A NEW APPROACH TO THE

ICE (Institution of Civil Engineers)

CONDITIONS OF CONTRACT

By the same author

A New Approach to the Standard Form of Building Contract
(Lancaster: MTP, 1972; The Construction Press Ltd., 1974)

A NEW APPROACH TO THE ICE (Institution of Civil Engineers) CONDITIONS OF CONTRACT

VOLUME 1

Glyn P. Jones MSc AIQS AIOB

Senior Lecturer, Quantity Surveying, Oxford Polytechnic

The Construction Press Ltd Hornby Lancaster 1975

ISBN 0 904406 07 5

Published by:

The Construction Press Ltd.,
Lunesdale House,
Hornby,
Lancaster,
LA2 8NB

CONTENTS

PREFACE AND ACKNOWLEDGEMENTS

This book is concerned with a systematic approach to the analysis of the new ICE Conditions of Contract (Fifth Edition). These flowcharts have been compiled for Engineers, Contractors, Employers, their Site Staff, and Quantity Surveyors. They are designed for those who are required to comply with the rules of this complex document, and for those who are required to solve contractual problems that inevitably arise in construction projects.

Chapter 1 examines some of the reasons why draftsmen write their clauses in ways which fail to communicate. It also points out many of the reasons preventing ordinary people from understanding those devilishly intricate legal sentences. Communications Psychologists will confirm that sentences with an unnatural arrangement soon confuse any educated person. It is therefore necessary to analyse long unnatural sentences and to rebuild each chunk to be understood, into a more natural step by step form. This way of breaking down prose (or numerical) problems, and rebuilding the results into a logical sequence of questioned pieces, is referred to as an algorithmic (or flowchart) approach.

This approach has already been used by the author to analyse the JCT Standard Form of Building Contract. Those practitioners who are therefore accustomed to those previously published flowcharts will not need the aid of Chapter 2 which merely explains the principles of constructing the charts contained in Chapter 3.

Since 1967 considerable interest has been aroused in the use of algorithmic methods. Her Majesty's Stationery Office has published two excellent papers (CAS nos. 2 & 12) on behalf of the Civil Service, concerning the application of these techniques to create flowcharts for decoding certain tax and legislative rules. Systems analysts and those involved with programmed learning are also actively developing the theory that surrounds these mechanistic ways, which endeavour to create simplicity out of complexity.

The Institution of Civil Engineers have kindly granted permission to reproduce the extracts of the Conditions interspersed in the text of this book.

Finally, I must make special acknowledgement of the work of my colleague John Hamwee BA (Oxon) Lecturer in Law, Oxford Polytechnic. He has acted as my constant guide and mentor in the construction of these flowcharts, and always expressed his legal expertise to me with patience, clarity and candour. If therefore there are any errors or obscurities which remain, they are entirely my own work.

Glyn P. Jones

Oxford
January 1975

1 INTRODUCTION AND OBJECTIVES

Ordinary people are now increasingly required to interpret complex legal language and possibly play quite important semi-legal roles whilst practising their professions. They now control and decide matters that were once the domain of the few. The majority of practitioners in the construction industry are required to know the details of several forms of contract, certain regulations, legislation concerning industrial relations, safety, health, and welfare requirements, and a minority are even required to delve into planning controls or tax strictures.

All of these important control documents communicate their contents in the same kind of nineteenth-century jargon about which R. Graham Page, former government minister, and a lawyer by profession, wrote:

> . . . The language of the parliamentary draftsman . . . is forced on to the statute book . . . not because the Minister (or anyone else for that matter) knows what it means but because it is quicker that way.[1]

Somehow or other practitioners are using the legal profession's system of communication, and amazingly are doing so without arousing reaction or resorting to widespread contractual litigation. The problem of identifying injustices, of course, inhibits the argument that our litigation record is no reflection of the efficiency of legal language in our control documents. Can it be that we have on the one hand a high-level formal communication system which no-one really understands, and on the other hand we actually use an unwritten low-level system of controlling matters, based upon the maxim of compromise?

Lawyers and psychologists[2] can rightly claim that man is a miserable component in *any* communications system. He is biased, he understands and sees only that which he wishes to, he has a short span of immediate memory and absolute judgement, he expects to decode legal jargon without formal training, and resorts to all sorts of low-level compromise in the firm belief that it is generally better that way, than the ultimate high-level legal way originally intended by draftsmen.

Draftsmen maintain they cannot create simplicity out of inherent complexity. If we choose to live in a highly developed community then our rules of conduct will proliferate and be complex. There will also be complicated interactions between commercial contracts and our social laws.

The draftsman does what he is told to do, using the only techniques he knows of which achieve comprehensive unambiguous communication. He has to ensure that what he writes means precisely what he intended. In so doing he is not concerned if his words do not yield his meaning on *first* reading. His only real concern is to prevent others from ever deciding against the meaning he originally intended.

This chapter sets out to examine some of the reasons for our inability to cope with legal language. The new ICE Conditions of Contract, like any other rules entrusted to legal communicants, do not yield their draftsman's meaning to an educated majority on first reading. We hardly ever discuss this important communications problem and seem to accept as inevitable the resultant monopoly in instant understanding granted to a few.

The aim here is to identify those areas of drafting that cause most difficulty and to question whether traditional contract form can be changed to bring about an improvement in our ability to understand.

Language Interpretation

Legal Expressions

The language used in contracts is undoubtedly the language of one expert to another; nevertheless, experts still express differences of opinion[3] about the meanings and precision of their own jargon. This stems from the fact that our English language is an imperfect code of communication and we must accept that it is futile to expect the draftsman to provide absolute answers to the meaning of any word. It is only when people agree on the object to which their words refer that minds meet and communication or agreement can occur.

The courts of law recently rejected an appeal that a 'warehouse' could contain a 'supermarket' for the purposes of the Town and Country Planning Act 1971. Mr Justice O'Connor apparently[4] sought the aid of his *Shorter Oxford English Dictionary* to assist him in deciding this question, but a dictionary is almost the last place in which to find the meaning of a word taken out of its context. Even if groups of synonyms are used to reinforce a word they cannot define that word con-

1. *The Times,* 6 April 1970.

2. Miller, G. A., *The Psychology of Communication* (London: Penguin Press, 1968).

3. Harris, Sir William, and Gardam, D., QC, 'Clearing the Critics' Confusion', *New Civil Engineer,* 20/27 December 1973.

4. Linnel, R., 'Legal', *Building Design,* 12 April 1974.

clusively. There remains ambiguity until we can point to an object, see it, touch it, then place a label on it to say: 'This is what I mean when I say "warehouse".'

It is not even safe to assume for example the simple labels 'dog', 'chair', 'pencil', 'pig', can only be placed upon one particular object. The dictionary informs us of some but not all of the objects to which we can attach these labels. We could conceivably state: 'If that pig drops a pig on my pig I will call a pig.' This is grammatically correct and carries lots of meaning. The referent list for 'pig' includes:

animals with curly tails,
a block of iron,
uncouth husbands,
rough policemen.

The draftsman therefore sees no harm in using the word 'determine' in one clause to mean 'decide', and in an adjacent clause to mean 'terminate'. In context the words will never be interpreted incorrectly by the expert he is addressing. But this habit confuses laymen who would prefer draftsmen to change their words if they are changing their meaning.

It is in fact the order and relationship (syntax) of words which decides the meaning intended. The interpretation of each word is governed by the words surrounding that word.

The legal profession has developed a system of syntax using a combination of well tried and preferably tested words, in the same formalized way that Engineers have developed their own symbolic language to achieve universal precision. Legal formalism cannot claim the precision of mathematical text since their symbols are words, not digits. Incidentally, business English similarly uses just about 800 words[5]. with relatively high frequency to produce a formalized cliché language, and statistical studies (in 1930) of telephonic speech showed that 96 per cent of conversations consisted of 737 words. Not surprisingly then, this cliché-ridden legal language has flourished in the fields of commercial contract, statutory rules, and social legislation.

5. Cherry, C., *On Human Communication* (Massachusetts: MIT Press, 1957).

The Changing Meaning of Words

The words used by draftsmen are those which have weathered the social upheavals which change our colloquial and written language. English is changing and the tempo of change is quickening. Words drop out and others are admitted. The word 'blurb' for instance (born in America about 1924) appears in the *Shorter Oxford English Dictionary* with only one succinct meaning, but draftsmen would instinctively steer clear of such a newcomer to our collection of 500 000 English words, until public opinion agrees it is the very best word to use in particular circumstances. Words obsolete in natural speech may therefore still appear in a draftsman's collection. We no longer actually say 'pursuant to' as in Clause 7(3) or 'importing' as in Clause 1(2). People no longer have a 'calling' (as in Clause 16), to become a bricklayer or carpenter. The angelic ethereal apparitions that presumably hovered over a carpenter's cot have been displaced by a spectre named the 'lump', but this word stands no chance of ever appearing in Clause 4 even though as an objective label it innocently and colloquially only refers to groups of self-employed persons.

When we now say 'watching' as in Clause 19 do we include closed-circuit TV if required by the Engineer? Are organized picket lines a 'supervening event' in the context of Clause 64? What was meant in 1874 is not meant today. We cannot expect the meaning of written words to survive extensive changes in culture, and draftsmen cannot anticipate every change in our ways or beliefs. The draftsman therefore chooses labels such as 'supervening event' simply because this label can be placed upon many and ever-changing things. In context the words can be interpreted correctly but we definitely cannot, and do not wish to, give these words a once-and-for-all meaning. When there is judicial interpretation of these kind of words lawyers then have a little more legal theory to add to their huge store called 'precedent'.

Abstract Words

Abstract words are particularly troublesome and liable to misunderstanding unless the layman realizes that the draftsman particularly choses such words to enable the minds of Engineers and Contractors to come together in agreement that those general words refer to a particular object or situation that they are concerned with. For this reason Clause 31, for instance, purposely refers to 'all reasonable facilities' and Clause 13 similarly uses the wide term 'arrangements'.

It is impossible and unwise to avoid the use of abstract words. The well-worn word 'reasonable' is indispensable. This particular word is the cornerstone of legal language. It is the litmus paper used to test every implied right or obligation.

Emotive Words

The statement 'All bad girls must be subject to a curfew' is liable to misinterpretation. Every person would interpret the word 'bad' differently. It is an emotive word. A 'bad girl' $_{1949}$ is not a 'bad girl'$_{1974}$. Somewhere between 1949 and 1974 there was presumably a social upheaval. Draftsmen do not use emotive words, and for this reason the word 'flagrantly' to be found in the 4th Edition (Clause 63) has now been removed from the new Conditions.

Words of the Same Category *(ejusdem generis)*

Draftsmen may use a list of precise words followed with general words. The general words take their meaning from the preceding precise words. This rule of connection is referred to as *ejusdem generis.*

Example 1 illustrates this unwritten connection between the last four general words with the preceding more precise words. The words 'usurped power' will therefore cover all situations of the same category, in the same context, as conveyed by the preceding precise words.

By this rule the ICE Conditions would regard damage that may have been caused during a lone bombing raid by say the Welsh Air Force as an Excepted Risk under Clause 20(3). This rule of *ejusdem generis* is clearly a useful one permitting correct flexibility of interpretation in a drafting situation that would otherwise lead to an infinite

Example 1

Excepted Risks

20. (3) The 'Excepted Risks' are riot war invasion act of foreign enemies hostilities (whether war be declared or not) civil war rebellion revolution insurrection or military or usurped power.

list of precise words, or to not including unforeseeable risks which ought rightly to be excepted ones.

The Meaning of Words in Their Context

If we begin to argue about the meanings of words (labels) it is because we do not agree that a certain label should be placed upon a certain object or situation. There has therefore been developed a drafting habit of not changing a word or syntax throughout a particular document unless the draftsman intends to change his meaning. Unfortunately this habit is not slavishly adhered to. If the draftsman decides that identical words can in different contexts have different meanings (eg determine = decide; determine = terminate) then he will sensibly use the same label to mean different things. On the other hand if he decides that different words can in context have the same meaning then he may use his draftsman's vagarious licence, as illustrated in Example 2. A caprice of this kind can cause considerable confusion to laymen.

Example 2

The ICE Conditions use the following words to mean the same thing. (The *SOED* gives the words *B* for the words *A.)*

A	*B*
order	an instruction, a direction
instruction	a direction, an order
direction	an instruction, an order
to order	to direct
to require	to demand
to demand	to order

The words do not comprehensively and conclusively tell practitioners how to recognize an 'order', etc., when we see one. To obviate any argument in civil engineering circles as to whether a drawing issued by the Engineer constitutes an 'order' the draftsman has added further words into Clause 7(3). But this then leaves us the problem of deciding whether a sketch on a cement bag is a 'drawing' or not. The draftsman to a certain extent anticipates this kind of iterative problem by providing, as he has done in Clause 1(1)(g) his own special definition of the word 'drawing', in the context of this Contract, which he considers goes far enough towards precision without excluding unforeseeable but nonetheless acceptable alternatives.

Definitions

To state what an English word means is an onerous task for word-mongers. They have no option but to use other words to convey their meaning. These other words may then themselves require defining, so theoretically leading to a never-ending activity. Fig. 1 illustrates the whirligig effect of using some defined words with which to define words. Draftsmen therefore rightly reserve definitions for objects or situations in which it is *vital* to determine the limits or boundaries of meaning intended.

Words defined by the draftsman and particular words, as opposed to general words, are given upper-case letters throughout contract documents. This subtle signal can change the meaning of words. A 'Contract' is therefore not a contract, a 'Contractor' is not a contractor, 'Plant' is not plant, and so forth. More confusingly, for the layman, 'Constructional Plant' defined in Clause 1 is not the 'Plant' defined in Clause 53 and the definitions (Example 3) given in Clause 70 have their source elsewhere outside the contract documents. (Incidentally the word 'invoice' is redundant anyway as it does not actually appear in Clause 70.) This use of upper-case letters, if confined to defined words, would effectively signal this fact to any reader. Unfortunately however the upper-case signal is used to denote special words which *have* been carefully defined and those which *have not* been defined at all, as evidenced in the bracketed words of clause 41 (Example 4).

There is further complication (Clause 1 concerning 'Definitions and Interpretation') surrounding

Example 3

Value Added Tax

70. (1) In this Clause 'exempt supply' *'invoice'* 'tax' 'taxable person' and 'taxable supply' have the same meanings as in Part I of the Finance Act 1972.

Example 4

Commencement of Works

41. The Contractor shall commence the Works on or as soon as is reasonably possible after the [Date for Commencement] of the Works to be notified by the Engineer in writing which date shall be within a reasonable time after the date of acceptance of the [Tender]. Thereafter the Contractor shall proceed with the Works with due expedition and without delay in accordance with the Contract.

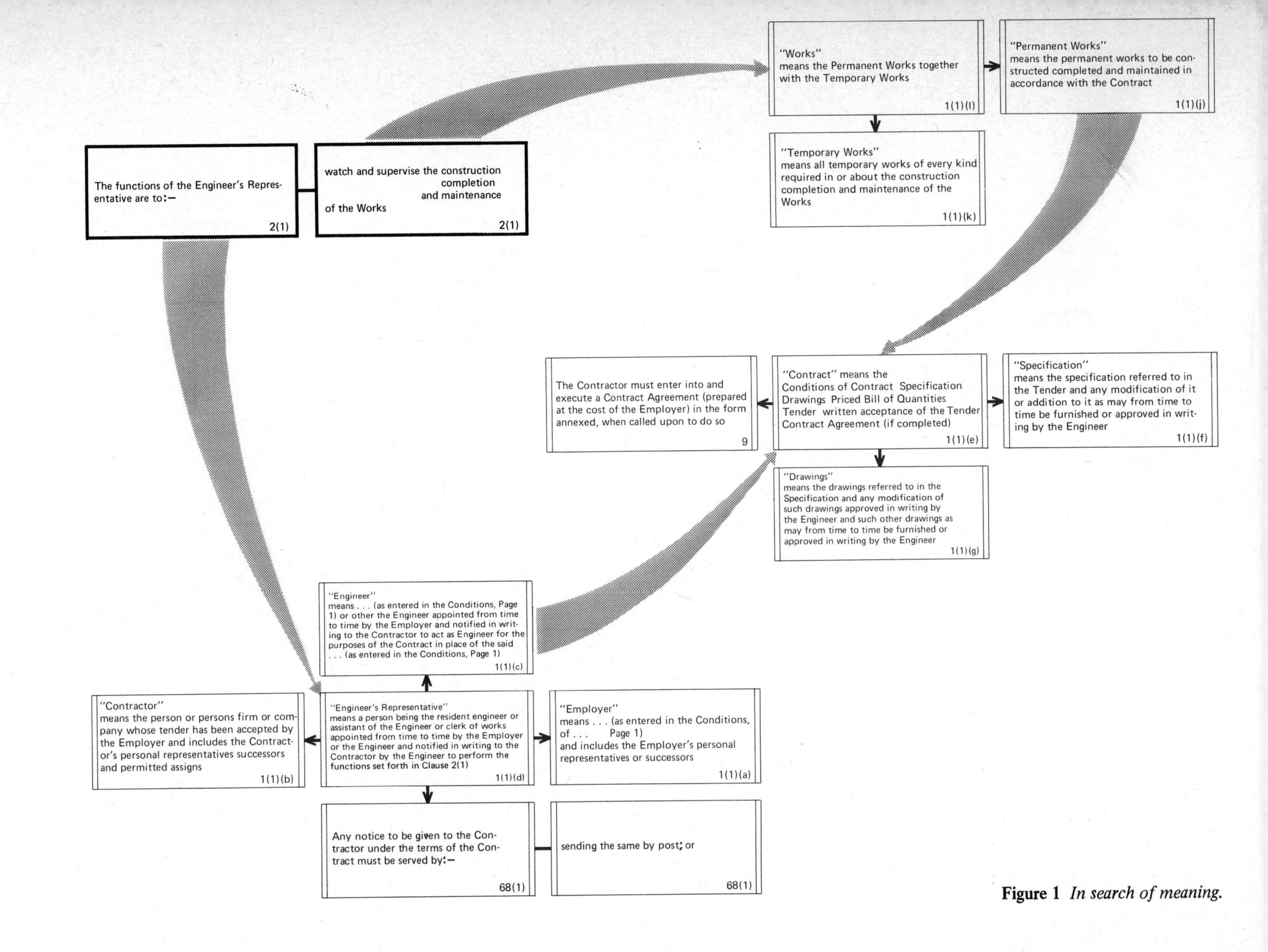

Figure 1 *In search of meaning.*

one word of great importance to the Conditions. The word is 'cost' and at first glance seems to have been defined in Clause 1(5) as shown in Example 5. The Joint Committee in fact compound this impression.[3] Upon closer scrutiny it can be seen however that the statement really only states the following:

> The word 'cost' (whatever that means) is **deemed** to include on Site and off Site overhead costs (whatever that means).

The word cost is not one-valued, it is multi-valued and can take on as many values as there are contexts. Neither the dictionary nor the statement in Clause 1(5) is of any assistance in deciding what we really mean when we say 'cost'. It is an extremely obtuse way of saying 'thou shalt not profit in certain circumstances'.

Example 5

Cost

1. (5) The word 'cost' when used in the Conditions of Contract shall be deemed to include overhead costs whether on or off the Site except where the contrary is expressly stated.

Example 6

Responsibility for Reinstatement

20. (2) In case any damage loss or injury from any cause whatsoever (save and except the Excepted Risks as defined in sub-clause (3) of this Clause) shall happen to the Works or any part thereof while the Contractor shall be responsible for the care thereof the Contractor shall at his own cost repair and make good the same so that at completion the Permanent Works shall be in good order and condition and in conformity in every respect with the requirements of the Contract and the Engineer's instructions.

Implied Meanings

One statement written down can imply another without actually saying so. 'Jack ordered a pint' implies that Jack has a legal right to enter a pub and order beer. Similarly Clause 63 in the ICE Conditions states the parties rights and obligations in the event of forfeiture and this statement implies, without actually saying so, that the parties have common law rights in respect of rescission or repudiation.

A contract may seem suspiciously silent on many matters that appear to be important. For instance, the consequences of acting corruptly to obtain, or in the execution of, a local authority building contract are spelt out in the 'RIBA' form of contract, but the new ICE Contract makes no mention of such matters. This is not to say civil engineering contractors and engineers can entice each other to abandon their virtues. The Prevention of Corruption Act provides stiff penalties for he who corrupts and he who has been corrupted. If the rules of this Act are combined with the rules of local authority standing orders (which usually require the termination of all contracts with any Contractor who has acted corruptly) then the draftsman can consider it unnecessary to spell this combination out. Contracts cannot oust criminal law, as Shylock discovered to his dismay. Compliance with all our law is implicit in every contract between man and man and needs no mention. But if any compliance with an Act is a precondition of a contract then the draftsman has to mention this. In this respect Clause 27 of the ICE Conditions refers for instance to the requirements of the Public Utilities Street Works Act in civil engineering work.

Both the 'RIBA' form of building contract and the ICE Conditions for civil engineering work stand on common ground in many respects. Vagaries in their drafting unfortunately cloud matters unnecessarily for those who have to daily deal with both these forms of contract.

Understanding Sentences

A contract is a declaration of a complex process involving numerous variables and an infinite number of interactions. Communications theorists are able to predict[2] when people will begin to make errors of judgement, whether handling objects in the span of attention, digits or words in the span of immediate memory, or when handling alternative choices which condition their judgement. We cannot think simultaneously about everything we know and our capacity to remember limits our intelligence. Our innate ability to understand is seriously diminished by some of the methods used to convey the information contained in a legal contract.

The long legal sentence is probably the heaviest cross the layman has to bear when identifying the benefits to be derived from conforming to all a contract's directions, or the undesirable consequences of ignoring an obligation. But any sentence laid down in an unnatural way is even more likely to strain the layman's innate ability to understand. Example 6 illustrates the effects of both a long (87-word) sentence and an unnatural arrangement in that the main clause is hidden away behind the 44 words that precede it, which all have to be read, understood, and stored in our low-capacity memory without receiving the all important main clause context clues as to meaning.

The draftsman paradoxically creates his unnatural sentences in an attempt to achieve certainty. Example 7 illustrates how this can happen.

The last 15 words of this conundrum are crucial, so are the first 8 which express the main clause. These words regarding the £15 till contents are really the most important within this process declaration. The process cannot commence unless

2. Miller, G. A., *The Psychology of Communication* (London: Penguin Press, 1968).

3. Harris, Sir William, and Gardam, D., QC, 'Clearing the Critics' Confusion', *New Civil Engineer*, 20/27 December 1973, p. 35.

Example 7

> You may on credit to regular customers only serve Wondabeer Brew except if any cider is available except if the cider available is more than 16p per pint and only if at all times there is no less than £15 in the till.

Example 8

> **ENGINEER'S REPRESENTATIVE**
>
> **Functions and Powers of Engineer's Representative**
>
> 2. (1) The functions of the Engineer's Representative are to watch and supervise the construction completion and maintenance of the Works. He shall have no authority to relieve the Contractor of any of his duties or obligations under the Contract nor except as expressly provided hereunder to order any work involving delay or any extra payment by the Employer nor to make any variation of or in the Works.
>
> **Appointment of Assistants**
>
> (2) The Engineer or the Engineer's Representative may appoint any number of persons to assist the Engineer's Representative in the exercise of his functions under sub-clause (1) of this Clause. He shall notify to the . . .

Example 9

> **Delay in Issue**
>
> 7. (3) If by reason of any failure or inability of the Engineer to issue at a time reasonable in all the circumstances drawings or instructions requested by the Contractor and considered necessary by the Engineer in accordance with sub-clause (1) of this Clause the Contractor suffers delay or incurs cost then the Engineer shall take such delay into account in determining any extension of time to which the Contractor is entitled under Clause 44 and the Contractor shall subject to Clause 52 (4) be paid in accordance with Clause 60 the amount of such cost as may be reasonable. If such drawings . . .

Example 10

> **Delay and Extra Cost**
>
> 12. (3) To the extent that the Engineer shall decide that the whole or some part of the said physical conditions or artificial obstructions could not reasonably have been foreseen by an experienced contractor the Engineer shall take any delay suffered by the Contractor as a result of such conditions or obstructions into account in determining any extension of time to which the Contractor is entitled under Clause 44 and the Contractor shall subject to Clause 52(4) (notwithstanding that the Engineer may not have given any instructions or orders pursuant to sub-clause (2) of this Clause) be paid in accordance with Clause . . .

there is the stipulated sum in the till. On this basis a draftsman traditionally brings forward this part of the statement to the start point of the sentence but in so doing would create a screen of 15 words before the operative part of the main clause.

The linguist Noam Chomsky reckons it is perfectly possible to construct a mile-long grammatical English sentence. Politicians speak using such sentences and we can understand them (even though we may not believe them). The difference between a long natural political sentence and a long unnatural legal sentence is that the legal sentence violates the principles of sentence construction in a natural language. This accounts for our failure to analyse the otherwise grammatically correct legal statement.

The draftsman is not long-winded even though his sentences are. His objectives are compressed clarity for his fellow experts. He does, however, have an unnatural fear of punctuation resulting in a tendency to treat the paragraph as a non-stop journey.

Take five simple sentences:

1. The contract was applauded by the critics.
2. The drafting made the contract.
3. The joint committee's deliberations eventually led to the drafting.
4. The chairman attended all the deliberations.
5. We all thanked the chairman.

The above sentences can be compressed into one statement:

> We all thanked the chairman who attended all the deliberations that eventually led to the drafting that made the contract that was applauded by the critics.

This statement, although unfamiliar, is intelligible and unambiguous. Furthermore we have achieved a saving in words of 19 per cent. Brevity is appreciated by the legal profession. Their environment is crowded with 20 000 000 statutory words, the contents of 250 000 test cases, and the outpourings from Parliament. They therefore speak succinctly to one another, their drafts are renowned for their compressed style. But it is this increased amount of information contained within an unnatural and compressed syntax that governs our understanding.

More on Understanding

Most practitioners would agree that the first 100 words of clause 2 of the ICE Conditions are quite easily understood on first reading (Example 8).

But the first 100 words of clause 7(3) (Example 9), or of clause 12(3) (Example 10), are not likely to yield their meaning on *first* reading. If we apply the formula of R. Gunning to measure the readability of the samples we have the following results:

Clause 2 = Index 14 (50 per cent monosyllabic).
Clause 7(3) = Index 24 (55 per cent monosyllabic).
Clause 12(3) = Index 26 (55 per cent monosyllabic).

The above indices can be compared with the following, in a table of relative readability:

Index	
6- 8	Children's books.
8-10	Popular papers and light fiction.
10-12	Quality papers and middle-brow fiction.
12-15	Reports in general and difficult fiction.
15-20	Scientific and technical reports.
20+	Treatises, especially scientific.

If the first 100 words of Clause 22(1) (Example 11) are measured for readability we find it has an index of 26 (45 per cent monosyllabic). Now most people would find the words sample in Example 11 less intelligible than the words sample in Example 10 even though they have the same index and only a slight monosyllabic difference.

Looking back for reasons we find Clause 2 scores well by virtue of its short sentences. Clause 7(3) is marginally better than Clause 12(3) because it contains one sentence, whereas Clause 12(3) is non-stop.

Readability is one thing, intelligibility is another. Clause 12(3) and Clause 22(1) are both as readable as each other but Clause 12(3) is easier to understand. The answer to this lies in the legal cliché 'except if and so far as . . .' contained in Clause 22(1) and placed as a string of qualifiers between the subject of the sentence ('The Contractor'), and the statement made about the subject. This legal cliché, unlike a natural statement, expresses alternative choice conditioning our judgement, contains two ideas, and strains our capacity by being positioned between the subject and main part of the sentence. Research by Wason and Jones (1965-6) indicates that[6] 'strings of qualifiers ("providing", "unless", "except", etc.) are known to be troublesome'. The effect of a negative qualifier in a sentence is also confusing. Example 14 in Chapter 2 illustrates the confusion caused when a string of qualifiers is used combined with a negative qualifier.

Formal experiments[2] (using 22 word sentences, and Harvard students) have been carried out to establish how well we drop into subroutines to analyse certain sentences. The tests involved taking 5 sentences each containing 1 idea, which could be connected and combined into 1 test sentence of 22 words. When the subjects of the test sentence were separated from each successive predicate (the statement made about a subject) by 1 idea everyone could handle the sentence easily but when 2 ideas were introduced some could not handle the sentence, and when 3 ideas were introduced between the subject and its predicate everyone had difficulty. Now this experiment involved 22 words whereas the Clause 22(1) word sample contains 100 words, and although the draftsman has endeavoured to aid the reader by placing brackets around the offending cliché, the reader is still loath to jump over any groups of words in search of the main clause, and so overloads his capacity with the information taken in strict reading sequence. Recorded[2] eye movements indicate that 'frantic' eye movements occur if readers cannot readily associate verbs with their subject.

Clause 65(6)(c) of the new ICE Conditions (Example 12) contains every one of these obstacles known to inhibit understanding and to cause mistakes. The sentence forcibly illustrates the effect of a 251-word sentence containing 19 verbs (or participles), placed well away from their subject, and the slotting of numerous ideas between subjects and successive predicates. The phrase between the asterisks ** even uses the same label to mean different things, almost in the way of our previous conundrum: 'If that pig drops a pig on my pig I will call a pig'.

There are unfortunately other more important parts of the Conditions which in the same way have been compressed into incomprehensible verbiage. The long legal sentence usually contains too many ideas for us to handle in our low capacity immediate memory. Most people become confused if more than about 6 ideas condition our judgement.[2] This can be best illustrated if the reader has any difficulty on first reading the statement contained in Example 13A. This statement is

Example 11

Damage to Persons and Property

22. (1) The Contractor shall (except if and so far as the Contract otherwise provides) indemnify and keep indemnified the Employer against all losses and claims for injuries or damage to any person or property whatsoever (other than the Works for which insurance is required under Clause 21 but including surface or other damage to land being the Site suffered by any persons in beneficial occupation of such land) which may arise out of or in consequence of the construction and maintenance of the Works and against all claims demands proceedings damages costs charges and expenses whatsoever in respect thereof or in . . .

2. Miller, G. A., *The Psychology of Communication* (London: Penguin Press, 1968).

6. Lewis, B. N., *Decision Logic Tables for Algorithms and Logical Trees* (London: HMSO, 1970); Lewis B. N., Horabin, I. S., and Gane, C. P., *Flow Charts, Logical Trees, and Algorithms for Rules and Regulations* (London: HMSO, 1967).

neither long nor legal but it is unnatural and does contain numerous ideas. As stated earlier it is the increased amount of information contained within unnatural compressed syntax that governs our understanding, but precognition or advance information is equally important.

Half a dozen friendly engineers were at separate times asked by the author to handle the information contained in Example 13A and then to answer the questions shown in Example 13B. No time for completion was stipulated. No-one obtained all correct answers to the five questions in the 3 minutes originally allowed for this 11-plus poser.

Interestingly, the engineers explained afterwards that they were looking for either a trick or complexity in the statements. Advance information that they were being asked to process an innocent 11-plus poser would have provided the precognition which plays an important part in our ability to process a message correctly.

Laymen expect legal statements like conundrums to contain tricks or insurmountable complexity. A glance at the ICE Clause for Contract Price Fluctuations will confirm that a readily understood formula can, when converted into legal statements, become incomprehensible. But once we are told that the conundrum in question is a description of the Baxter Formula then this vital advance information plays its part in enabling us to decode the message. Draftsmen are reluctant to use formulae in numerical or alphabetic form even though those they are addressing revel in such symbols which condense facts more succinctly than any draftsman can when he uses words.

Conclusions

This chapter has endeavoured to illustrate some of the problems facing a draftsman whilst at the same time discussing the problems his resultant draft creates for the layman. Obviously a standard method of drafting would together with some standard phraseology lighten every laymen's load by sweeping away every unnecessary whim and eccentricity at present admissable in legal drafts. Contractors concerned with *both* building and

Example 12

Clause 65

(c) In the event that the Contract includes the Contract Price Fluctuations Clause the terms of that Clause shall continue to apply but if subsequent to the outbreak of war the index figures therein referred to shall cease to be published or in the event that the contract shall not include a Price Fluctuations Clause in that form the following paragraph shall have effect:–

If under decision of the Civil Engineering Construction Conciliation Board or of any other body recognised as an appropriate body for regulating the rates of wages in any trade or industry other than the Civil Engineering Construction Industry to which Contractors undertaking works of civil engineering construction give effect by agreement or in practice or by reason of any Statute or Statutory Instrument there shall during the currency of the Contract be any increase or decrease in the wages or the rates of wages or in the allowances or rates of allowances (including allowances in respect of holidays) payable to or in respect of labour of any kind prevailing at the date of outbreak of war as then fixed by the said Board or such other body as aforesaid or by Statute or Statutory Instrument or any increase in the amount payable by the Contractor by virtue or in respect of any Scheme of State Insurance or if there shall be any increase or decrease in the cost prevailing at the date of the said outbreak of war of any materials consumable stores fuel or power (and whether for permanent or temporary works) which *increase or increases decrease or decreases shall result in an increase or decrease* of cost to the Contractor in carrying out the Works the net increase or decrease of cost shall form an addition or deduction as the case may be to or from the Contract Price and be paid to or allowed by the Contractor accordingly.

Example 13A

Statements

Jack is twice as old as he was five years ago. His mother was then six times as old as he was. She is now 35 years old.

Example 13B

1. How old is Jack?
2. How old was his mother when Jack was 5?
3. How old will Jack be when his mother is 50?
4. How old was Jack's mother when he was born?
5. What is the difference between Jack's age and his mother's?

civil engineering contracts must in the meantime bear the brunt of their draftsman's vagaries.

There exists a phenomenal amount of *legal* theory, but no modern *drafting* theory. Piesse and Gilchrist-Smith[7] do throw some light upon the pitfalls of drafting, but in the main it is clearly an archaic intuitively based activity. So far as drafting is concerned the legal profession is in the same position the construction industry was in prior to the advent of critical path theory (or even prior to the agreement reached upon a standard method of measurement). Modern planning theory relies more and more upon a semi-pictorial network, or model, representing the logic of a complex process. It is rapidly reinforcing our intuitive ways of predicting results. Networks are of course more time-consuming and mentally rigorous but the results do present us with a rational basis for communication.

Aristotle (the father of formal logic)[8] thought a fly had eight legs and wrote it down. For centuries no-one really looked at a fly's legs and scholars were content to accept the authority of Aristotle. Have we really looked at legal drafting to see whether it is in fact the unambiguous, intelligible, logical, and precise format claimed, with authority, by the legal profession?

Lawyers will shake their heads at the notion that their legal language and format is unsuited to the majority who have to use it in their daily activities, that strictured mechanistic ways can improve matters, and inch us a little nearer the day when we can solve our own contract problems by following sign-posted ways of arriving at *the* only correct conclusion.

The next chapter examines ways of arranging legal drafts in such a way that we do reduce the load on memory, by organizing our facts into units, so that we can make a sequence of simpler judgements to arrive at the contract's conclusions.

7. Piesse, E. L., and Gilchrist-Smith, J., *The Elements of Drafting* (London: Stevens and Sons, 1965).

8. Chase, S., *The Tyranny of Words* (New York: Harcourt Brace & World Inc., 1937).

2 THE REASON FOR FLOWCHARTS

Low-level communications theory can be applied to the ICE Conditions of Contract to create some simplicity out of its complexity. People have always endeavoured to improve their methods of conveying information and reducing the ambiguity that is present in every method of communication. Methods of conveying military information have certainly improved since the day when an original message of 'Send reinforcements. Going to advance' was understood as 'Send three-and-fourpence. Going to a dance'.[9] Fig. 2 illustrates one entrepreneur's simple way of enabling people of all nationalities to understand his Spanish bills. Whether we use signs, muscle contractions, smoke signals, semaphore, drums, words, diagrams, digits, phrases, clauses, morse, or out-of-the-ether pictures, there will always be some ambiguity. But improvements can be made by developing methods of deduction, by organizing facts into units, by reducing the load on memory, by arranging the task in such a way that we make a sequence of several absolute judgements all in a row.

Such organizational improvements can be illustrated in a simple way by applying the above principles to enable us to process the data contained within one pack of ordinary playing cards. A pack of cards contains 52 pieces of information. The problem posed is can we reorganize this information first into manageable units, and then develop methods of deduction which will enable anyone to correctly conclude, in a methodical way, which card has been secretly chosen from the pack by asking no more questions than are necessary?

Figure 2

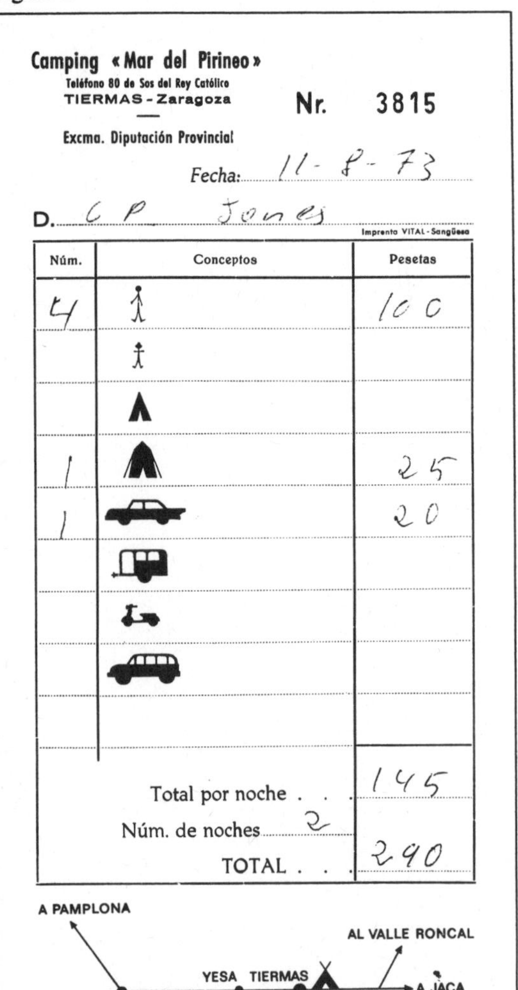

Camping «Mar del Pirineo»
Teléfono 80 de Sos del Rey Católico
TIERMAS - Zaragoza
Excma. Diputación Provincial

Nr. 3815

Fecha: 11-8-73

D. C P Jones

Imprenta VITAL - Sangüesa

Núm.	Conceptos	Pesetas
4	(person)	100
	(child)	
	(small tent)	
1	(large tent)	25
1	(car)	20
	(caravan)	
	(scooter)	
	(van)	

Total por noche . . . 145
Núm. de noches 2
TOTAL . . . 290

There is no simpler question than that which can be answered by a direct 'Yes', or 'No'. Only if the question is right can the answer be right. Using 'Yes/No', or two-way questions correctly positioned we find that 6 binary (Yes/No) questions will, each time, enable us to correctly identify the card chosen, as shown in Fig. 3.

We have here the basis for an algorithm. An algorithm is the word used to describe a logical procedure for the solution of any problem in a given class. It can be said that a problem is insoluble if it can be proved no algorithm can be given for solving it. An algorithmic approach to a problem uses a step by step procedure in a reasoned logical way. Fig. 3 could be arranged in the form of a logic table. Decision logic tables[6] display the permutation of possibilities in any given situation. All the possibilities can, however, prove a daunting display to those who are usually only interested in one of those possibilities, the rest prevent him from seeing the wood for trees.

Flowcharts are a logical network providing a pictorial step by step representation of a process (algorithm) indicating the arrangement and action of its parts by which it produces a given result. Flowcharting simplifies any plan process or procedure no matter how complex the problem.[10] Algorithms dispense with the strings of qualifiers with which draftsmen create their devilishly intricate sentences.

Fig. 4 sets out six basic symbols required to enable legal language to be decoded, organized into manageable units, arranged for sequential

6. Lewis, B. N., *Decision Logic Tables for Algorithms and Logical Trees* (London: HMSO, 1970); Lewis, B. N., Horabin, I. S., and Gane, C. P., *Flow Charts, Logical Trees, and Algorithms for Rules and Regulations* (London: HMSO, 1967).

9. Young, J., *Information Theory* (London: Butterworth, 1971).

10. Jones, Glyn P., *A New Approach to the Standard Form of Building Contract* (Lancaster: MTP, 1972; Construction Press, 1974).

Figure 3

Figure 4

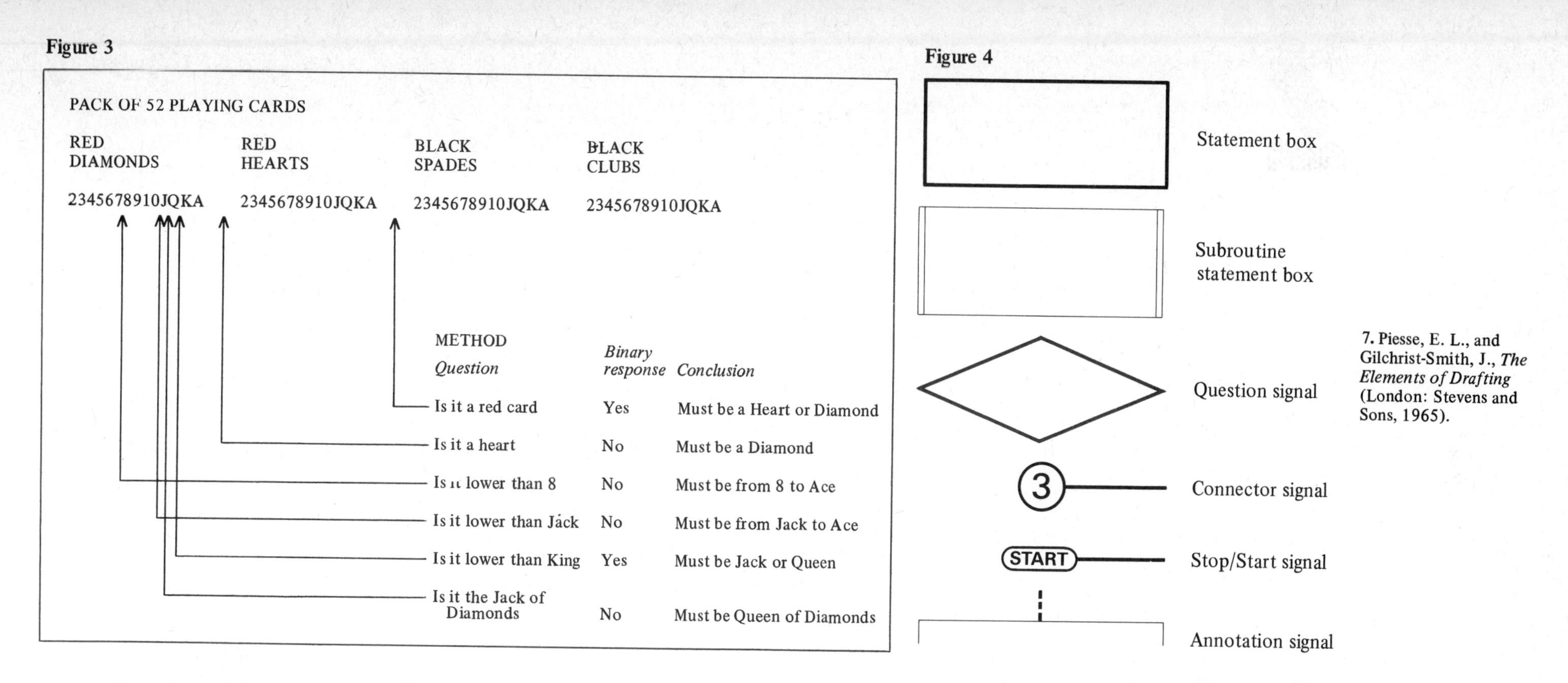

questioning, so that we arrive in a logical step by step way at the only correct conclusion. These basic symbols are taken from BS 4058 which sets out signals internationally agreed for use in data processing, problem definition, and problem analysis.

Draftsmen may recoil at the suggestion that their statements can be rearranged to upset their rigid syntactical rules they claim inherent in their compositions. Statement 1 below demonstrates one reason for their concern. However, Statement 2, indicates five rearrangements of one statement without any impairment of meaning.

Statement 1

1. I wanted to survive Adolph Hitler.
2. I wanted Adolph Hitler to survive.

Statement 2

1. Home they brought her warrior dead.
2. They brought her dead warrior home.
3. They brought her warrior home dead.
4. They brought home her dead warrior.
5. They brought home her warrior dead.

We can all see clearly that ordinary statements *can* be reorganized without impairing meaning. If great care is taken to preserve traditional syntactical chunks, it is also possible to rearrange legal statements, without impairing the draftsman's meaning. Certain words spell trouble for the layman. The word 'shall' is one such word. It even proves troublesome to lawyers and has done so throughout the ages. Coode[7] clarified some of the drafting problems caused by this word in the early nineteenth century. A glance in any dictionary only heightens the confusion and confirms that this word should now be declared unfit to serve. The words 'must' or 'may' are more positive in denoting the appropriate obligation to act as conveyed by the words surrounding 'shall' in the original draft. The chartist can confidently dump aged words such as 'pursuant to' and can for instance choose to preserve the word 'determine' when used to mean 'terminate', but drop the same word when it is used to mean making a decision, in favour of the word 'decide'.

Digits or the signals on playing cards carry much less information than words. Simple English sentences on average contain about 20 words and only 1 idea. Legal sentences (for example see Clause 49 (5)(a) or worse still Clause 65(6)(c)) can contain

7. Piesse, E. L., and Gilchrist-Smith, J., *The Elements of Drafting* (London: Stevens and Sons, 1965).

ten times this number. Each idea should therefore be questioned in the mechanistic way that we worked through the 52 reorganized ideas contained in a pack of cards. Each question enables us to arrive at the information next needed to predict the next step to be taken correctly from a set of steps available to eventually arrive at the only correct conclusion. For equally probably alternatives the Log_2 of the number of alternatives will equal the number of questions needed to make all correct decisions. A legal sentence containing 6 ideas is in effect presenting the reader with 64 alternatives. This is one way of quantifying the information content of a statement. There is, alas, no way of likewise measuring its meaning.

Method Rules

To create consistent flowcharts certain rules emerge:

1. Simplify (or decode) the problem to form statements.

2. Reorganize the facts into statements which are manageable.

3. Arrange the statements into a logical sequence.

4. Question each statement.

5. If any predefined process or routine has to be followed at any particular point in the sequence, a subroutine 'obey' signal is required.

6. If any additional information is essential, to aid the reader in his understanding, this can be tagged onto any statement or subroutine by the use of an annotation signal.
This same signal can be used to confirm a logical conclusion.

These rules, when followed, frequently expose inherent anomalies or ambiguity. Example 14 illustrates all the above rules in action, including the exposure of anomaly and ambiguity, which is dealt with by using the annotation signal to convey information to aid the reader, as stated in rule 6 above.

The use of annotation provides the reader with the advance information or precognition referred to in Chapter 1 (see Examples 13A and B) which plays such an important part in our ability to process a message correctly.

Example 14

Assume a landlord goes away on holiday and leaves his barman the following message:

You may on credit to regular customers only serve Wondabeer Brew except if any cider is available except if the cider available is more than 16p per pint and only if at all times there is no less than £15 in the till.

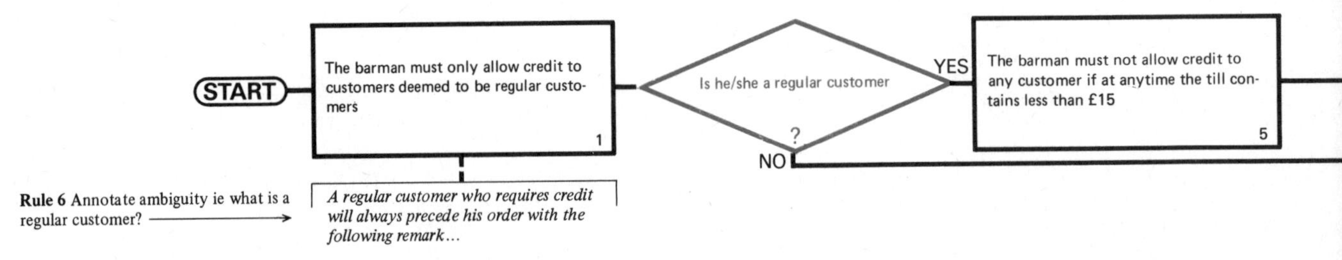

Figure 5 *Rule 1. Simplification*

There are 5 constraints placed upon the barman allowing credit. These constraints create 32 alternatives.

Simplified statements	1	2	3	4	5	6	7	8	9	10	11	12	13	14	15	16	17	18	19	20	21	22	23	24	25	26	27	28	29	30	31	32
1. Credit customers?	0	0	0	0	0	0	0	0	0	0	0	0	0	0	0	0	1	1	1	1	1	1	1	1	1	1	1	1	1	1	1	1
2. Serve Wondabeer?	0	0	0	0	1	0	1	1	0	0	1	1	0	1	1	1	1	1	1	1	0	1	0	0	1	1	0	0	1	0	0	0
3. Cider available?	0	0	0	1	1	1	1	1	0	1	0	0	1	0	0	1	1	1	1	0	0	0	0	0	1	0	1	1	0	1	1	0
4. Cider > 16p per pint?	0	0	1	1	1	0	0	1	1	1	0	1	0	0	1	0	1	1	0	0	0	1	1	0	0	0	1	0	1	1	0	1
5. Till contains ⩾ £15?	0	1	1	1	1	0	0	0	0	0	0	0	1	1	1	1	1	0	0	0	0	1	1	1	1	1	1	1	0	0	0	0

Key 0 = NO 1 = YES

Figure 6 *Rule 2. Form statements*

The barman must only allow credit to customers deemed to be regular customers — 1

The barman may serve Wondabeer on credit to the regular customer — 2

The barman must check if any cider is available — 3

The barman must check whether the price per pint is more than 16p — 4

The barman must not allow credit to any customer if at anytime the till contains less than £15 — 5

Figure 7 *Rules 3 and 4. Arrange the statements into a logical sequence and question each statement*

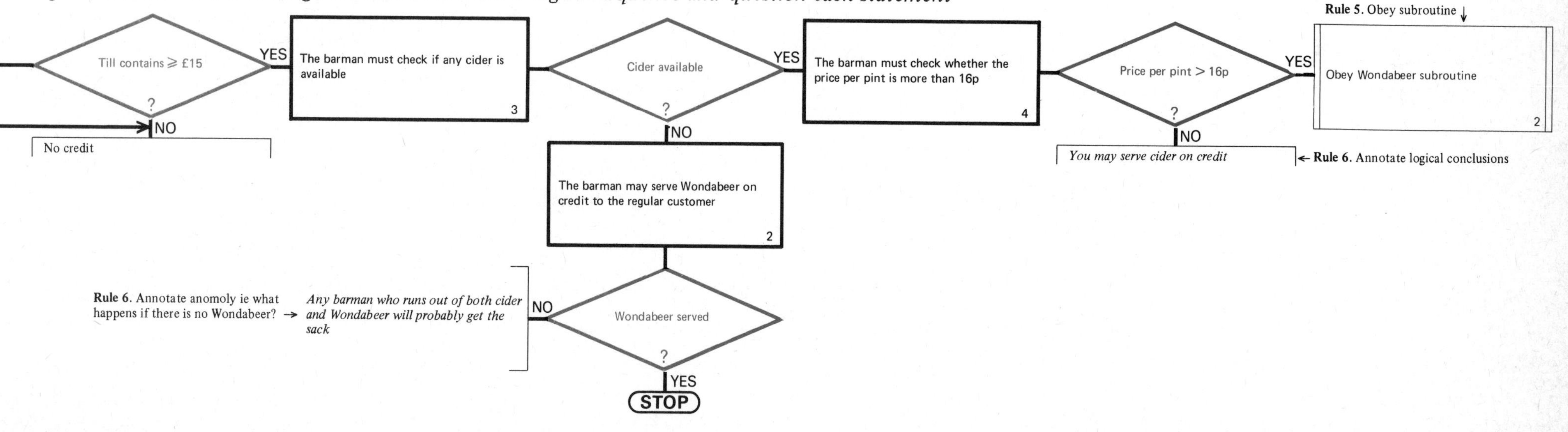

Flowcharting the ICE Conditions

The principles of flowcharting can be applied to legal text to create a logical network to represent the process of sequential judgements required to arrive at the contract's only correct conclusion.

Fig. 8 (pp 16-17) illustrates all the rules of flowcharting applied to the ICE Conditions of Contract (clause 26).

Using the Flowcharts

The flowcharts in Chapter 3 follow the principles outlined in this chapter. The reader is required to read the charts from top to bottom and left to right, in that order. Arrowheads to indicate the direction of flow have not been necessary except in isolated situations where the reader is required either to move directly to a certain position in the charts or to move against the normal flow of top to bottom and left to right.

The legal source of a statement is indicated in the bottom right-hand corner of each box. The numerical code in the bottom left-hand corner is a reference aid for simplifying the use of subroutines enabling readers to readily find the exact point at which to enter an 'obey' routine. A dot (•) under a subroutine box signifies that the particular subroutine is to be published in volume 2. Example 15 illustrates both of these points.

The flowcharts should be self-explanatory although two signals used at the end of certain statements perhaps require explanation. The sign **:—** is used to break information up into manageable units or signifies there are *variable* endings to be fitted to such a statement. The sign **;** is also used to break information up into manageable chunks and indicates that *additional* text follows. Example 16 illustrates both of these points.

Obviously every question demands an objective answer. When disputes arise the parties should proceed step by step to the question point at which one states 'yes' and the other 'no'. This then pinpoints where their disagreement lies, and the flowchart will indicate to the parties their remedies or the contractual consequences.

Example 15

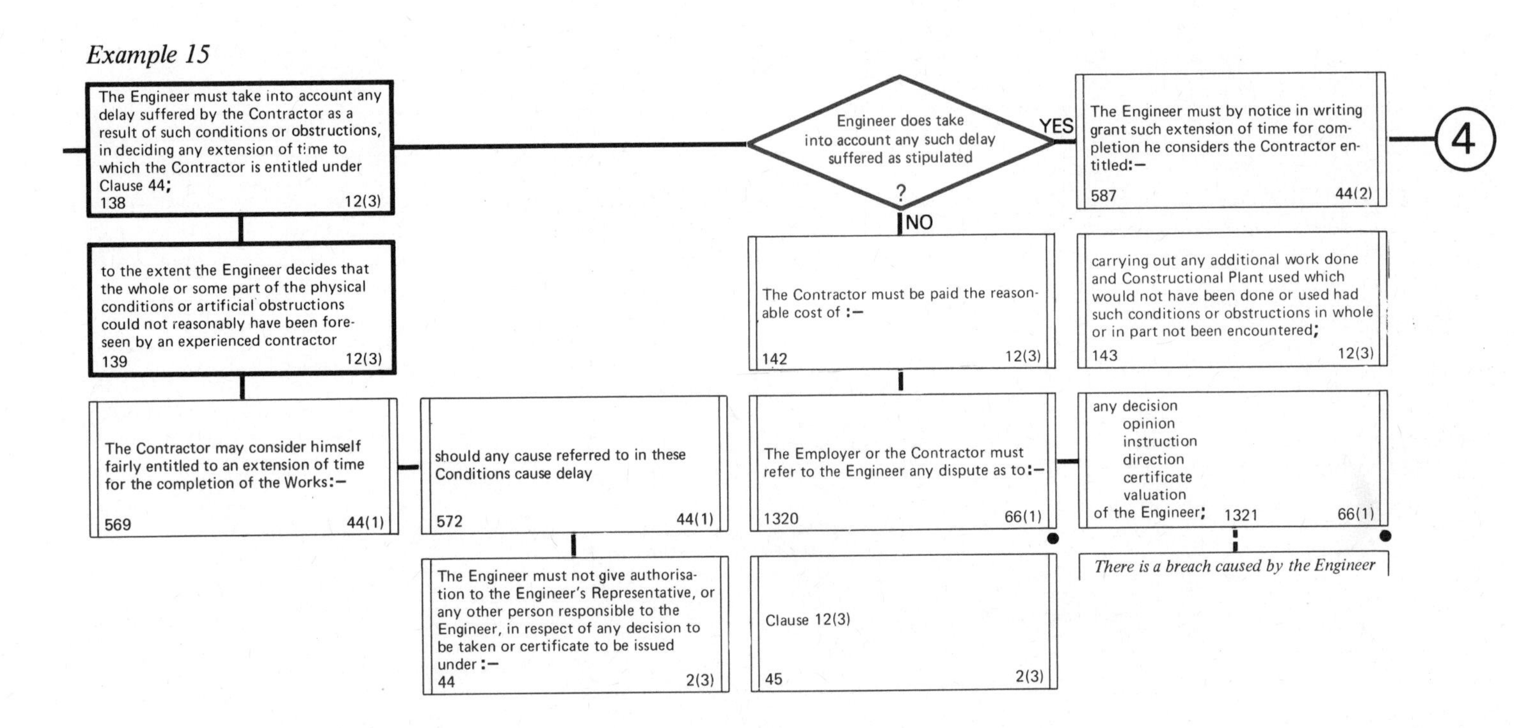

Example 16

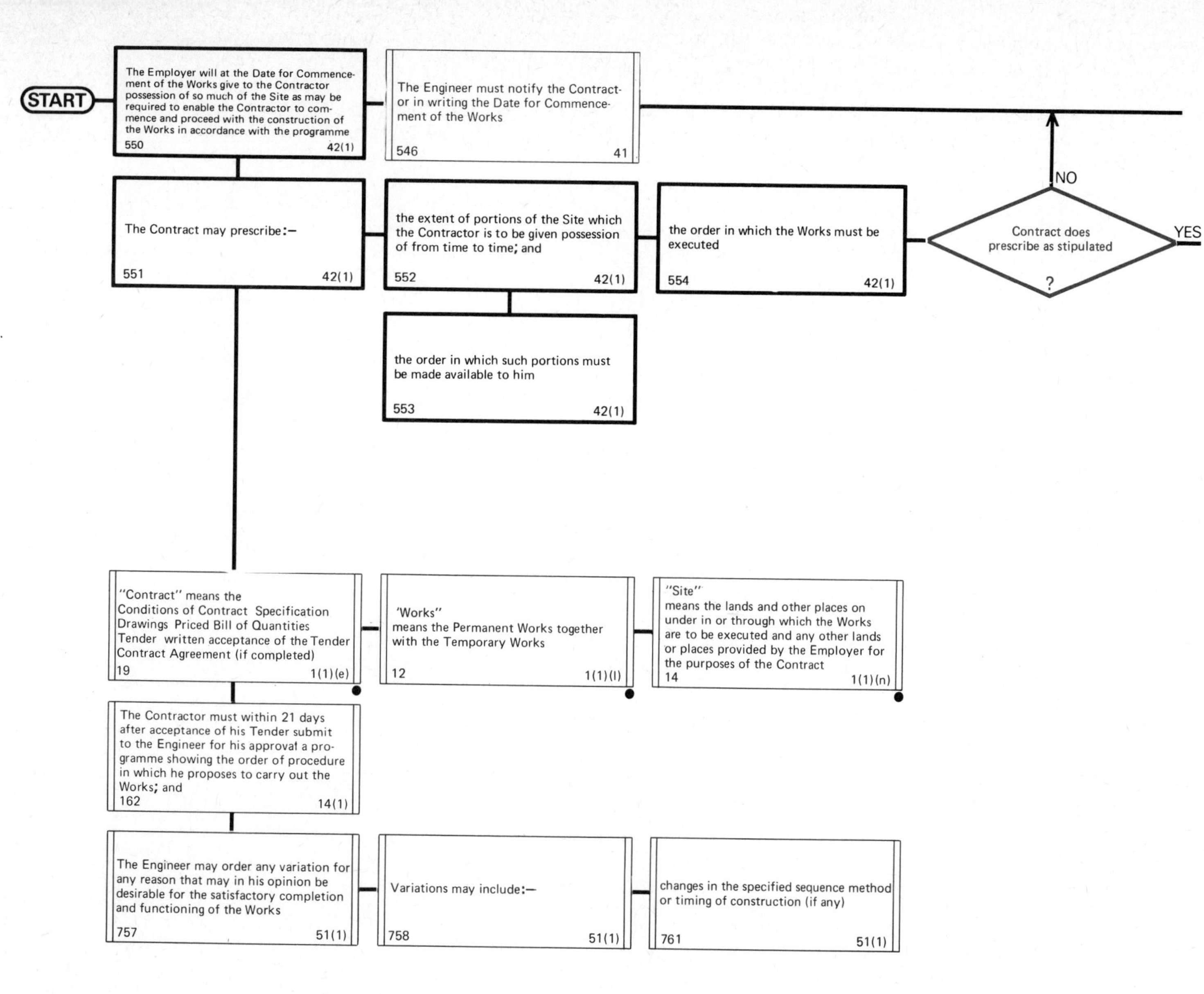

START
The Employer will at the Date for Commencement of the Works give to the Contractor possession of so much of the Site as may be required to enable the Contractor to commence and proceed with the construction of the Works in accordance with the programme
550 42(1)
The Engineer must notify the Contractor in writing the Date for Commencement of the Works
546 41
The Contract may prescribe:—
551 42(1)
the extent of portions of the Site which the Contractor is to be given possession of from time to time; and
552 42(1)
the order in which the Works must be executed
554 42(1)
Contract does prescribe as stipulated
?
NO
YES
the order in which such portions must be made available to him
553 42(1)
"Contract" means the Conditions of Contract Specification Drawings Priced Bill of Quantities Tender written acceptance of the Tender Contract Agreement (if completed)
19 1(1)(e)
'Works" means the Permanent Works together with the Temporary Works
12 1(1)(l)
"Site" means the lands and other places on under in or through which the Works are to be executed and any other lands or places provided by the Employer for the purposes of the Contract
14 1(1)(n)
The Contractor must within 21 days after acceptance of his Tender submit to the Engineer for his approval a programme showing the order of procedure in which he proposes to carry out the Works; and
162 14(1)
The Engineer may order any variation for any reason that may in his opinion be desirable for the satisfactory completion and functioning of the Works
757 51(1)
Variations may include:—
758 51(1)
changes in the specified sequence method or timing of construction (if any)
761 51(1)

Figure 8

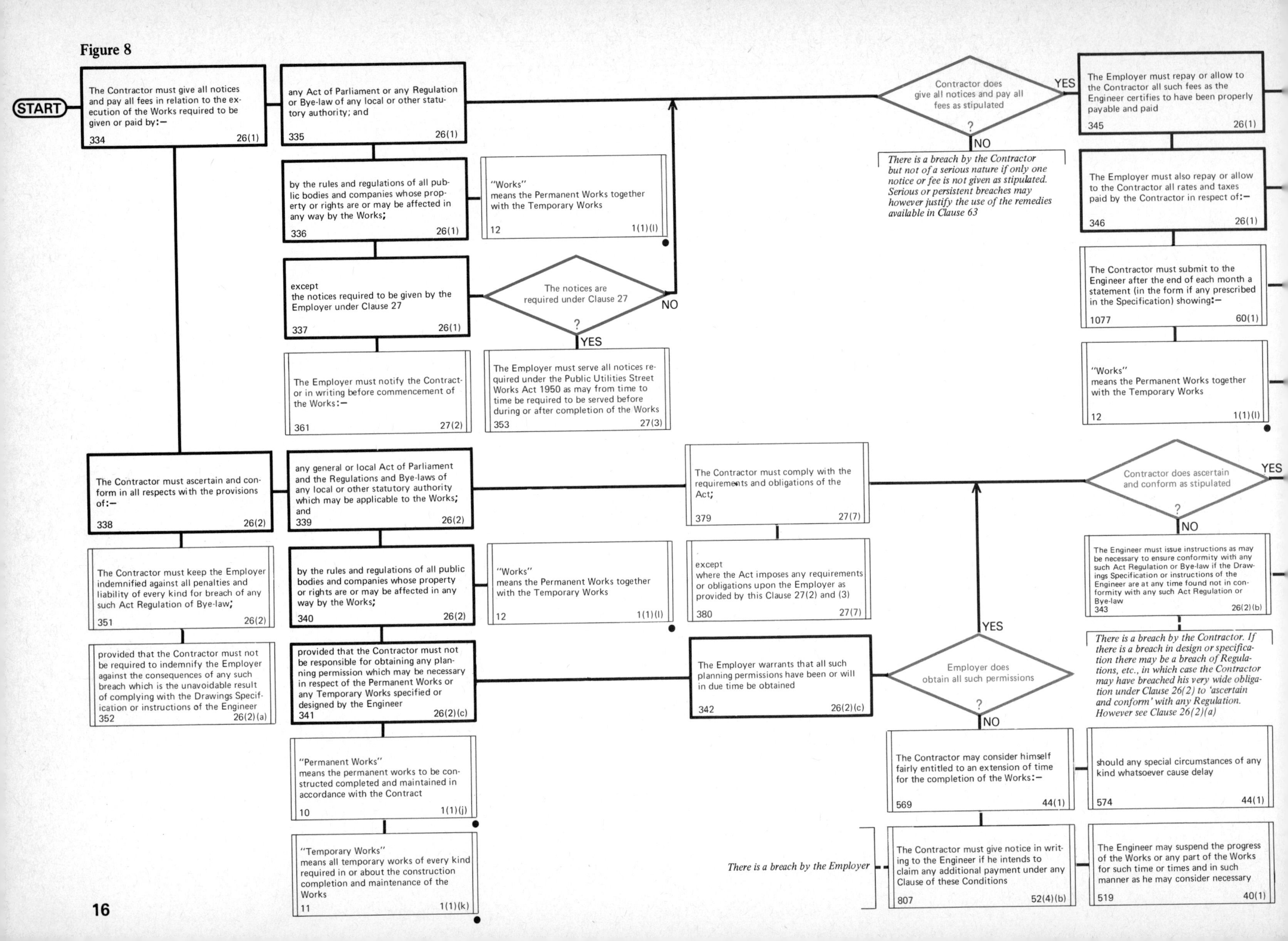

START
The Contractor must give all notices and pay all fees in relation to the execution of the Works required to be given or paid by:— 334 26(1)
any Act of Parliament or any Regulation or Bye-law of any local or other statutory authority; and 335 26(1)
by the rules and regulations of all public bodies and companies whose property or rights are or may be affected in any way by the Works; 336 26(1)
"Works" means the Permanent Works together with the Temporary Works 12 1(1)(l)
except the notices required to be given by the Employer under Clause 27 337 26(1)
The notices are required under Clause 27 ?
NO
YES
The Employer must notify the Contractor in writing before commencement of the Works:— 361 27(2)
The Employer must serve all notices required under the Public Utilities Street Works Act 1950 as may from time to time be required to be served before during or after completion of the Works 353 27(3)
Contractor does give all notices and pay all fees as stipulated ?
YES
NO
There is a breach by the Contractor but not of a serious nature if only one notice or fee is not given as stipulated. Serious or persistent breaches may however justify the use of the remedies available in Clause 63
The Employer must repay or allow to the Contractor all such fees as the Engineer certifies to have been properly payable and paid 345 26(1)
The Employer must also repay or allow to the Contractor all rates and taxes paid by the Contractor in respect of:— 346 26(1)
The Contractor must submit to the Engineer after the end of each month a statement (in the form if any prescribed in the Specification) showing:— 1077 60(1)
"Works" means the Permanent Works together with the Temporary Works 12 1(1)(l)
The Contractor must ascertain and conform in all respects with the provisions of:— 338 26(2)
any general or local Act of Parliament and the Regulations and Bye-laws of any local or other statutory authority which may be applicable to the Works; and 339 26(2)
The Contractor must comply with the requirements and obligations of the Act; 379 27(7)
Contractor does ascertain and conform as stipulated ?
YES
NO
The Engineer must issue instructions as may be necessary to ensure conformity with any such Act Regulation or Bye-law if the Drawings Specification or instructions of the Engineer are at any time found not in conformity with any such Act Regulation or Bye-law 343 26(2)(b)
There is a breach by the Contractor. If there is a breach in design or specification there may be a breach of Regulations, etc., in which case the Contractor may have breached his very wide obligation under Clause 26(2) to 'ascertain and conform' with any Regulation. However see Clause 26(2)(a)
The Contractor must keep the Employer indemnified against all penalties and liability of every kind for breach of any such Act Regulation of Bye-law; 351 26(2)
by the rules and regulations of all public bodies and companies whose property or rights are or may be affected in any way by the Works; 340 26(2)
"Works" means the Permanent Works together with the Temporary Works 12 1(1)(l)
except where the Act imposes any requirements or obligations upon the Employer as provided by this Clause 27(2) and (3) 380 27(7)
provided that the Contractor must not be required to indemnify the Employer against the consequences of any such breach which is the unavoidable result of complying with the Drawings Specification or instructions of the Engineer 352 26(2)(a)
provided that the Contractor must not be responsible for obtaining any planning permission which may be necessary in respect of the Permanent Works or any Temporary Works specified or designed by the Engineer 341 26(2)(c)
The Employer warrants that all such planning permissions have been or will in due time be obtained 342 26(2)(c)
Employer does obtain all such permissions ?
YES
NO
"Permanent Works" means the permanent works to be constructed completed and maintained in accordance with the Contract 10 1(1)(j)
"Temporary Works" means all temporary works of every kind required in or about the construction completion and maintenance of the Works 11 1(1)(k)
The Contractor may consider himself fairly entitled to an extension of time for the completion of the Works:— 569 44(1)
should any special circumstances of any kind whatsoever cause delay 574 44(1)
There is a breach by the Employer
The Contractor must give notice in writing to the Engineer if he intends to claim any additional payment under any Clause of these Conditions 807 52(4)(b)
The Engineer may suspend the progress of the Works or any part of the Works for such time or times and in such manner as he may consider necessary 519 40(1)

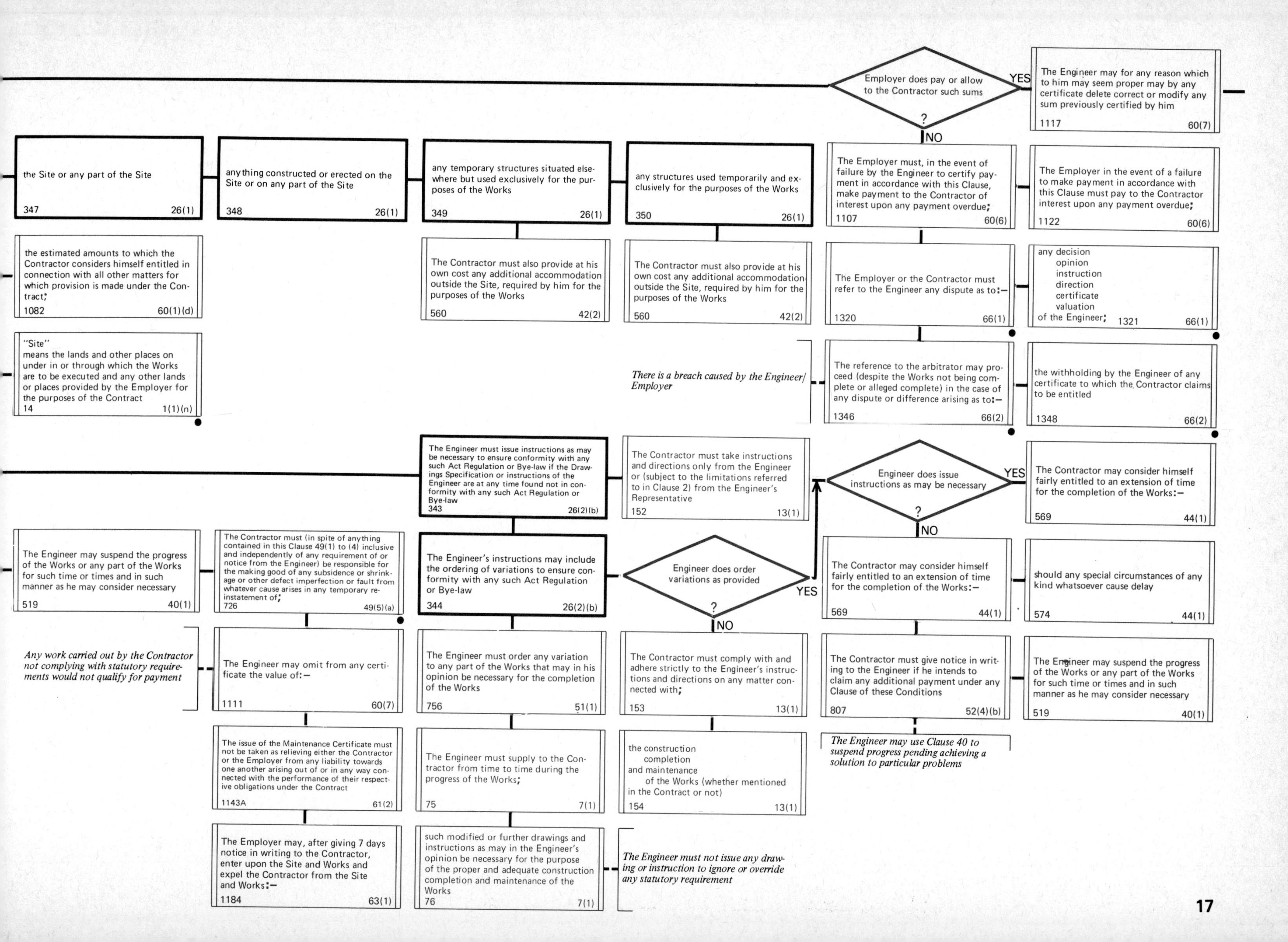
Employer does pay or allow to the Contractor such sums ?
YES
NO
The Engineer may for any reason which to him may seem proper may by any certificate delete correct or modify any sum previously certified by him 1117 60(7)
the Site or any part of the Site 347 26(1)
anything constructed or erected on the Site or on any part of the Site 348 26(1)
any temporary structures situated elsewhere but used exclusively for the purposes of the Works 349 26(1)
any structures used temporarily and exclusively for the purposes of the Works 350 26(1)
The Employer must, in the event of failure by the Engineer to certify payment in accordance with this Clause, make payment to the Contractor of interest upon any payment overdue; 1107 60(6)
The Employer in the event of a failure to make payment in accordance with this Clause must pay to the Contractor interest upon any payment overdue; 1122 60(6)
the estimated amounts to which the Contractor considers himself entitled in connection with all other matters for which provision is made under the Contract; 1082 60(1)(d)
The Contractor must also provide at his own cost any additional accommodation outside the Site, required by him for the purposes of the Works 560 42(2)
The Contractor must also provide at his own cost any additional accommodation outside the Site, required by him for the purposes of the Works 560 42(2)
The Employer or the Contractor must refer to the Engineer any dispute as to:– 1320 66(1)
any decision opinion instruction direction certificate valuation of the Engineer; 1321 66(1)
"Site" means the lands and other places on under in or through which the Works are to be executed and any other lands or places provided by the Employer for the purposes of the Contract 14 1(1)(n)
There is a breach caused by the Engineer/ Employer
The reference to the arbitrator may proceed (despite the Works not being complete or alleged complete) in the case of any dispute or difference arising as to:– 1346 66(2)
the withholding by the Engineer of any certificate to which the Contractor claims to be entitled 1348 66(2)
The Engineer must issue instructions as may be necessary to ensure conformity with any such Act Regulation or Bye-law if the Drawings Specification or instructions of the Engineer are at any time found not in conformity with any such Act Regulation or Bye-law 343 26(2)(b)
The Contractor must take instructions and directions only from the Engineer or (subject to the limitations referred to in Clause 2) from the Engineer's Representative 152 13(1)
Engineer does issue instructions as may be necessary ?
YES
NO
The Contractor may consider himself fairly entitled to an extension of time for the completion of the Works:– 569 44(1)
The Engineer may suspend the progress of the Works or any part of the Works for such time or times and in such manner as he may consider necessary 519 40(1)
The Contractor must (in spite of anything contained in this Clause 49(1) to (4) inclusive and independently of any requirement of or notice from the Engineer) be responsible for the making good of any subsidence or shrinkage or other defect imperfection or fault from whatever cause arises in any temporary reinstatement of; 726 49(5)(a)
The Engineer's instructions may include the ordering of variations to ensure conformity with any such Act Regulation or Bye-law 344 26(2)(b)
Engineer does order variations as provided ?
YES
NO
The Contractor may consider himself fairly entitled to an extension of time for the completion of the Works:– 569 44(1)
should any special circumstances of any kind whatsoever cause delay 574 44(1)
Any work carried out by the Contractor not complying with statutory requirements would not qualify for payment
The Engineer may omit from any certificate the value of:– 1111 60(7)
The Engineer must order any variation to any part of the Works that may in his opinion be necessary for the completion of the Works 756 51(1)
The Contractor must comply with and adhere strictly to the Engineer's instructions and directions on any matter connected with; 153 13(1)
The Contractor must give notice in writing to the Engineer if he intends to claim any additional payment under any Clause of these Conditions 807 52(4)(b)
The Engineer may suspend the progress of the Works or any part of the Works for such time or times and in such manner as he may consider necessary 519 40(1)
The Engineer may use Clause 40 to suspend progress pending achieving a solution to particular problems
The issue of the Maintenance Certificate must not be taken as relieving either the Contractor or the Employer from any liability towards one another arising out of or in any way connected with the performance of their respective obligations under the Contract 1143A 61(2)
The Engineer must supply to the Contractor from time to time during the progress of the Works; 75 7(1)
the construction completion and maintenance of the Works (whether mentioned in the Contract or not) 154 13(1)
The Employer may, after giving 7 days notice in writing to the Contractor, enter upon the Site and Works and expel the Contractor from the Site and Works:– 1184 63(1)
such modified or further drawings and instructions as may in the Engineer's opinion be necessary for the purpose of the proper and adequate construction completion and maintenance of the Works 76 7(1)
The Engineer must not issue any drawing or instruction to ignore or override any statutory requirement

FLOWCHARTS

Table of Flowcharts

Index to Flowcharts

Clause 2 ENGINEER'S REPRESENTATIVE

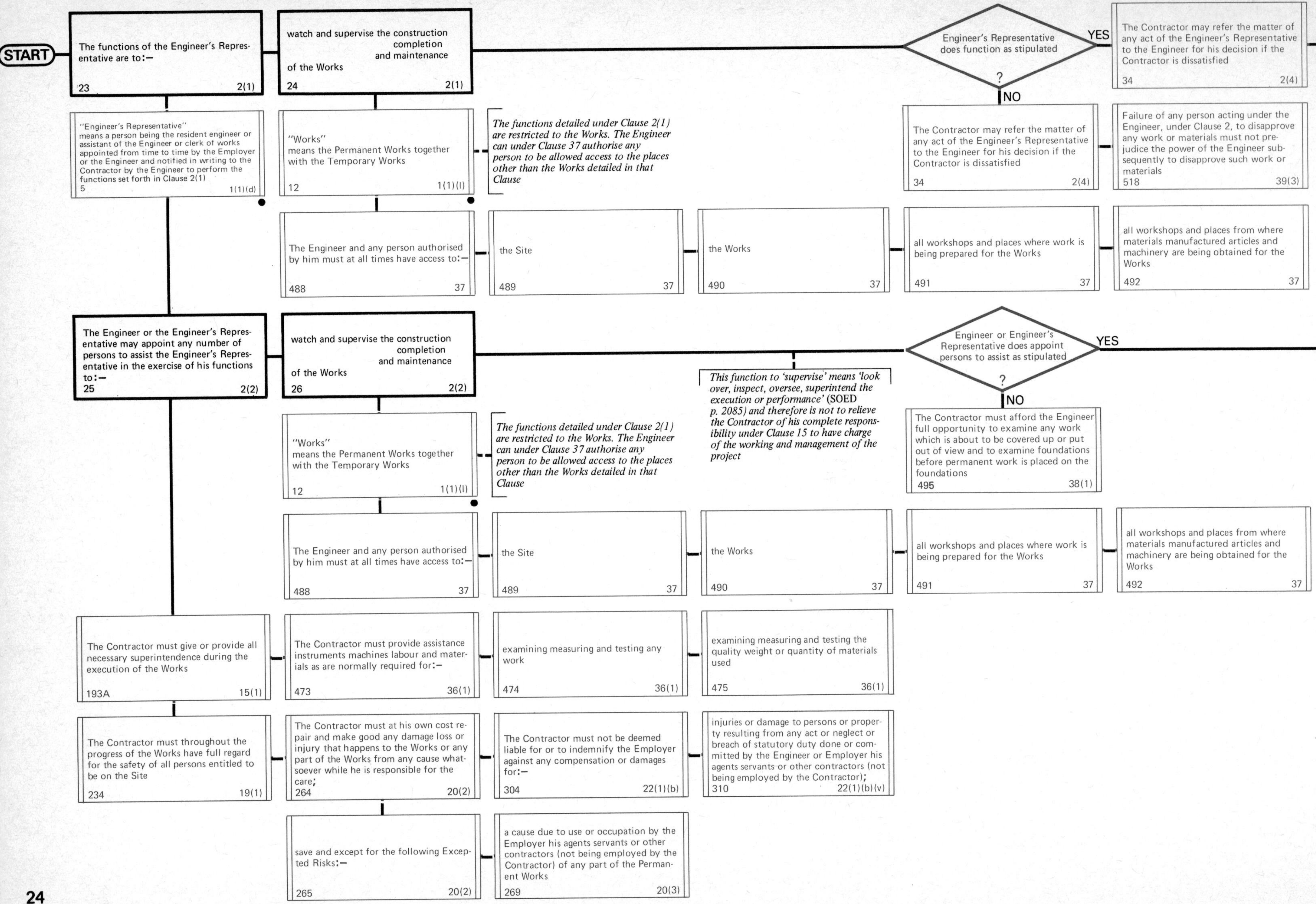

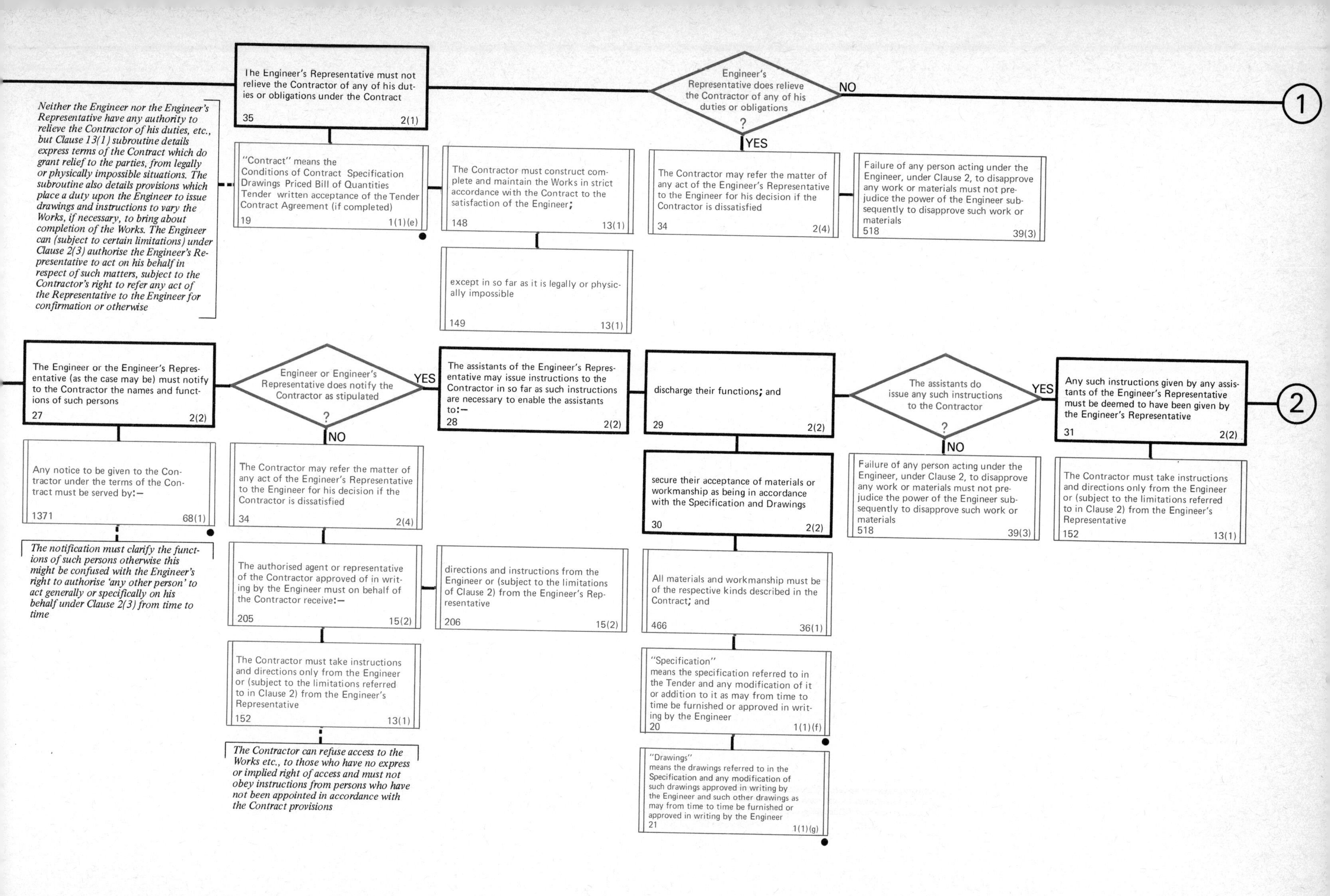

The Engineer's Representative must not relieve the Contractor of any of his duties or obligations under the Contract
35 2(1)
Engineer's Representative does relieve the Contractor of any of his duties or obligations ?
NO
YES
1
Neither the Engineer nor the Engineer's Representative have any authority to relieve the Contractor of his duties, etc., but Clause 13(1) subroutine details express terms of the Contract which do grant relief to the parties, from legally or physically impossible situations. The subroutine also details provisions which place a duty upon the Engineer to issue drawings and instructions to vary the Works, if necessary, to bring about completion of the Works. The Engineer can (subject to certain limitations) under Clause 2(3) authorise the Engineer's Representative to act on his behalf in respect of such matters, subject to the Contractor's right to refer any act of the Representative to the Engineer for confirmation or otherwise
"Contract" means the Conditions of Contract Specification Drawings Priced Bill of Quantities Tender written acceptance of the Tender Contract Agreement (if completed)
19 1(1)(e)
The Contractor must construct complete and maintain the Works in strict accordance with the Contract to the satisfaction of the Engineer;
148 13(1)
except in so far as it is legally or physically impossible
149 13(1)
The Contractor may refer the matter of any act of the Engineer's Representative to the Engineer for his decision if the Contractor is dissatisfied
34 2(4)
Failure of any person acting under the Engineer, under Clause 2, to disapprove any work or materials must not prejudice the power of the Engineer subsequently to disapprove such work or materials
518 39(3)
The Engineer or the Engineer's Representative (as the case may be) must notify to the Contractor the names and functions of such persons
27 2(2)
Engineer or Engineer's Representative does notify the Contractor as stipulated ?
YES
NO
The assistants of the Engineer's Representative may issue instructions to the Contractor in so far as such instructions are necessary to enable the assistants to:—
28 2(2)
discharge their functions; and
29 2(2)
The assistants do issue any such instructions to the Contractor ?
YES
NO
Any such instructions given by any assistants of the Engineer's Representative must be deemed to have been given by the Engineer's Representative
31 2(2)
2
Any notice to be given to the Contractor under the terms of the Contract must be served by:—
1371 68(1)
The notification must clarify the functions of such persons otherwise this might be confused with the Engineer's right to authorise 'any other person' to act generally or specifically on his behalf under Clause 2(3) from time to time
The Contractor may refer the matter of any act of the Engineer's Representative to the Engineer for his decision if the Contractor is dissatisfied
34 2(4)
The authorised agent or representative of the Contractor approved of in writing by the Engineer must on behalf of the Contractor receive:—
205 15(2)
directions and instructions from the Engineer or (subject to the limitations of Clause 2) from the Engineer's Representative
206 15(2)
The Contractor must take instructions and directions only from the Engineer or (subject to the limitations referred to in Clause 2) from the Engineer's Representative
152 13(1)
The Contractor can refuse access to the Works etc., to those who have no express or implied right of access and must not obey instructions from persons who have not been appointed in accordance with the Contract provisions
secure their acceptance of materials or workmanship as being in accordance with the Specification and Drawings
30 2(2)
All materials and workmanship must be of the respective kinds described in the Contract; and
466 36(1)
"Specification" means the specification referred to in the Tender and any modification of it or addition to it as may from time to time be furnished or approved in writing by the Engineer
20 1(1)(f)
"Drawings" means the drawings referred to in the Specification and any modification of such drawings approved in writing by the Engineer and such other drawings as may from time to time be furnished or approved in writing by the Engineer
21 1(1)(g)
Failure of any person acting under the Engineer, under Clause 2, to disapprove any work or materials must not prejudice the power of the Engineer subsequently to disapprove such work or materials
518 39(3)
The Contractor must take instructions and directions only from the Engineer or (subject to the limitations referred to in Clause 2) from the Engineer's Representative
152 13(1)

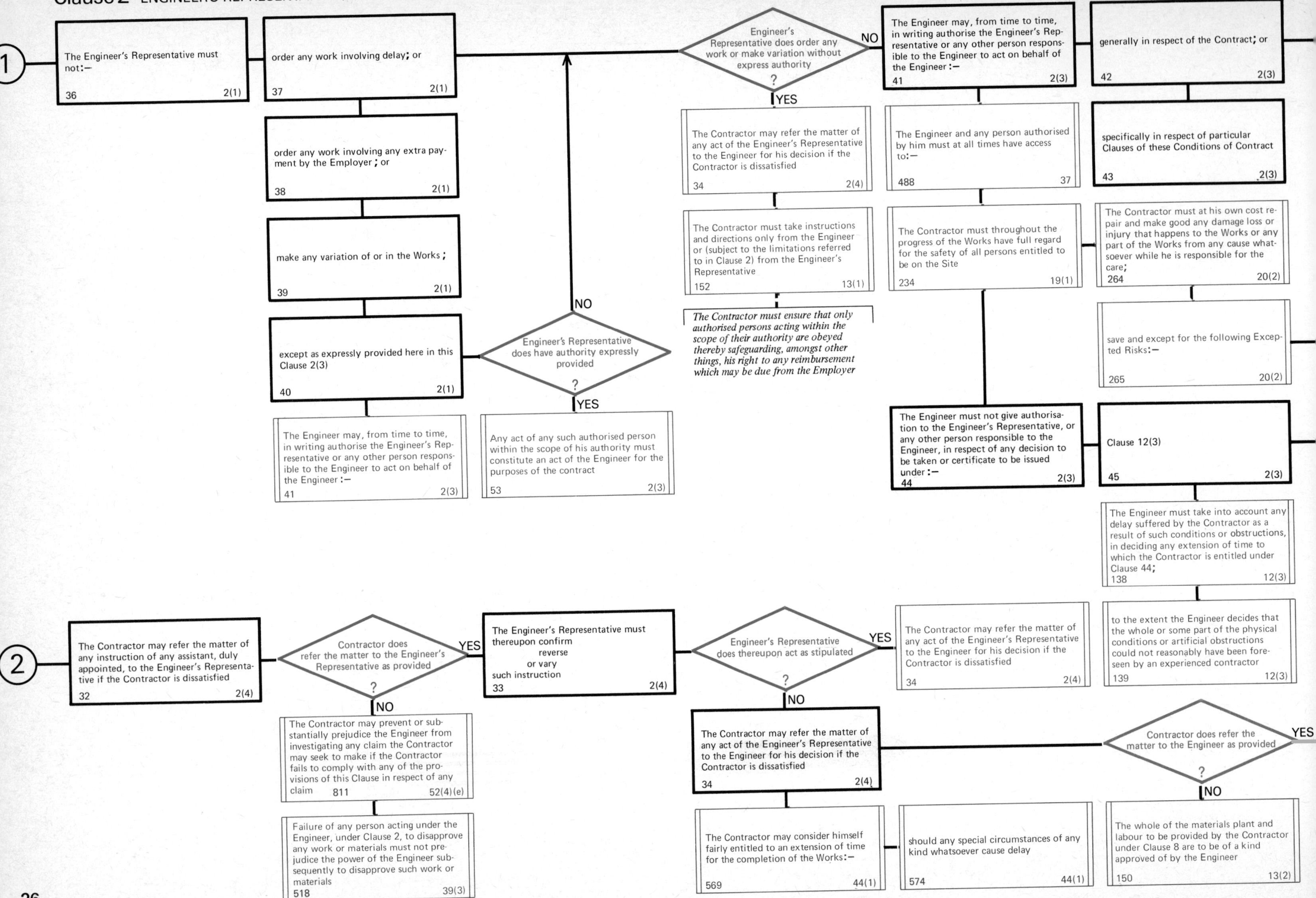
1
The Engineer's Representative must not:— 36 2(1)
order any work involving delay; or 37 2(1)
order any work involving any extra payment by the Employer; or 38 2(1)
make any variation of or in the Works; 39 2(1)
except as expressly provided here in this Clause 2(3) 40 2(1)
The Engineer may, from time to time, in writing authorise the Engineer's Representative or any other person responsible to the Engineer to act on behalf of the Engineer:— 41 2(3)
Engineer's Representative does have authority expressly provided ?
YES
NO
Any act of any such authorised person within the scope of his authority must constitute an act of the Engineer for the purposes of the contract 53 2(3)
Engineer's Representative does order any work or make variation without express authority ?
YES
NO
The Contractor may refer the matter of any act of the Engineer's Representative to the Engineer for his decision if the Contractor is dissatisfied 34 2(4)
The Contractor must take instructions and directions only from the Engineer or (subject to the limitations referred to in Clause 2) from the Engineer's Representative 152 13(1)
The Contractor must ensure that only authorised persons acting within the scope of their authority are obeyed thereby safeguarding, amongst other things, his right to any reimbursement which may be due from the Employer
The Engineer may, from time to time, in writing authorise the Engineer's Representative or any other person responsible to the Engineer to act on behalf of the Engineer:— 41 2(3)
generally in respect of the Contract; or 42 2(3)
specifically in respect of particular Clauses of these Conditions of Contract 43 2(3)
The Engineer and any person authorised by him must at all times have access to:— 488 37
The Contractor must throughout the progress of the Works have full regard for the safety of all persons entitled to be on the Site 234 19(1)
The Contractor must at his own cost repair and make good any damage loss or injury that happens to the Works or any part of the Works from any cause whatsoever while he is responsible for the care; 264 20(2)
save and except for the following Excepted Risks:— 265 20(2)
The Engineer must not give authorisation to the Engineer's Representative, or any other person responsible to the Engineer, in respect of any decision to be taken or certificate to be issued under:— 44 2(3)
Clause 12(3) 45 2(3)
The Engineer must take into account any delay suffered by the Contractor as a result of such conditions or obstructions, in deciding any extension of time to which the Contractor is entitled under Clause 44; 138 12(3)
to the extent the Engineer decides that the whole or some part of the physical conditions or artificial obstructions could not reasonably have been foreseen by an experienced contractor 139 12(3)
2
The Contractor may refer the matter of any instruction of any assistant, duly appointed, to the Engineer's Representative if the Contractor is dissatisfied 32 2(4)
Contractor does refer the matter to the Engineer's Representative as provided ?
YES
NO
The Contractor may prevent or substantially prejudice the Engineer from investigating any claim the Contractor may seek to make if the Contractor fails to comply with any of the provisions of this Clause in respect of any claim 811 52(4)(e)
Failure of any person acting under the Engineer, under Clause 2, to disapprove any work or materials must not prejudice the power of the Engineer subsequently to disapprove such work or materials 518 39(3)
The Engineer's Representative must thereupon confirm reverse or vary such instruction 33 2(4)
Engineer's Representative does thereupon act as stipulated ?
YES
NO
The Contractor may refer the matter of any act of the Engineer's Representative to the Engineer for his decision if the Contractor is dissatisfied 34 2(4)
The Contractor may refer the matter of any act of the Engineer's Representative to the Engineer for his decision if the Contractor is dissatisfied 34 2(4)
The Contractor may consider himself fairly entitled to an extension of time for the completion of the Works:— 569 44(1)
should any special circumstances of any kind whatsoever cause delay 574 44(1)
Contractor does refer the matter to the Engineer as provided ?
YES
NO
The whole of the materials plant and labour to be provided by the Contractor under Clause 8 are to be of a kind approved of by the Engineer 150 13(2)

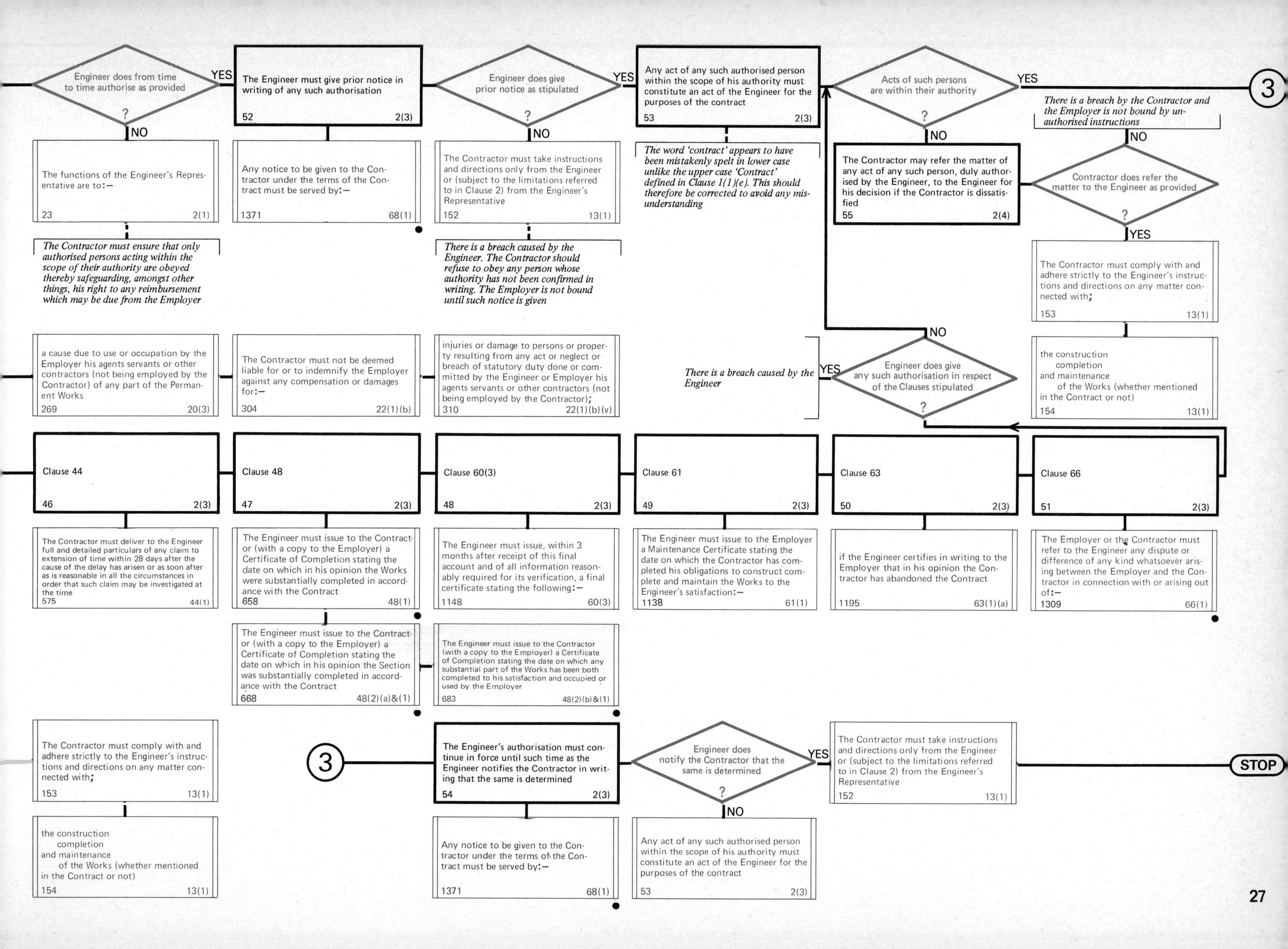

Engineer does from time to time authorise as provided ?
YES
NO
The Engineer must give prior notice in writing of any such authorisation
52 2(3)
Engineer does give prior notice as stipulated ?
YES
NO
Any act of any such authorised person within the scope of his authority must constitute an act of the Engineer for the purposes of the contract
53 2(3)
Acts of such persons are within their authority ?
YES
NO
3
There is a breach by the Contractor and the Employer is not bound by unauthorised instructions
The functions of the Engineer's Representative are to:—
23 2(1)
Any notice to be given to the Contractor under the terms of the Contract must be served by:—
1371 68(1)
The Contractor must take instructions and directions only from the Engineer or (subject to the limitations referred to in Clause 2) from the Engineer's Representative
152 13(1)
The word 'contract' appears to have been mistakenly spelt in lower case unlike the upper case 'Contract' defined in Clause 1(1)(e). This should therefore be corrected to avoid any misunderstanding
The Contractor may refer the matter of any act of any such person, duly authorised by the Engineer, to the Engineer for his decision if the Contractor is dissatisfied
55 2(4)
Contractor does refer the matter to the Engineer as provided ?
YES
The Contractor must ensure that only authorised persons acting within the scope of their authority are obeyed thereby safeguarding, amongst other things, his right to any reimbursement which may be due from the Employer
There is a breach caused by the Engineer. The Contractor should refuse to obey any person whose authority has not been confirmed in writing. The Employer is not bound until such notice is given
The Contractor must comply with and adhere strictly to the Engineer's instructions and directions on any matter connected with;
153 13(1)
a cause due to use or occupation by the Employer his agents servants or other contractors (not being employed by the Contractor) of any part of the Permanent Works
269 20(3)
The Contractor must not be deemed liable for or to indemnify the Employer against any compensation or damages for:—
304 22(1)(b)
injuries or damage to persons or property resulting from any act or neglect or breach of statutory duty done or committed by the Engineer or Employer his agents servants or other contractors (not being employed by the Contractor);
310 22(1)(b)(v)
There is a breach caused by the Engineer
YES
Engineer does give any such authorisation in respect of the Clauses stipulated ?
NO
the construction
completion
and maintenance
of the Works (whether mentioned in the Contract or not)
154 13(1)
Clause 44
46 2(3)
Clause 48
47 2(3)
Clause 60(3)
48 2(3)
Clause 61
49 2(3)
Clause 63
50 2(3)
Clause 66
51 2(3)
The Contractor must deliver to the Engineer full and detailed particulars of any claim to extension of time within 28 days after the cause of the delay has arisen or as soon after as is reasonable in all the circumstances in order that such claim may be investigated at the time
575 44(1)
The Engineer must issue to the Contractor (with a copy to the Employer) a Certificate of Completion stating the date on which in his opinion the Works were substantially completed in accordance with the Contract
658 48(1)
The Engineer must issue, within 3 months after receipt of this final account and of all information reasonably required for its verification, a final certificate stating the following:—
1148 60(3)
The Engineer must issue to the Employer a Maintenance Certificate stating the date on which the Contractor has completed his obligations to construct complete and maintain the Works to the Engineer's satisfaction:—
1138 61(1)
if the Engineer certifies in writing to the Employer that in his opinion the Contractor has abandoned the Contract
1195 63(1)(a)
The Employer or the Contractor must refer to the Engineer any dispute or difference of any kind whatsoever arising between the Employer and the Contractor in connection with or arising out of:—
1309 66(1)
The Engineer must issue to the Contractor (with a copy to the Employer) a Certificate of Completion stating the date on which in his opinion the Section was substantially completed in accordance with the Contract
668 48(2)(a)&(1)
The Engineer must issue to the Contractor (with a copy to the Employer) a Certificate of Completion stating the date on which any substantial part of the Works has been both completed to his satisfaction and occupied or used by the Employer
683 48(2)(b)&(1)
The Contractor must comply with and adhere strictly to the Engineer's instructions and directions on any matter connected with;
153 13(1)
3
The Engineer's authorisation must continue in force until such time as the Engineer notifies the Contractor in writing that the same is determined
54 2(3)
Engineer does notify the Contractor that the same is determined ?
YES
NO
The Contractor must take instructions and directions only from the Engineer or (subject to the limitations referred to in Clause 2) from the Engineer's Representative
152 13(1)
STOP
the construction
completion
and maintenance
of the Works (whether mentioned in the Contract or not)
154 13(1)
Any notice to be given to the Contractor under the terms of the Contract must be served by:—
1371 68(1)
Any act of any such authorised person within the scope of his authority must constitute an act of the Engineer for the purposes of the contract
53 2(3)

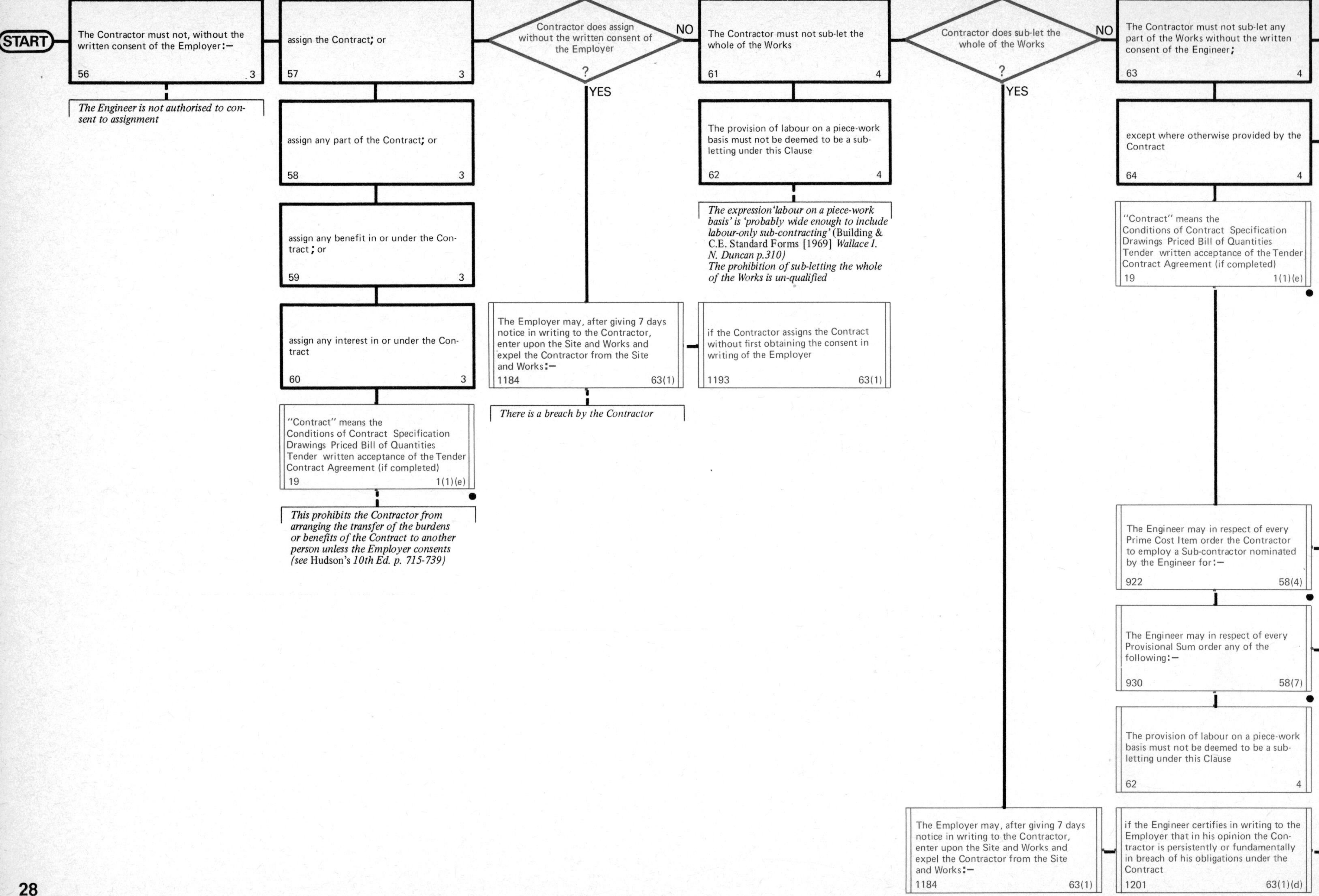
START
The Contractor must not, without the written consent of the Employer:—
56 3
The Engineer is not authorised to consent to assignment
assign the Contract; or
57 3
assign any part of the Contract; or
58 3
assign any benefit in or under the Contract; or
59 3
assign any interest in or under the Contract
60 3
"Contract" means the Conditions of Contract Specification Drawings Priced Bill of Quantities Tender written acceptance of the Tender Contract Agreement (if completed)
19 1(1)(e)
This prohibits the Contractor from arranging the transfer of the burdens or benefits of the Contract to another person unless the Employer consents (see Hudson's 10th Ed. p. 715-739)
Contractor does assign without the written consent of the Employer
?
NO
YES
The Employer may, after giving 7 days notice in writing to the Contractor, enter upon the Site and Works and expel the Contractor from the Site and Works:—
1184 63(1)
There is a breach by the Contractor
The Contractor must not sub-let the whole of the Works
61 4
The provision of labour on a piece-work basis must not be deemed to be a sub-letting under this Clause
62 4
The expression 'labour on a piece-work basis' is 'probably wide enough to include labour-only sub-contracting' (Building & C.E. Standard Forms [1969] Wallace I. N. Duncan p.310)
The prohibition of sub-letting the whole of the Works is un-qualified
if the Contractor assigns the Contract without first obtaining the consent in writing of the Employer
1193 63(1)
Contractor does sub-let the whole of the Works
?
NO
YES
The Employer may, after giving 7 days notice in writing to the Contractor, enter upon the Site and Works and expel the Contractor from the Site and Works:—
1184 63(1)
The Contractor must not sub-let any part of the Works without the written consent of the Engineer;
63 4
except where otherwise provided by the Contract
64 4
"Contract" means the Conditions of Contract Specification Drawings Priced Bill of Quantities Tender written acceptance of the Tender Contract Agreement (if completed)
19 1(1)(e)
The Engineer may in respect of every Prime Cost Item order the Contractor to employ a Sub-contractor nominated by the Engineer for:—
922 58(4)
The Engineer may in respect of every Provisional Sum order any of the following:—
930 58(7)
The provision of labour on a piece-work basis must not be deemed to be a sub-letting under this Clause
62 4
if the Engineer certifies in writing to the Employer that in his opinion the Contractor is persistently or fundamentally in breach of his obligations under the Contract
1201 63(1)(d)

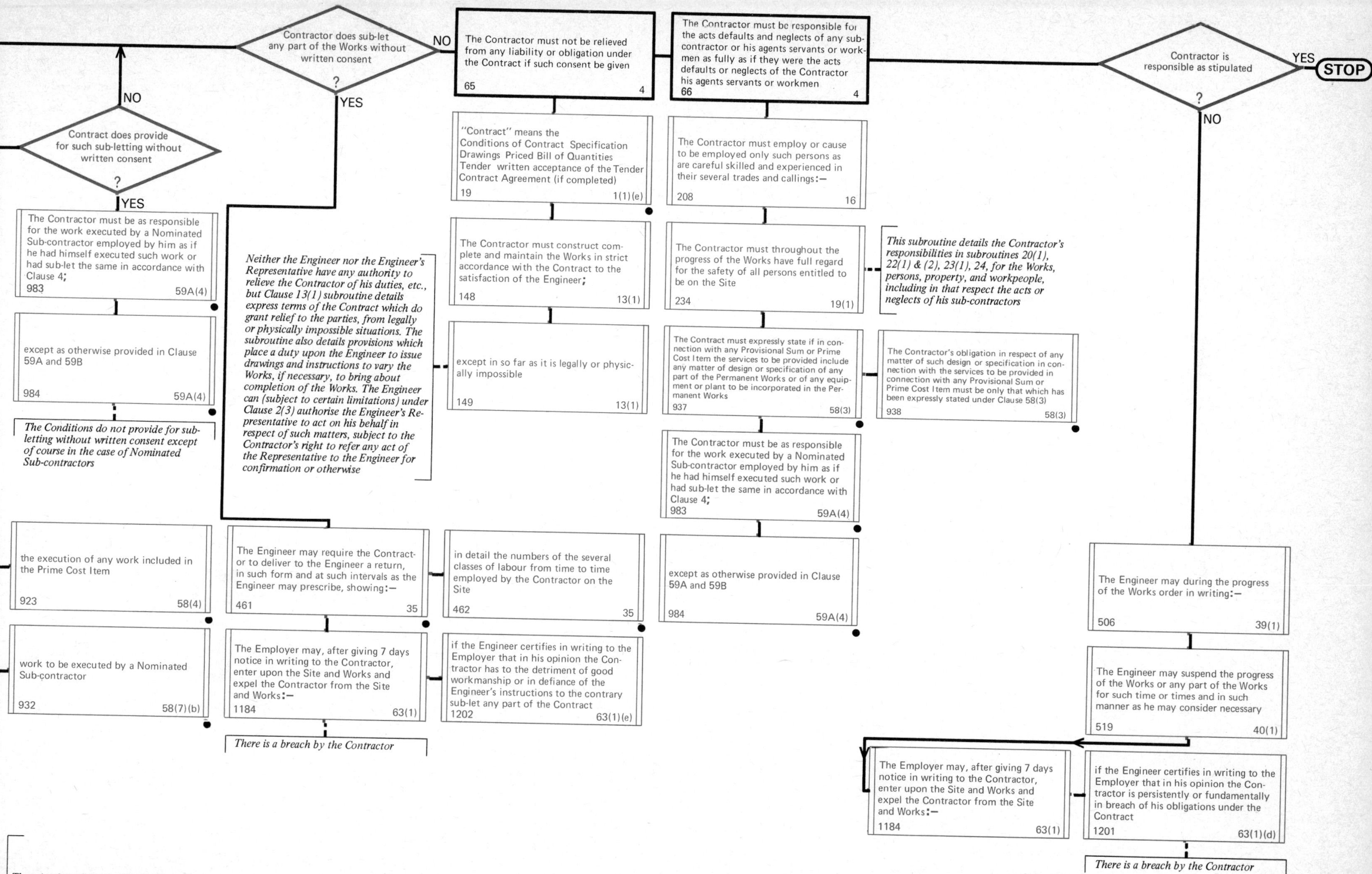
Contractor does sub-let any part of the Works without written consent
?
NO
YES
The Contractor must not be relieved from any liability or obligation under the Contract if such consent be given
65
4
The Contractor must be responsible for the acts defaults and neglects of any sub-contractor or his agents servants or workmen as fully as if they were the acts defaults or neglects of the Contractor his agents servants or workmen
66
4
Contractor is responsible as stipulated
?
YES
STOP
NO
Contract does provide for such sub-letting without written consent
?
NO
YES
The Contractor must be as responsible for the work executed by a Nominated Sub-contractor employed by him as if he had himself executed such work or had sub-let the same in accordance with Clause 4;
983
59A(4)
except as otherwise provided in Clause 59A and 59B
984
59A(4)
The Conditions do not provide for sub-letting without written consent except of course in the case of Nominated Sub-contractors
Neither the Engineer nor the Engineer's Representative have any authority to relieve the Contractor of his duties, etc., but Clause 13(1) subroutine details express terms of the Contract which do grant relief to the parties, from legally or physically impossible situations. The subroutine also details provisions which place a duty upon the Engineer to issue drawings and instructions to vary the Works, if necessary, to bring about completion of the Works. The Engineer can (subject to certain limitations) under Clause 2(3) authorise the Engineer's Representative to act on his behalf in respect of such matters, subject to the Contractor's right to refer any act of the Representative to the Engineer for confirmation or otherwise
"Contract" means the Conditions of Contract Specification Drawings Priced Bill of Quantities Tender written acceptance of the Tender Contract Agreement (if completed)
19
1(1)(e)
The Contractor must construct complete and maintain the Works in strict accordance with the Contract to the satisfaction of the Engineer;
148
13(1)
except in so far as it is legally or physically impossible
149
13(1)
The Contractor must employ or cause to be employed only such persons as are careful skilled and experienced in their several trades and callings:—
208
16
The Contractor must throughout the progress of the Works have full regard for the safety of all persons entitled to be on the Site
234
19(1)
This subroutine details the Contractor's responsibilities in subroutines 20(1), 22(1) & (2), 23(1), 24, for the Works, persons, property, and workpeople, including in that respect the acts or neglects of his sub-contractors
The Contract must expressly state if in connection with any Provisional Sum or Prime Cost Item the services to be provided include any matter of design or specification of any part of the Permanent Works or of any equipment or plant to be incorporated in the Permanent Works
937
58(3)
The Contractor's obligation in respect of any matter of such design or specification in connection with the services to be provided in connection with any Provisional Sum or Prime Cost Item must be only that which has been expressly stated under Clause 58(3)
938
58(3)
The Contractor must be as responsible for the work executed by a Nominated Sub-contractor employed by him as if he had himself executed such work or had sub-let the same in accordance with Clause 4;
983
59A(4)
except as otherwise provided in Clause 59A and 59B
984
59A(4)
the execution of any work included in the Prime Cost Item
923
58(4)
work to be executed by a Nominated Sub-contractor
932
58(7)(b)
The Engineer may require the Contractor to deliver to the Engineer a return, in such form and at such intervals as the Engineer may prescribe, showing:—
461
35
in detail the numbers of the several classes of labour from time to time employed by the Contractor on the Site
462
35
The Employer may, after giving 7 days notice in writing to the Contractor, enter upon the Site and Works and expel the Contractor from the Site and Works:—
1184
63(1)
if the Engineer certifies in writing to the Employer that in his opinion the Contractor has to the detriment of good workmanship or in defiance of the Engineer's instructions to the contrary sub-let any part of the Contract
1202
63(1)(e)
There is a breach by the Contractor
The Engineer may during the progress of the Works order in writing:—
506
39(1)
The Engineer may suspend the progress of the Works or any part of the Works for such time or times and in such manner as he may consider necessary
519
40(1)
The Employer may, after giving 7 days notice in writing to the Contractor, enter upon the Site and Works and expel the Contractor from the Site and Works:—
1184
63(1)
if the Engineer certifies in writing to the Employer that in his opinion the Contractor is persistently or fundamentally in breach of his obligations under the Contract
1201
63(1)(d)
There is a breach by the Contractor
There is a breach by the Contractor

Clauses 5-7 CONTRACT DOCUMENTS

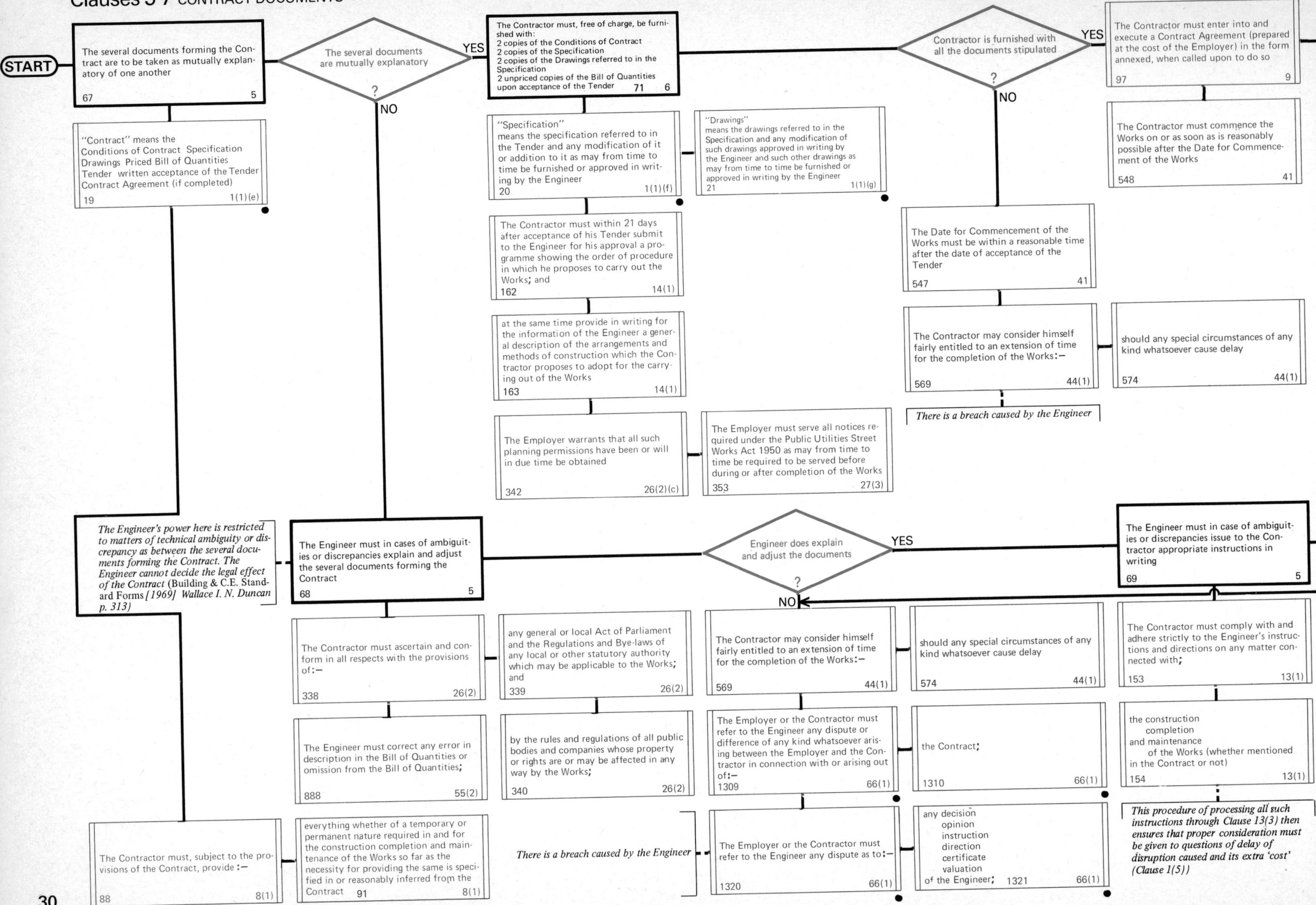

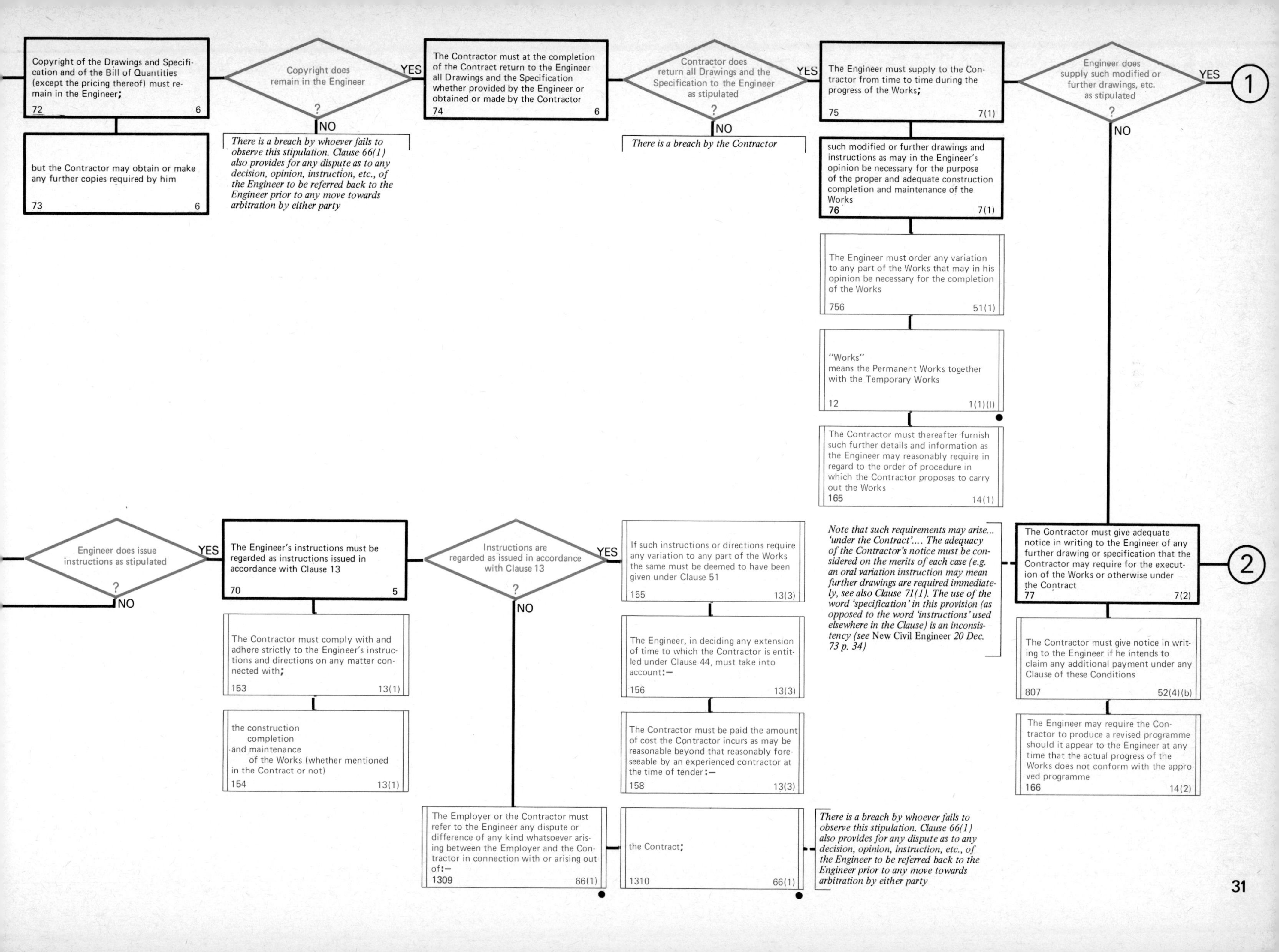

Copyright of the Drawings and Specification and of the Bill of Quantities (except the pricing thereof) must remain in the Engineer;
72 6
but the Contractor may obtain or make any further copies required by him
73 6
Copyright does remain in the Engineer ?
YES
NO
There is a breach by whoever fails to observe this stipulation. Clause 66(1) also provides for any dispute as to any decision, opinion, instruction, etc., of the Engineer to be referred back to the Engineer prior to any move towards arbitration by either party
The Contractor must at the completion of the Contract return to the Engineer all Drawings and the Specification whether provided by the Engineer or obtained or made by the Contractor
74 6
Contractor does return all Drawings and the Specification to the Engineer as stipulated ?
YES
NO
There is a breach by the Contractor
The Engineer must supply to the Contractor from time to time during the progress of the Works;
75 7(1)
such modified or further drawings and instructions as may in the Engineer's opinion be necessary for the purpose of the proper and adequate construction completion and maintenance of the Works
76 7(1)
The Engineer must order any variation to any part of the Works that may in his opinion be necessary for the completion of the Works
756 51(1)
"Works" means the Permanent Works together with the Temporary Works
12 1(1)(l)
The Contractor must thereafter furnish such further details and information as the Engineer may reasonably require in regard to the order of procedure in which the Contractor proposes to carry out the Works
165 14(1)
Engineer does supply such modified or further drawings, etc. as stipulated ?
YES
NO
1
Engineer does issue instructions as stipulated ?
YES
NO
The Engineer's instructions must be regarded as instructions issued in accordance with Clause 13
70 5
The Contractor must comply with and adhere strictly to the Engineer's instructions and directions on any matter connected with;
153 13(1)
the construction completion and maintenance of the Works (whether mentioned in the Contract or not)
154 13(1)
Instructions are regarded as issued in accordance with Clause 13 ?
YES
NO
If such instructions or directions require any variation to any part of the Works the same must be deemed to have been given under Clause 51
155 13(3)
The Engineer, in deciding any extension of time to which the Contractor is entitled under Clause 44, must take into account:—
156 13(3)
The Contractor must be paid the amount of cost the Contractor incurs as may be reasonable beyond that reasonably foreseeable by an experienced contractor at the time of tender:—
158 13(3)
The Employer or the Contractor must refer to the Engineer any dispute or difference of any kind whatsoever arising between the Employer and the Contractor in connection with or arising out of:—
1309 66(1)
the Contract;
1310 66(1)
There is a breach by whoever fails to observe this stipulation. Clause 66(1) also provides for any dispute as to any decision, opinion, instruction, etc., of the Engineer to be referred back to the Engineer prior to any move towards arbitration by either party
Note that such requirements may arise... 'under the Contract'.... The adequacy of the Contractor's notice must be considered on the merits of each case (e.g. an oral variation instruction may mean further drawings are required immediately, see also Clause 71(1). The use of the word 'specification' in this provision (as opposed to the word 'instructions' used elsewhere in the Clause) is an inconsistency (see New Civil Engineer 20 Dec. 73 p. 34)
The Contractor must give adequate notice in writing to the Engineer of any further drawing or specification that the Contractor may require for the execution of the Works or otherwise under the Contract
77 7(2)
2
The Contractor must give notice in writing to the Engineer if he intends to claim any additional payment under any Clause of these Conditions
807 52(4)(b)
The Engineer may require the Contractor to produce a revised programme should it appear to the Engineer at any time that the actual progress of the Works does not conform with the approved programme
166 14(2)

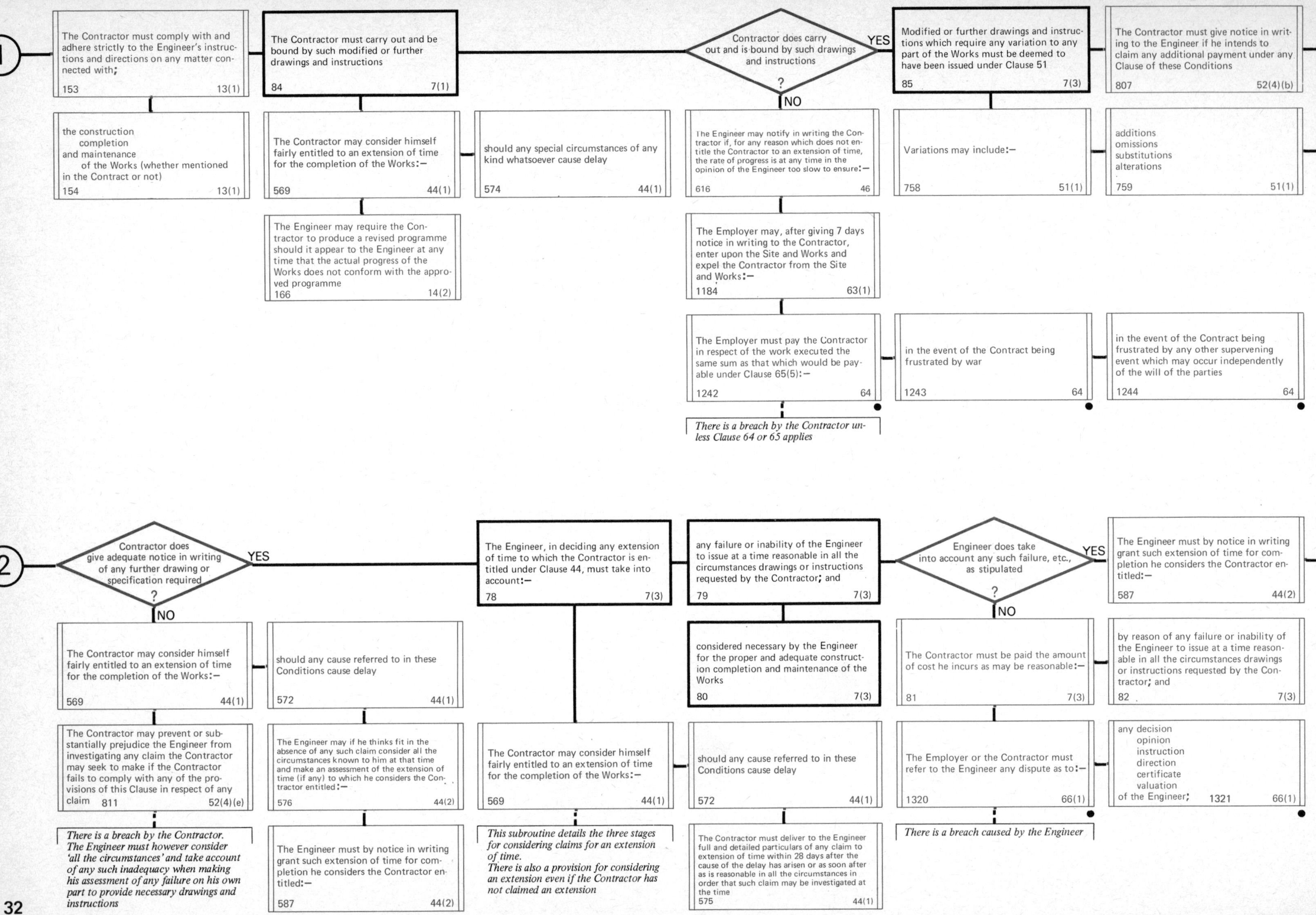
1
The Contractor must comply with and adhere strictly to the Engineer's instructions and directions on any matter connected with;
153 13(1)
The Contractor must carry out and be bound by such modified or further drawings and instructions
84 7(1)
Contractor does carry out and is bound by such drawings and instructions ?
YES
NO
Modified or further drawings and instructions which require any variation to any part of the Works must be deemed to have been issued under Clause 51
85 7(3)
The Contractor must give notice in writing to the Engineer if he intends to claim any additional payment under any Clause of these Conditions
807 52(4)(b)
the construction completion and maintenance of the Works (whether mentioned in the Contract or not)
154 13(1)
The Contractor may consider himself fairly entitled to an extension of time for the completion of the Works:—
569 44(1)
should any special circumstances of any kind whatsoever cause delay
574 44(1)
The Engineer may notify in writing the Contractor if, for any reason which does not entitle the Contractor to an extension of time, the rate of progress is at any time in the opinion of the Engineer too slow to ensure:—
616 46
Variations may include:—
758 51(1)
additions omissions substitutions alterations
759 51(1)
The Engineer may require the Contractor to produce a revised programme should it appear to the Engineer at any time that the actual progress of the Works does not conform with the approved programme
166 14(2)
The Employer may, after giving 7 days notice in writing to the Contractor, enter upon the Site and Works and expel the Contractor from the Site and Works:—
1184 63(1)
The Employer must pay the Contractor in respect of the work executed the same sum as that which would be payable under Clause 65(5):—
1242 64
in the event of the Contract being frustrated by war
1243 64
in the event of the Contract being frustrated by any other supervening event which may occur independently of the will of the parties
1244 64
There is a breach by the Contractor unless Clause 64 or 65 applies
2
Contractor does give adequate notice in writing of any further drawing or specification required ?
YES
NO
The Engineer, in deciding any extension of time to which the Contractor is entitled under Clause 44, must take into account:—
78 7(3)
any failure or inability of the Engineer to issue at a time reasonable in all the circumstances drawings or instructions requested by the Contractor; and
79 7(3)
Engineer does take into account any such failure, etc., as stipulated ?
YES
NO
The Engineer must by notice in writing grant such extension of time for completion he considers the Contractor entitled:—
587 44(2)
The Contractor may consider himself fairly entitled to an extension of time for the completion of the Works:—
569 44(1)
should any cause referred to in these Conditions cause delay
572 44(1)
considered necessary by the Engineer for the proper and adequate construction completion and maintenance of the Works
80 7(3)
The Contractor must be paid the amount of cost he incurs as may be reasonable:—
81 7(3)
by reason of any failure or inability of the Engineer to issue at a time reasonable in all the circumstances drawings or instructions requested by the Contractor; and
82 7(3)
The Contractor may prevent or substantially prejudice the Engineer from investigating any claim the Contractor may seek to make if the Contractor fails to comply with any of the provisions of this Clause in respect of any claim 811 52(4)(e)
The Engineer may if he thinks fit in the absence of any such claim consider all the circumstances known to him at that time and make an assessment of the extension of time (if any) to which he considers the Contractor entitled:—
576 44(2)
The Contractor may consider himself fairly entitled to an extension of time for the completion of the Works:—
569 44(1)
should any cause referred to in these Conditions cause delay
572 44(1)
The Employer or the Contractor must refer to the Engineer any dispute as to:—
1320 66(1)
any decision opinion instruction direction certificate valuation of the Engineer; 1321 66(1)
There is a breach by the Contractor. The Engineer must however consider 'all the circumstances' and take account of any such inadequacy when making his assessment of any failure on his own part to provide necessary drawings and instructions
The Engineer must by notice in writing grant such extension of time for completion he considers the Contractor entitled:—
587 44(2)
This subroutine details the three stages for considering claims for an extension of time.
There is also a provision for considering an extension even if the Contractor has not claimed an extension
The Contractor must deliver to the Engineer full and detailed particulars of any claim to extension of time within 28 days after the cause of the delay has arisen or as soon after as is reasonable in all the circumstances in order that such claim may be investigated at the time
575 44(1)
There is a breach caused by the Engineer

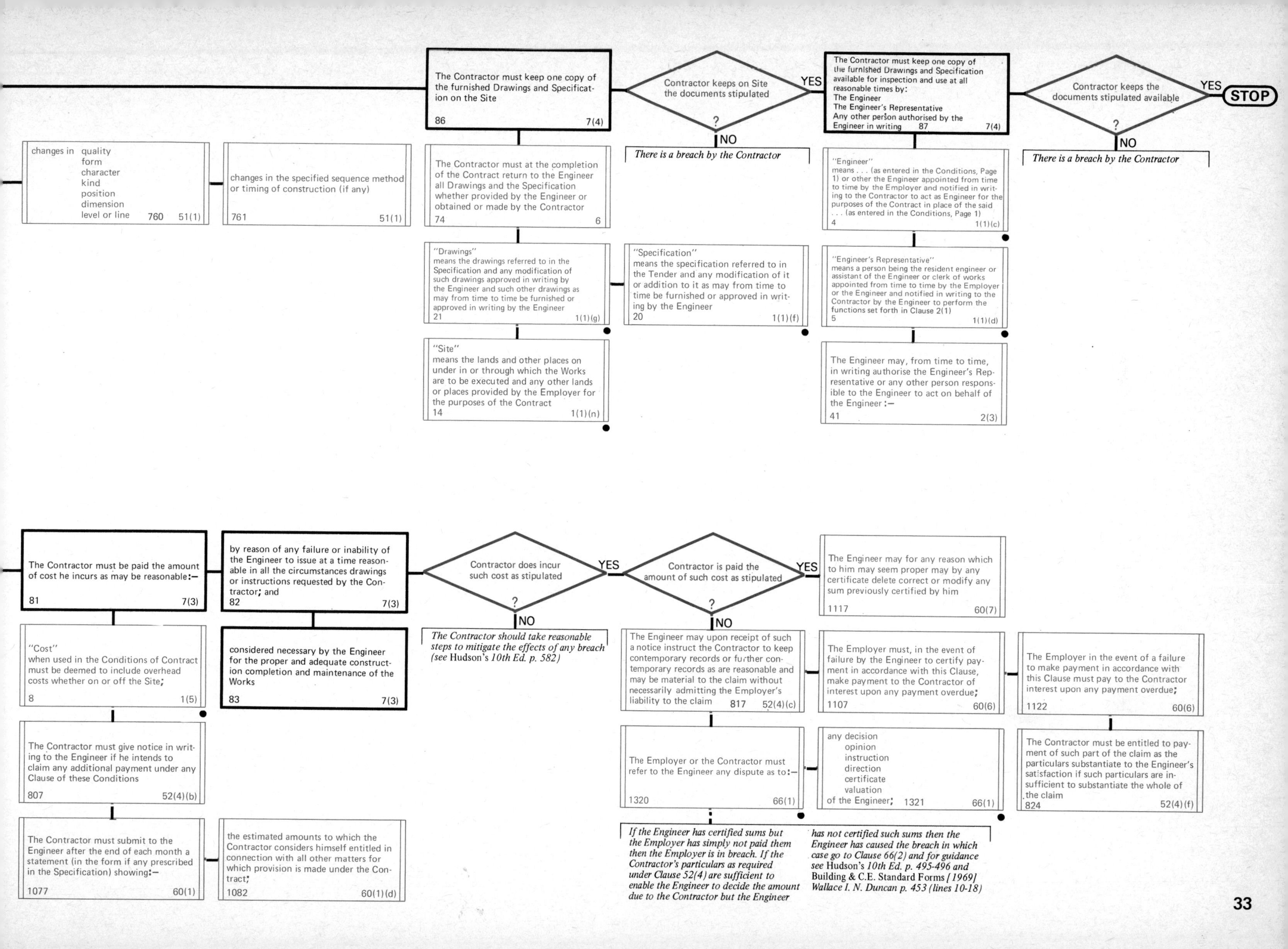

The Contractor must keep one copy of the furnished Drawings and Specification on the Site
86 7(4)
Contractor keeps on Site the documents stipulated
?
YES
NO
There is a breach by the Contractor
The Contractor must keep one copy of the furnished Drawings and Specification available for inspection and use at all reasonable times by:
The Engineer
The Engineer's Representative
Any other person authorised by the Engineer in writing 87 7(4)
Contractor keeps the documents stipulated available
?
YES
STOP
NO
There is a breach by the Contractor
changes in quality
form
character
kind
position
dimension
level or line 760 51(1)
changes in the specified sequence method or timing of construction (if any)
761 51(1)
The Contractor must at the completion of the Contract return to the Engineer all Drawings and the Specification whether provided by the Engineer or obtained or made by the Contractor
74 6
"Engineer"
means . . . (as entered in the Conditions, Page 1) or other the Engineer appointed from time to time by the Employer and notified in writing to the Contractor to act as Engineer for the purposes of the Contract in place of the said . . . (as entered in the Conditions, Page 1)
4 1(1)(c)
"Drawings"
means the drawings referred to in the Specification and any modification of such drawings approved in writing by the Engineer and such other drawings as may from time to time be furnished or approved in writing by the Engineer
21 1(1)(g)
"Specification"
means the specification referred to in the Tender and any modification of it or addition to it as may from time to time be furnished or approved in writing by the Engineer
20 1(1)(f)
"Engineer's Representative"
means a person being the resident engineer or assistant of the Engineer or clerk of works appointed from time to time by the Employer or the Engineer and notified in writing to the Contractor by the Engineer to perform the functions set forth in Clause 2(1)
5 1(1)(d)
"Site"
means the lands and other places on under in or through which the Works are to be executed and any other lands or places provided by the Employer for the purposes of the Contract
14 1(1)(n)
The Engineer may, from time to time, in writing authorise the Engineer's Representative or any other person responsible to the Engineer to act on behalf of the Engineer :—
41 2(3)
The Contractor must be paid the amount of cost he incurs as may be reasonable:—
81 7(3)
by reason of any failure or inability of the Engineer to issue at a time reasonable in all the circumstances drawings or instructions requested by the Contractor; and
82 7(3)
Contractor does incur such cost as stipulated
?
YES
NO
Contractor is paid the amount of such cost as stipulated
?
YES
NO
The Engineer may for any reason which to him may seem proper may by any certificate delete correct or modify any sum previously certified by him
1117 60(7)
"Cost"
when used in the Conditions of Contract must be deemed to include overhead costs whether on or off the Site;
8 1(5)
considered necessary by the Engineer for the proper and adequate construction completion and maintenance of the Works
83 7(3)
The Contractor should take reasonable steps to mitigate the effects of any breach (see Hudson's 10th Ed. p. 582)
The Engineer may upon receipt of such a notice instruct the Contractor to keep contemporary records or further contemporary records as are reasonable and may be material to the claim without necessarily admitting the Employer's liability to the claim 817 52(4)(c)
The Employer must, in the event of failure by the Engineer to certify payment in accordance with this Clause, make payment to the Contractor of interest upon any payment overdue;
1107 60(6)
The Employer in the event of a failure to make payment in accordance with this Clause must pay to the Contractor interest upon any payment overdue;
1122 60(6)
The Contractor must give notice in writing to the Engineer if he intends to claim any additional payment under any Clause of these Conditions
807 52(4)(b)
The Employer or the Contractor must refer to the Engineer any dispute as to:—
1320 66(1)
any decision
opinion
instruction
direction
certificate
valuation
of the Engineer; 1321 66(1)
The Contractor must be entitled to payment of such part of the claim as the particulars substantiate to the Engineer's satisfaction if such particulars are insufficient to substantiate the whole of the claim
824 52(4)(f)
The Contractor must submit to the Engineer after the end of each month a statement (in the form if any prescribed in the Specification) showing:—
1077 60(1)
the estimated amounts to which the Contractor considers himself entitled in connection with all other matters for which provision is made under the Contract;
1082 60(1)(d)
If the Engineer has certified sums but the Employer has simply not paid them then the Employer is in breach. If the Contractor's particulars as required under Clause 52(4) are sufficient to enable the Engineer to decide the amount due to the Contractor but the Engineer has not certified such sums then the Engineer has caused the breach in which case go to Clause 66(2) and for guidance see Hudson's 10th Ed. p. 495-496 and Building & C.E. Standard Forms [1969] Wallace I. N. Duncan p. 453 (lines 10-18)

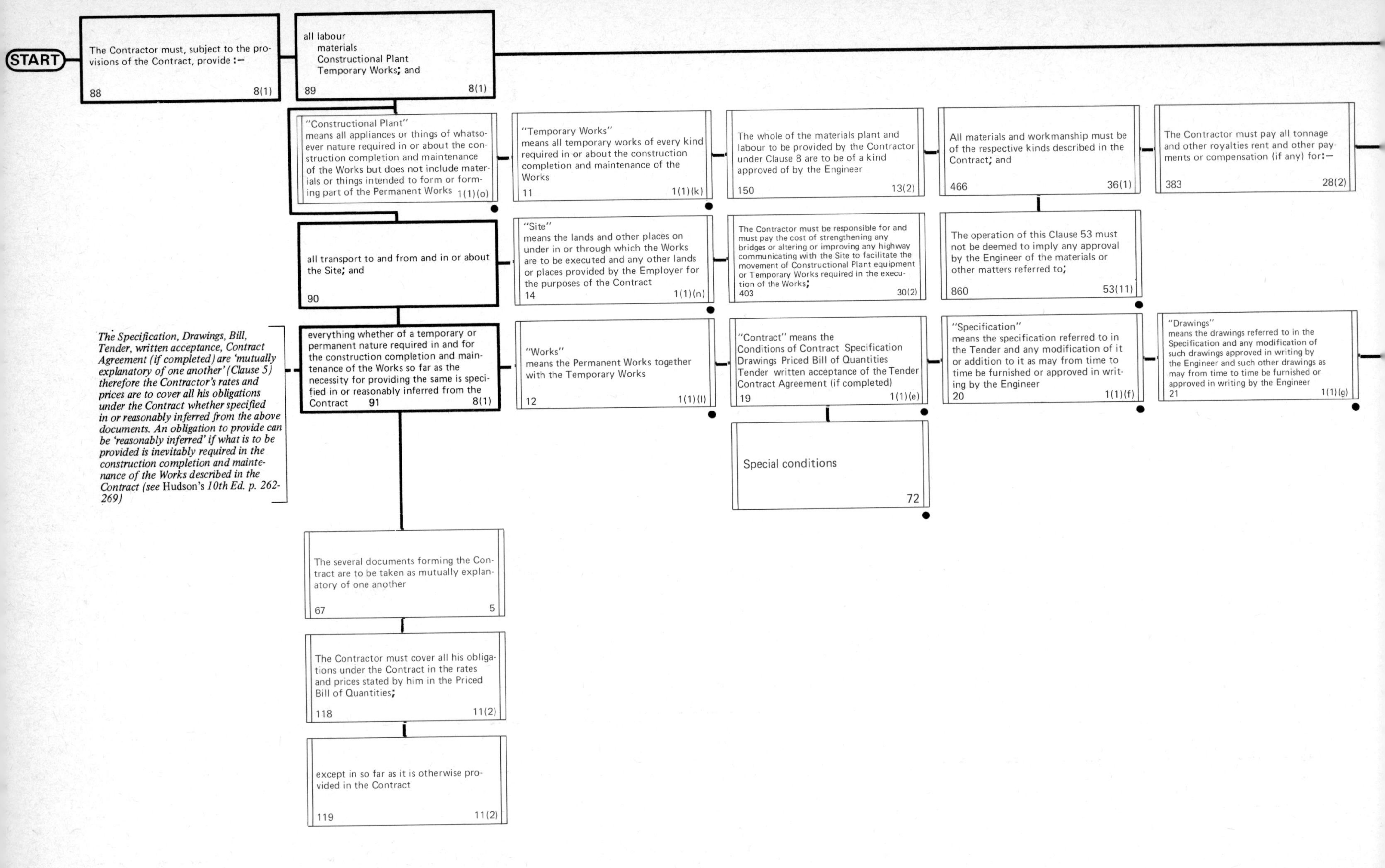
START
The Contractor must, subject to the provisions of the Contract, provide :—
88 8(1)
all labour
materials
Constructional Plant
Temporary Works; and
89 8(1)
"Constructional Plant"
means all appliances or things of whatsoever nature required in or about the construction completion and maintenance of the Works but does not include materials or things intended to form or forming part of the Permanent Works 1(1)(o)
"Temporary Works"
means all temporary works of every kind required in or about the construction completion and maintenance of the Works
11 1(1)(k)
The whole of the materials plant and labour to be provided by the Contractor under Clause 8 are to be of a kind approved of by the Engineer
150 13(2)
All materials and workmanship must be of the respective kinds described in the Contract; and
466 36(1)
The Contractor must pay all tonnage and other royalties rent and other payments or compensation (if any) for:—
383 28(2)
all transport to and from and in or about the Site; and
90
"Site"
means the lands and other places on under in or through which the Works are to be executed and any other lands or places provided by the Employer for the purposes of the Contract
14 1(1)(n)
The Contractor must be responsible for and must pay the cost of strengthening any bridges or altering or improving any highway communicating with the Site to facilitate the movement of Constructional Plant equipment or Temporary Works required in the execution of the Works;
403 30(2)
The operation of this Clause 53 must not be deemed to imply any approval by the Engineer of the materials or other matters referred to;
860 53(11)
The Specification, Drawings, Bill, Tender, written acceptance, Contract Agreement (if completed) are 'mutually explanatory of one another' (Clause 5) therefore the Contractor's rates and prices are to cover all his obligations under the Contract whether specified in or reasonably inferred from the above documents. An obligation to provide can be 'reasonably inferred' if what is to be provided is inevitably required in the construction completion and maintenance of the Works described in the Contract (see Hudson's 10th Ed. p. 262-269)
everything whether of a temporary or permanent nature required in and for the construction completion and maintenance of the Works so far as the necessity for providing the same is specified in or reasonably inferred from the Contract 91 8(1)
"Works"
means the Permanent Works together with the Temporary Works
12 1(1)(l)
"Contract" means the Conditions of Contract Specification Drawings Priced Bill of Quantities Tender written acceptance of the Tender Contract Agreement (if completed)
19 1(1)(e)
"Specification"
means the specification referred to in the Tender and any modification of it or addition to it as may from time to time be furnished or approved in writing by the Engineer
20 1(1)(f)
"Drawings"
means the drawings referred to in the Specification and any modification of such drawings approved in writing by the Engineer and such other drawings as may from time to time be furnished or approved in writing by the Engineer
21 1(1)(g)
Special conditions
72
The several documents forming the Contract are to be taken as mutually explanatory of one another
67 5
The Contractor must cover all his obligations under the Contract in the rates and prices stated by him in the Priced Bill of Quantities;
118 11(2)
except in so far as it is otherwise provided in the Contract
119 11(2)

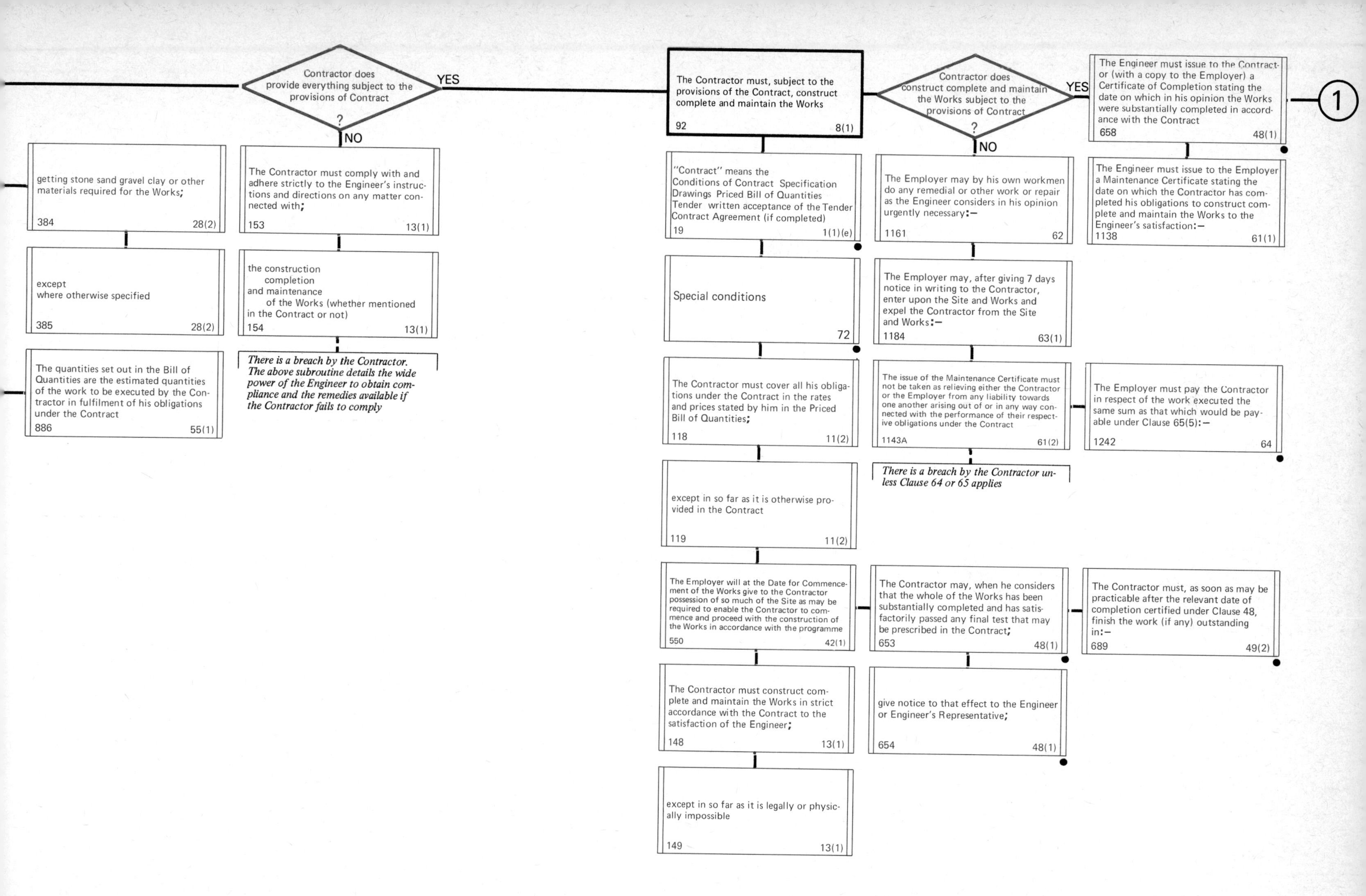
Contractor does provide everything subject to the provisions of Contract ?
YES
NO
getting stone sand gravel clay or other materials required for the Works;
384 28(2)
except where otherwise specified
385 28(2)
The quantities set out in the Bill of Quantities are the estimated quantities of the work to be executed by the Contractor in fulfilment of his obligations under the Contract
886 55(1)
The Contractor must comply with and adhere strictly to the Engineer's instructions and directions on any matter connected with;
153 13(1)
the construction completion and maintenance of the Works (whether mentioned in the Contract or not)
154 13(1)
There is a breach by the Contractor. The above subroutine details the wide power of the Engineer to obtain compliance and the remedies available if the Contractor fails to comply
The Contractor must, subject to the provisions of the Contract, construct complete and maintain the Works
92 8(1)
"Contract" means the Conditions of Contract Specification Drawings Priced Bill of Quantities Tender written acceptance of the Tender Contract Agreement (if completed)
19 1(1)(e)
Special conditions
72
The Contractor must cover all his obligations under the Contract in the rates and prices stated by him in the Priced Bill of Quantities;
118 11(2)
except in so far as it is otherwise provided in the Contract
119 11(2)
The Employer will at the Date for Commencement of the Works give to the Contractor possession of so much of the Site as may be required to enable the Contractor to commence and proceed with the construction of the Works in accordance with the programme
550 42(1)
The Contractor must construct complete and maintain the Works in strict accordance with the Contract to the satisfaction of the Engineer;
148 13(1)
except in so far as it is legally or physically impossible
149 13(1)
Contractor does construct complete and maintain the Works subject to the provisions of Contract ?
YES
NO
The Employer may by his own workmen do any remedial or other work or repair as the Engineer considers in his opinion urgently necessary:—
1161 62
The Employer may, after giving 7 days notice in writing to the Contractor, enter upon the Site and Works and expel the Contractor from the Site and Works:—
1184 63(1)
The issue of the Maintenance Certificate must not be taken as relieving either the Contractor or the Employer from any liability towards one another arising out of or in any way connected with the performance of their respective obligations under the Contract
1143A 61(2)
There is a breach by the Contractor unless Clause 64 or 65 applies
The Contractor may, when he considers that the whole of the Works has been substantially completed and has satisfactorily passed any final test that may be prescribed in the Contract;
653 48(1)
give notice to that effect to the Engineer or Engineer's Representative;
654 48(1)
The Engineer must issue to the Contractor (with a copy to the Employer) a Certificate of Completion stating the date on which in his opinion the Works were substantially completed in accordance with the Contract
658 48(1)
1
The Engineer must issue to the Employer a Maintenance Certificate stating the date on which the Contractor has completed his obligations to construct complete and maintain the Works to the Engineer's satisfaction:—
1138 61(1)
The Employer must pay the Contractor in respect of the work executed the same sum as that which would be payable under Clause 65(5):—
1242 64
The Contractor must, as soon as may be practicable after the relevant date of completion certified under Clause 48, finish the work (if any) outstanding in:—
689 49(2)

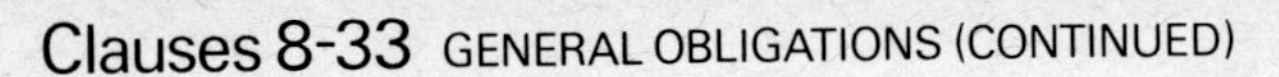

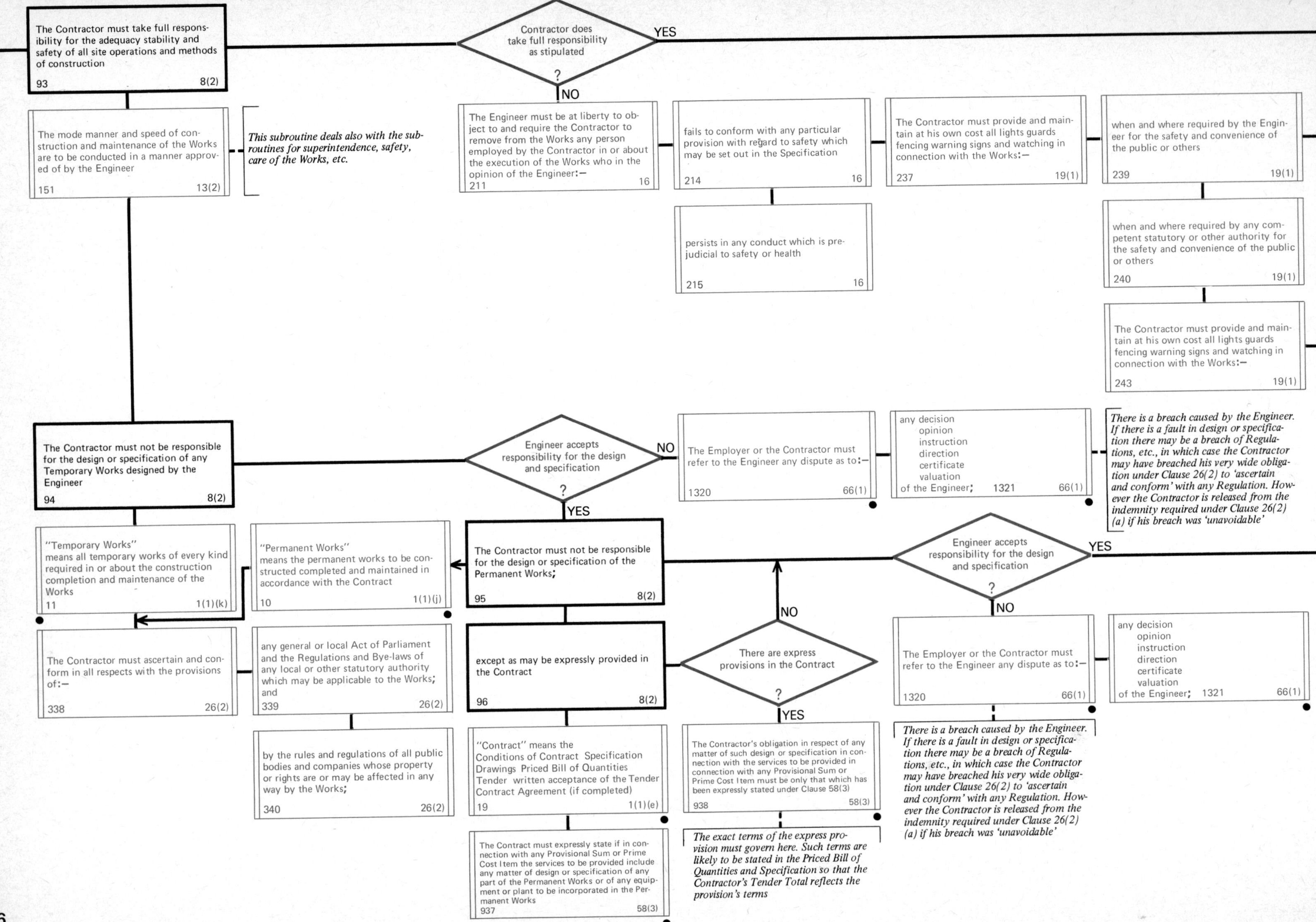
1
The Contractor must take full responsibility for the adequacy stability and safety of all site operations and methods of construction
93 8(2)
Contractor does take full responsibility as stipulated ?
YES
NO
The mode manner and speed of construction and maintenance of the Works are to be conducted in a manner approved of by the Engineer
151 13(2)
This subroutine deals also with the subroutines for superintendence, safety, care of the Works, etc.
The Engineer must be at liberty to object to and require the Contractor to remove from the Works any person employed by the Contractor in or about the execution of the Works who in the opinion of the Engineer:—
211 16
fails to conform with any particular provision with regard to safety which may be set out in the Specification
214 16
persists in any conduct which is prejudicial to safety or health
215 16
The Contractor must provide and maintain at his own cost all lights guards fencing warning signs and watching in connection with the Works:—
237 19(1)
when and where required by the Engineer for the safety and convenience of the public or others
239 19(1)
when and where required by any competent statutory or other authority for the safety and convenience of the public or others
240 19(1)
The Contractor must provide and maintain at his own cost all lights guards fencing warning signs and watching in connection with the Works:—
243 19(1)
The Contractor must not be responsible for the design or specification of any Temporary Works designed by the Engineer
94 8(2)
Engineer accepts responsibility for the design and specification ?
NO
YES
The Employer or the Contractor must refer to the Engineer any dispute as to:—
1320 66(1)
any decision opinion instruction direction certificate valuation of the Engineer; 1321 66(1)
There is a breach caused by the Engineer. If there is a fault in design or specification there may be a breach of Regulations, etc., in which case the Contractor may have breached his very wide obligation under Clause 26(2) to 'ascertain and conform' with any Regulation. However the Contractor is released from the indemnity required under Clause 26(2) (a) if his breach was 'unavoidable'
"Temporary Works" means all temporary works of every kind required in or about the construction completion and maintenance of the Works
11 1(1)(k)
"Permanent Works" means the permanent works to be constructed completed and maintained in accordance with the Contract
10 1(1)(j)
The Contractor must not be responsible for the design or specification of the Permanent Works;
95 8(2)
Engineer accepts responsibility for the design and specification ?
YES
NO
The Contractor must ascertain and conform in all respects with the provisions of:—
338 26(2)
any general or local Act of Parliament and the Regulations and Bye-laws of any local or other statutory authority which may be applicable to the Works; and
339 26(2)
except as may be expressly provided in the Contract
96 8(2)
NO
There are express provisions in the Contract ?
YES
The Employer or the Contractor must refer to the Engineer any dispute as to:—
1320 66(1)
any decision opinion instruction direction certificate valuation of the Engineer; 1321 66(1)
by the rules and regulations of all public bodies and companies whose property or rights are or may be affected in any way by the Works;
340 26(2)
"Contract" means the Conditions of Contract Specification Drawings Priced Bill of Quantities Tender written acceptance of the Tender Contract Agreement (if completed)
19 1(1)(e)
The Contractor's obligation in respect of any matter of such design or specification in connection with the services to be provided in connection with any Provisional Sum or Prime Cost Item must be only that which has been expressly stated under Clause 58(3)
938 58(3)
There is a breach caused by the Engineer. If there is a fault in design or specification there may be a breach of Regulations, etc., in which case the Contractor may have breached his very wide obligation under Clause 26(2) to 'ascertain and conform' with any Regulation. However the Contractor is released from the indemnity required under Clause 26(2) (a) if his breach was 'unavoidable'
The Contract must expressly state if in connection with any Provisional Sum or Prime Cost Item the services to be provided include any matter of design or specification of any part of the Permanent Works or of any equipment or plant to be incorporated in the Permanent Works
937 58(3)
The exact terms of the express provision must govern here. Such terms are likely to be stated in the Priced Bill of Quantities and Specification so that the Contractor's Tender Total reflects the provision's terms

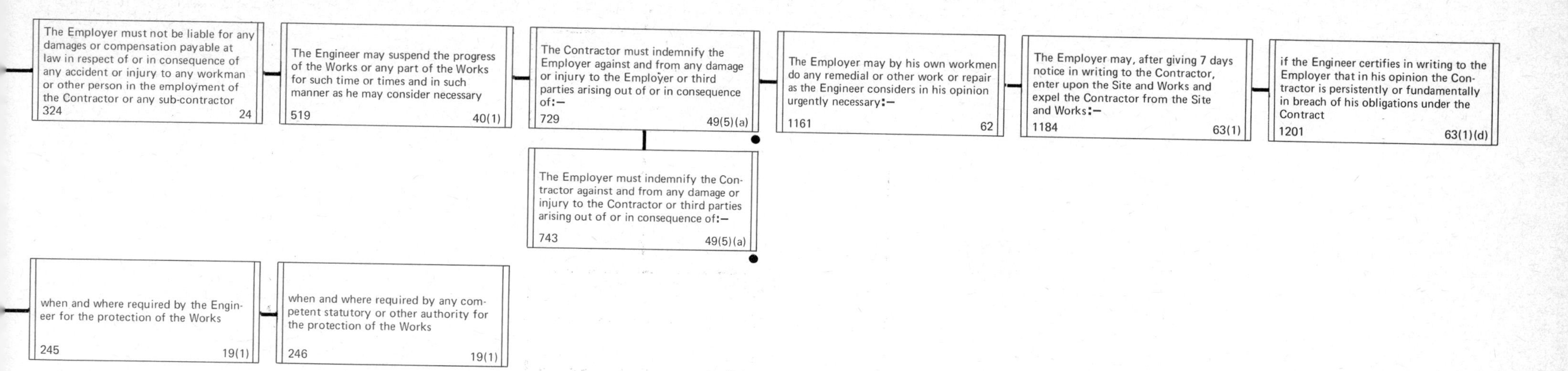
STOP
The Employer must not be liable for any damages or compensation payable at law in respect of or in consequence of any accident or injury to any workman or other person in the employment of the Contractor or any sub-contractor
324 24
The Engineer may suspend the progress of the Works or any part of the Works for such time or times and in such manner as he may consider necessary
519 40(1)
The Contractor must indemnify the Employer against and from any damage or injury to the Employer or third parties arising out of or in consequence of:—
729 49(5)(a)
The Employer must indemnify the Contractor against and from any damage or injury to the Contractor or third parties arising out of or in consequence of:—
743 49(5)(a)
The Employer may by his own workmen do any remedial or other work or repair as the Engineer considers in his opinion urgently necessary:—
1161 62
The Employer may, after giving 7 days notice in writing to the Contractor, enter upon the Site and Works and expel the Contractor from the Site and Works:—
1184 63(1)
if the Engineer certifies in writing to the Employer that in his opinion the Contractor is persistently or fundamentally in breach of his obligations under the Contract
1201 63(1)(d)
when and where required by the Engineer for the protection of the Works
245 19(1)
when and where required by any competent statutory or other authority for the protection of the Works
246 19(1)

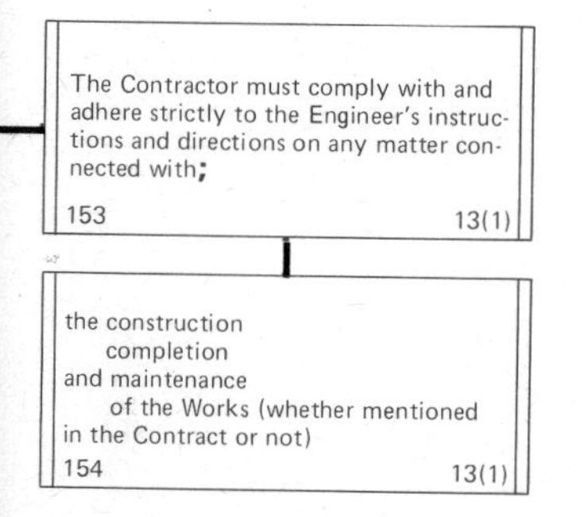
The Contractor must comply with and adhere strictly to the Engineer's instructions and directions on any matter connected with;
153 13(1)
the construction
completion
and maintenance
of the Works (whether mentioned in the Contract or not)
154 13(1)

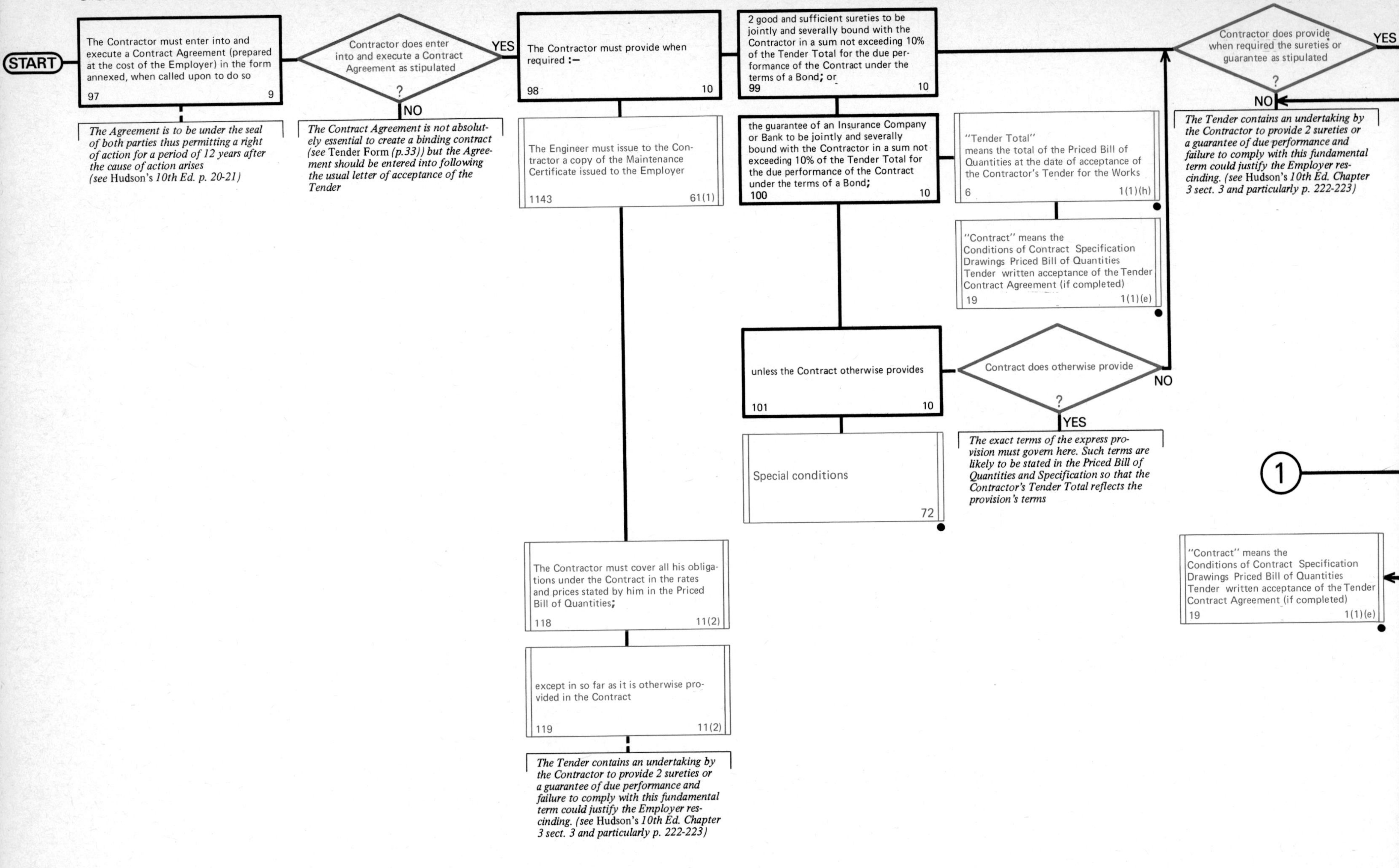
START
The Contractor must enter into and execute a Contract Agreement (prepared at the cost of the Employer) in the form annexed, when called upon to do so
97 9
The Agreement is to be under the seal of both parties thus permitting a right of action for a period of 12 years after the cause of action arises (see Hudson's 10th Ed. p. 20-21)
Contractor does enter into and execute a Contract Agreement as stipulated ?
YES
NO
The Contract Agreement is not absolutely essential to create a binding contract (see Tender Form (p.33)) but the Agreement should be entered into following the usual letter of acceptance of the Tender
The Contractor must provide when required :—
98 10
The Engineer must issue to the Contractor a copy of the Maintenance Certificate issued to the Employer
1143 61(1)
The Contractor must cover all his obligations under the Contract in the rates and prices stated by him in the Priced Bill of Quantities;
118 11(2)
except in so far as it is otherwise provided in the Contract
119 11(2)
The Tender contains an undertaking by the Contractor to provide 2 sureties or a guarantee of due performance and failure to comply with this fundamental term could justify the Employer rescinding. (see Hudson's 10th Ed. Chapter 3 sect. 3 and particularly p. 222-223)
2 good and sufficient sureties to be jointly and severally bound with the Contractor in a sum not exceeding 10% of the Tender Total for the due performance of the Contract under the terms of a Bond; or
99 10
the guarantee of an Insurance Company or Bank to be jointly and severally bound with the Contractor in a sum not exceeding 10% of the Tender Total for the due performance of the Contract under the terms of a Bond;
100 10
unless the Contract otherwise provides
101 10
Special conditions
72
"Tender Total" means the total of the Priced Bill of Quantities at the date of acceptance of the Contractor's Tender for the Works
6 1(1)(h)
"Contract" means the Conditions of Contract Specification Drawings Priced Bill of Quantities Tender written acceptance of the Tender Contract Agreement (if completed)
19 1(1)(e)
Contract does otherwise provide ?
NO
YES
The exact terms of the express provision must govern here. Such terms are likely to be stated in the Priced Bill of Quantities and Specification so that the Contractor's Tender Total reflects the provision's terms
Contractor does provide when required the sureties or guarantee as stipulated ?
YES
NO
The Tender contains an undertaking by the Contractor to provide 2 sureties or a guarantee of due performance and failure to comply with this fundamental term could justify the Employer rescinding. (see Hudson's 10th Ed. Chapter 3 sect. 3 and particularly p. 222-223)
1
"Contract" means the Conditions of Contract Specification Drawings Priced Bill of Quantities Tender written acceptance of the Tender Contract Agreement (if completed)
19 1(1)(e)

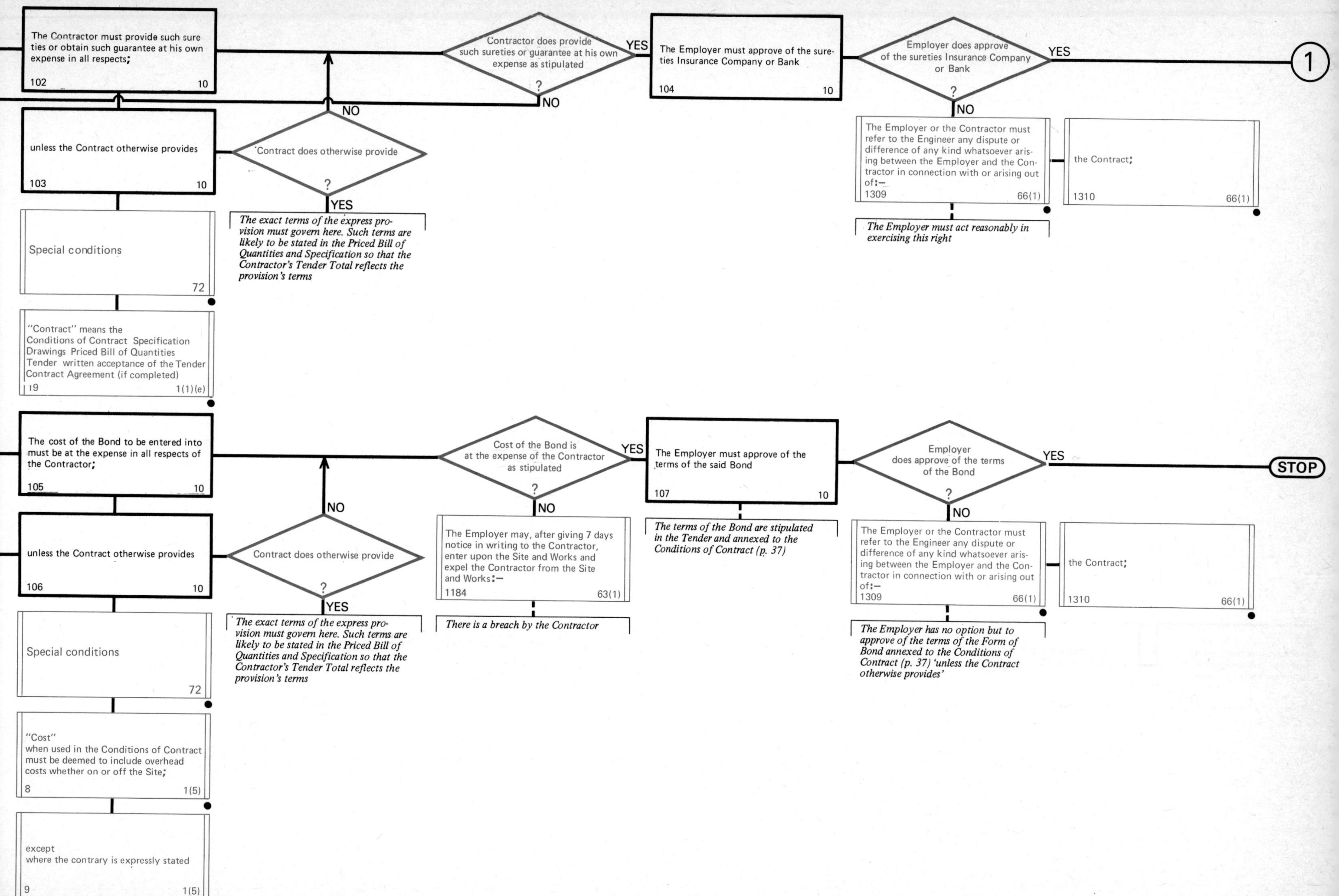

The Contractor must provide such sureties or obtain such guarantee at his own expense in all respects;
102
10
Contractor does provide such sureties or guarantee at his own expense as stipulated
?
YES
NO
The Employer must approve of the sureties Insurance Company or Bank
104
10
Employer does approve of the sureties Insurance Company or Bank
?
YES
NO
1
unless the Contract otherwise provides
103
10
Contract does otherwise provide
?
NO
YES
The exact terms of the express provision must govern here. Such terms are likely to be stated in the Priced Bill of Quantities and Specification so that the Contractor's Tender Total reflects the provision's terms
Special conditions
72
"Contract" means the Conditions of Contract Specification Drawings Priced Bill of Quantities Tender written acceptance of the Tender Contract Agreement (if completed)
19
1(1)(e)
The Employer or the Contractor must refer to the Engineer any dispute or difference of any kind whatsoever arising between the Employer and the Contractor in connection with or arising out of:—
1309
66(1)
the Contract;
1310
66(1)
The Employer must act reasonably in exercising this right
The cost of the Bond to be entered into must be at the expense in all respects of the Contractor;
105
10
Cost of the Bond is at the expense of the Contractor as stipulated
?
YES
NO
The Employer must approve of the terms of the said Bond
107
10
Employer does approve of the terms of the Bond
?
YES
NO
STOP
unless the Contract otherwise provides
106
10
Contract does otherwise provide
?
NO
YES
The exact terms of the express provision must govern here. Such terms are likely to be stated in the Priced Bill of Quantities and Specification so that the Contractor's Tender Total reflects the provision's terms
The Employer may, after giving 7 days notice in writing to the Contractor, enter upon the Site and Works and expel the Contractor from the Site and Works:—
1184
63(1)
There is a breach by the Contractor
The terms of the Bond are stipulated in the Tender and annexed to the Conditions of Contract (p. 37)
The Employer or the Contractor must refer to the Engineer any dispute or difference of any kind whatsoever arising between the Employer and the Contractor in connection with or arising out of:—
1309
66(1)
the Contract;
1310
66(1)
The Employer has no option but to approve of the terms of the Form of Bond annexed to the Conditions of Contract (p. 37) 'unless the Contract otherwise provides'
Special conditions
72
"Cost" when used in the Conditions of Contract must be deemed to include overhead costs whether on or off the Site;
8
1(5)
except where the contrary is expressly stated
9
1(5)

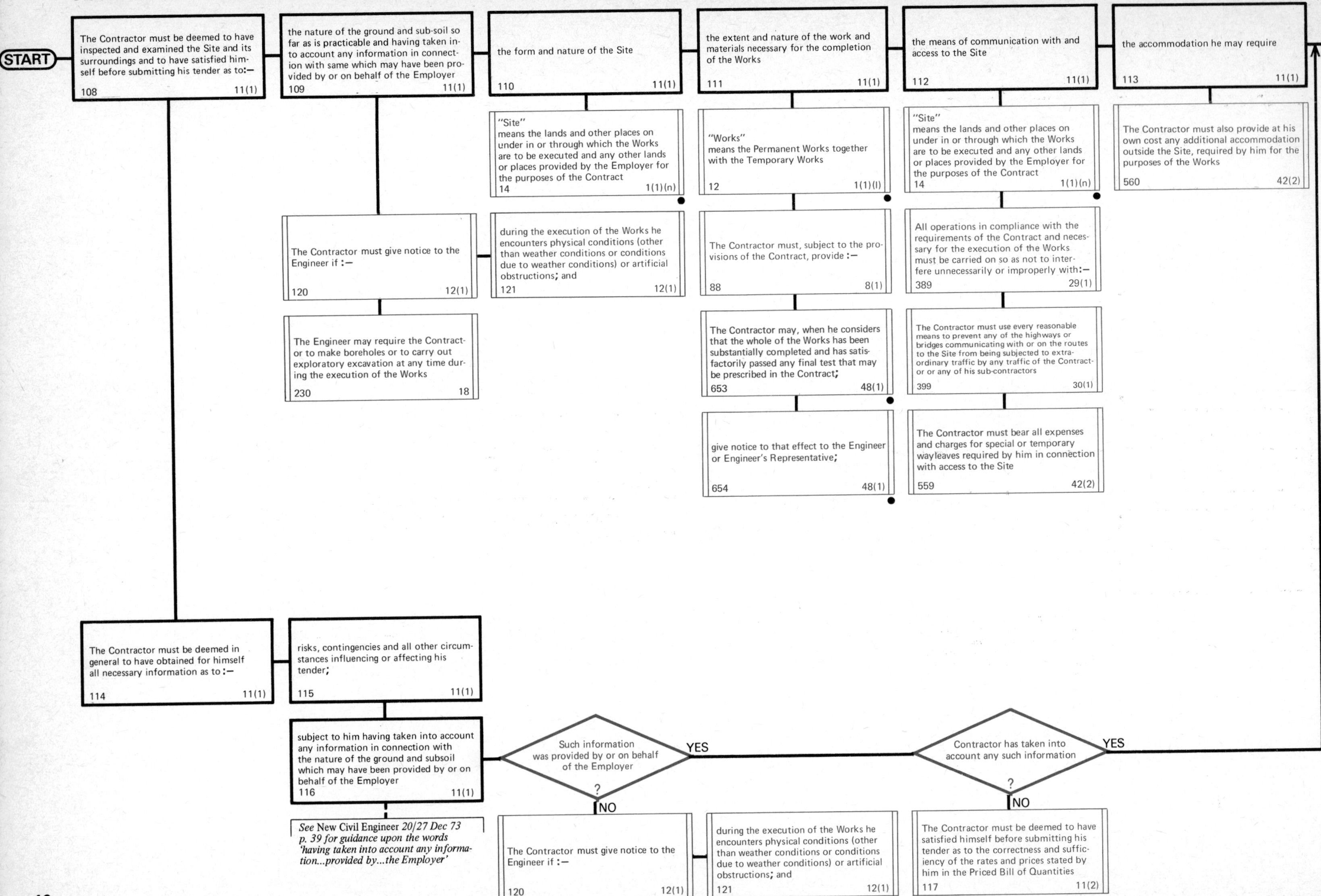
START
The Contractor must be deemed to have inspected and examined the Site and its surroundings and to have satisfied himself before submitting his tender as to:— 108 11(1)
the nature of the ground and sub-soil so far as is practicable and having taken into account any information in connection with same which may have been provided by or on behalf of the Employer 109 11(1)
the form and nature of the Site 110 11(1)
the extent and nature of the work and materials necessary for the completion of the Works 111 11(1)
the means of communication with and access to the Site 112 11(1)
the accommodation he may require 113 11(1)
"Site" means the lands and other places on under in or through which the Works are to be executed and any other lands or places provided by the Employer for the purposes of the Contract 14 1(1)(n)
"Works" means the Permanent Works together with the Temporary Works 12 1(1)(l)
"Site" means the lands and other places on under in or through which the Works are to be executed and any other lands or places provided by the Employer for the purposes of the Contract 14 1(1)(n)
The Contractor must also provide at his own cost any additional accommodation outside the Site, required by him for the purposes of the Works 560 42(2)
The Contractor must give notice to the Engineer if :— 120 12(1)
during the execution of the Works he encounters physical conditions (other than weather conditions or conditions due to weather conditions) or artificial obstructions; and 121 12(1)
The Contractor must, subject to the provisions of the Contract, provide :— 88 8(1)
All operations in compliance with the requirements of the Contract and necessary for the execution of the Works must be carried on so as not to interfere unnecessarily or improperly with:— 389 29(1)
The Engineer may require the Contractor to make boreholes or to carry out exploratory excavation at any time during the execution of the Works 230 18
The Contractor may, when he considers that the whole of the Works has been substantially completed and has satisfactorily passed any final test that may be prescribed in the Contract; 653 48(1)
The Contractor must use every reasonable means to prevent any of the highways or bridges communicating with or on the routes to the Site from being subjected to extraordinary traffic by any traffic of the Contractor or any of his sub-contractors 399 30(1)
give notice to that effect to the Engineer or Engineer's Representative; 654 48(1)
The Contractor must bear all expenses and charges for special or temporary wayleaves required by him in connection with access to the Site 559 42(2)
The Contractor must be deemed in general to have obtained for himself all necessary information as to :— 114 11(1)
risks, contingencies and all other circumstances influencing or affecting his tender; 115 11(1)
subject to him having taken into account any information in connection with the nature of the ground and subsoil which may have been provided by or on behalf of the Employer 116 11(1)
Such information was provided by or on behalf of the Employer ?
YES
NO
Contractor has taken into account any such information ?
YES
NO
See New Civil Engineer 20/27 Dec 73 p. 39 for guidance upon the words 'having taken into account any information...provided by...the Employer'
The Contractor must give notice to the Engineer if :— 120 12(1)
during the execution of the Works he encounters physical conditions (other than weather conditions or conditions due to weather conditions) or artificial obstructions; and 121 12(1)
The Contractor must be deemed to have satisfied himself before submitting his tender as to the correctness and sufficiency of the rates and prices stated by him in the Priced Bill of Quantities 117 11(2)

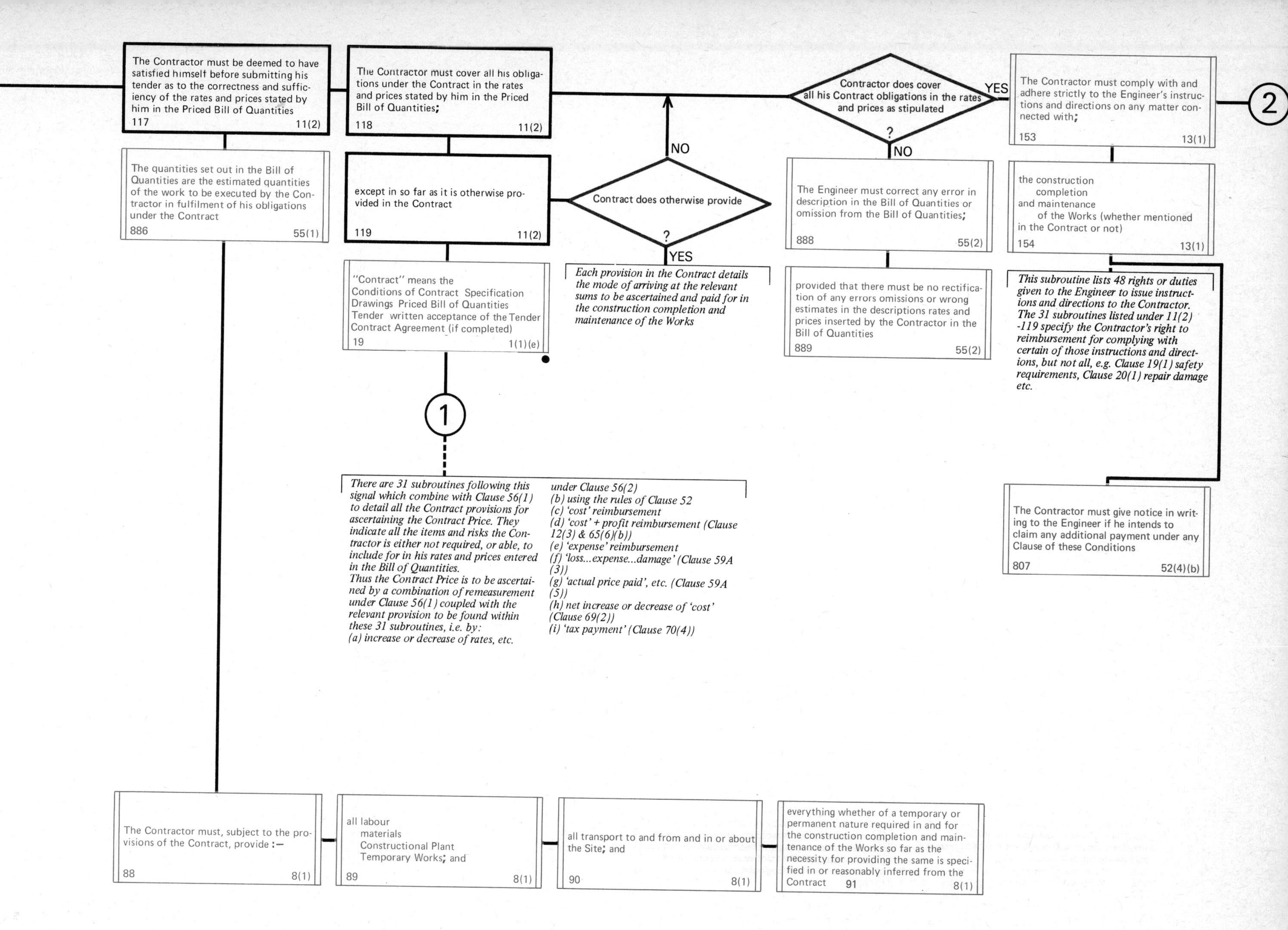

The Contractor must be deemed to have satisfied himself before submitting his tender as to the correctness and sufficiency of the rates and prices stated by him in the Priced Bill of Quantities
117 11(2)
The Contractor must cover all his obligations under the Contract in the rates and prices stated by him in the Priced Bill of Quantities;
118 11(2)
Contractor does cover all his Contract obligations in the rates and prices as stipulated
?
YES
NO
The Contractor must comply with and adhere strictly to the Engineer's instructions and directions on any matter connected with;
153 13(1)
2
The quantities set out in the Bill of Quantities are the estimated quantities of the work to be executed by the Contractor in fulfilment of his obligations under the Contract
886 55(1)
except in so far as it is otherwise provided in the Contract
119 11(2)
Contract does otherwise provide
?
NO
YES
The Engineer must correct any error in description in the Bill of Quantities or omission from the Bill of Quantities;
888 55(2)
the construction completion and maintenance of the Works (whether mentioned in the Contract or not)
154 13(1)
"Contract" means the Conditions of Contract Specification Drawings Priced Bill of Quantities Tender written acceptance of the Tender Contract Agreement (if completed)
19 1(1)(e)
Each provision in the Contract details the mode of arriving at the relevant sums to be ascertained and paid for in the construction completion and maintenance of the Works
provided that there must be no rectification of any errors omissions or wrong estimates in the descriptions rates and prices inserted by the Contractor in the Bill of Quantities
889 55(2)
This subroutine lists 48 rights or duties given to the Engineer to issue instructions and directions to the Contractor. The 31 subroutines listed under 11(2) -119 specify the Contractor's right to reimbursement for complying with certain of those instructions and directions, but not all, e.g. Clause 19(1) safety requirements, Clause 20(1) repair damage etc.
1
There are 31 subroutines following this signal which combine with Clause 56(1) to detail all the Contract provisions for ascertaining the Contract Price. They indicate all the items and risks the Contractor is either not required, or able, to include for in his rates and prices entered in the Bill of Quantities.
Thus the Contract Price is to be ascertained by a combination of remeasurement under Clause 56(1) coupled with the relevant provision to be found within these 31 subroutines, i.e. by:
(a) increase or decrease of rates, etc. under Clause 56(2)
(b) using the rules of Clause 52
(c) 'cost' reimbursement
(d) 'cost' + profit reimbursement (Clause 12(3) & 65(6)(b))
(e) 'expense' reimbursement
(f) 'loss...expense...damage' (Clause 59A (3))
(g) 'actual price paid', etc. (Clause 59A (5))
(h) net increase or decrease of 'cost' (Clause 69(2))
(i) 'tax payment' (Clause 70(4))
The Contractor must give notice in writing to the Engineer if he intends to claim any additional payment under any Clause of these Conditions
807 52(4)(b)
The Contractor must, subject to the provisions of the Contract, provide :—
88 8(1)
all labour materials Constructional Plant Temporary Works; and
89 8(1)
all transport to and from and in or about the Site; and
90 8(1)
everything whether of a temporary or permanent nature required in and for the construction completion and maintenance of the Works so far as the necessity for providing the same is specified in or reasonably inferred from the Contract
91 8(1)

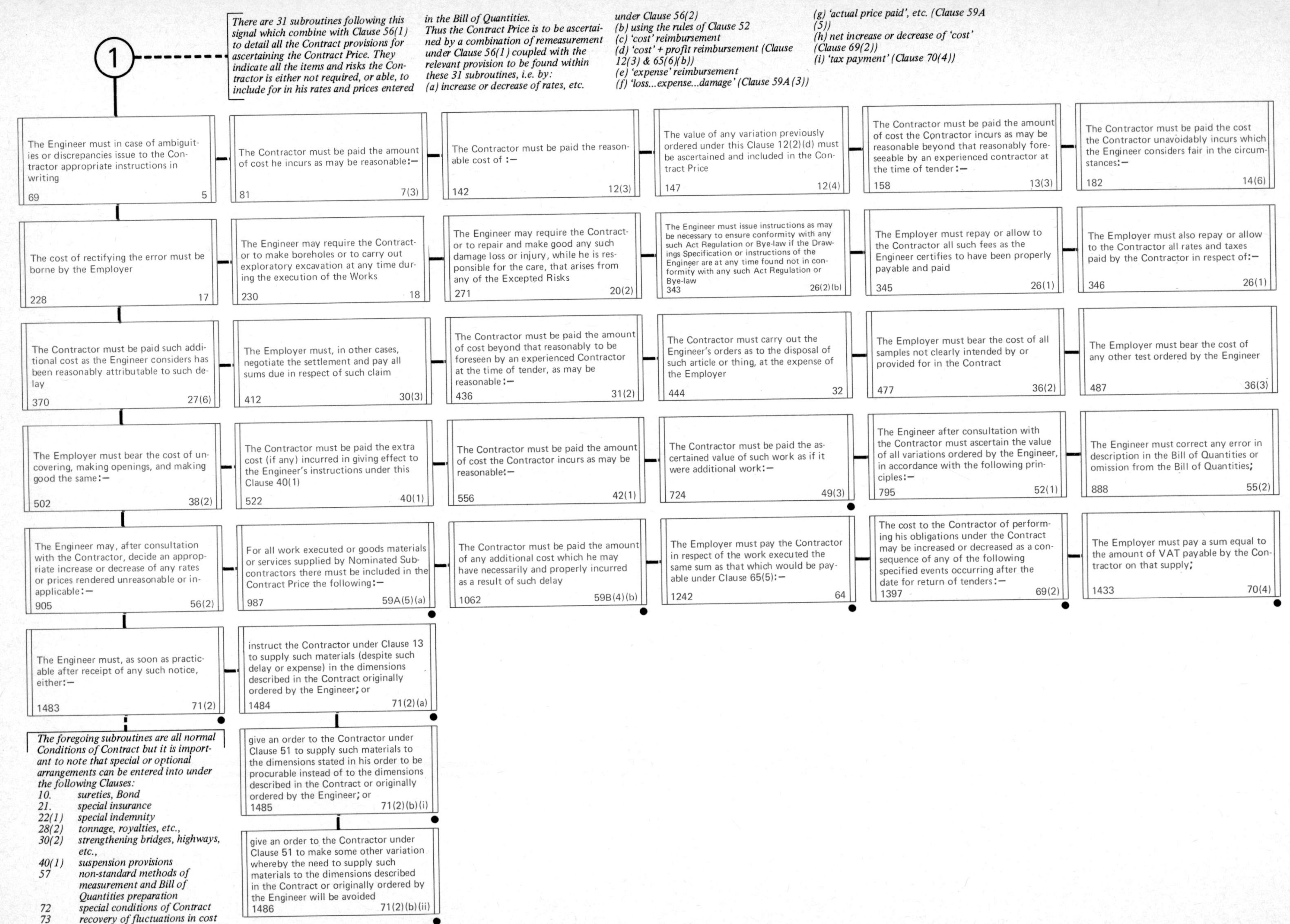
1
There are 31 subroutines following this signal which combine with Clause 56(1) to detail all the Contract provisions for ascertaining the Contract Price. They indicate all the items and risks the Contractor is either not required, or able, to include for in his rates and prices entered in the Bill of Quantities. Thus the Contract Price is to be ascertained by a combination of remeasurement under Clause 56(1) coupled with the relevant provision to be found within these 31 subroutines, i.e. by: (a) increase or decrease of rates, etc. under Clause 56(2) (b) using the rules of Clause 52 (c) 'cost' reimbursement (d) 'cost' + profit reimbursement (Clause 12(3) & 65(6)(b)) (e) 'expense' reimbursement (f) 'loss...expense...damage' (Clause 59A(3)) (g) 'actual price paid', etc. (Clause 59A(5)) (h) net increase or decrease of 'cost' (Clause 69(2)) (i) 'tax payment' (Clause 70(4))
The Engineer must in case of ambiguities or discrepancies issue to the Contractor appropriate instructions in writing
69 5
The Contractor must be paid the amount of cost he incurs as may be reasonable:—
81 7(3)
The Contractor must be paid the reasonable cost of :—
142 12(3)
The value of any variation previously ordered under this Clause 12(2)(d) must be ascertained and included in the Contract Price
147 12(4)
The Contractor must be paid the amount of cost the Contractor incurs as may be reasonable beyond that reasonably foreseeable by an experienced contractor at the time of tender:—
158 13(3)
The Contractor must be paid the cost the Contractor unavoidably incurs which the Engineer considers fair in the circumstances:—
182 14(6)
The cost of rectifying the error must be borne by the Employer
228 17
The Engineer may require the Contractor to make boreholes or to carry out exploratory excavation at any time during the execution of the Works
230 18
The Engineer may require the Contractor to repair and make good any such damage loss or injury, while he is responsible for the care, that arises from any of the Excepted Risks
271 20(2)
The Engineer must issue instructions as may be necessary to ensure conformity with any such Act Regulation or Bye-law if the Drawings Specification or instructions of the Engineer are at any time found not in conformity with any such Act Regulation or Bye-law
343 26(2)(b)
The Employer must repay or allow to the Contractor all such fees as the Engineer certifies to have been properly payable and paid
345 26(1)
The Employer must also repay or allow to the Contractor all rates and taxes paid by the Contractor in respect of:—
346 26(1)
The Contractor must be paid such additional cost as the Engineer considers has been reasonably attributable to such delay
370 27(6)
The Employer must, in other cases, negotiate the settlement and pay all sums due in respect of such claim
412 30(3)
The Contractor must be paid the amount of cost beyond that reasonably to be foreseen by an experienced Contractor at the time of tender, as may be reasonable:—
436 31(2)
The Contractor must carry out the Engineer's orders as to the disposal of such article or thing, at the expense of the Employer
444 32
The Employer must bear the cost of all samples not clearly intended by or provided for in the Contract
477 36(2)
The Employer must bear the cost of any other test ordered by the Engineer
487 36(3)
The Employer must bear the cost of uncovering, making openings, and making good the same:—
502 38(2)
The Contractor must be paid the extra cost (if any) incurred in giving effect to the Engineer's instructions under this Clause 40(1)
522 40(1)
The Contractor must be paid the amount of cost the Contractor incurs as may be reasonable:—
556 42(1)
The Contractor must be paid the ascertained value of such work as if it were additional work:—
724 49(3)
The Engineer after consultation with the Contractor must ascertain the value of all variations ordered by the Engineer, in accordance with the following principles:—
795 52(1)
The Engineer must correct any error in description in the Bill of Quantities or omission from the Bill of Quantities;
888 55(2)
The Engineer may, after consultation with the Contractor, decide an appropriate increase or decrease of any rates or prices rendered unreasonable or inapplicable:—
905 56(2)
For all work executed or goods materials or services supplied by Nominated Sub-contractors there must be included in the Contract Price the following:—
987 59A(5)(a)
The Contractor must be paid the amount of any additional cost which he may have necessarily and properly incurred as a result of such delay
1062 59B(4)(b)
The Employer must pay the Contractor in respect of the work executed the same sum as that which would be payable under Clause 65(5):—
1242 64
The cost to the Contractor of performing his obligations under the Contract may be increased or decreased as a consequence of any of the following specified events occurring after the date for return of tenders:—
1397 69(2)
The Employer must pay a sum equal to the amount of VAT payable by the Contractor on that supply;
1433 70(4)
The Engineer must, as soon as practicable after receipt of any such notice, either:—
1483 71(2)
instruct the Contractor under Clause 13 to supply such materials (despite such delay or expense) in the dimensions described in the Contract originally ordered by the Engineer; or
1484 71(2)(a)
give an order to the Contractor under Clause 51 to supply such materials to the dimensions stated in his order to be procurable instead of to the dimensions described in the Contract or originally ordered by the Engineer; or
1485 71(2)(b)(i)
give an order to the Contractor under Clause 51 to make some other variation whereby the need to supply such materials to the dimensions described in the Contract or originally ordered by the Engineer will be avoided
1486 71(2)(b)(ii)
The foregoing subroutines are all normal Conditions of Contract but it is important to note that special or optional arrangements can be entered into under the following Clauses:
10. sureties, Bond
21. special insurance
22(1) special indemnity
28(2) tonnage, royalties, etc.,
30(2) strengthening bridges, highways, etc.,
40(1) suspension provisions
57 non-standard methods of measurement and Bill of Quantities preparation
72 special conditions of Contract
73 recovery of fluctuations in cost

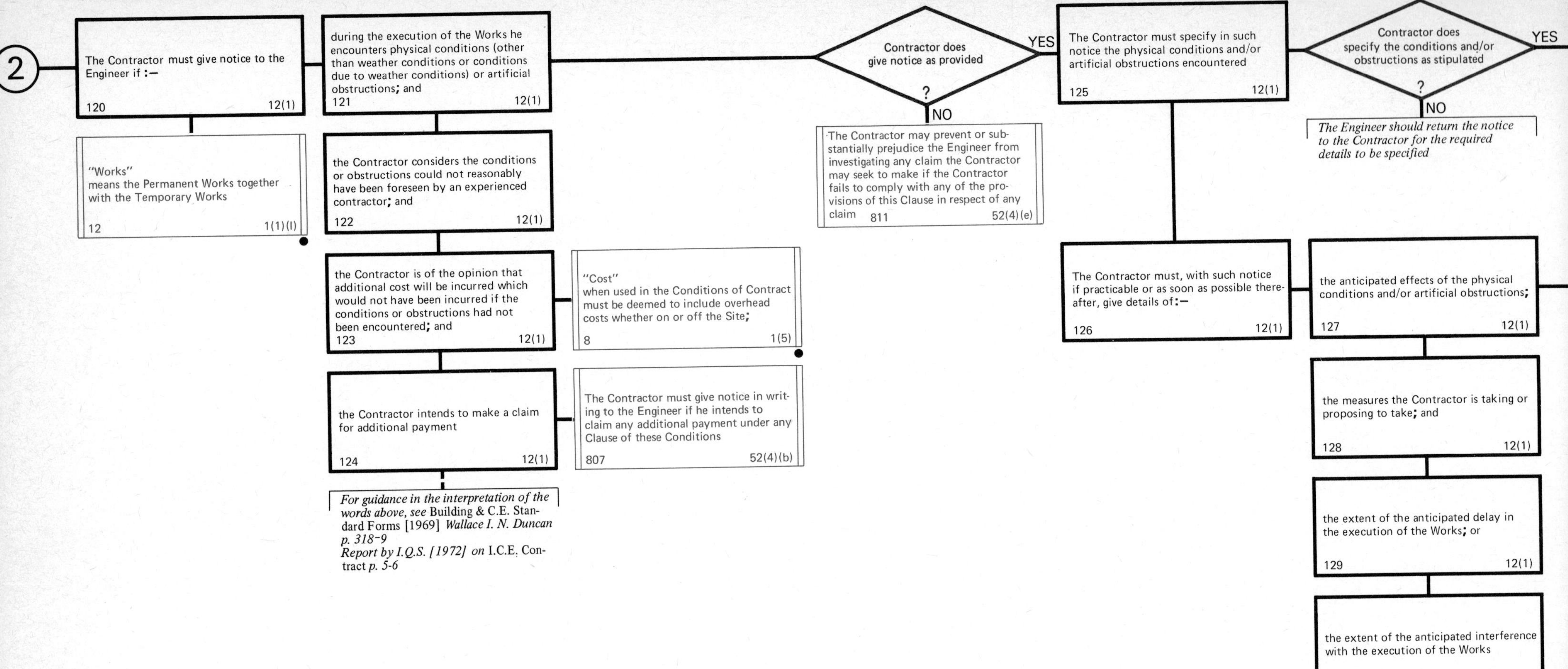
2
The Contractor must give notice to the Engineer if :—
120 12(1)
"Works"
means the Permanent Works together with the Temporary Works
12 1(1)(l)
during the execution of the Works he encounters physical conditions (other than weather conditions or conditions due to weather conditions) or artificial obstructions; and
121 12(1)
the Contractor considers the conditions or obstructions could not reasonably have been foreseen by an experienced contractor; and
122 12(1)
the Contractor is of the opinion that additional cost will be incurred which would not have been incurred if the conditions or obstructions had not been encountered; and
123 12(1)
"Cost"
when used in the Conditions of Contract must be deemed to include overhead costs whether on or off the Site;
8 1(5)
the Contractor intends to make a claim for additional payment
124 12(1)
The Contractor must give notice in writing to the Engineer if he intends to claim any additional payment under any Clause of these Conditions
807 52(4)(b)
For guidance in the interpretation of the words above, see Building & C.E. Standard Forms [1969] Wallace I. N. Duncan p. 318-9
Report by I.Q.S. [1972] on I.C.E. Contract p. 5-6
Contractor does give notice as provided
?
YES
NO
The Contractor may prevent or substantially prejudice the Engineer from investigating any claim the Contractor may seek to make if the Contractor fails to comply with any of the provisions of this Clause in respect of any claim 811 52(4)(e)
The Contractor must specify in such notice the physical conditions and/or artificial obstructions encountered
125 12(1)
Contractor does specify the conditions and/or obstructions as stipulated
?
YES
NO
The Engineer should return the notice to the Contractor for the required details to be specified
The Contractor must, with such notice if practicable or as soon as possible thereafter, give details of:—
126 12(1)
the anticipated effects of the physical conditions and/or artificial obstructions;
127 12(1)
the measures the Contractor is taking or proposing to take; and
128 12(1)
the extent of the anticipated delay in the execution of the Works; or
129 12(1)
the extent of the anticipated interference with the execution of the Works
130 12(1)
"Works"
means the Permanent Works together with the Temporary Works
12 1(1)(l)

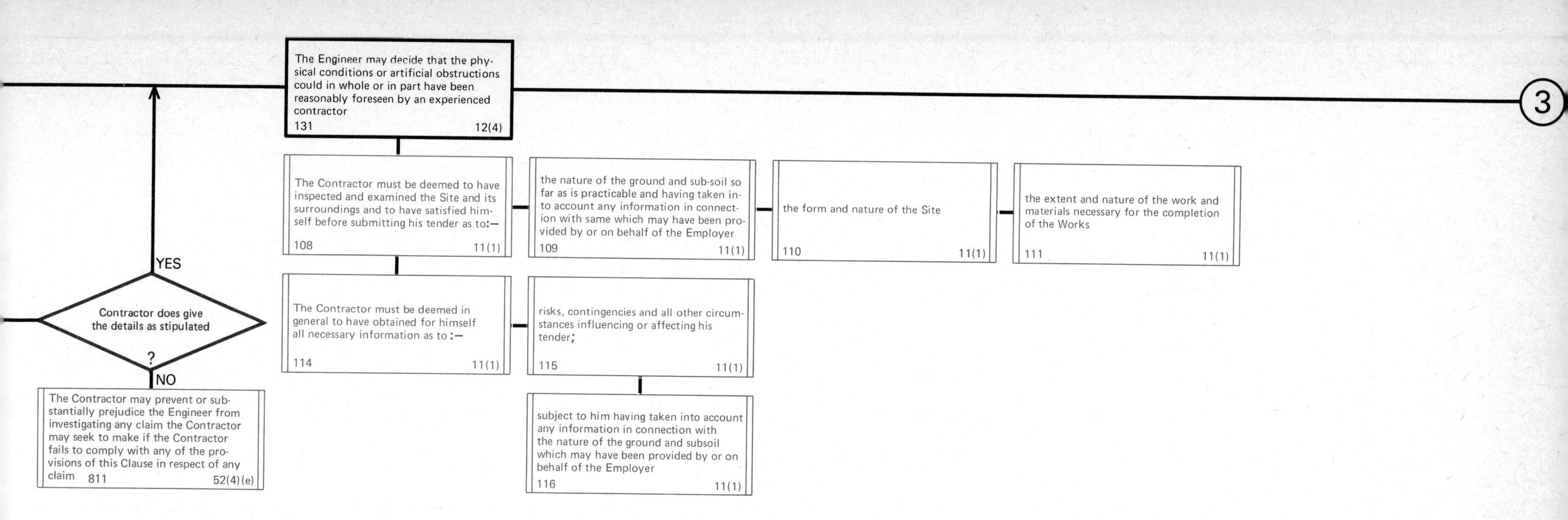
The Engineer may decide that the physical conditions or artificial obstructions could in whole or in part have been reasonably foreseen by an experienced contractor
131 12(4)
3
YES
Contractor does give the details as stipulated
?
NO
The Contractor may prevent or substantially prejudice the Engineer from investigating any claim the Contractor may seek to make if the Contractor fails to comply with any of the provisions of this Clause in respect of any claim 811 52(4)(e)
The Contractor must be deemed to have inspected and examined the Site and its surroundings and to have satisfied himself before submitting his tender as to:—
108 11(1)
the nature of the ground and sub-soil so far as is practicable and having taken into account any information in connection with same which may have been provided by or on behalf of the Employer
109 11(1)
the form and nature of the Site
110 11(1)
the extent and nature of the work and materials necessary for the completion of the Works
111 11(1)
The Contractor must be deemed in general to have obtained for himself all necessary information as to:—
114 11(1)
risks, contingencies and all other circumstances influencing or affecting his tender;
115 11(1)
subject to him having taken into account any information in connection with the nature of the ground and subsoil which may have been provided by or on behalf of the Employer
116 11(1)

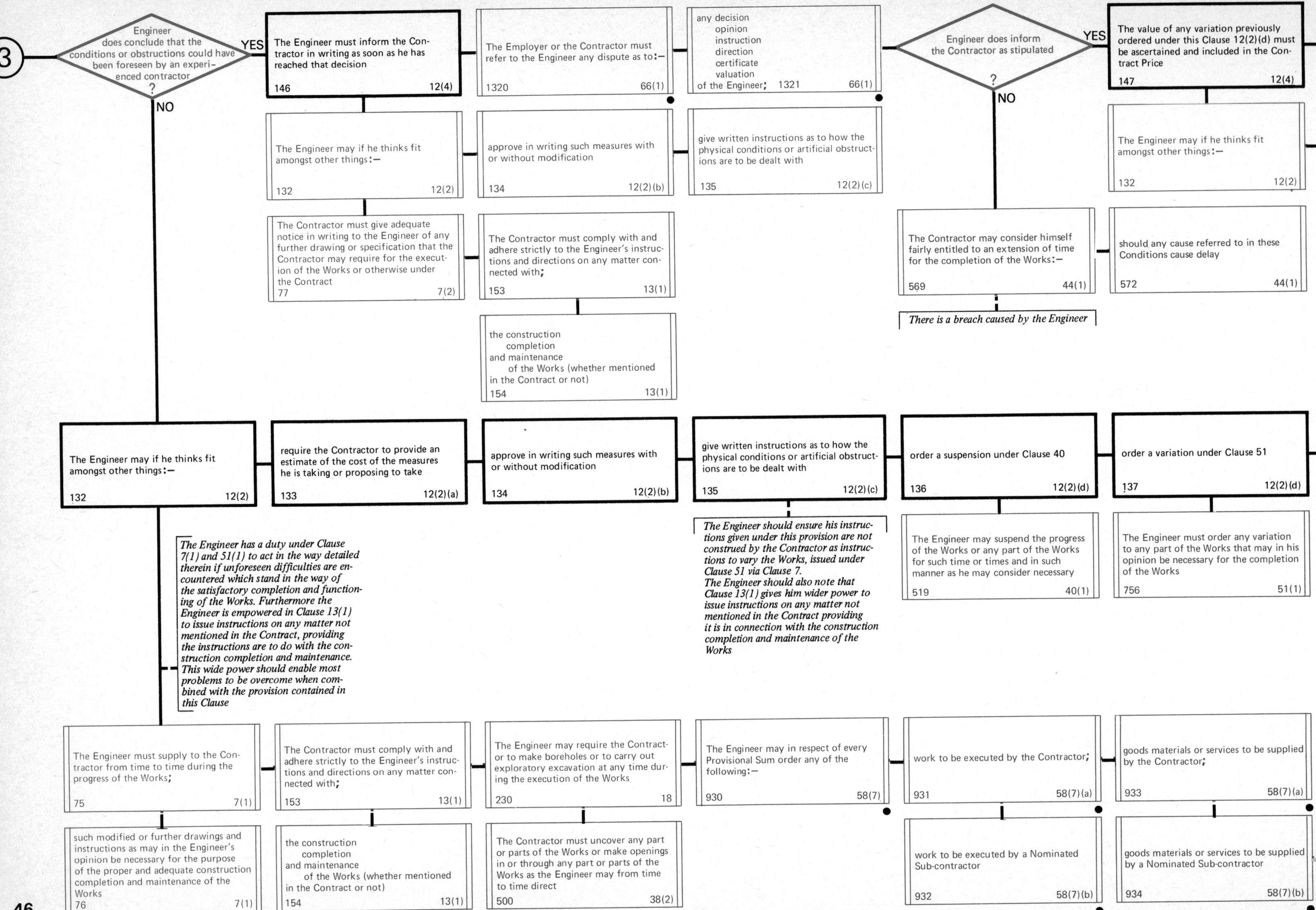
3
Engineer does conclude that the conditions or obstructions could have been foreseen by an experienced contractor ?
YES
NO
The Engineer must inform the Contractor in writing as soon as he has reached that decision 146 12(4)
The Employer or the Contractor must refer to the Engineer any dispute as to:— 1320 66(1)
any decision opinion instruction direction certificate valuation of the Engineer; 1321 66(1)
Engineer does inform the Contractor as stipulated ?
YES
NO
The value of any variation previously ordered under this Clause 12(2)(d) must be ascertained and included in the Contract Price 147 12(4)
The Engineer may if he thinks fit amongst other things:— 132 12(2)
approve in writing such measures with or without modification 134 12(2)(b)
give written instructions as to how the physical conditions or artificial obstructions are to be dealt with 135 12(2)(c)
The Engineer may if he thinks fit amongst other things:— 132 12(2)
The Contractor must give adequate notice in writing to the Engineer of any further drawing or specification that the Contractor may require for the execution of the Works or otherwise under the Contract 77 7(2)
The Contractor must comply with and adhere strictly to the Engineer's instructions and directions on any matter connected with; 153 13(1)
The Contractor may consider himself fairly entitled to an extension of time for the completion of the Works:— 569 44(1)
should any cause referred to in these Conditions cause delay 572 44(1)
There is a breach caused by the Engineer
the construction completion and maintenance of the Works (whether mentioned in the Contract or not) 154 13(1)
The Engineer may if he thinks fit amongst other things:— 132 12(2)
require the Contractor to provide an estimate of the cost of the measures he is taking or proposing to take 133 12(2)(a)
approve in writing such measures with or without modification 134 12(2)(b)
give written instructions as to how the physical conditions or artificial obstructions are to be dealt with 135 12(2)(c)
order a suspension under Clause 40 136 12(2)(d)
order a variation under Clause 51 137 12(2)(d)
The Engineer has a duty under Clause 7(1) and 51(1) to act in the way detailed therein if unforeseen difficulties are encountered which stand in the way of the satisfactory completion and functioning of the Works. Furthermore the Engineer is empowered in Clause 13(1) to issue instructions on any matter not mentioned in the Contract, providing the instructions are to do with the construction completion and maintenance. This wide power should enable most problems to be overcome when combined with the provision contained in this Clause
The Engineer should ensure his instructions given under this provision are not construed by the Contractor as instructions to vary the Works, issued under Clause 51 via Clause 7. The Engineer should also note that Clause 13(1) gives him wider power to issue instructions on any matter not mentioned in the Contract providing it is in connection with the construction completion and maintenance of the Works
The Engineer may suspend the progress of the Works or any part of the Works for such time or times and in such manner as he may consider necessary 519 40(1)
The Engineer must order any variation to any part of the Works that may in his opinion be necessary for the completion of the Works 756 51(1)
The Engineer must supply to the Contractor from time to time during the progress of the Works; 75 7(1)
The Contractor must comply with and adhere strictly to the Engineer's instructions and directions on any matter connected with; 153 13(1)
The Engineer may require the Contractor to make boreholes or to carry out exploratory excavation at any time during the execution of the Works 230 18
The Engineer may in respect of every Provisional Sum order any of the following:— 930 58(7)
work to be executed by the Contractor; 931 58(7)(a)
goods materials or services to be supplied by the Contractor; 933 58(7)(a)
such modified or further drawings and instructions as may in the Engineer's opinion be necessary for the purpose of the proper and adequate construction completion and maintenance of the Works 76 7(1)
the construction completion and maintenance of the Works (whether mentioned in the Contract or not) 154 13(1)
The Contractor must uncover any part or parts of the Works or make openings in or through any part or parts of the Works as the Engineer may from time to time direct 500 38(2)
work to be executed by a Nominated Sub-contractor 932 58(7)(b)
goods materials or services to be supplied by a Nominated Sub-contractor 934 58(7)(b)

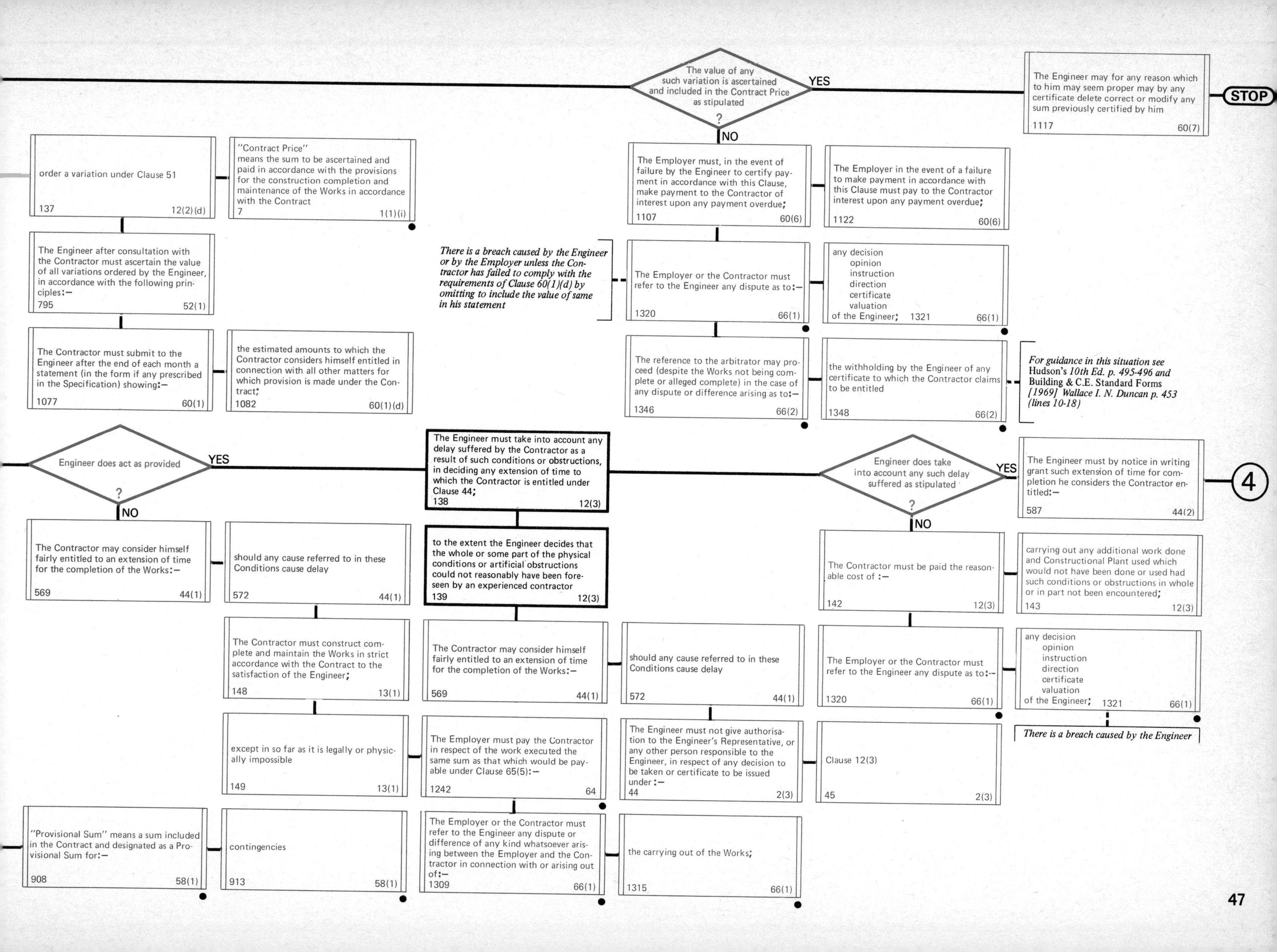
The value of any such variation is ascertained and included in the Contract Price as stipulated ?
YES
NO
The Engineer may for any reason which to him may seem proper may by any certificate delete correct or modify any sum previously certified by him 1117 60(7)
STOP
order a variation under Clause 51 137 12(2)(d)
"Contract Price" means the sum to be ascertained and paid in accordance with the provisions for the construction completion and maintenance of the Works in accordance with the Contract 7 1(1)(i)
The Employer must, in the event of failure by the Engineer to certify payment in accordance with this Clause, make payment to the Contractor of interest upon any payment overdue; 1107 60(6)
The Employer in the event of a failure to make payment in accordance with this Clause must pay to the Contractor interest upon any payment overdue; 1122 60(6)
The Engineer after consultation with the Contractor must ascertain the value of all variations ordered by the Engineer, in accordance with the following principles:— 795 52(1)
There is a breach caused by the Engineer or by the Employer unless the Contractor has failed to comply with the requirements of Clause 60(1)(d) by omitting to include the value of same in his statement
The Employer or the Contractor must refer to the Engineer any dispute as to:— 1320 66(1)
any decision opinion instruction direction certificate valuation of the Engineer; 1321 66(1)
The Contractor must submit to the Engineer after the end of each month a statement (in the form if any prescribed in the Specification) showing:— 1077 60(1)
the estimated amounts to which the Contractor considers himself entitled in connection with all other matters for which provision is made under the Contract; 1082 60(1)(d)
The reference to the arbitrator may proceed (despite the Works not being complete or alleged complete) in the case of any dispute or difference arising as to:— 1346 66(2)
the withholding by the Engineer of any certificate to which the Contractor claims to be entitled 1348 66(2)
For guidance in this situation see Hudson's 10th Ed. p. 495-496 and Building & C.E. Standard Forms [1969] Wallace I. N. Duncan p. 453 (lines 10-18)
Engineer does act as provided ?
YES
NO
The Engineer must take into account any delay suffered by the Contractor as a result of such conditions or obstructions, in deciding any extension of time to which the Contractor is entitled under Clause 44; 138 12(3)
Engineer does take into account any such delay suffered as stipulated ?
YES
NO
The Engineer must by notice in writing grant such extension of time for completion he considers the Contractor entitled:— 587 44(2)
4
The Contractor may consider himself fairly entitled to an extension of time for the completion of the Works:— 569 44(1)
should any cause referred to in these Conditions cause delay 572 44(1)
to the extent the Engineer decides that the whole or some part of the physical conditions or artificial obstructions could not reasonably have been foreseen by an experienced contractor 139 12(3)
The Contractor must be paid the reasonable cost of :— 142 12(3)
carrying out any additional work done and Constructional Plant used which would not have been done or used had such conditions or obstructions in whole or in part not been encountered; 143 12(3)
The Contractor must construct complete and maintain the Works in strict accordance with the Contract to the satisfaction of the Engineer; 148 13(1)
The Contractor may consider himself fairly entitled to an extension of time for the completion of the Works:— 569 44(1)
should any cause referred to in these Conditions cause delay 572 44(1)
The Employer or the Contractor must refer to the Engineer any dispute as to:— 1320 66(1)
any decision opinion instruction direction certificate valuation of the Engineer; 1321 66(1)
There is a breach caused by the Engineer
except in so far as it is legally or physically impossible 149 13(1)
The Employer must pay the Contractor in respect of the work executed the same sum as that which would be payable under Clause 65(5):— 1242 64
The Engineer must not give authorisation to the Engineer's Representative, or any other person responsible to the Engineer, in respect of any decision to be taken or certificate to be issued under:— 44 2(3)
Clause 12(3) 45 2(3)
"Provisional Sum" means a sum included in the Contract and designated as a Provisional Sum for:— 908 58(1)
contingencies 913 58(1)
The Employer or the Contractor must refer to the Engineer any dispute or difference of any kind whatsoever arising between the Employer and the Contractor in connection with or arising out of:— 1309 66(1)
the carrying out of the Works; 1315 66(1)

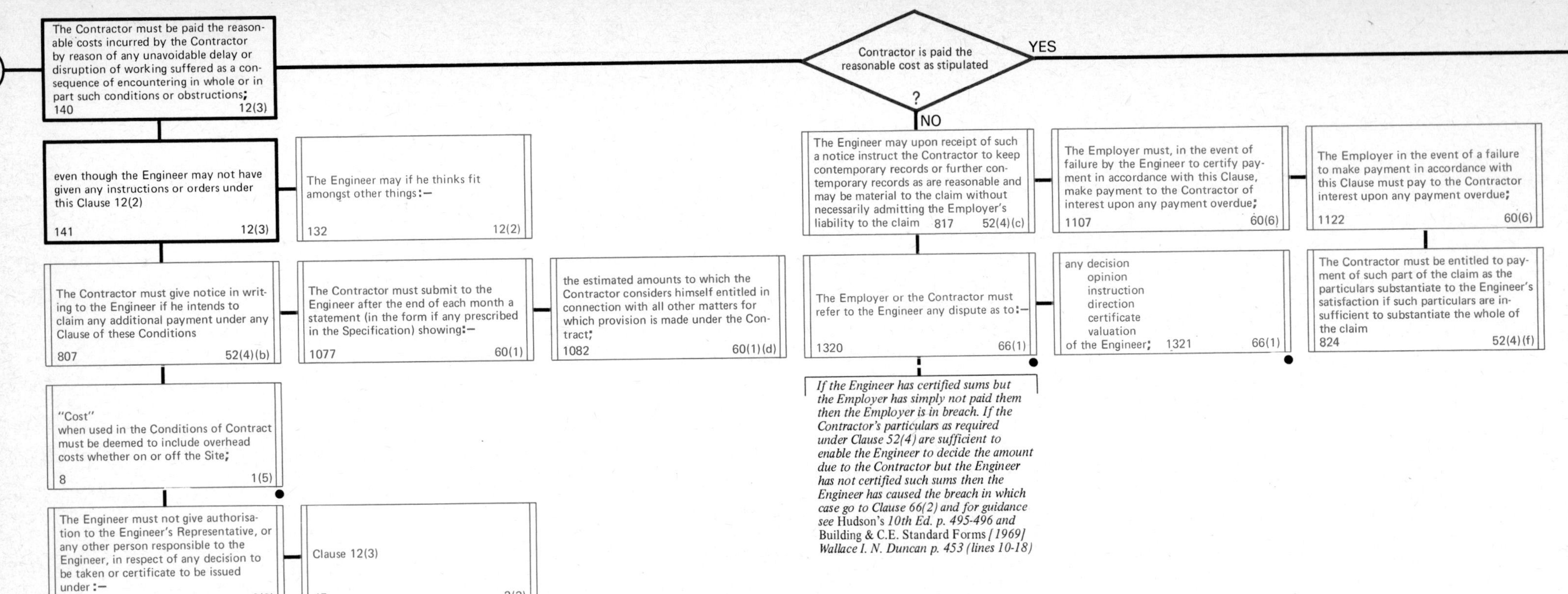
4
The Contractor must be paid the reasonable costs incurred by the Contractor by reason of any unavoidable delay or disruption of working suffered as a consequence of encountering in whole or in part such conditions or obstructions;
140 12(3)
even though the Engineer may not have given any instructions or orders under this Clause 12(2)
141 12(3)
The Engineer may if he thinks fit amongst other things:—
132 12(2)
The Contractor must give notice in writing to the Engineer if he intends to claim any additional payment under any Clause of these Conditions
807 52(4)(b)
The Contractor must submit to the Engineer after the end of each month a statement (in the form if any prescribed in the Specification) showing:—
1077 60(1)
the estimated amounts to which the Contractor considers himself entitled in connection with all other matters for which provision is made under the Contract;
1082 60(1)(d)
"Cost"
when used in the Conditions of Contract must be deemed to include overhead costs whether on or off the Site;
8 1(5)
The Engineer must not give authorisation to the Engineer's Representative, or any other person responsible to the Engineer, in respect of any decision to be taken or certificate to be issued under:—
44 2(3)
Clause 12(3)
45 2(3)
Contractor is paid the reasonable cost as stipulated
?
YES
NO
The Engineer may upon receipt of such a notice instruct the Contractor to keep contemporary records or further contemporary records as are reasonable and may be material to the claim without necessarily admitting the Employer's liability to the claim 817 52(4)(c)
The Employer must, in the event of failure by the Engineer to certify payment in accordance with this Clause, make payment to the Contractor of interest upon any payment overdue;
1107 60(6)
The Employer in the event of a failure to make payment in accordance with this Clause must pay to the Contractor interest upon any payment overdue;
1122 60(6)
The Employer or the Contractor must refer to the Engineer any dispute as to:—
1320 66(1)
any decision
opinion
instruction
direction
certificate
valuation
of the Engineer; 1321 66(1)
The Contractor must be entitled to payment of such part of the claim as the particulars substantiate to the Engineer's satisfaction if such particulars are insufficient to substantiate the whole of the claim
824 52(4)(f)
If the Engineer has certified sums but the Employer has simply not paid them then the Employer is in breach. If the Contractor's particulars as required under Clause 52(4) are sufficient to enable the Engineer to decide the amount due to the Contractor but the Engineer has not certified such sums then the Engineer has caused the breach in which case go to Clause 66(2) and for guidance see Hudson's 10th Ed. p. 495-496 and Building & C.E. Standard Forms [1969] Wallace I. N. Duncan p. 453 (lines 10-18)

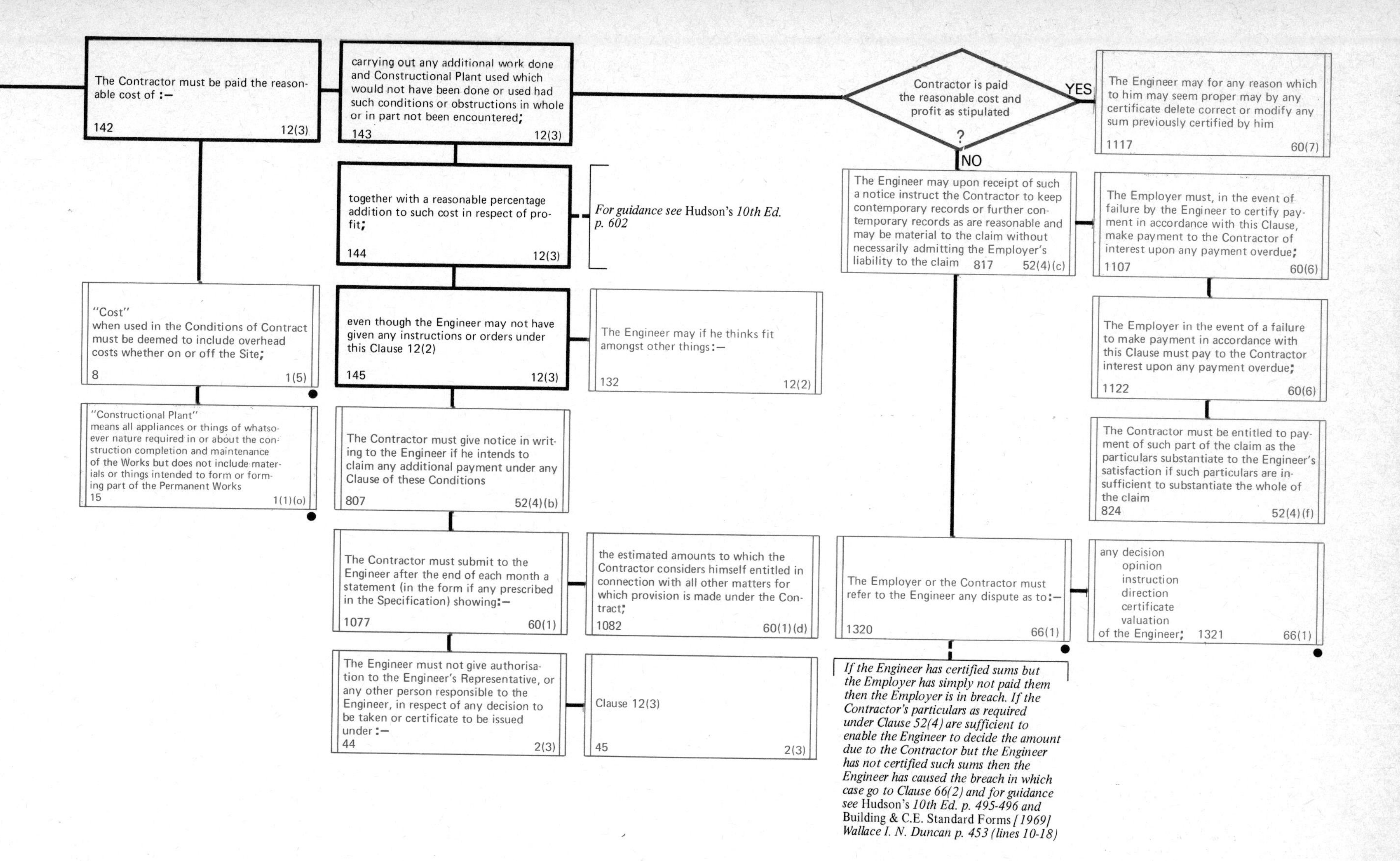

The Contractor must be paid the reasonable cost of :—
142 12(3)
carrying out any additional work done and Constructional Plant used which would not have been done or used had such conditions or obstructions in whole or in part not been encountered;
143 12(3)
together with a reasonable percentage addition to such cost in respect of profit;
144 12(3)
For guidance see Hudson's 10th Ed. p. 602
even though the Engineer may not have given any instructions or orders under this Clause 12(2)
145 12(3)
The Engineer may if he thinks fit amongst other things:—
132 12(2)
"Cost"
when used in the Conditions of Contract must be deemed to include overhead costs whether on or off the Site;
8 1(5)
"Constructional Plant"
means all appliances or things of whatsoever nature required in or about the construction completion and maintenance of the Works but does not include materials or things intended to form or forming part of the Permanent Works
15 1(1)(o)
The Contractor must give notice in writing to the Engineer if he intends to claim any additional payment under any Clause of these Conditions
807 52(4)(b)
The Contractor must submit to the Engineer after the end of each month a statement (in the form if any prescribed in the Specification) showing:—
1077 60(1)
the estimated amounts to which the Contractor considers himself entitled in connection with all other matters for which provision is made under the Contract;
1082 60(1)(d)
The Engineer must not give authorisation to the Engineer's Representative, or any other person responsible to the Engineer, in respect of any decision to be taken or certificate to be issued under:—
44 2(3)
Clause 12(3)
45 2(3)
Contractor is paid the reasonable cost and profit as stipulated ?
YES
NO
The Engineer may for any reason which to him may seem proper may by any certificate delete correct or modify any sum previously certified by him
1117 60(7)
The Engineer may upon receipt of such a notice instruct the Contractor to keep contemporary records or further contemporary records as are reasonable and may be material to the claim without necessarily admitting the Employer's liability to the claim 817 52(4)(c)
The Employer must, in the event of failure by the Engineer to certify payment in accordance with this Clause, make payment to the Contractor of interest upon any payment overdue;
1107 60(6)
The Employer in the event of a failure to make payment in accordance with this Clause must pay to the Contractor interest upon any payment overdue;
1122 60(6)
The Contractor must be entitled to payment of such part of the claim as the particulars substantiate to the Engineer's satisfaction if such particulars are insufficient to substantiate the whole of the claim
824 52(4)(f)
The Employer or the Contractor must refer to the Engineer any dispute as to:—
1320 66(1)
any decision
opinion
instruction
direction
certificate
valuation
of the Engineer; 1321 66(1)
If the Engineer has certified sums but the Employer has simply not paid them then the Employer is in breach. If the Contractor's particulars as required under Clause 52(4) are sufficient to enable the Engineer to decide the amount due to the Contractor but the Engineer has not certified such sums then the Engineer has caused the breach in which case go to Clause 66(2) and for guidance see Hudson's 10th Ed. p. 495-496 and Building & C.E. Standard Forms [1969] Wallace I. N. Duncan p. 453 (lines 10-18)

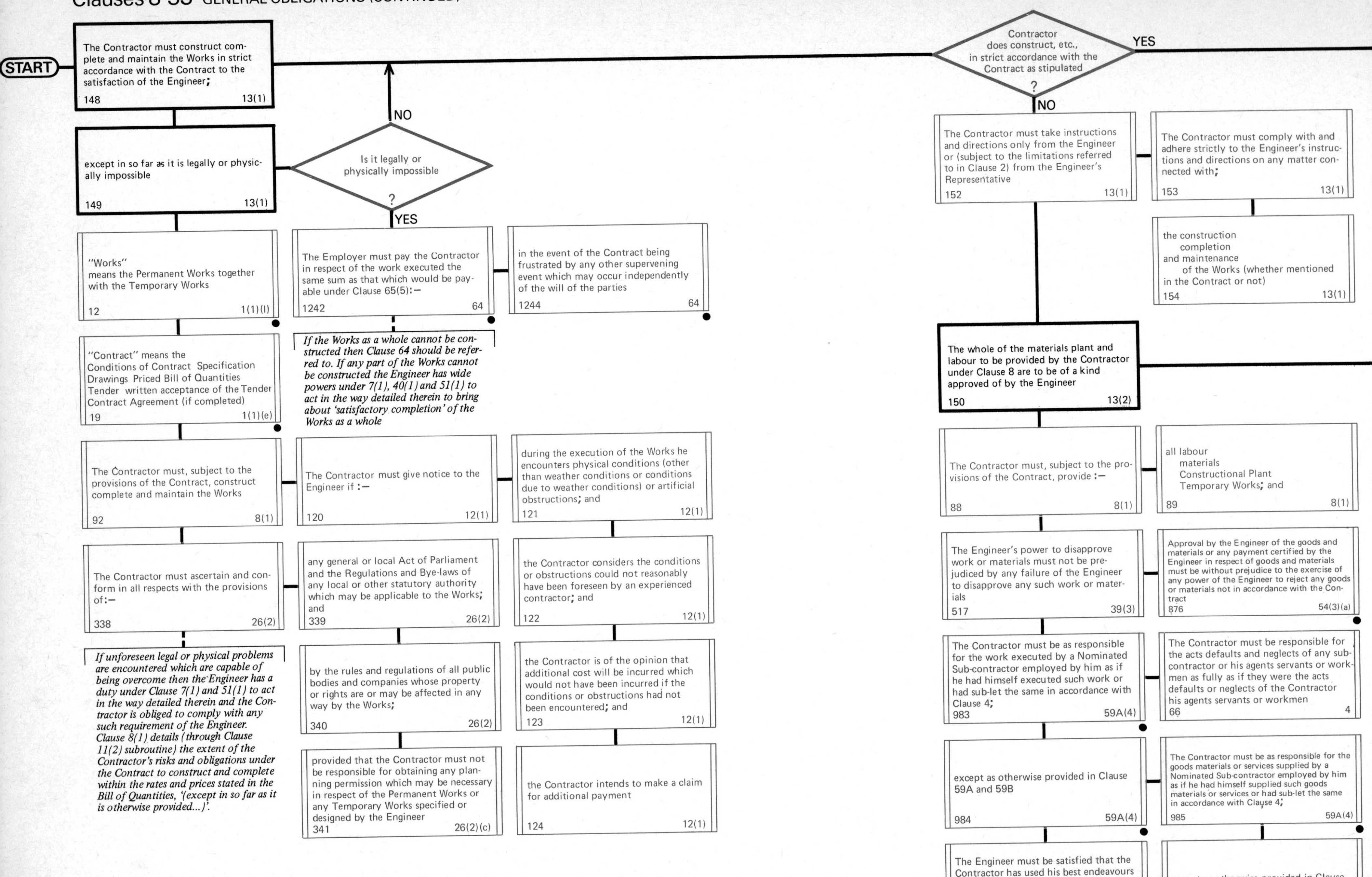
START
The Contractor must construct complete and maintain the Works in strict accordance with the Contract to the satisfaction of the Engineer;
148 13(1)
except in so far as it is legally or physically impossible
149 13(1)
Is it legally or physically impossible ?
NO
YES
"Works" means the Permanent Works together with the Temporary Works
12 1(1)(l)
"Contract" means the Conditions of Contract Specification Drawings Priced Bill of Quantities Tender written acceptance of the Tender Contract Agreement (if completed)
19 1(1)(e)
The Employer must pay the Contractor in respect of the work executed the same sum as that which would be payable under Clause 65(5):—
1242 64
in the event of the Contract being frustrated by any other supervening event which may occur independently of the will of the parties
1244 64
If the Works as a whole cannot be constructed then Clause 64 should be referred to. If any part of the Works cannot be constructed the Engineer has wide powers under 7(1), 40(1) and 51(1) to act in the way detailed therein to bring about 'satisfactory completion' of the Works as a whole
The Contractor must, subject to the provisions of the Contract, construct complete and maintain the Works
92 8(1)
The Contractor must give notice to the Engineer if :—
120 12(1)
during the execution of the Works he encounters physical conditions (other than weather conditions or conditions due to weather conditions) or artificial obstructions; and
121 12(1)
The Contractor must ascertain and conform in all respects with the provisions of:—
338 26(2)
any general or local Act of Parliament and the Regulations and Bye-laws of any local or other statutory authority which may be applicable to the Works; and
339 26(2)
the Contractor considers the conditions or obstructions could not reasonably have been foreseen by an experienced contractor; and
122 12(1)
If unforeseen legal or physical problems are encountered which are capable of being overcome then the Engineer has a duty under Clause 7(1) and 51(1) to act in the way detailed therein and the Contractor is obliged to comply with any such requirement of the Engineer. Clause 8(1) details (through Clause 11(2) subroutine) the extent of the Contractor's risks and obligations under the Contract to construct and complete within the rates and prices stated in the Bill of Quantities, '(except in so far as it is otherwise provided...)'.
by the rules and regulations of all public bodies and companies whose property or rights are or may be affected in any way by the Works;
340 26(2)
the Contractor is of the opinion that additional cost will be incurred which would not have been incurred if the conditions or obstructions had not been encountered; and
123 12(1)
provided that the Contractor must not be responsible for obtaining any planning permission which may be necessary in respect of the Permanent Works or any Temporary Works specified or designed by the Engineer
341 26(2)(c)
the Contractor intends to make a claim for additional payment
124 12(1)
Contractor does construct, etc., in strict accordance with the Contract as stipulated ?
YES
NO
The Contractor must take instructions and directions only from the Engineer or (subject to the limitations referred to in Clause 2) from the Engineer's Representative
152 13(1)
The Contractor must comply with and adhere strictly to the Engineer's instructions and directions on any matter connected with;
153 13(1)
the construction
completion
and maintenance
of the Works (whether mentioned in the Contract or not)
154 13(1)
The whole of the materials plant and labour to be provided by the Contractor under Clause 8 are to be of a kind approved of by the Engineer
150 13(2)
The Contractor must, subject to the provisions of the Contract, provide :—
88 8(1)
all labour
materials
Constructional Plant
Temporary Works; and
89 8(1)
The Engineer's power to disapprove work or materials must not be prejudiced by any failure of the Engineer to disapprove any such work or materials
517 39(3)
Approval by the Engineer of the goods and materials or any payment certified by the Engineer in respect of goods and materials must be without prejudice to the exercise of any power of the Engineer to reject any goods or materials not in accordance with the Contract
876 54(3)(a)
The Contractor must be as responsible for the work executed by a Nominated Sub-contractor employed by him as if he had himself executed such work or had sub-let the same in accordance with Clause 4;
983 59A(4)
The Contractor must be responsible for the acts defaults and neglects of any sub-contractor or his agents servants or workmen as fully as if they were the acts defaults or neglects of the Contractor his agents servants or workmen
66 4
except as otherwise provided in Clause 59A and 59B
984 59A(4)
The Contractor must be as responsible for the goods materials or services supplied by a Nominated Sub-contractor employed by him as if he had himself supplied such goods materials or services or had sub-let the same in accordance with Clause 4;
985 59A(4)
The Engineer must be satisfied that the Contractor has used his best endeavours to obtain materials to the dimensions described in the Contract or ordered by the Engineer; and
1480 71(2)
except as otherwise provided in Clause 59A and 59B
986 59A(4)

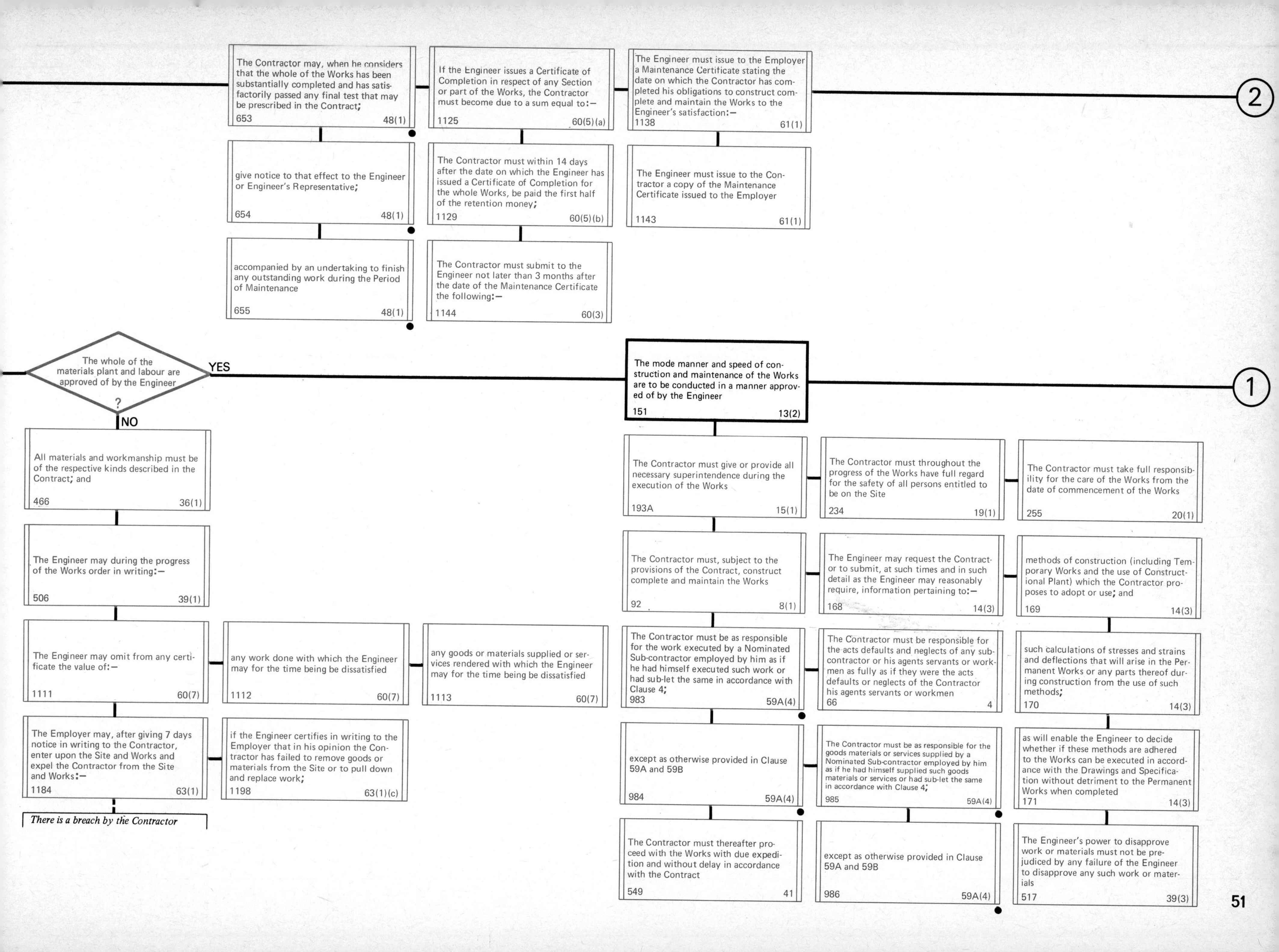

The Contractor may, when he considers that the whole of the Works has been substantially completed and has satisfactorily passed any final test that may be prescribed in the Contract;
653 48(1)
give notice to that effect to the Engineer or Engineer's Representative;
654 48(1)
accompanied by an undertaking to finish any outstanding work during the Period of Maintenance
655 48(1)
If the Engineer issues a Certificate of Completion in respect of any Section or part of the Works, the Contractor must become due to a sum equal to:—
1125 60(5)(a)
The Contractor must within 14 days after the date on which the Engineer has issued a Certificate of Completion for the whole Works, be paid the first half of the retention money;
1129 60(5)(b)
The Contractor must submit to the Engineer not later than 3 months after the date of the Maintenance Certificate the following:—
1144 60(3)
The Engineer must issue to the Employer a Maintenance Certificate stating the date on which the Contractor has completed his obligations to construct complete and maintain the Works to the Engineer's satisfaction:—
1138 61(1)
The Engineer must issue to the Contractor a copy of the Maintenance Certificate issued to the Employer
1143 61(1)
2
The whole of the materials plant and labour are approved of by the Engineer ?
YES
NO
The mode manner and speed of construction and maintenance of the Works are to be conducted in a manner approved of by the Engineer
151 13(2)
1
All materials and workmanship must be of the respective kinds described in the Contract; and
466 36(1)
The Engineer may during the progress of the Works order in writing:—
506 39(1)
The Engineer may omit from any certificate the value of:—
1111 60(7)
any work done with which the Engineer may for the time being be dissatisfied
1112 60(7)
any goods or materials supplied or services rendered with which the Engineer may for the time being be dissatisfied
1113 60(7)
The Employer may, after giving 7 days notice in writing to the Contractor, enter upon the Site and Works and expel the Contractor from the Site and Works:—
1184 63(1)
if the Engineer certifies in writing to the Employer that in his opinion the Contractor has failed to remove goods or materials from the Site or to pull down and replace work;
1198 63(1)(c)
There is a breach by the Contractor
The Contractor must give or provide all necessary superintendence during the execution of the Works
193A 15(1)
The Contractor must throughout the progress of the Works have full regard for the safety of all persons entitled to be on the Site
234 19(1)
The Contractor must take full responsibility for the care of the Works from the date of commencement of the Works
255 20(1)
The Contractor must, subject to the provisions of the Contract, construct complete and maintain the Works
92 8(1)
The Engineer may request the Contractor to submit, at such times and in such detail as the Engineer may reasonably require, information pertaining to:—
168 14(3)
methods of construction (including Temporary Works and the use of Constructional Plant) which the Contractor proposes to adopt or use; and
169 14(3)
The Contractor must be as responsible for the work executed by a Nominated Sub-contractor employed by him as if he had himself executed such work or had sub-let the same in accordance with Clause 4;
983 59A(4)
The Contractor must be responsible for the acts defaults and neglects of any sub-contractor or his agents servants or workmen as fully as if they were the acts defaults or neglects of the Contractor his agents servants or workmen
66 4
such calculations of stresses and strains and deflections that will arise in the Permanent Works or any parts thereof during construction from the use of such methods;
170 14(3)
except as otherwise provided in Clause 59A and 59B
984 59A(4)
The Contractor must be as responsible for the goods materials or services supplied by a Nominated Sub-contractor employed by him as if he had himself supplied such goods materials or services or had sub-let the same in accordance with Clause 4;
985 59A(4)
as will enable the Engineer to decide whether if these methods are adhered to the Works can be executed in accordance with the Drawings and Specification without detriment to the Permanent Works when completed
171 14(3)
The Contractor must thereafter proceed with the Works with due expedition and without delay in accordance with the Contract
549 41
except as otherwise provided in Clause 59A and 59B
986 59A(4)
The Engineer's power to disapprove work or materials must not be prejudiced by any failure of the Engineer to disapprove any such work or materials
517 39(3)

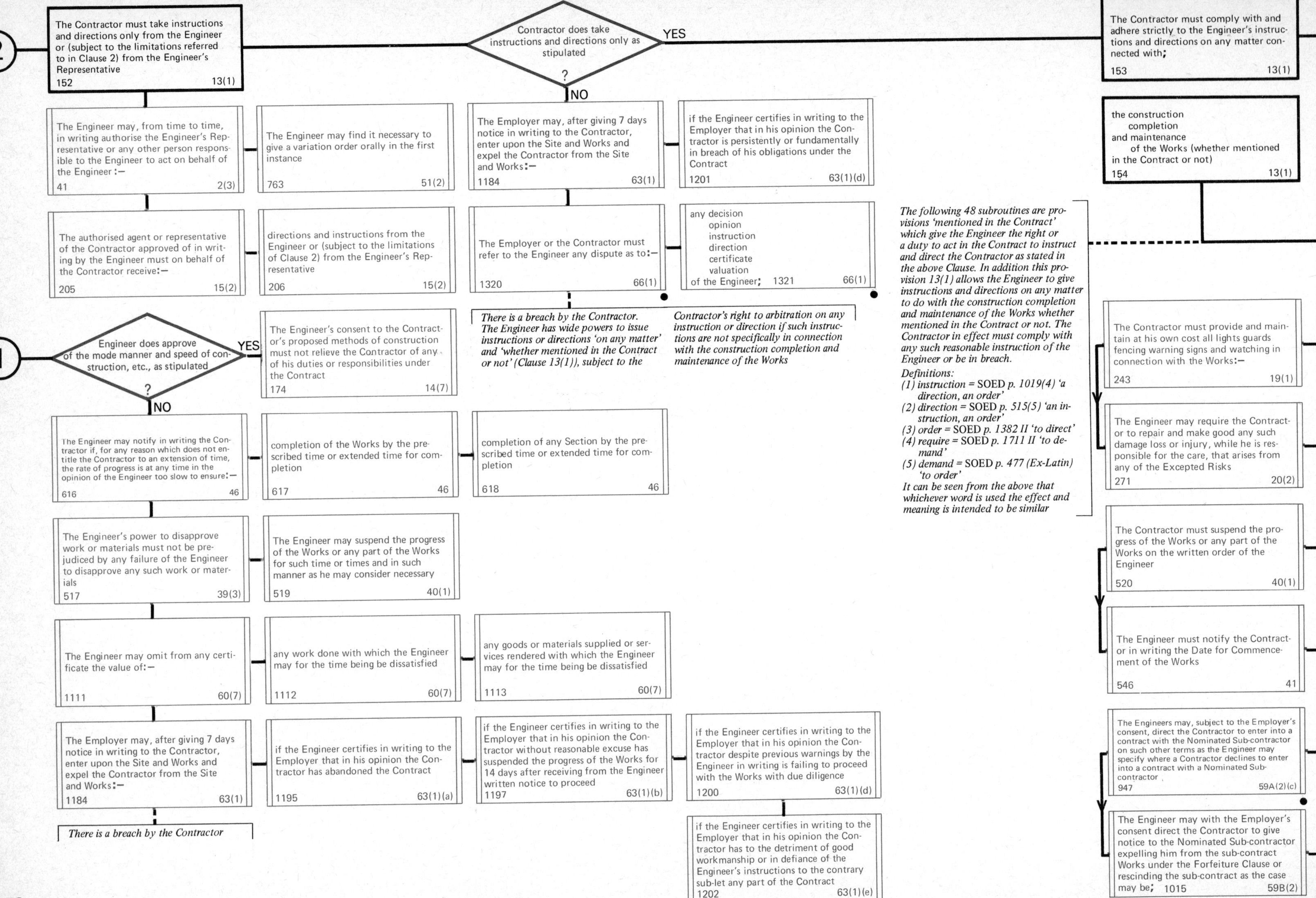
2
The Contractor must take instructions and directions only from the Engineer or (subject to the limitations referred to in Clause 2) from the Engineer's Representative
152 13(1)
Contractor does take instructions and directions only as stipulated
?
YES
NO
The Contractor must comply with and adhere strictly to the Engineer's instructions and directions on any matter connected with;
153 13(1)
the construction
completion
and maintenance
of the Works (whether mentioned in the Contract or not)
154 13(1)
The Engineer may, from time to time, in writing authorise the Engineer's Representative or any other person responsible to the Engineer to act on behalf of the Engineer:—
41 2(3)
The Engineer may find it necessary to give a variation order orally in the first instance
763 51(2)
The Employer may, after giving 7 days notice in writing to the Contractor, enter upon the Site and Works and expel the Contractor from the Site and Works:—
1184 63(1)
if the Engineer certifies in writing to the Employer that in his opinion the Contractor is persistently or fundamentally in breach of his obligations under the Contract
1201 63(1)(d)
The authorised agent or representative of the Contractor approved of in writing by the Engineer must on behalf of the Contractor receive:—
205 15(2)
directions and instructions from the Engineer or (subject to the limitations of Clause 2) from the Engineer's Representative
206 15(2)
The Employer or the Contractor must refer to the Engineer any dispute as to:—
1320 66(1)
any decision
opinion
instruction
direction
certificate
valuation
of the Engineer; 1321 66(1)
There is a breach by the Contractor. The Engineer has wide powers to issue instructions or directions 'on any matter' and 'whether mentioned in the Contract or not' (Clause 13(1)), subject to the
Contractor's right to arbitration on any instruction or direction if such instructions are not specifically in connection with the construction completion and maintenance of the Works
The following 48 subroutines are provisions 'mentioned in the Contract' which give the Engineer the right or a duty to act in the Contract to instruct and direct the Contractor as stated in the above Clause. In addition this provision 13(1) allows the Engineer to give instructions and directions on any matter to do with the construction completion and maintenance of the Works whether mentioned in the Contract or not. The Contractor in effect must comply with any such reasonable instruction of the Engineer or be in breach.
Definitions:
(1) instruction = SOED p. 1019(4) 'a direction, an order'
(2) direction = SOED p. 515(5) 'an instruction, an order'
(3) order = SOED p. 1382 II 'to direct'
(4) require = SOED p. 1711 II 'to demand'
(5) demand = SOED p. 477 (Ex-Latin) 'to order'
It can be seen from the above that whichever word is used the effect and meaning is intended to be similar
1
Engineer does approve of the mode manner and speed of construction, etc., as stipulated
?
YES
NO
The Engineer's consent to the Contractor's proposed methods of construction must not relieve the Contractor of any of his duties or responsibilities under the Contract
174 14(7)
The Contractor must provide and maintain at his own cost all lights guards fencing warning signs and watching in connection with the Works:—
243 19(1)
The Engineer may require the Contractor to repair and make good any such damage loss or injury, while he is responsible for the care, that arises from any of the Excepted Risks
271 20(2)
The Engineer may notify in writing the Contractor if, for any reason which does not entitle the Contractor to an extension of time, the rate of progress is at any time in the opinion of the Engineer too slow to ensure:—
616 46
completion of the Works by the prescribed time or extended time for completion
617 46
completion of any Section by the prescribed time or extended time for completion
618 46
The Engineer's power to disapprove work or materials must not be prejudiced by any failure of the Engineer to disapprove any such work or materials
517 39(3)
The Engineer may suspend the progress of the Works or any part of the Works for such time or times and in such manner as he may consider necessary
519 40(1)
The Contractor must suspend the progress of the Works or any part of the Works on the written order of the Engineer
520 40(1)
The Engineer must notify the Contractor in writing the Date for Commencement of the Works
546 41
The Engineer may omit from any certificate the value of:—
1111 60(7)
any work done with which the Engineer may for the time being be dissatisfied
1112 60(7)
any goods or materials supplied or services rendered with which the Engineer may for the time being be dissatisfied
1113 60(7)
The Employer may, after giving 7 days notice in writing to the Contractor, enter upon the Site and Works and expel the Contractor from the Site and Works:—
1184 63(1)
if the Engineer certifies in writing to the Employer that in his opinion the Contractor has abandoned the Contract
1195 63(1)(a)
if the Engineer certifies in writing to the Employer that in his opinion the Contractor without reasonable excuse has suspended the progress of the Works for 14 days after receiving from the Engineer written notice to proceed
1197 63(1)(b)
if the Engineer certifies in writing to the Employer that in his opinion the Contractor despite previous warnings by the Engineer in writing is failing to proceed with the Works with due diligence
1200 63(1)(d)
The Engineers may, subject to the Employer's consent, direct the Contractor to enter into a contract with the Nominated Sub-contractor on such other terms as the Engineer may specify where a Contractor declines to enter into a contract with a Nominated Sub-contractor
947 59A(2)(c)
There is a breach by the Contractor
if the Engineer certifies in writing to the Employer that in his opinion the Contractor has to the detriment of good workmanship or in defiance of the Engineer's instructions to the contrary sub-let any part of the Contract
1202 63(1)(e)
The Engineer may with the Employer's consent direct the Contractor to give notice to the Nominated Sub-contractor expelling him from the sub-contract Works under the Forfeiture Clause or rescinding the sub-contract as the case may be; 1015 59B(2)

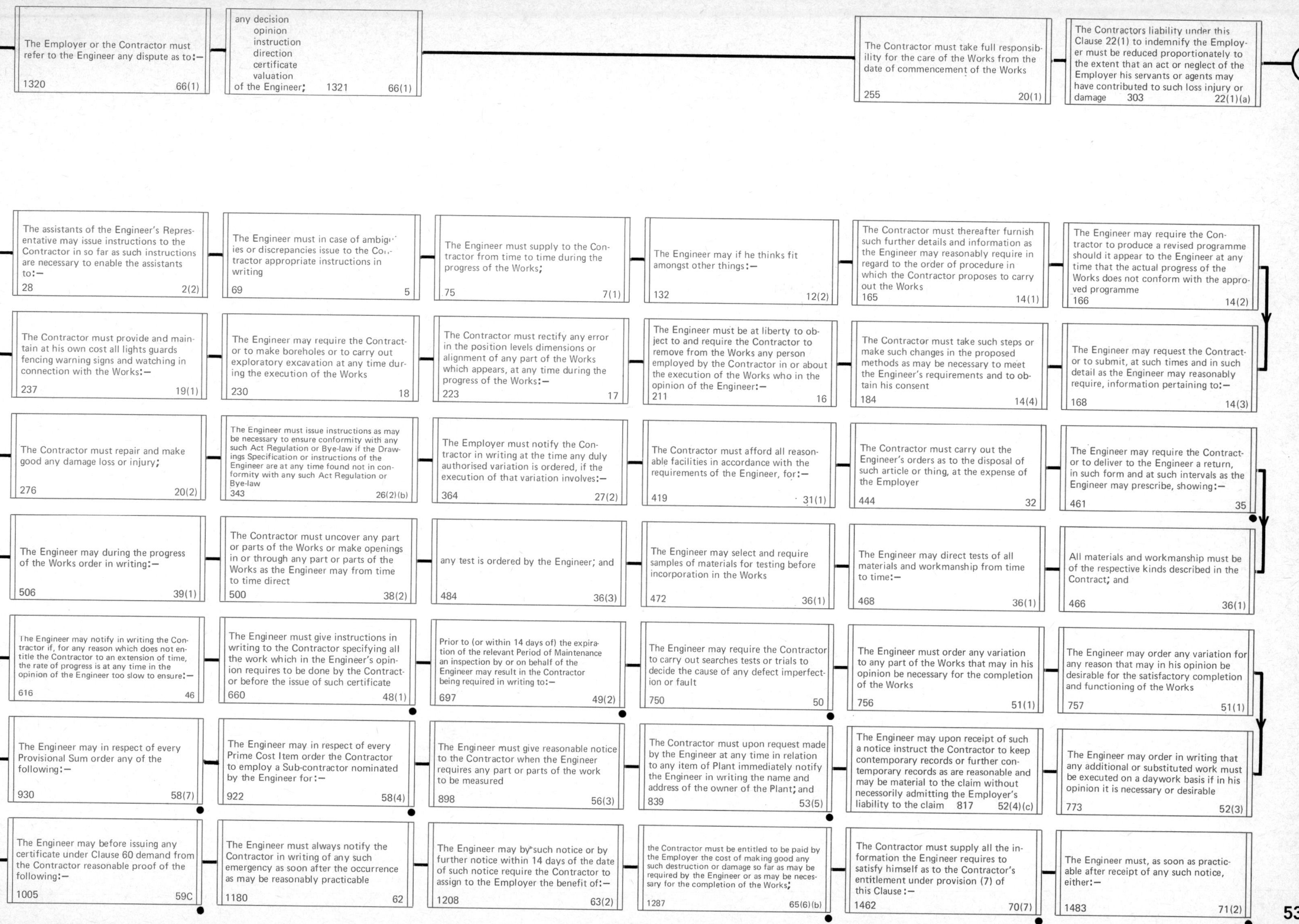

The Employer or the Contractor must refer to the Engineer any dispute as to:— 1320 66(1)
any decision opinion instruction direction certificate valuation of the Engineer; 1321 66(1)
The Contractor must take full responsibility for the care of the Works from the date of commencement of the Works 255 20(1)
The Contractors liability under this Clause 22(1) to indemnify the Employer must be reduced proportionately to the extent that an act or neglect of the Employer his servants or agents may have contributed to such loss injury or damage 303 22(1)(a)
3
The assistants of the Engineer's Representative may issue instructions to the Contractor in so far as such instructions are necessary to enable the assistants to:— 28 2(2)
The Engineer must in case of ambiguities or discrepancies issue to the Contractor appropriate instructions in writing 69 5
The Engineer must supply to the Contractor from time to time during the progress of the Works; 75 7(1)
The Engineer may if he thinks fit amongst other things:— 132 12(2)
The Contractor must thereafter furnish such further details and information as the Engineer may reasonably require in regard to the order of procedure in which the Contractor proposes to carry out the Works 165 14(1)
The Engineer may require the Contractor to produce a revised programme should it appear to the Engineer at any time that the actual progress of the Works does not conform with the approved programme 166 14(2)
The Contractor must provide and maintain at his own cost all lights guards fencing warning signs and watching in connection with the Works:— 237 19(1)
The Engineer may require the Contractor to make boreholes or to carry out exploratory excavation at any time during the execution of the Works 230 18
The Contractor must rectify any error in the position levels dimensions or alignment of any part of the Works which appears, at any time during the progress of the Works:— 223 17
The Engineer must be at liberty to object to and require the Contractor to remove from the Works any person employed by the Contractor in or about the execution of the Works who in the opinion of the Engineer:— 211 16
The Contractor must take such steps or make such changes in the proposed methods as may be necessary to meet the Engineer's requirements and to obtain his consent 184 14(4)
The Engineer may request the Contractor to submit, at such times and in such detail as the Engineer may reasonably require, information pertaining to:— 168 14(3)
The Contractor must repair and make good any damage loss or injury; 276 20(2)
The Engineer must issue instructions as may be necessary to ensure conformity with any such Act Regulation or Bye-law if the Drawings Specification or instructions of the Engineer are at any time found not in conformity with any such Act Regulation or Bye-law 343 26(2)(b)
The Employer must notify the Contractor in writing at the time any duly authorised variation is ordered, if the execution of that variation involves:— 364 27(2)
The Contractor must afford all reasonable facilities in accordance with the requirements of the Engineer, for:— 419 31(1)
The Contractor must carry out the Engineer's orders as to the disposal of such article or thing, at the expense of the Employer 444 32
The Engineer may require the Contractor to deliver to the Engineer a return, in such form and at such intervals as the Engineer may prescribe, showing:— 461 35
The Engineer may during the progress of the Works order in writing:— 506 39(1)
The Contractor must uncover any part or parts of the Works or make openings in or through any part or parts of the Works as the Engineer may from time to time direct 500 38(2)
any test is ordered by the Engineer; and 484 36(3)
The Engineer may select and require samples of materials for testing before incorporation in the Works 472 36(1)
The Engineer may direct tests of all materials and workmanship from time to time:— 468 36(1)
All materials and workmanship must be of the respective kinds described in the Contract; and 466 36(1)
The Engineer may notify in writing the Contractor if, for any reason which does not entitle the Contractor to an extension of time, the rate of progress is at any time in the opinion of the Engineer too slow to ensure:— 616 46
The Engineer must give instructions in writing to the Contractor specifying all the work which in the Engineer's opinion requires to be done by the Contractor or before the issue of such certificate 660 48(1)
Prior to (or within 14 days of) the expiration of the relevant Period of Maintenance an inspection by or on behalf of the Engineer may result in the Contractor being required in writing to:— 697 49(2)
The Engineer may require the Contractor to carry out searches tests or trials to decide the cause of any defect imperfection or fault 750 50
The Engineer must order any variation to any part of the Works that may in his opinion be necessary for the completion of the Works 756 51(1)
The Engineer may order any variation for any reason that may in his opinion be desirable for the satisfactory completion and functioning of the Works 757 51(1)
The Engineer may in respect of every Provisional Sum order any of the following:— 930 58(7)
The Engineer may in respect of every Prime Cost Item order the Contractor to employ a Sub-contractor nominated by the Engineer for:— 922 58(4)
The Engineer must give reasonable notice to the Contractor when the Engineer requires any part or parts of the work to be measured 898 56(3)
The Contractor must upon request made by the Engineer at any time in relation to any item of Plant immediately notify the Engineer in writing the name and address of the owner of the Plant; and 839 53(5)
The Engineer may upon receipt of such a notice instruct the Contractor to keep contemporary records or further contemporary records as are reasonable and may be material to the claim without necessorily admitting the Employer's liability to the claim 817 52(4)(c)
The Engineer may order in writing that any additional or substituted work must be executed on a daywork basis if in his opinion it is necessary or desirable 773 52(3)
The Engineer may before issuing any certificate under Clause 60 demand from the Contractor reasonable proof of the following:— 1005 59C
The Engineer must always notify the Contractor in writing of any such emergency as soon after the occurrence as may be reasonably practicable 1180 62
The Engineer may by such notice or by further notice within 14 days of the date of such notice require the Contractor to assign to the Employer the benefit of:— 1208 63(2)
the Contractor must be entitled to be paid by the Employer the cost of making good any such destruction or damage so far as may be required by the Engineer or as may be necessary for the completion of the Works; 1287 65(6)(b)
The Contractor must supply all the information the Engineer requires to satisfy himself as to the Contractor's entitlement under provision (7) of this Clause:— 1462 70(7)
The Engineer must, as soon as practicable after receipt of any such notice, either:— 1483 71(2)

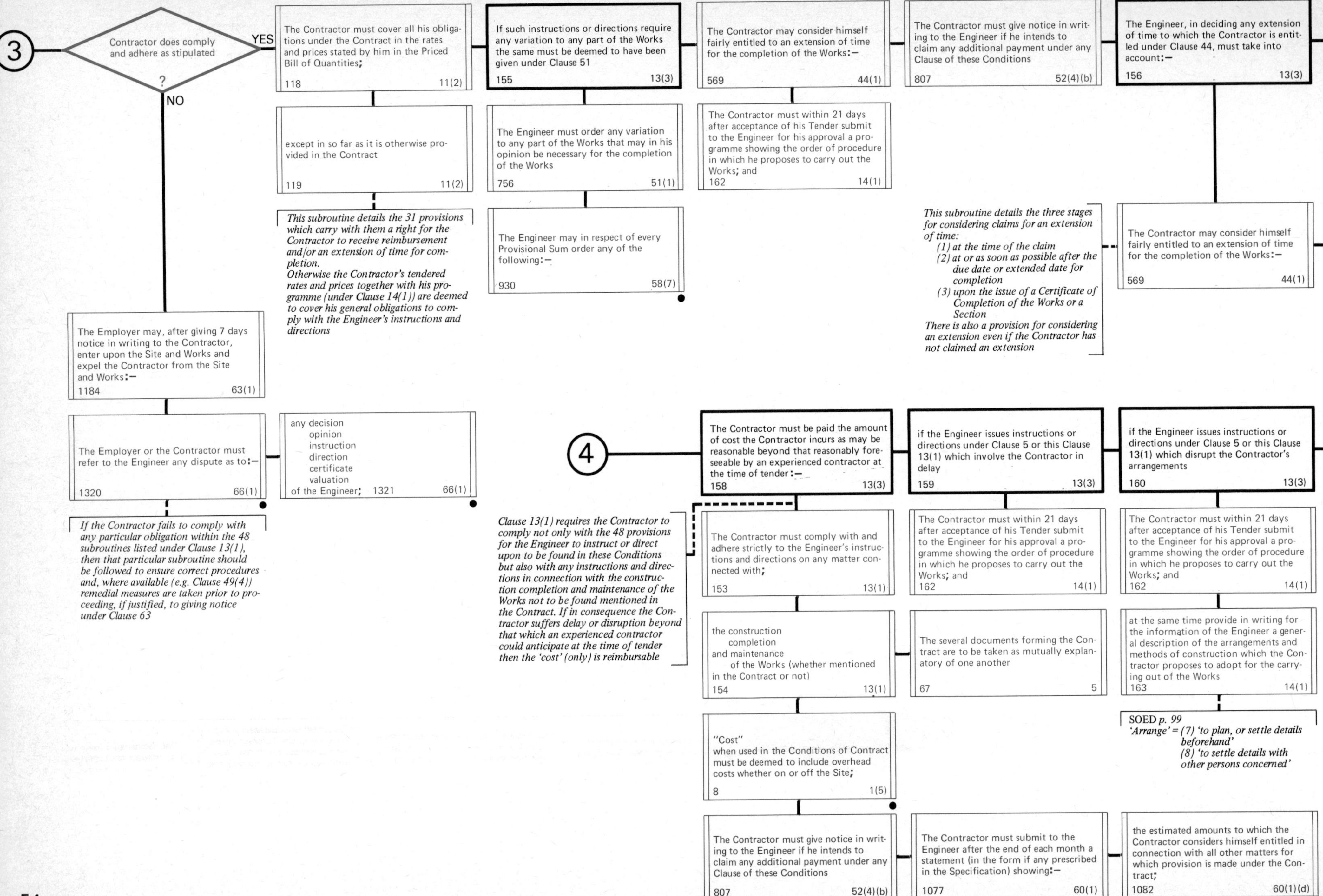
3
Contractor does comply and adhere as stipulated ?
YES
NO
The Contractor must cover all his obligations under the Contract in the rates and prices stated by him in the Priced Bill of Quantities; 118 11(2)
except in so far as it is otherwise provided in the Contract 119 11(2)
This subroutine details the 31 provisions which carry with them a right for the Contractor to receive reimbursement and/or an extension of time for completion. Otherwise the Contractor's tendered rates and prices together with his programme (under Clause 14(1)) are deemed to cover his general obligations to comply with the Engineer's instructions and directions
If such instructions or directions require any variation to any part of the Works the same must be deemed to have been given under Clause 51 155 13(3)
The Engineer must order any variation to any part of the Works that may in his opinion be necessary for the completion of the Works 756 51(1)
The Engineer may in respect of every Provisional Sum order any of the following:— 930 58(7)
The Contractor may consider himself fairly entitled to an extension of time for the completion of the Works:— 569 44(1)
The Contractor must within 21 days after acceptance of his Tender submit to the Engineer for his approval a programme showing the order of procedure in which he proposes to carry out the Works; and 162 14(1)
The Contractor must give notice in writing to the Engineer if he intends to claim any additional payment under any Clause of these Conditions 807 52(4)(b)
The Engineer, in deciding any extension of time to which the Contractor is entitled under Clause 44, must take into account:— 156 13(3)
This subroutine details the three stages for considering claims for an extension of time:
(1) at the time of the claim
(2) at or as soon as possible after the due date or extended date for completion
(3) upon the issue of a Certificate of Completion of the Works or a Section
There is also a provision for considering an extension even if the Contractor has not claimed an extension
The Contractor may consider himself fairly entitled to an extension of time for the completion of the Works:— 569 44(1)
The Employer may, after giving 7 days notice in writing to the Contractor, enter upon the Site and Works and expel the Contractor from the Site and Works:— 1184 63(1)
The Employer or the Contractor must refer to the Engineer any dispute as to:— 1320 66(1)
any decision
opinion
instruction
direction
certificate
valuation
of the Engineer; 1321 66(1)
If the Contractor fails to comply with any particular obligation within the 48 subroutines listed under Clause 13(1), then that particular subroutine should be followed to ensure correct procedures and, where available (e.g. Clause 49(4)) remedial measures are taken prior to proceeding, if justified, to giving notice under Clause 63
4
The Contractor must be paid the amount of cost the Contractor incurs as may be reasonable beyond that reasonably foreseeable by an experienced contractor at the time of tender:— 158 13(3)
if the Engineer issues instructions or directions under Clause 5 or this Clause 13(1) which involve the Contractor in delay 159 13(3)
if the Engineer issues instructions or directions under Clause 5 or this Clause 13(1) which disrupt the Contractor's arrangements 160 13(3)
Clause 13(1) requires the Contractor to comply not only with the 48 provisions for the Engineer to instruct or direct upon to be found in these Conditions but also with any instructions and directions in connection with the construction completion and maintenance of the Works not to be found mentioned in the Contract. If in consequence the Contractor suffers delay or disruption beyond that which an experienced contractor could anticipate at the time of tender then the 'cost' (only) is reimbursable
The Contractor must comply with and adhere strictly to the Engineer's instructions and directions on any matter connected with; 153 13(1)
the construction
completion
and maintenance
of the Works (whether mentioned in the Contract or not) 154 13(1)
"Cost" when used in the Conditions of Contract must be deemed to include overhead costs whether on or off the Site; 8 1(5)
The Contractor must give notice in writing to the Engineer if he intends to claim any additional payment under any Clause of these Conditions 807 52(4)(b)
The Contractor must within 21 days after acceptance of his Tender submit to the Engineer for his approval a programme showing the order of procedure in which he proposes to carry out the Works; and 162 14(1)
The several documents forming the Contract are to be taken as mutually explanatory of one another 67 5
The Contractor must submit to the Engineer after the end of each month a statement (in the form if any prescribed in the Specification) showing:— 1077 60(1)
The Contractor must within 21 days after acceptance of his Tender submit to the Engineer for his approval a programme showing the order of procedure in which he proposes to carry out the Works; and 162 14(1)
at the same time provide in writing for the information of the Engineer a general description of the arrangements and methods of construction which the Contractor proposes to adopt for the carrying out of the Works 163 14(1)
SOED p. 99
'Arrange' = (7) 'to plan, or settle details beforehand'
(8) 'to settle details with other persons concerned'
the estimated amounts to which the Contractor considers himself entitled in connection with all other matters for which provision is made under the Contract; 1082 60(1)(d)

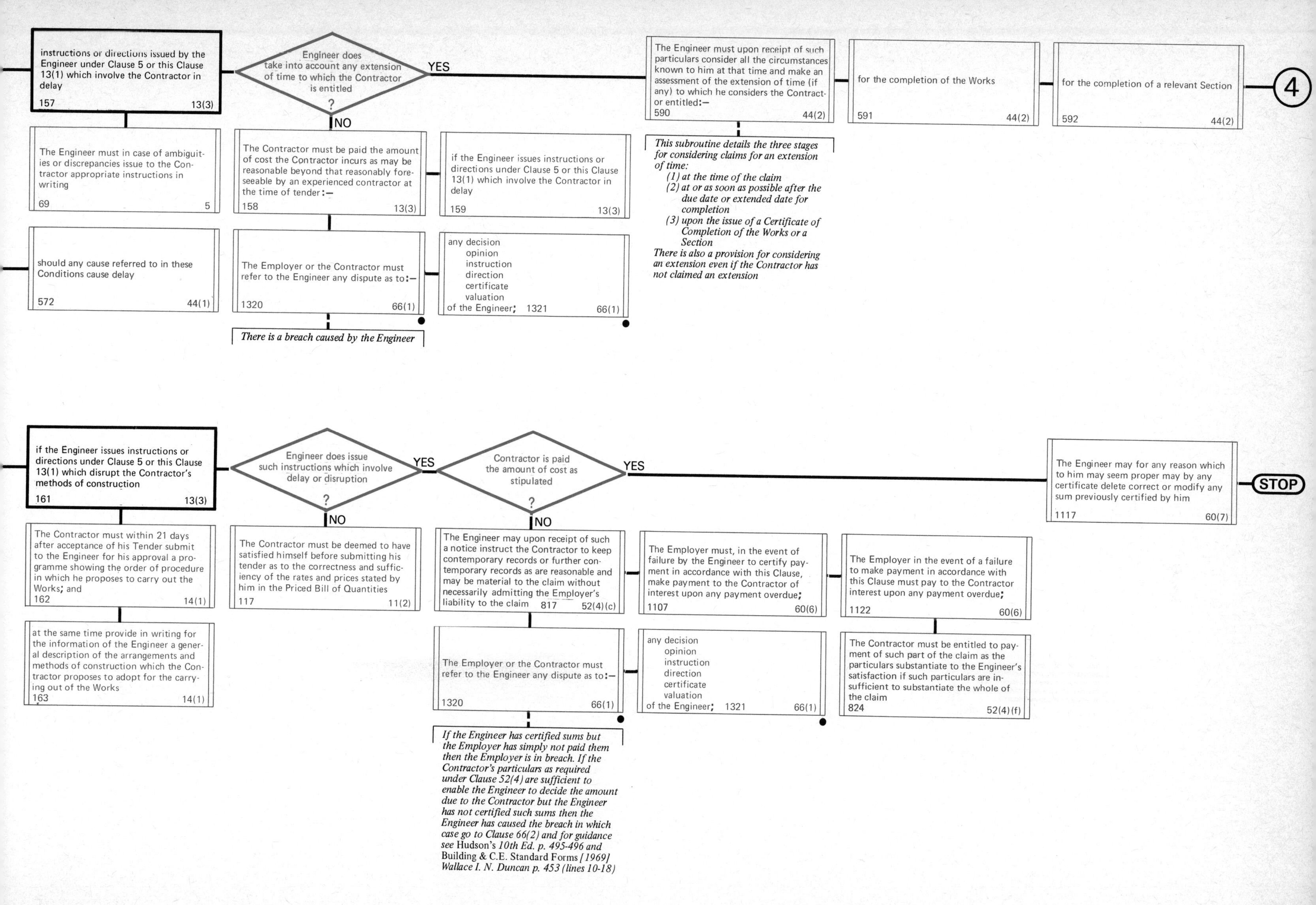
instructions or directions issued by the Engineer under Clause 5 or this Clause 13(1) which involve the Contractor in delay
157 13(3)
Engineer does take into account any extension of time to which the Contractor is entitled ?
YES
NO
The Engineer must upon receipt of such particulars consider all the circumstances known to him at that time and make an assessment of the extension of time (if any) to which he considers the Contractor entitled:—
590 44(2)
for the completion of the Works
591 44(2)
for the completion of a relevant Section
592 44(2)
4
The Engineer must in case of ambiguities or discrepancies issue to the Contractor appropriate instructions in writing
69 5
The Contractor must be paid the amount of cost the Contractor incurs as may be reasonable beyond that reasonably foreseeable by an experienced contractor at the time of tender:—
158 13(3)
if the Engineer issues instructions or directions under Clause 5 or this Clause 13(1) which involve the Contractor in delay
159 13(3)
This subroutine details the three stages for considering claims for an extension of time:
(1) at the time of the claim
(2) at or as soon as possible after the due date or extended date for completion
(3) upon the issue of a Certificate of Completion of the Works or a Section
There is also a provision for considering an extension even if the Contractor has not claimed an extension
should any cause referred to in these Conditions cause delay
572 44(1)
The Employer or the Contractor must refer to the Engineer any dispute as to:—
1320 66(1)
any decision
opinion
instruction
direction
certificate
valuation
of the Engineer; 1321 66(1)
There is a breach caused by the Engineer
if the Engineer issues instructions or directions under Clause 5 or this Clause 13(1) which disrupt the Contractor's methods of construction
161 13(3)
Engineer does issue such instructions which involve delay or disruption ?
YES
NO
Contractor is paid the amount of cost as stipulated ?
YES
NO
The Engineer may for any reason which to him may seem proper may by any certificate delete correct or modify any sum previously certified by him
1117 60(7)
STOP
The Contractor must within 21 days after acceptance of his Tender submit to the Engineer for his approval a programme showing the order of procedure in which he proposes to carry out the Works; and
162 14(1)
The Contractor must be deemed to have satisfied himself before submitting his tender as to the correctness and sufficiency of the rates and prices stated by him in the Priced Bill of Quantities
117 11(2)
The Engineer may upon receipt of such a notice instruct the Contractor to keep contemporary records or further contemporary records as are reasonable and may be material to the claim without necessarily admitting the Employer's liability to the claim 817 52(4)(c)
The Employer must, in the event of failure by the Engineer to certify payment in accordance with this Clause, make payment to the Contractor of interest upon any payment overdue;
1107 60(6)
The Employer in the event of a failure to make payment in accordance with this Clause must pay to the Contractor interest upon any payment overdue;
1122 60(6)
at the same time provide in writing for the information of the Engineer a general description of the arrangements and methods of construction which the Contractor proposes to adopt for the carrying out of the Works
163 14(1)
The Employer or the Contractor must refer to the Engineer any dispute as to:—
1320 66(1)
any decision
opinion
instruction
direction
certificate
valuation
of the Engineer; 1321 66(1)
The Contractor must be entitled to payment of such part of the claim as the particulars substantiate to the Engineer's satisfaction if such particulars are insufficient to substantiate the whole of the claim
824 52(4)(f)
If the Engineer has certified sums but the Employer has simply not paid them then the Employer is in breach. If the Contractor's particulars as required under Clause 52(4) are sufficient to enable the Engineer to decide the amount due to the Contractor but the Engineer has not certified such sums then the Engineer has caused the breach in which case go to Clause 66(2) and for guidance see Hudson's 10th Ed. p. 495-496 and Building & C.E. Standard Forms [1969] Wallace I. N. Duncan p. 453 (lines 10-18)

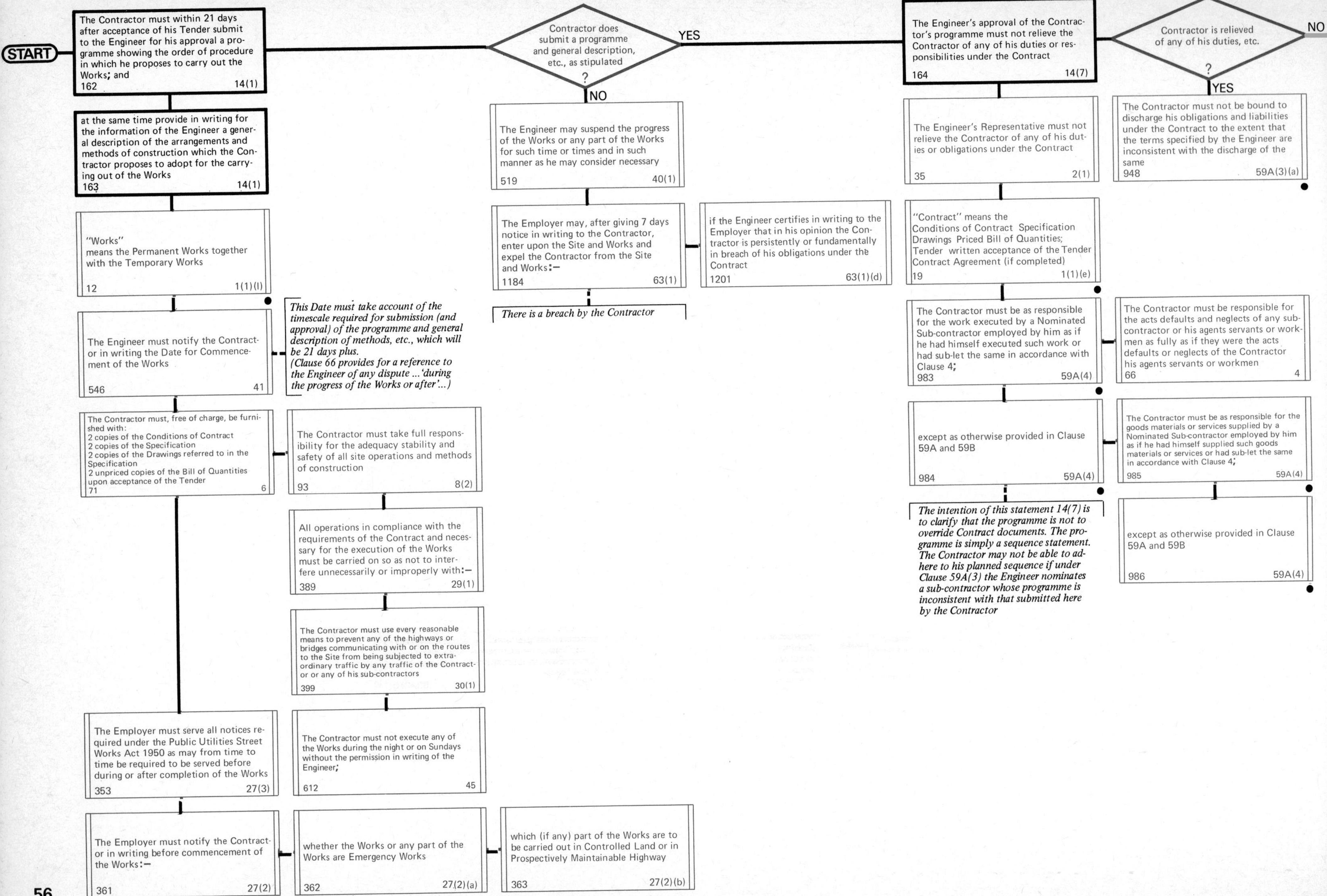
START
The Contractor must within 21 days after acceptance of his Tender submit to the Engineer for his approval a programme showing the order of procedure in which he proposes to carry out the Works; and 162 14(1)
at the same time provide in writing for the information of the Engineer a general description of the arrangements and methods of construction which the Contractor proposes to adopt for the carrying out of the Works 163 14(1)
"Works" means the Permanent Works together with the Temporary Works 12 1(1)(l)
The Engineer must notify the Contractor in writing the Date for Commencement of the Works 546 41
This Date must take account of the timescale required for submission (and approval) of the programme and general description of methods, etc., which will be 21 days plus. (Clause 66 provides for a reference to the Engineer of any dispute ... 'during the progress of the Works or after'...)
The Contractor must, free of charge, be furnished with: 2 copies of the Conditions of Contract 2 copies of the Specification 2 copies of the Drawings referred to in the Specification 2 unpriced copies of the Bill of Quantities upon acceptance of the Tender 71 6
The Contractor must take full responsibility for the adequacy stability and safety of all site operations and methods of construction 93 8(2)
All operations in compliance with the requirements of the Contract and necessary for the execution of the Works must be carried on so as not to interfere unnecessarily or improperly with:— 389 29(1)
The Contractor must use every reasonable means to prevent any of the highways or bridges communicating with or on the routes to the Site from being subjected to extraordinary traffic by any traffic of the Contractor or any of his sub-contractors 399 30(1)
The Contractor must not execute any of the Works during the night or on Sundays without the permission in writing of the Engineer; 612 45
The Employer must serve all notices required under the Public Utilities Street Works Act 1950 as may from time to time be required to be served before during or after completion of the Works 353 27(3)
The Employer must notify the Contractor in writing before commencement of the Works:— 361 27(2)
whether the Works or any part of the Works are Emergency Works 362 27(2)(a)
which (if any) part of the Works are to be carried out in Controlled Land or in Prospectively Maintainable Highway 363 27(2)(b)
Contractor does submit a programme and general description, etc., as stipulated ?
YES
NO
The Engineer may suspend the progress of the Works or any part of the Works for such time or times and in such manner as he may consider necessary 519 40(1)
The Employer may, after giving 7 days notice in writing to the Contractor, enter upon the Site and Works and expel the Contractor from the Site and Works:— 1184 63(1)
if the Engineer certifies in writing to the Employer that in his opinion the Contractor is persistently or fundamentally in breach of his obligations under the Contract 1201 63(1)(d)
There is a breach by the Contractor
The Engineer's approval of the Contractor's programme must not relieve the Contractor of any of his duties or responsibilities under the Contract 164 14(7)
The Engineer's Representative must not relieve the Contractor of any of his duties or obligations under the Contract 35 2(1)
"Contract" means the Conditions of Contract Specification Drawings Priced Bill of Quantities; Tender written acceptance of the Tender Contract Agreement (if completed) 19 1(1)(e)
The Contractor must be as responsible for the work executed by a Nominated Sub-contractor employed by him as if he had himself executed such work or had sub-let the same in accordance with Clause 4; 983 59A(4)
except as otherwise provided in Clause 59A and 59B 984 59A(4)
The intention of this statement 14(7) is to clarify that the programme is not to override Contract documents. The programme is simply a sequence statement. The Contractor may not be able to adhere to his planned sequence if under Clause 59A(3) the Engineer nominates a sub-contractor whose programme is inconsistent with that submitted here by the Contractor
Contractor is relieved of any of his duties, etc. ?
NO
YES
The Contractor must not be bound to discharge his obligations and liabilities under the Contract to the extent that the terms specified by the Engineer are inconsistent with the discharge of the same 948 59A(3)(a)
The Contractor must be responsible for the acts defaults and neglects of any sub-contractor or his agents servants or workmen as fully as if they were the acts defaults or neglects of the Contractor his agents servants or workmen 66 4
The Contractor must be as responsible for the goods materials or services supplied by a Nominated Sub-contractor employed by him as if he had himself supplied such goods materials or services or had sub-let the same in accordance with Clause 4; 985 59A(4)
except as otherwise provided in Clause 59A and 59B 986 59A(4)

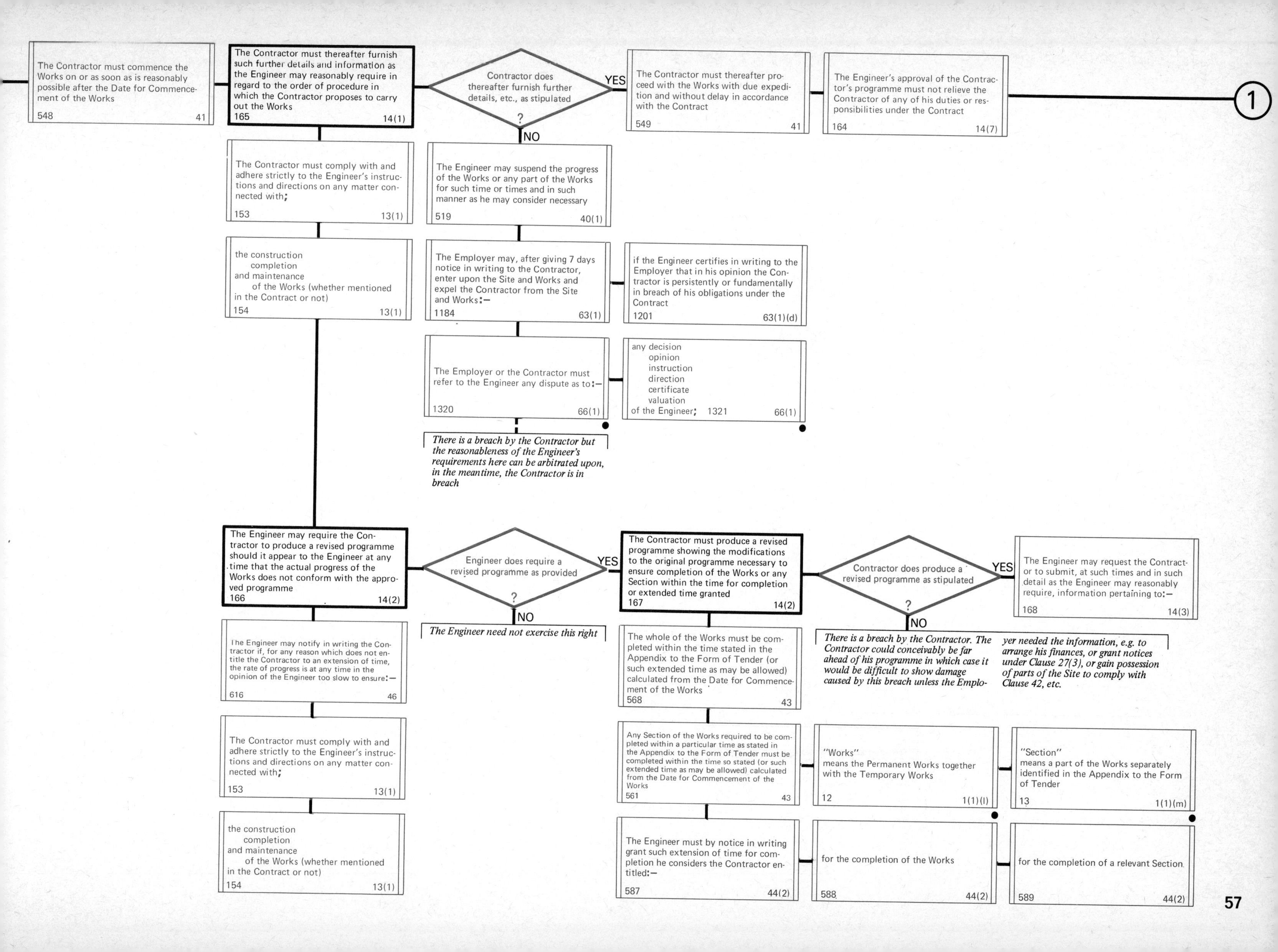

The Contractor must commence the Works on or as soon as is reasonably possible after the Date for Commencement of the Works
548 41
The Contractor must thereafter furnish such further details and information as the Engineer may reasonably require in regard to the order of procedure in which the Contractor proposes to carry out the Works
165 14(1)
Contractor does thereafter furnish further details, etc., as stipulated
?
YES
NO
The Contractor must thereafter proceed with the Works with due expedition and without delay in accordance with the Contract
549 41
The Engineer's approval of the Contractor's programme must not relieve the Contractor of any of his duties or responsibilities under the Contract
164 14(7)
1
The Contractor must comply with and adhere strictly to the Engineer's instructions and directions on any matter connected with;
153 13(1)
the construction completion and maintenance of the Works (whether mentioned in the Contract or not)
154 13(1)
The Engineer may suspend the progress of the Works or any part of the Works for such time or times and in such manner as he may consider necessary
519 40(1)
The Employer may, after giving 7 days notice in writing to the Contractor, enter upon the Site and Works and expel the Contractor from the Site and Works:—
1184 63(1)
if the Engineer certifies in writing to the Employer that in his opinion the Contractor is persistently or fundamentally in breach of his obligations under the Contract
1201 63(1)(d)
The Employer or the Contractor must refer to the Engineer any dispute as to:—
1320 66(1)
any decision opinion instruction direction certificate valuation of the Engineer; 1321 66(1)
There is a breach by the Contractor but the reasonableness of the Engineer's requirements here can be arbitrated upon, in the meantime, the Contractor is in breach
The Engineer may require the Contractor to produce a revised programme should it appear to the Engineer at any time that the actual progress of the Works does not conform with the approved programme
166 14(2)
Engineer does require a revised programme as provided
?
YES
NO
The Engineer need not exercise this right
The Contractor must produce a revised programme showing the modifications to the original programme necessary to ensure completion of the Works or any Section within the time for completion or extended time granted
167 14(2)
Contractor does produce a revised programme as stipulated
?
YES
NO
The Engineer may request the Contractor to submit, at such times and in such detail as the Engineer may reasonably require, information pertaining to:—
168 14(3)
There is a breach by the Contractor. The Contractor could conceivably be far ahead of his programme in which case it would be difficult to show damage caused by this breach unless the Employer needed the information, e.g. to arrange his finances, or grant notices under Clause 27(3), or gain possession of parts of the Site to comply with Clause 42, etc.
The Engineer may notify in writing the Contractor if, for any reason which does not entitle the Contractor to an extension of time, the rate of progress is at any time in the opinion of the Engineer too slow to ensure:—
616 46
The Contractor must comply with and adhere strictly to the Engineer's instructions and directions on any matter connected with;
153 13(1)
the construction completion and maintenance of the Works (whether mentioned in the Contract or not)
154 13(1)
The whole of the Works must be completed within the time stated in the Appendix to the Form of Tender (or such extended time as may be allowed) calculated from the Date for Commencement of the Works
568 43
Any Section of the Works required to be completed within a particular time as stated in the Appendix to the Form of Tender must be completed within the time so stated (or such extended time as may be allowed) calculated from the Date for Commencement of the Works
561 43
"Works" means the Permanent Works together with the Temporary Works
12 1(1)(l)
"Section" means a part of the Works separately identified in the Appendix to the Form of Tender
13 1(1)(m)
The Engineer must by notice in writing grant such extension of time for completion he considers the Contractor entitled:—
587 44(2)
for the completion of the Works
588 44(2)
for the completion of a relevant Section
589 44(2)

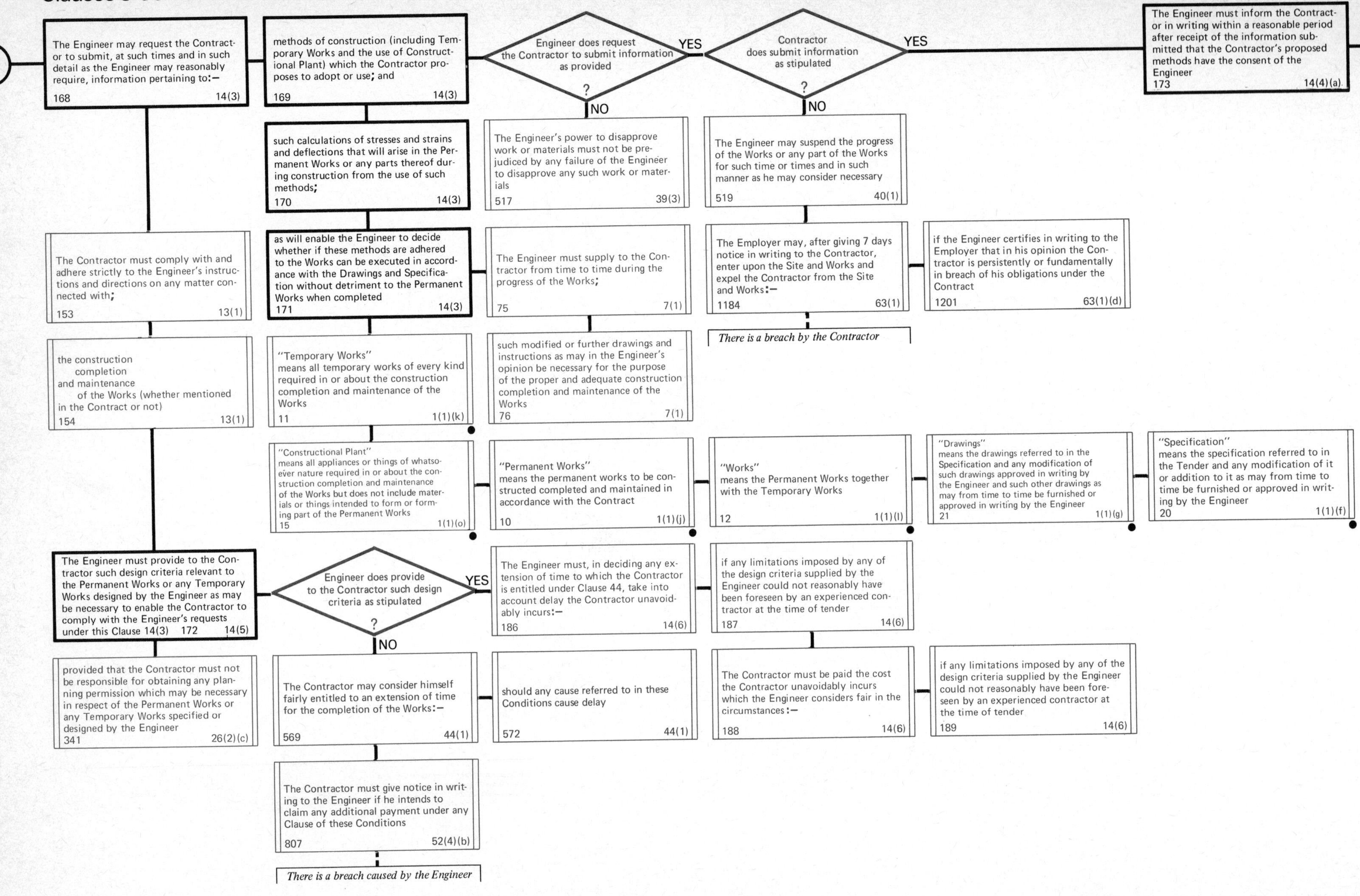
1
The Engineer may request the Contractor to submit, at such times and in such detail as the Engineer may reasonably require, information pertaining to:— 168 14(3)
methods of construction (including Temporary Works and the use of Constructional Plant) which the Contractor proposes to adopt or use; and 169 14(3)
Engineer does request the Contractor to submit information as provided ?
YES
NO
Contractor does submit information as stipulated ?
YES
NO
The Engineer must inform the Contractor in writing within a reasonable period after receipt of the information submitted that the Contractor's proposed methods have the consent of the Engineer 173 14(4)(a)
such calculations of stresses and strains and deflections that will arise in the Permanent Works or any parts thereof during construction from the use of such methods; 170 14(3)
The Engineer's power to disapprove work or materials must not be prejudiced by any failure of the Engineer to disapprove any such work or materials 517 39(3)
The Engineer may suspend the progress of the Works or any part of the Works for such time or times and in such manner as he may consider necessary 519 40(1)
The Contractor must comply with and adhere strictly to the Engineer's instructions and directions on any matter connected with; 153 13(1)
as will enable the Engineer to decide whether if these methods are adhered to the Works can be executed in accordance with the Drawings and Specification without detriment to the Permanent Works when completed 171 14(3)
The Engineer must supply to the Contractor from time to time during the progress of the Works; 75 7(1)
The Employer may, after giving 7 days notice in writing to the Contractor, enter upon the Site and Works and expel the Contractor from the Site and Works:— 1184 63(1)
if the Engineer certifies in writing to the Employer that in his opinion the Contractor is persistently or fundamentally in breach of his obligations under the Contract 1201 63(1)(d)
There is a breach by the Contractor
the construction completion and maintenance of the Works (whether mentioned in the Contract or not) 154 13(1)
"Temporary Works" means all temporary works of every kind required in or about the construction completion and maintenance of the Works 11 1(1)(k)
such modified or further drawings and instructions as may in the Engineer's opinion be necessary for the purpose of the proper and adequate construction completion and maintenance of the Works 76 7(1)
"Constructional Plant" means all appliances or things of whatsoever nature required in or about the construction completion and maintenance of the Works but does not include materials or things intended to form or forming part of the Permanent Works 15 1(1)(o)
"Permanent Works" means the permanent works to be constructed completed and maintained in accordance with the Contract 10 1(1)(j)
"Works" means the Permanent Works together with the Temporary Works 12 1(1)(l)
"Drawings" means the drawings referred to in the Specification and any modification of such drawings approved in writing by the Engineer and such other drawings as may from time to time be furnished or approved in writing by the Engineer 21 1(1)(g)
"Specification" means the specification referred to in the Tender and any modification of it or addition to it as may from time to time be furnished or approved in writing by the Engineer 20 1(1)(f)
The Engineer must provide to the Contractor such design criteria relevant to the Permanent Works or any Temporary Works designed by the Engineer as may be necessary to enable the Contractor to comply with the Engineer's requests under this Clause 14(3) 172 14(5)
Engineer does provide to the Contractor such design criteria as stipulated ?
YES
NO
The Engineer must, in deciding any extension of time to which the Contractor is entitled under Clause 44, take into account delay the Contractor unavoidably incurs:— 186 14(6)
if any limitations imposed by any of the design criteria supplied by the Engineer could not reasonably have been foreseen by an experienced contractor at the time of tender 187 14(6)
provided that the Contractor must not be responsible for obtaining any planning permission which may be necessary in respect of the Permanent Works or any Temporary Works specified or designed by the Engineer 341 26(2)(c)
The Contractor may consider himself fairly entitled to an extension of time for the completion of the Works:— 569 44(1)
should any cause referred to in these Conditions cause delay 572 44(1)
The Contractor must be paid the cost the Contractor unavoidably incurs which the Engineer considers fair in the circumstances:— 188 14(6)
if any limitations imposed by any of the design criteria supplied by the Engineer could not reasonably have been foreseen by an experienced contractor at the time of tender 189 14(6)
The Contractor must give notice in writing to the Engineer if he intends to claim any additional payment under any Clause of these Conditions 807 52(4)(b)
There is a breach caused by the Engineer

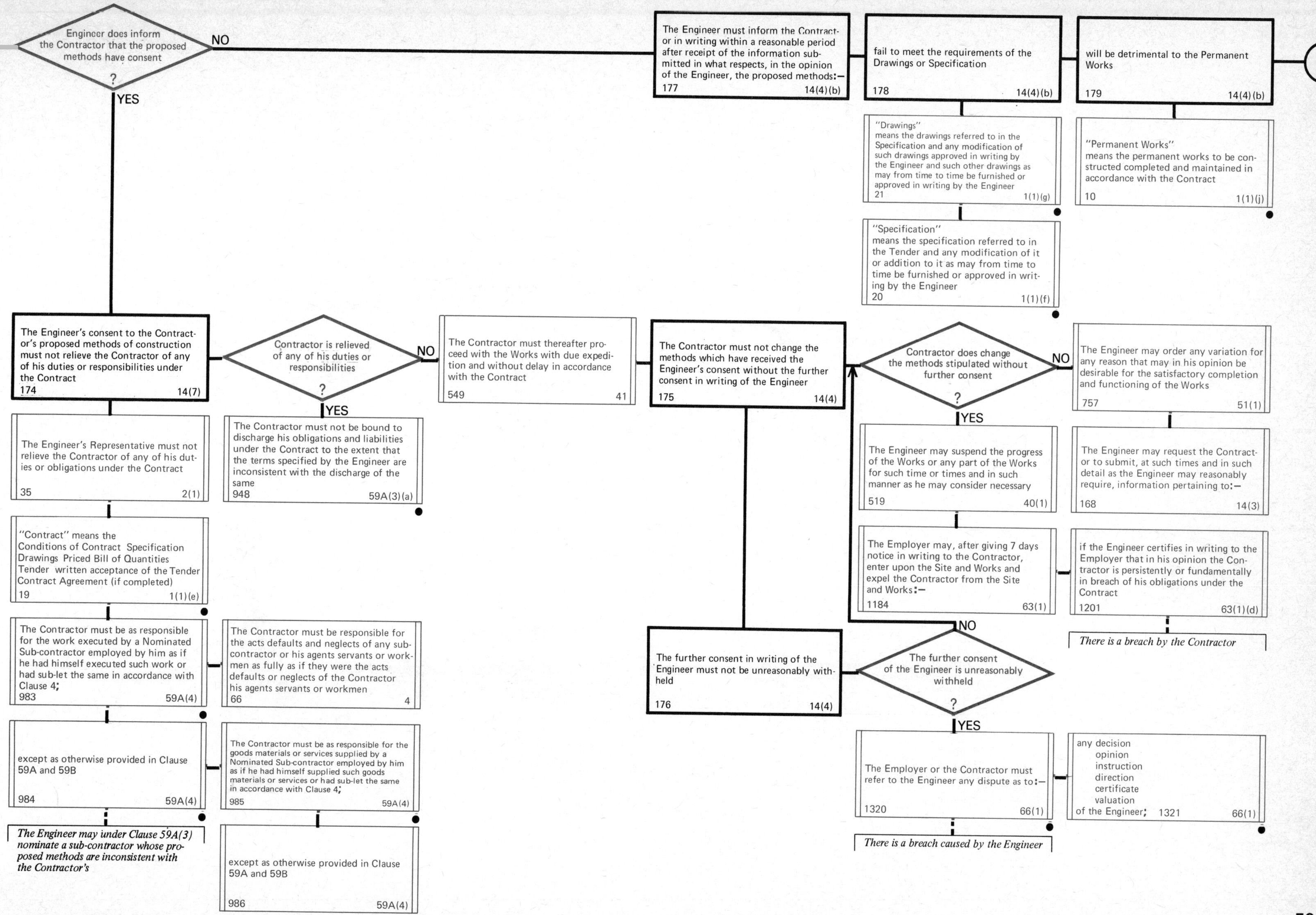

Engineer does inform the Contractor that the proposed methods have consent ?
NO
YES
The Engineer must inform the Contractor in writing within a reasonable period after receipt of the information submitted in what respects, in the opinion of the Engineer, the proposed methods:— 177 14(4)(b)
fail to meet the requirements of the Drawings or Specification 178 14(4)(b)
will be detrimental to the Permanent Works 179 14(4)(b)
2
"Drawings" means the drawings referred to in the Specification and any modification of such drawings approved in writing by the Engineer and such other drawings as may from time to time be furnished or approved in writing by the Engineer 21 1(1)(g)
"Permanent Works" means the permanent works to be constructed completed and maintained in accordance with the Contract 10 1(1)(j)
"Specification" means the specification referred to in the Tender and any modification of it or addition to it as may from time to time be furnished or approved in writing by the Engineer 20 1(1)(f)
The Engineer's consent to the Contractor's proposed methods of construction must not relieve the Contractor of any of his duties or responsibilities under the Contract 174 14(7)
Contractor is relieved of any of his duties or responsibilities ?
NO
YES
The Contractor must thereafter proceed with the Works with due expedition and without delay in accordance with the Contract 549 41
The Contractor must not change the methods which have received the Engineer's consent without the further consent in writing of the Engineer 175 14(4)
Contractor does change the methods stipulated without further consent ?
NO
YES
The Engineer may order any variation for any reason that may in his opinion be desirable for the satisfactory completion and functioning of the Works 757 51(1)
The Engineer's Representative must not relieve the Contractor of any of his duties or obligations under the Contract 35 2(1)
The Contractor must not be bound to discharge his obligations and liabilities under the Contract to the extent that the terms specified by the Engineer are inconsistent with the discharge of the same 948 59A(3)(a)
The Engineer may suspend the progress of the Works or any part of the Works for such time or times and in such manner as he may consider necessary 519 40(1)
The Engineer may request the Contractor to submit, at such times and in such detail as the Engineer may reasonably require, information pertaining to:— 168 14(3)
"Contract" means the Conditions of Contract Specification Drawings Priced Bill of Quantities Tender written acceptance of the Tender Contract Agreement (if completed) 19 1(1)(e)
The Employer may, after giving 7 days notice in writing to the Contractor, enter upon the Site and Works and expel the Contractor from the Site and Works:— 1184 63(1)
if the Engineer certifies in writing to the Employer that in his opinion the Contractor is persistently or fundamentally in breach of his obligations under the Contract 1201 63(1)(d)
There is a breach by the Contractor
The Contractor must be as responsible for the work executed by a Nominated Sub-contractor employed by him as if he had himself executed such work or had sub-let the same in accordance with Clause 4; 983 59A(4)
The Contractor must be responsible for the acts defaults and neglects of any sub-contractor or his agents servants or workmen as fully as if they were the acts defaults or neglects of the Contractor his agents servants or workmen 66 4
NO
The further consent in writing of the Engineer must not be unreasonably withheld 176 14(4)
The further consent of the Engineer is unreasonably withheld ?
YES
except as otherwise provided in Clause 59A and 59B 984 59A(4)
The Contractor must be as responsible for the goods materials or services supplied by a Nominated Sub-contractor employed by him as if he had himself supplied such goods materials or services or had sub-let the same in accordance with Clause 4; 985 59A(4)
The Employer or the Contractor must refer to the Engineer any dispute as to:— 1320 66(1)
any decision opinion instruction direction certificate valuation of the Engineer; 1321 66(1)
The Engineer may under Clause 59A(3) nominate a sub-contractor whose proposed methods are inconsistent with the Contractor's
There is a breach caused by the Engineer
except as otherwise provided in Clause 59A and 59B 986 59A(4)

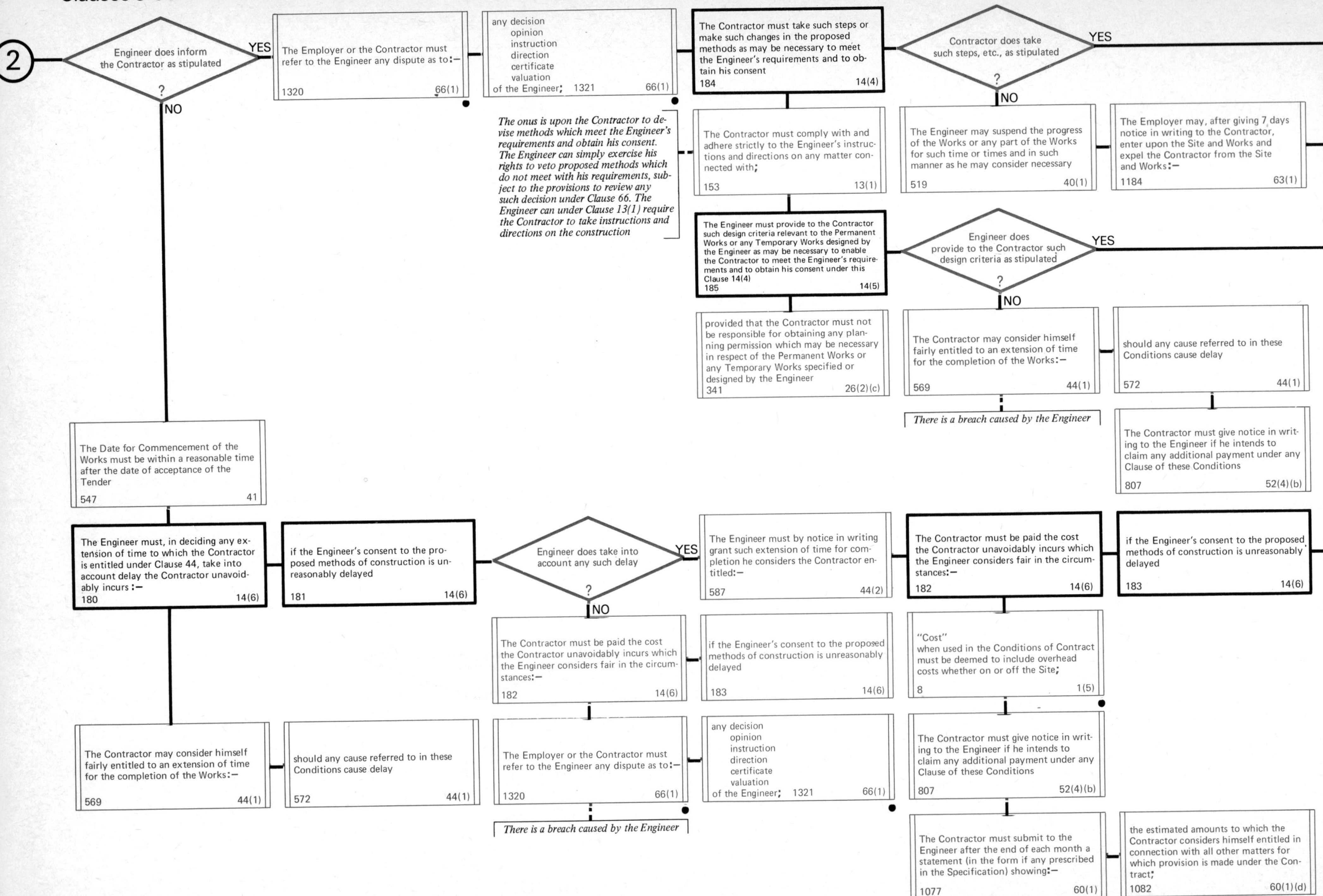
2
Engineer does inform the Contractor as stipulated ?
YES
NO
The Employer or the Contractor must refer to the Engineer any dispute as to:—
1320 66(1)
any decision opinion instruction direction certificate valuation of the Engineer; 1321 66(1)
The Contractor must take such steps or make such changes in the proposed methods as may be necessary to meet the Engineer's requirements and to obtain his consent
184 14(4)
Contractor does take such steps, etc., as stipulated ?
YES
NO
The onus is upon the Contractor to devise methods which meet the Engineer's requirements and obtain his consent. The Engineer can simply exercise his rights to veto proposed methods which do not meet with his requirements, subject to the provisions to review any such decision under Clause 66. The Engineer can under Clause 13(1) require the Contractor to take instructions and directions on the construction
The Contractor must comply with and adhere strictly to the Engineer's instructions and directions on any matter connected with;
153 13(1)
The Engineer may suspend the progress of the Works or any part of the Works for such time or times and in such manner as he may consider necessary
519 40(1)
The Employer may, after giving 7 days notice in writing to the Contractor, enter upon the Site and Works and expel the Contractor from the Site and Works:—
1184 63(1)
The Engineer must provide to the Contractor such design criteria relevant to the Permanent Works or any Temporary Works designed by the Engineer as may be necessary to enable the Contractor to meet the Engineer's requirements and to obtain his consent under this Clause 14(4)
185 14(5)
Engineer does provide to the Contractor such design criteria as stipulated ?
YES
NO
provided that the Contractor must not be responsible for obtaining any planning permission which may be necessary in respect of the Permanent Works or any Temporary Works specified or designed by the Engineer
341 26(2)(c)
The Contractor may consider himself fairly entitled to an extension of time for the completion of the Works:—
569 44(1)
should any cause referred to in these Conditions cause delay
572 44(1)
There is a breach caused by the Engineer
The Contractor must give notice in writing to the Engineer if he intends to claim any additional payment under any Clause of these Conditions
807 52(4)(b)
The Date for Commencement of the Works must be within a reasonable time after the date of acceptance of the Tender
547 41
The Engineer must, in deciding any extension of time to which the Contractor is entitled under Clause 44, take into account delay the Contractor unavoidably incurs:—
180 14(6)
if the Engineer's consent to the proposed methods of construction is unreasonably delayed
181 14(6)
Engineer does take into account any such delay ?
YES
NO
The Engineer must by notice in writing grant such extension of time for completion he considers the Contractor entitled:—
587 44(2)
The Contractor must be paid the cost the Contractor unavoidably incurs which the Engineer considers fair in the circumstances:—
182 14(6)
if the Engineer's consent to the proposed methods of construction is unreasonably delayed
183 14(6)
The Contractor must be paid the cost the Contractor unavoidably incurs which the Engineer considers fair in the circumstances:—
182 14(6)
if the Engineer's consent to the proposed methods of construction is unreasonably delayed
183 14(6)
"Cost" when used in the Conditions of Contract must be deemed to include overhead costs whether on or off the Site;
8 1(5)
The Contractor may consider himself fairly entitled to an extension of time for the completion of the Works:—
569 44(1)
should any cause referred to in these Conditions cause delay
572 44(1)
The Employer or the Contractor must refer to the Engineer any dispute as to:—
1320 66(1)
any decision opinion instruction direction certificate valuation of the Engineer; 1321 66(1)
There is a breach caused by the Engineer
The Contractor must give notice in writing to the Engineer if he intends to claim any additional payment under any Clause of these Conditions
807 52(4)(b)
The Contractor must submit to the Engineer after the end of each month a statement (in the form if any prescribed in the Specification) showing:—
1077 60(1)
the estimated amounts to which the Contractor considers himself entitled in connection with all other matters for which provision is made under the Contract;
1082 60(1)(d)

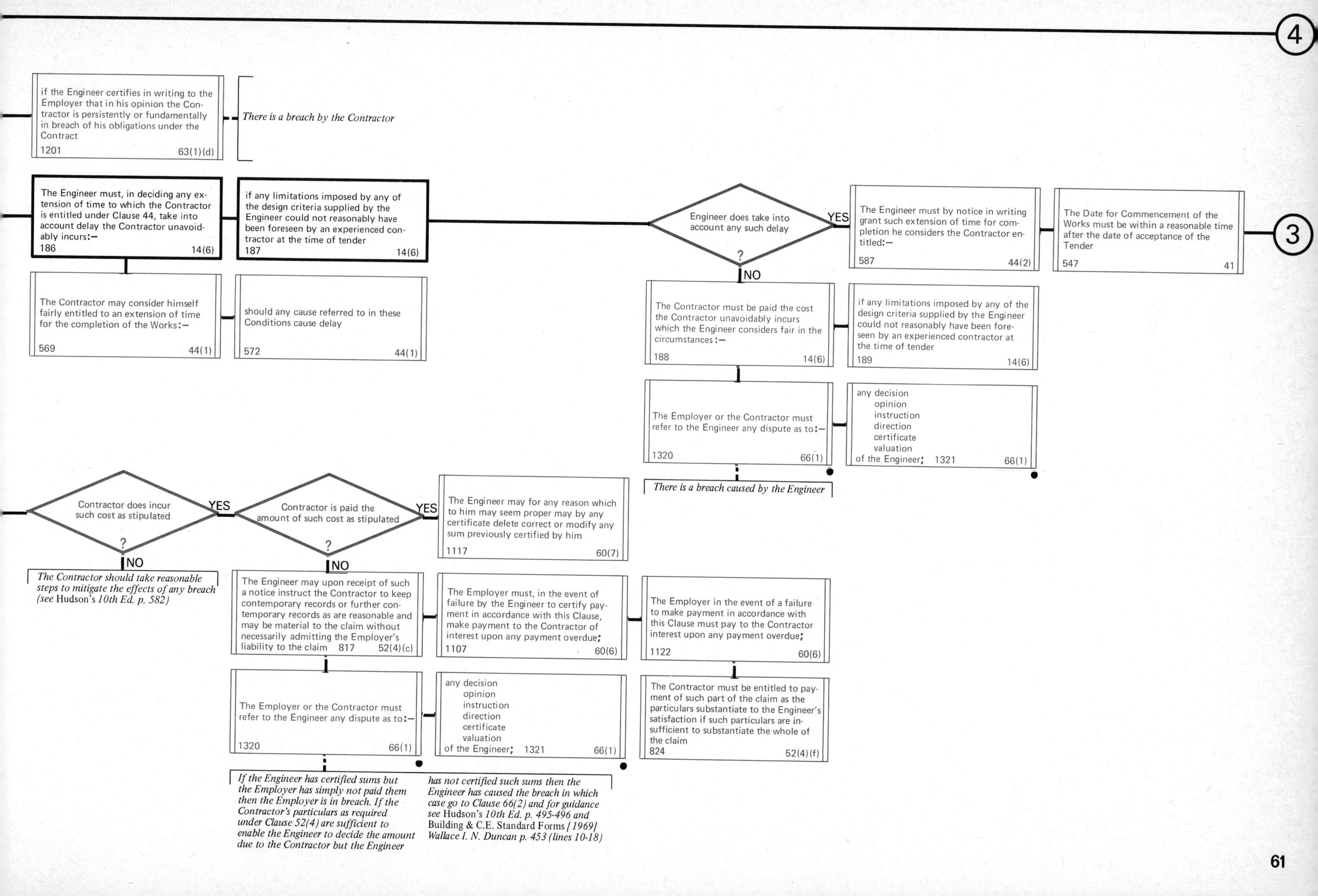
4
if the Engineer certifies in writing to the Employer that in his opinion the Contractor is persistently or fundamentally in breach of his obligations under the Contract
1201 63(1)(d)
There is a breach by the Contractor
The Engineer must, in deciding any extension of time to which the Contractor is entitled under Clause 44, take into account delay the Contractor unavoidably incurs:—
186 14(6)
if any limitations imposed by any of the design criteria supplied by the Engineer could not reasonably have been foreseen by an experienced contractor at the time of tender
187 14(6)
Engineer does take into account any such delay
?
YES
NO
The Engineer must by notice in writing grant such extension of time for completion he considers the Contractor entitled:—
587 44(2)
The Date for Commencement of the Works must be within a reasonable time after the date of acceptance of the Tender
547 41
3
The Contractor may consider himself fairly entitled to an extension of time for the completion of the Works:—
569 44(1)
should any cause referred to in these Conditions cause delay
572 44(1)
The Contractor must be paid the cost the Contractor unavoidably incurs which the Engineer considers fair in the circumstances:—
188 14(6)
if any limitations imposed by any of the design criteria supplied by the Engineer could not reasonably have been foreseen by an experienced contractor at the time of tender
189 14(6)
The Employer or the Contractor must refer to the Engineer any dispute as to:—
1320 66(1)
any decision opinion instruction direction certificate valuation of the Engineer; 1321 66(1)
There is a breach caused by the Engineer
Contractor does incur such cost as stipulated
?
YES
NO
Contractor is paid the amount of such cost as stipulated
?
YES
NO
The Engineer may for any reason which to him may seem proper may by any certificate delete correct or modify any sum previously certified by him
1117 60(7)
The Contractor should take reasonable steps to mitigate the effects of any breach (see Hudson's 10th Ed. p. 582)
The Engineer may upon receipt of such a notice instruct the Contractor to keep contemporary records or further contemporary records as are reasonable and may be material to the claim without necessarily admitting the Employer's liability to the claim 817 52(4)(c)
The Employer must, in the event of failure by the Engineer to certify payment in accordance with this Clause, make payment to the Contractor of interest upon any payment overdue;
1107 60(6)
The Employer in the event of a failure to make payment in accordance with this Clause must pay to the Contractor interest upon any payment overdue;
1122 60(6)
The Employer or the Contractor must refer to the Engineer any dispute as to:—
1320 66(1)
any decision opinion instruction direction certificate valuation of the Engineer; 1321 66(1)
The Contractor must be entitled to payment of such part of the claim as the particulars substantiate to the Engineer's satisfaction if such particulars are insufficient to substantiate the whole of the claim
824 52(4)(f)
If the Engineer has certified sums but the Employer has simply not paid them then the Employer is in breach. If the Contractor's particulars as required under Clause 52(4) are sufficient to enable the Engineer to decide the amount due to the Contractor but the Engineer has not certified such sums then the Engineer has caused the breach in which case go to Clause 66(2) and for guidance see Hudson's 10th Ed. p. 495-496 and Building & C.E. Standard Forms [1969] Wallace I. N. Duncan p. 453 (lines 10-18)

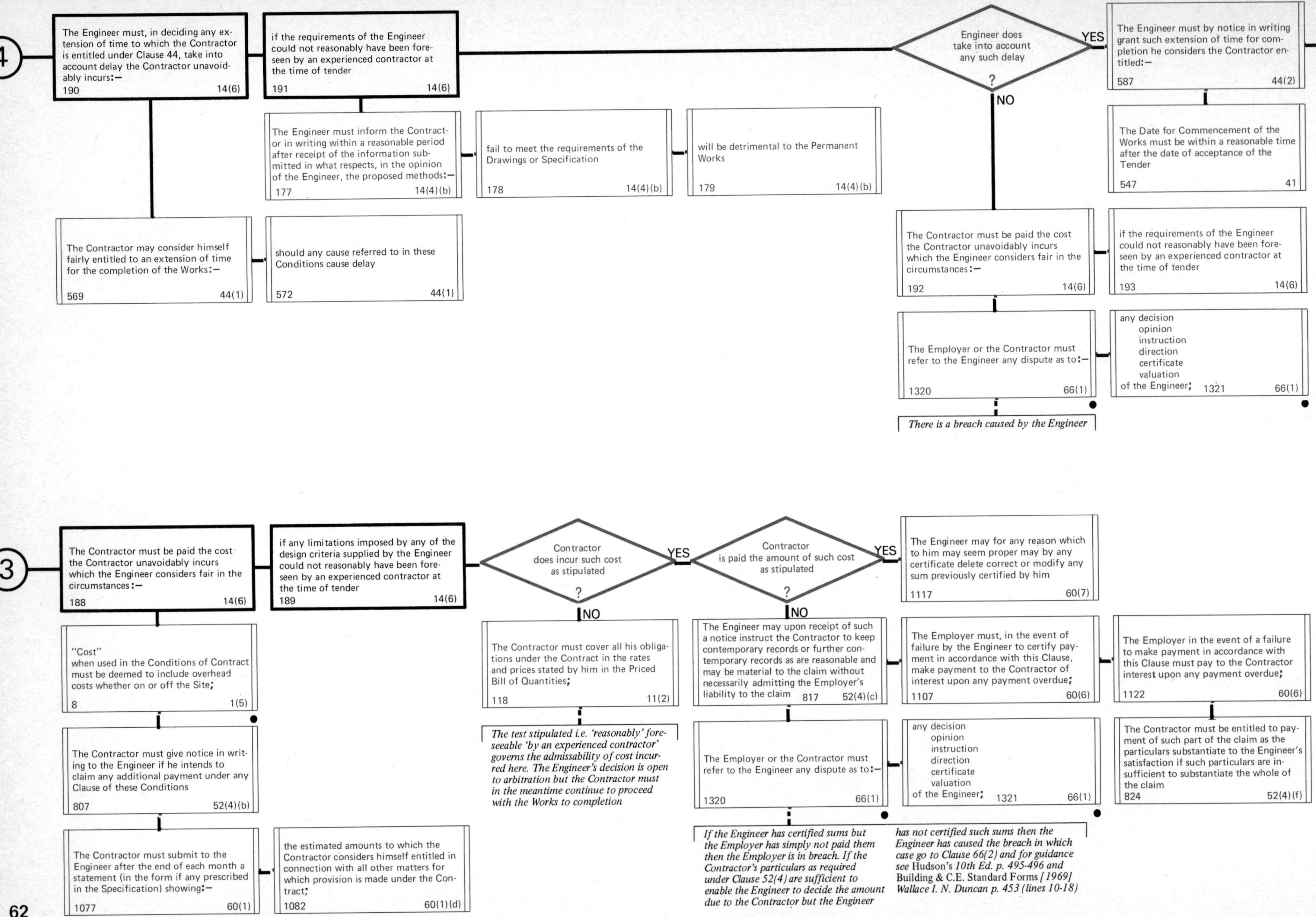
4
The Engineer must, in deciding any extension of time to which the Contractor is entitled under Clause 44, take into account delay the Contractor unavoidably incurs:— 190 14(6)
if the requirements of the Engineer could not reasonably have been foreseen by an experienced contractor at the time of tender 191 14(6)
Engineer does take into account any such delay ?
YES
NO
The Engineer must by notice in writing grant such extension of time for completion he considers the Contractor entitled:— 587 44(2)
The Date for Commencement of the Works must be within a reasonable time after the date of acceptance of the Tender 547 41
The Engineer must inform the Contractor in writing within a reasonable period after receipt of the information submitted in what respects, in the opinion of the Engineer, the proposed methods:— 177 14(4)(b)
fail to meet the requirements of the Drawings or Specification 178 14(4)(b)
will be detrimental to the Permanent Works 179 14(4)(b)
The Contractor may consider himself fairly entitled to an extension of time for the completion of the Works:— 569 44(1)
should any cause referred to in these Conditions cause delay 572 44(1)
The Contractor must be paid the cost the Contractor unavoidably incurs which the Engineer considers fair in the circumstances:— 192 14(6)
if the requirements of the Engineer could not reasonably have been foreseen by an experienced contractor at the time of tender 193 14(6)
The Employer or the Contractor must refer to the Engineer any dispute as to:— 1320 66(1)
any decision opinion instruction direction certificate valuation of the Engineer; 1321 66(1)
There is a breach caused by the Engineer
3
The Contractor must be paid the cost the Contractor unavoidably incurs which the Engineer considers fair in the circumstances:— 188 14(6)
if any limitations imposed by any of the design criteria supplied by the Engineer could not reasonably have been foreseen by an experienced contractor at the time of tender 189 14(6)
Contractor does incur such cost as stipulated ?
YES
NO
Contractor is paid the amount of such cost as stipulated ?
YES
NO
The Engineer may for any reason which to him may seem proper may by any certificate delete correct or modify any sum previously certified by him 1117 60(7)
"Cost" when used in the Conditions of Contract must be deemed to include overhead costs whether on or off the Site; 8 1(5)
The Contractor must give notice in writing to the Engineer if he intends to claim any additional payment under any Clause of these Conditions 807 52(4)(b)
The Contractor must submit to the Engineer after the end of each month a statement (in the form if any prescribed in the Specification) showing:— 1077 60(1)
the estimated amounts to which the Contractor considers himself entitled in connection with all other matters for which provision is made under the Contract; 1082 60(1)(d)
The Contractor must cover all his obligations under the Contract in the rates and prices stated by him in the Priced Bill of Quantities; 118 11(2)
The test stipulated i.e. 'reasonably' foreseeable 'by an experienced contractor' governs the admissability of cost incurred here. The Engineer's decision is open to arbitration but the Contractor must in the meantime continue to proceed with the Works to completion
The Engineer may upon receipt of such a notice instruct the Contractor to keep contemporary records or further contemporary records as are reasonable and may be material to the claim without necessarily admitting the Employer's liability to the claim 817 52(4)(c)
The Employer must, in the event of failure by the Engineer to certify payment in accordance with this Clause, make payment to the Contractor of interest upon any payment overdue; 1107 60(6)
The Employer in the event of a failure to make payment in accordance with this Clause must pay to the Contractor interest upon any payment overdue; 1122 60(6)
The Employer or the Contractor must refer to the Engineer any dispute as to:— 1320 66(1)
any decision opinion instruction direction certificate valuation of the Engineer; 1321 66(1)
The Contractor must be entitled to payment of such part of the claim as the particulars substantiate to the Engineer's satisfaction if such particulars are insufficient to substantiate the whole of the claim 824 52(4)(f)
If the Engineer has certified sums but the Employer has simply not paid them then the Employer is in breach. If the Contractor's particulars as required under Clause 52(4) are sufficient to enable the Engineer to decide the amount due to the Contractor but the Engineer has not certified such sums then the Engineer has caused the breach in which case go to Clause 66(2) and for guidance see Hudson's 10th Ed. p. 495-496 and Building & C.E. Standard Forms [1969] Wallace I. N. Duncan p. 453 (lines 10-18)

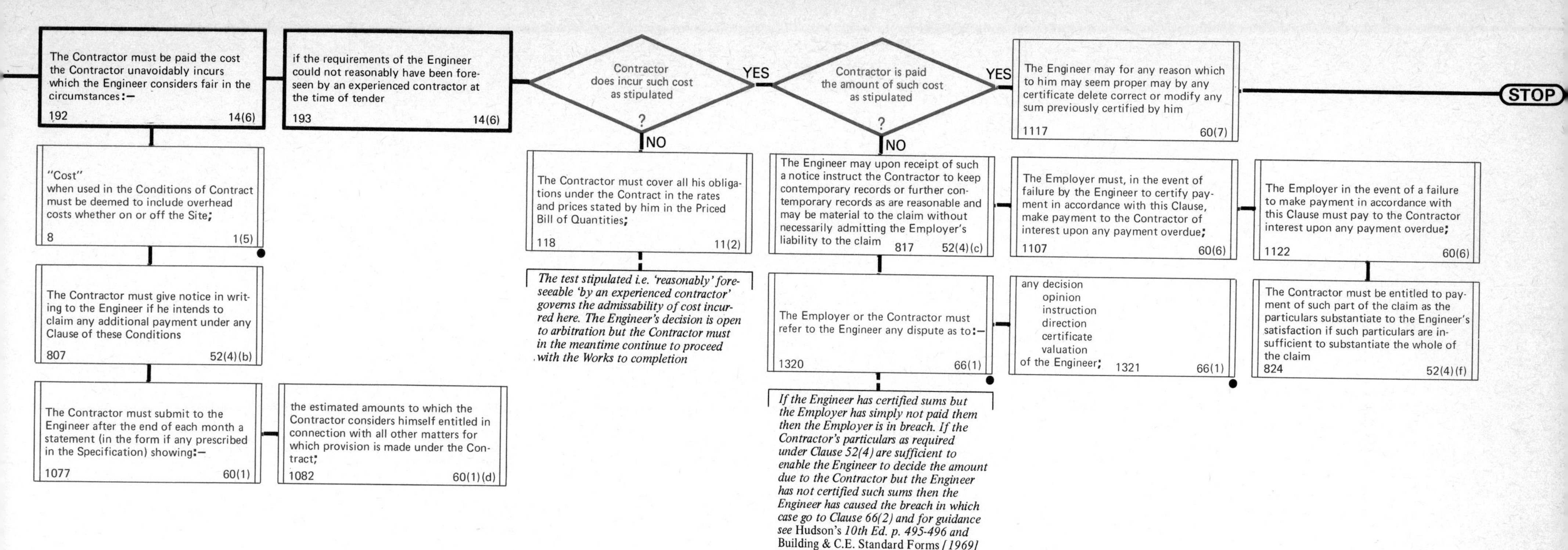
The Contractor must be paid the cost the Contractor unavoidably incurs which the Engineer considers fair in the circumstances:—
192 14(6)
if the requirements of the Engineer could not reasonably have been foreseen by an experienced contractor at the time of tender
193 14(6)
Contractor does incur such cost as stipulated ?
YES
NO
Contractor is paid the amount of such cost as stipulated ?
YES
NO
The Engineer may for any reason which to him may seem proper may by any certificate delete correct or modify any sum previously certified by him
1117 60(7)
STOP
"Cost" when used in the Conditions of Contract must be deemed to include overhead costs whether on or off the Site;
8 1(5)
The Contractor must cover all his obligations under the Contract in the rates and prices stated by him in the Priced Bill of Quantities;
118 11(2)
The Engineer may upon receipt of such a notice instruct the Contractor to keep contemporary records or further contemporary records as are reasonable and may be material to the claim without necessarily admitting the Employer's liability to the claim
817 52(4)(c)
The Employer must, in the event of failure by the Engineer to certify payment in accordance with this Clause, make payment to the Contractor of interest upon any payment overdue;
1107 60(6)
The Employer in the event of a failure to make payment in accordance with this Clause must pay to the Contractor interest upon any payment overdue;
1122 60(6)
The Contractor must give notice in writing to the Engineer if he intends to claim any additional payment under any Clause of these Conditions
807 52(4)(b)
The test stipulated i.e. 'reasonably' foreseeable 'by an experienced contractor' governs the admissability of cost incurred here. The Engineer's decision is open to arbitration but the Contractor must in the meantime continue to proceed with the Works to completion
The Employer or the Contractor must refer to the Engineer any dispute as to:—
1320 66(1)
any decision opinion instruction direction certificate valuation of the Engineer;
1321 66(1)
The Contractor must be entitled to payment of such part of the claim as the particulars substantiate to the Engineer's satisfaction if such particulars are insufficient to substantiate the whole of the claim
824 52(4)(f)
The Contractor must submit to the Engineer after the end of each month a statement (in the form if any prescribed in the Specification) showing:—
1077 60(1)
the estimated amounts to which the Contractor considers himself entitled in connection with all other matters for which provision is made under the Contract;
1082 60(1)(d)
If the Engineer has certified sums but the Employer has simply not paid them then the Employer is in breach. If the Contractor's particulars as required under Clause 52(4) are sufficient to enable the Engineer to decide the amount due to the Contractor but the Engineer has not certified such sums then the Engineer has caused the breach in which case go to Clause 66(2) and for guidance see Hudson's 10th Ed. p. 495-496 and Building & C.E. Standard Forms [1969] Wallace I. N. Duncan p. 453 (lines 10-18)

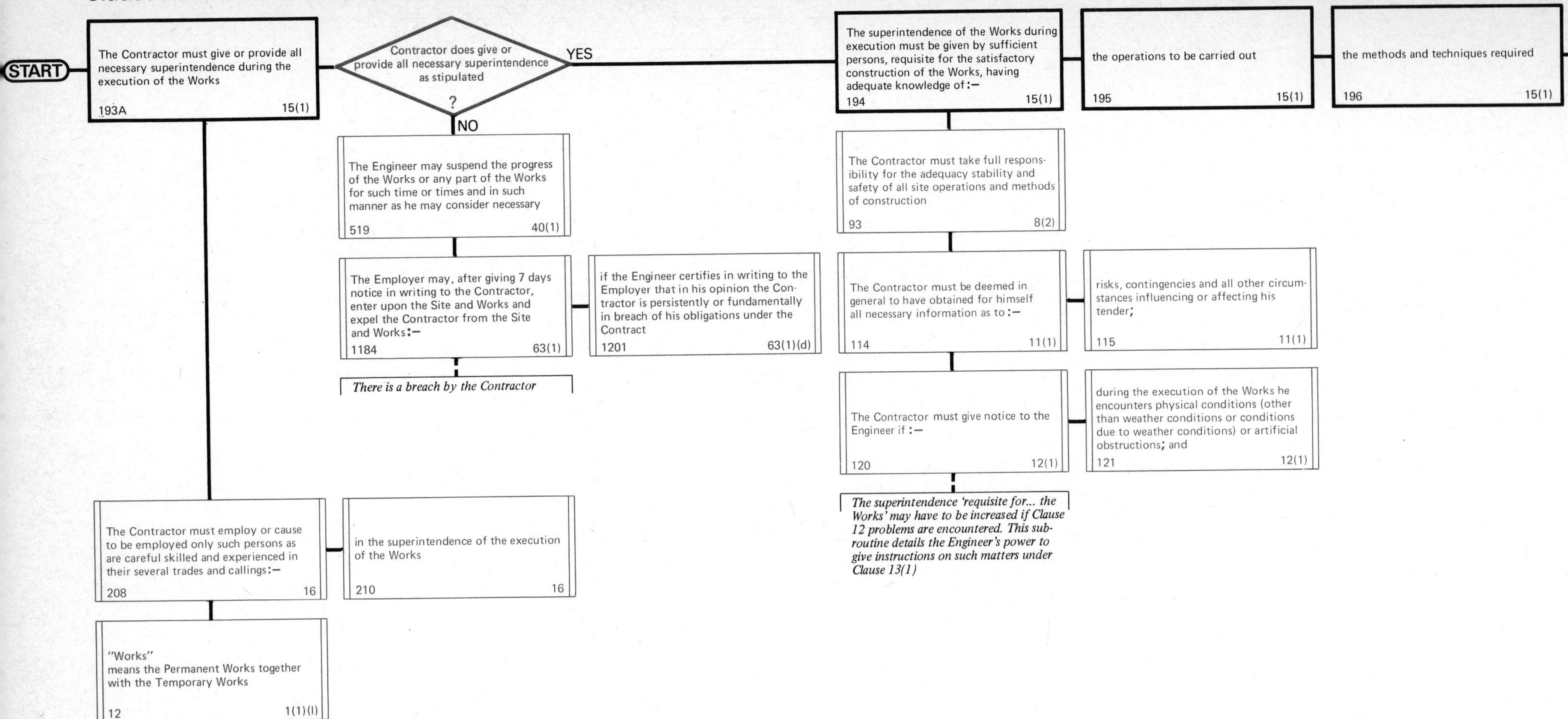
START
The Contractor must give or provide all necessary superintendence during the execution of the Works
193A 15(1)
Contractor does give or provide all necessary superintendence as stipulated
?
YES
NO
The superintendence of the Works during execution must be given by sufficient persons, requisite for the satisfactory construction of the Works, having adequate knowledge of :—
194 15(1)
the operations to be carried out
195 15(1)
the methods and techniques required
196 15(1)
The Engineer may suspend the progress of the Works or any part of the Works for such time or times and in such manner as he may consider necessary
519 40(1)
The Employer may, after giving 7 days notice in writing to the Contractor, enter upon the Site and Works and expel the Contractor from the Site and Works:—
1184 63(1)
if the Engineer certifies in writing to the Employer that in his opinion the Contractor is persistently or fundamentally in breach of his obligations under the Contract
1201 63(1)(d)
There is a breach by the Contractor
The Contractor must take full responsibility for the adequacy stability and safety of all site operations and methods of construction
93 8(2)
The Contractor must be deemed in general to have obtained for himself all necessary information as to :—
114 11(1)
risks, contingencies and all other circumstances influencing or affecting his tender;
115 11(1)
The Contractor must give notice to the Engineer if :—
120 12(1)
during the execution of the Works he encounters physical conditions (other than weather conditions or conditions due to weather conditions) or artificial obstructions; and
121 12(1)
The superintendence 'requisite for... the Works' may have to be increased if Clause 12 problems are encountered. This subroutine details the Engineer's power to give instructions on such matters under Clause 13(1)
The Contractor must employ or cause to be employed only such persons as are careful skilled and experienced in their several trades and callings:—
208 16
in the superintendence of the execution of the Works
210 16
"Works"
means the Permanent Works together with the Temporary Works
12 1(1)(l)

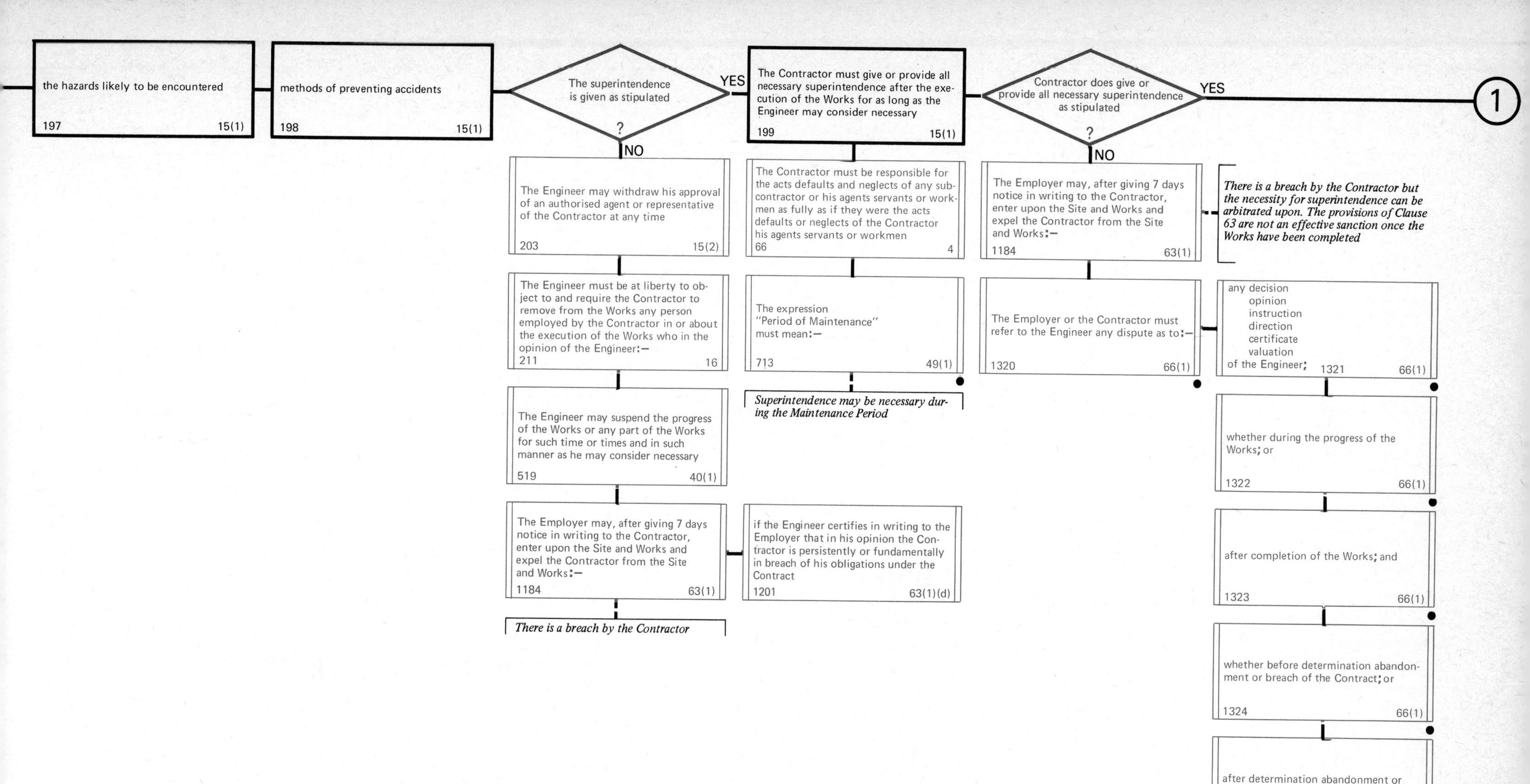

the hazards likely to be encountered
197
15(1)
methods of preventing accidents
198
15(1)
The superintendence is given as stipulated
?
YES
NO
The Contractor must give or provide all necessary superintendence after the execution of the Works for as long as the Engineer may consider necessary
199
15(1)
Contractor does give or provide all necessary superintendence as stipulated
?
YES
NO
1
The Engineer may withdraw his approval of an authorised agent or representative of the Contractor at any time
203
15(2)
The Contractor must be responsible for the acts defaults and neglects of any sub-contractor or his agents servants or workmen as fully as if they were the acts defaults or neglects of the Contractor his agents servants or workmen
66
4
The Employer may, after giving 7 days notice in writing to the Contractor, enter upon the Site and Works and expel the Contractor from the Site and Works:—
1184
63(1)
There is a breach by the Contractor but the necessity for superintendence can be arbitrated upon. The provisions of Clause 63 are not an effective sanction once the Works have been completed
The Engineer must be at liberty to object to and require the Contractor to remove from the Works any person employed by the Contractor in or about the execution of the Works who in the opinion of the Engineer:—
211
16
The expression "Period of Maintenance" must mean:—
713
49(1)
The Employer or the Contractor must refer to the Engineer any dispute as to:—
1320
66(1)
any decision opinion instruction direction certificate valuation of the Engineer;
1321
66(1)
Superintendence may be necessary during the Maintenance Period
The Engineer may suspend the progress of the Works or any part of the Works for such time or times and in such manner as he may consider necessary
519
40(1)
whether during the progress of the Works; or
1322
66(1)
The Employer may, after giving 7 days notice in writing to the Contractor, enter upon the Site and Works and expel the Contractor from the Site and Works:—
1184
63(1)
if the Engineer certifies in writing to the Employer that in his opinion the Contractor is persistently or fundamentally in breach of his obligations under the Contract
1201
63(1)(d)
after completion of the Works; and
1323
66(1)
There is a breach by the Contractor
whether before determination abandonment or breach of the Contract; or
1324
66(1)
after determination abandonment or breach of the Contract
1325
66(1)

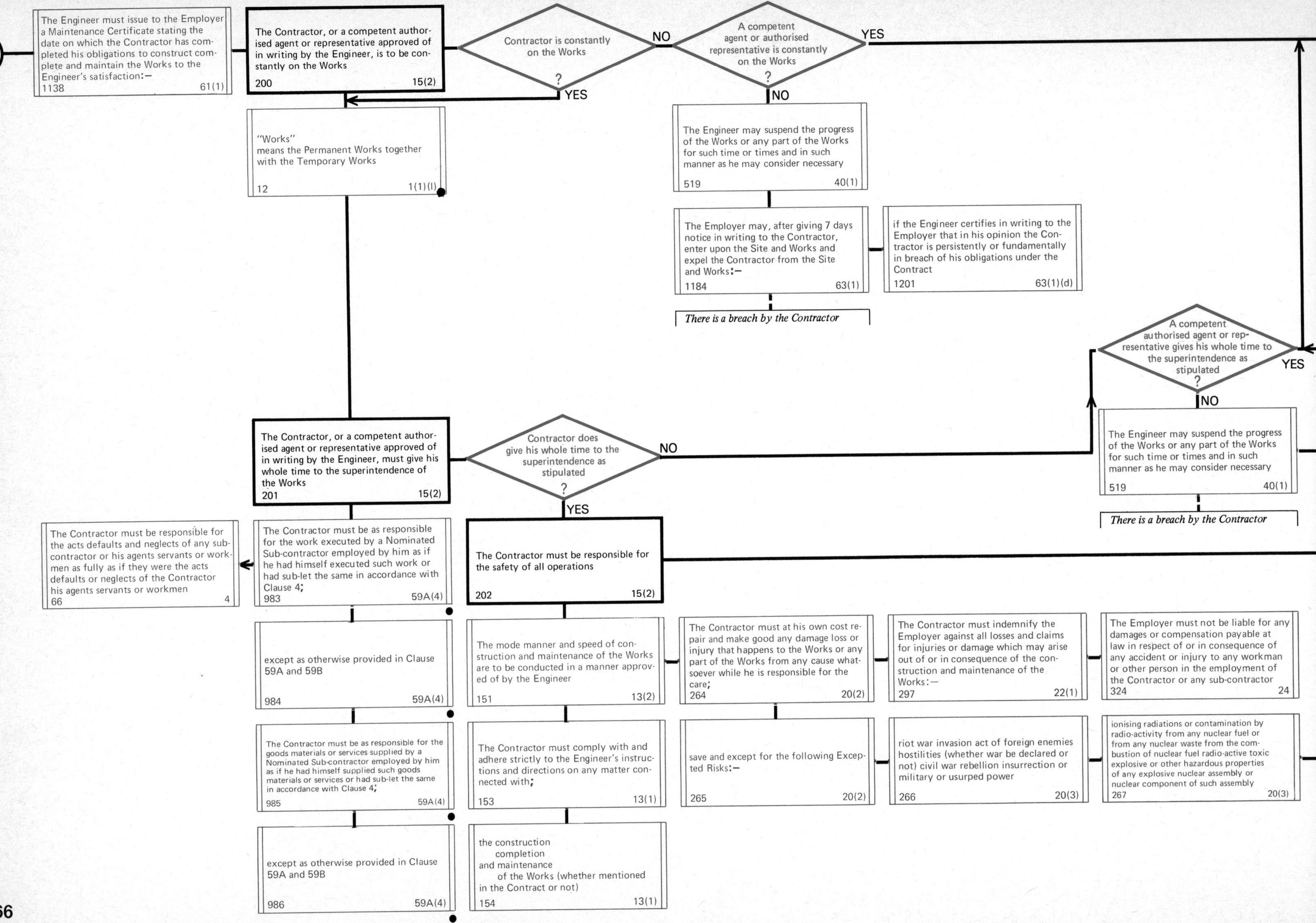
1
The Engineer must issue to the Employer a Maintenance Certificate stating the date on which the Contractor has completed his obligations to construct complete and maintain the Works to the Engineer's satisfaction:— 1138 61(1)
The Contractor, or a competent authorised agent or representative approved of in writing by the Engineer, is to be constantly on the Works 200 15(2)
Contractor is constantly on the Works ?
NO
YES
A competent agent or authorised representative is constantly on the Works ?
YES
NO
"Works" means the Permanent Works together with the Temporary Works 12 1(1)(l)
The Engineer may suspend the progress of the Works or any part of the Works for such time or times and in such manner as he may consider necessary 519 40(1)
The Employer may, after giving 7 days notice in writing to the Contractor, enter upon the Site and Works and expel the Contractor from the Site and Works:— 1184 63(1)
if the Engineer certifies in writing to the Employer that in his opinion the Contractor is persistently or fundamentally in breach of his obligations under the Contract 1201 63(1)(d)
There is a breach by the Contractor
A competent authorised agent or representative gives his whole time to the superintendence as stipulated ?
YES
NO
The Contractor, or a competent authorised agent or representative approved of in writing by the Engineer, must give his whole time to the superintendence of the Works 201 15(2)
Contractor does give his whole time to the superintendence as stipulated ?
NO
YES
The Engineer may suspend the progress of the Works or any part of the Works for such time or times and in such manner as he may consider necessary 519 40(1)
There is a breach by the Contractor
The Contractor must be responsible for the acts defaults and neglects of any sub-contractor or his agents servants or workmen as fully as if they were the acts defaults or neglects of the Contractor his agents servants or workmen 66 4
The Contractor must be as responsible for the work executed by a Nominated Sub-contractor employed by him as if he had himself executed such work or had sub-let the same in accordance with Clause 4; 983 59A(4)
The Contractor must be responsible for the safety of all operations 202 15(2)
except as otherwise provided in Clause 59A and 59B 984 59A(4)
The mode manner and speed of construction and maintenance of the Works are to be conducted in a manner approved of by the Engineer 151 13(2)
The Contractor must at his own cost repair and make good any damage loss or injury that happens to the Works or any part of the Works from any cause whatsoever while he is responsible for the care; 264 20(2)
The Contractor must indemnify the Employer against all losses and claims for injuries or damage which may arise out of or in consequence of the construction and maintenance of the Works:— 297 22(1)
The Employer must not be liable for any damages or compensation payable at law in respect of or in consequence of any accident or injury to any workman or other person in the employment of the Contractor or any sub-contractor 324 24
The Contractor must be as responsible for the goods materials or services supplied by a Nominated Sub-contractor employed by him as if he had himself supplied such goods materials or services or had sub-let the same in accordance with Clause 4; 985 59A(4)
The Contractor must comply with and adhere strictly to the Engineer's instructions and directions on any matter connected with; 153 13(1)
save and except for the following Excepted Risks:— 265 20(2)
riot war invasion act of foreign enemies hostilities (whether war be declared or not) civil war rebellion insurrection or military or usurped power 266 20(3)
ionising radiations or contamination by radio-activity from any nuclear fuel or from any nuclear waste from the combustion of nuclear fuel radio-active toxic explosive or other hazardous properties of any explosive nuclear assembly or nuclear component of such assembly 267 20(3)
except as otherwise provided in Clause 59A and 59B 986 59A(4)
the construction completion and maintenance of the Works (whether mentioned in the Contract or not) 154 13(1)

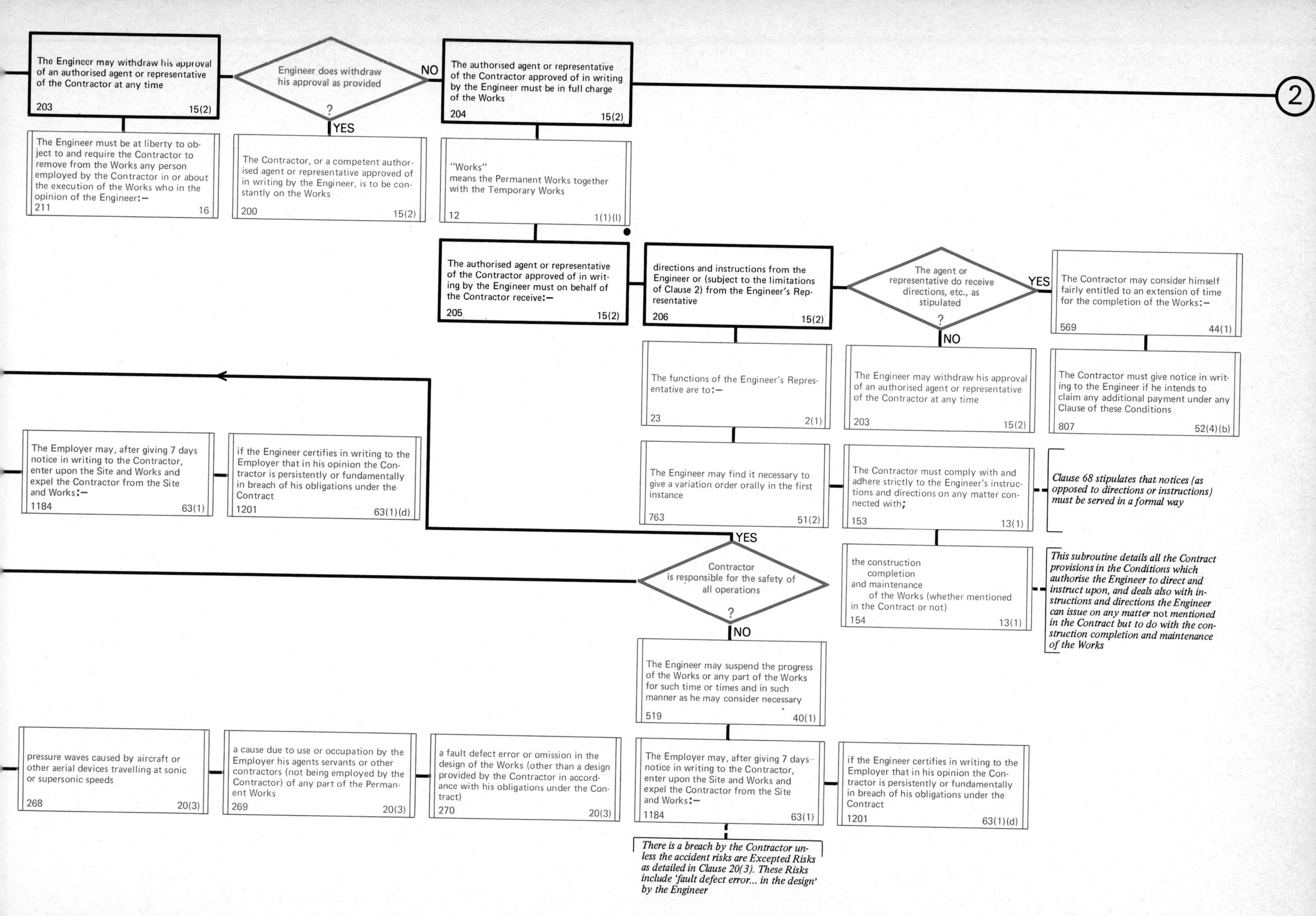

The Engineer may withdraw his approval of an authorised agent or representative of the Contractor at any time
203 15(2)
Engineer does withdraw his approval as provided ?
NO
YES
The authorised agent or representative of the Contractor approved of in writing by the Engineer must be in full charge of the Works
204 15(2)
2
The Engineer must be at liberty to object to and require the Contractor to remove from the Works any person employed by the Contractor in or about the execution of the Works who in the opinion of the Engineer:—
211 16
The Contractor, or a competent authorised agent or representative approved of in writing by the Engineer, is to be constantly on the Works
200 15(2)
"Works" means the Permanent Works together with the Temporary Works
12 1(1)(l)
The authorised agent or representative of the Contractor approved of in writing by the Engineer must on behalf of the Contractor receive:—
205 15(2)
directions and instructions from the Engineer or (subject to the limitations of Clause 2) from the Engineer's Representative
206 15(2)
The agent or representative do receive directions, etc., as stipulated ?
YES
NO
The Contractor may consider himself fairly entitled to an extension of time for the completion of the Works:—
569 44(1)
The functions of the Engineer's Representative are to:—
23 2(1)
The Engineer may withdraw his approval of an authorised agent or representative of the Contractor at any time
203 15(2)
The Contractor must give notice in writing to the Engineer if he intends to claim any additional payment under any Clause of these Conditions
807 52(4)(b)
The Employer may, after giving 7 days notice in writing to the Contractor, enter upon the Site and Works and expel the Contractor from the Site and Works:—
1184 63(1)
if the Engineer certifies in writing to the Employer that in his opinion the Contractor is persistently or fundamentally in breach of his obligations under the Contract
1201 63(1)(d)
The Engineer may find it necessary to give a variation order orally in the first instance
763 51(2)
The Contractor must comply with and adhere strictly to the Engineer's instructions and directions on any matter connected with;
153 13(1)
Clause 68 stipulates that notices (as opposed to directions or instructions) must be served in a formal way
YES
Contractor is responsible for the safety of all operations ?
NO
the construction completion and maintenance of the Works (whether mentioned in the Contract or not)
154 13(1)
This subroutine details all the Contract provisions in the Conditions which authorise the Engineer to direct and instruct upon, and deals also with instructions and directions the Engineer can issue on any matter not mentioned in the Contract but to do with the construction completion and maintenance of the Works
The Engineer may suspend the progress of the Works or any part of the Works for such time or times and in such manner as he may consider necessary
519 40(1)
pressure waves caused by aircraft or other aerial devices travelling at sonic or supersonic speeds
268 20(3)
a cause due to use or occupation by the Employer his agents servants or other contractors (not being employed by the Contractor) of any part of the Permanent Works
269 20(3)
a fault defect error or omission in the design of the Works (other than a design provided by the Contractor in accordance with his obligations under the Contract)
270 20(3)
The Employer may, after giving 7 days notice in writing to the Contractor, enter upon the Site and Works and expel the Contractor from the Site and Works:—
1184 63(1)
if the Engineer certifies in writing to the Employer that in his opinion the Contractor is persistently or fundamentally in breach of his obligations under the Contract
1201 63(1)(d)
There is a breach by the Contractor unless the accident risks are Excepted Risks as detailed in Clause 20(3). These Risks include 'fault defect error... in the design' by the Engineer

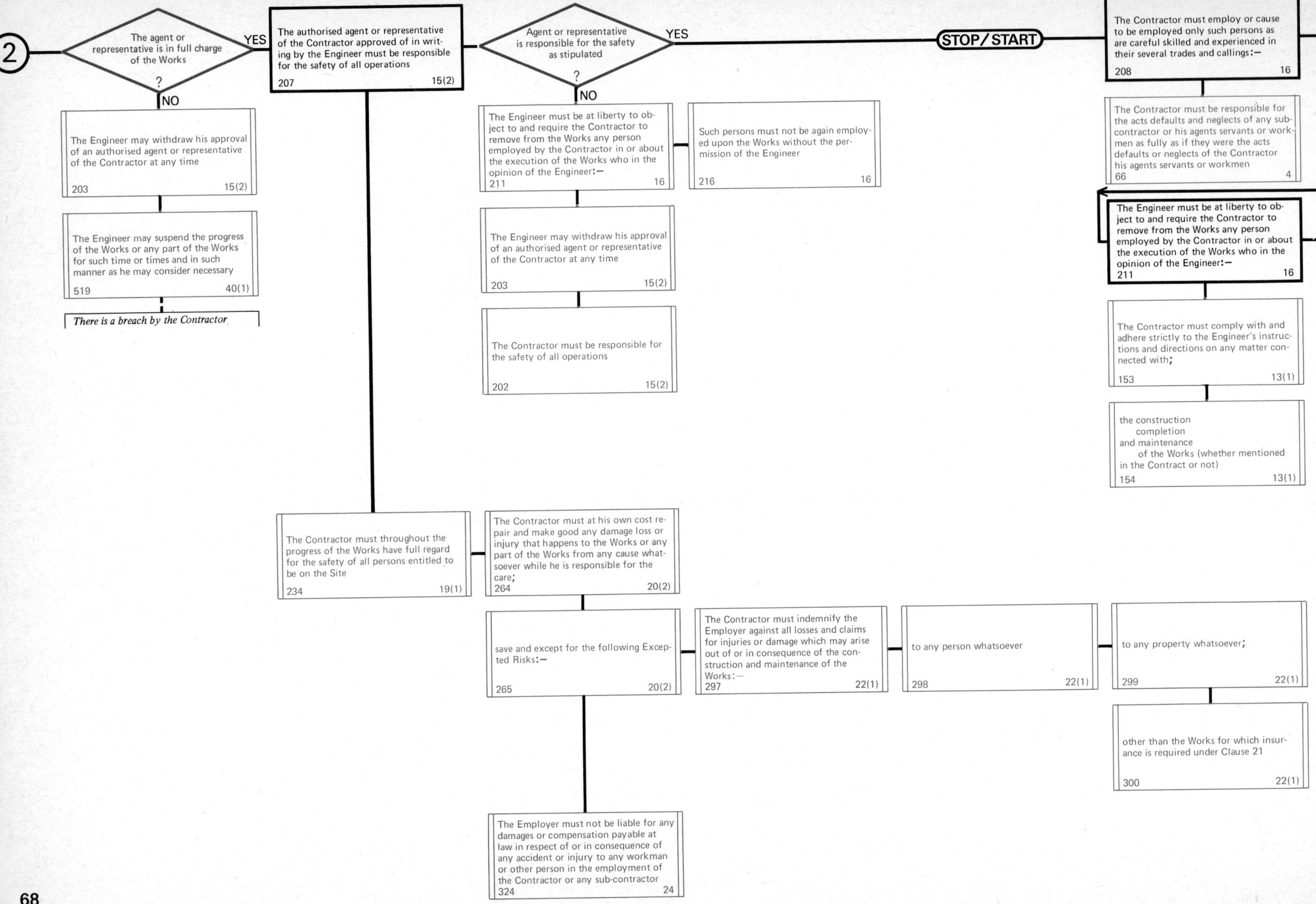
2
The agent or representative is in full charge of the Works ?
YES
NO
The authorised agent or representative of the Contractor approved of in writing by the Engineer must be responsible for the safety of all operations 207 15(2)
Agent or representative is responsible for the safety as stipulated ?
YES
NO
STOP/START
The Contractor must employ or cause to be employed only such persons as are careful skilled and experienced in their several trades and callings:— 208 16
The Engineer may withdraw his approval of an authorised agent or representative of the Contractor at any time 203 15(2)
The Engineer must be at liberty to object to and require the Contractor to remove from the Works any person employed by the Contractor in or about the execution of the Works who in the opinion of the Engineer:— 211 16
Such persons must not be again employed upon the Works without the permission of the Engineer 216 16
The Contractor must be responsible for the acts defaults and neglects of any sub-contractor or his agents servants or workmen as fully as if they were the acts defaults or neglects of the Contractor his agents servants or workmen 66 4
The Engineer must be at liberty to object to and require the Contractor to remove from the Works any person employed by the Contractor in or about the execution of the Works who in the opinion of the Engineer:— 211 16
The Engineer may suspend the progress of the Works or any part of the Works for such time or times and in such manner as he may consider necessary 519 40(1)
The Engineer may withdraw his approval of an authorised agent or representative of the Contractor at any time 203 15(2)
There is a breach by the Contractor
The Contractor must be responsible for the safety of all operations 202 15(2)
The Contractor must comply with and adhere strictly to the Engineer's instructions and directions on any matter connected with; 153 13(1)
the construction completion and maintenance of the Works (whether mentioned in the Contract or not) 154 13(1)
The Contractor must throughout the progress of the Works have full regard for the safety of all persons entitled to be on the Site 234 19(1)
The Contractor must at his own cost repair and make good any damage loss or injury that happens to the Works or any part of the Works from any cause whatsoever while he is responsible for the care; 264 20(2)
save and except for the following Excepted Risks:— 265 20(2)
The Contractor must indemnify the Employer against all losses and claims for injuries or damage which may arise out of or in consequence of the construction and maintenance of the Works:— 297 22(1)
to any person whatsoever 298 22(1)
to any property whatsoever; 299 22(1)
other than the Works for which insurance is required under Clause 21 300 22(1)
The Employer must not be liable for any damages or compensation payable at law in respect of or in consequence of any accident or injury to any workman or other person in the employment of the Contractor or any sub-contractor 324 24

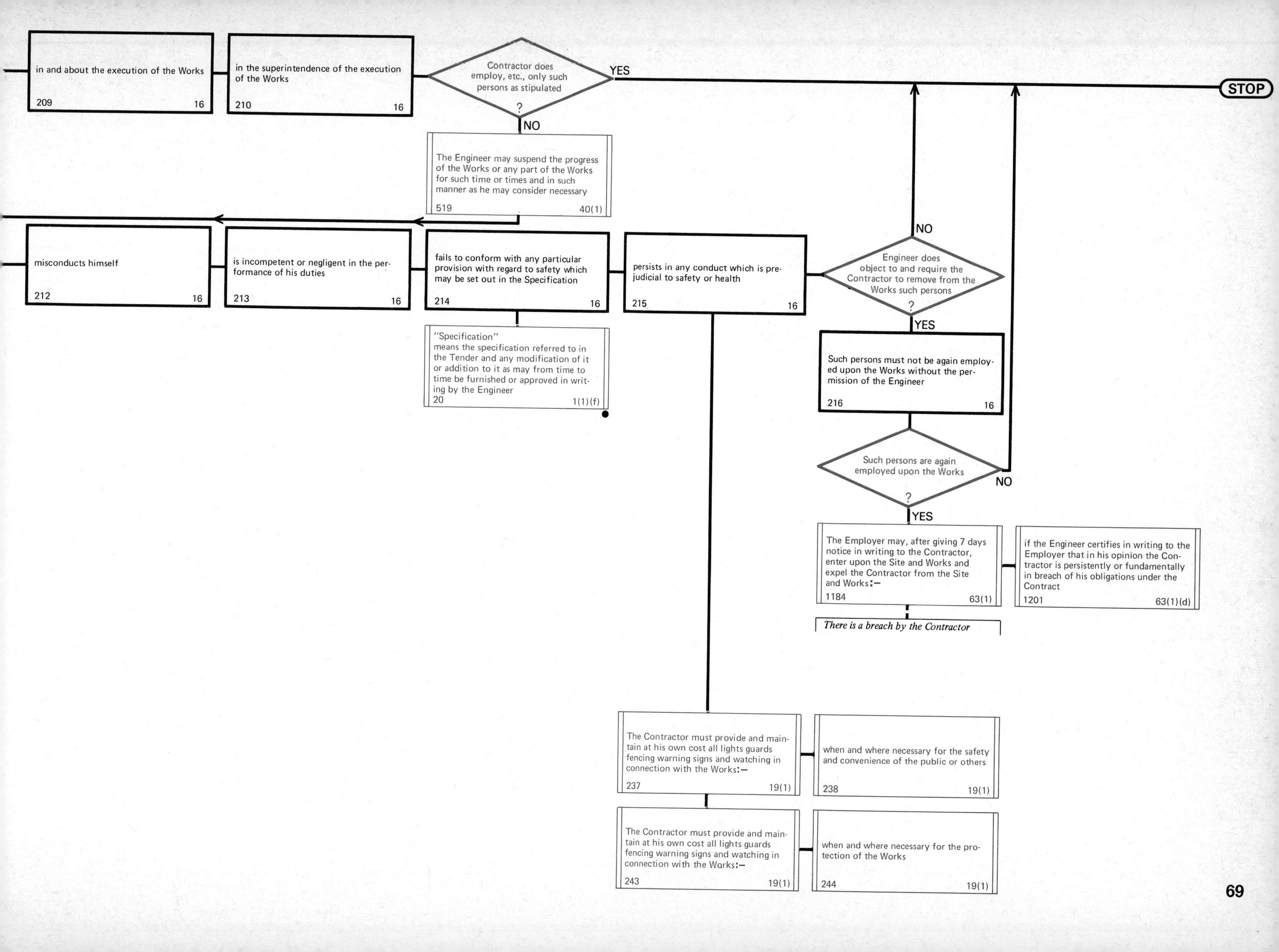

in and about the execution of the Works
209 16
in the superintendence of the execution of the Works
210 16
Contractor does employ, etc., only such persons as stipulated ?
YES
STOP
NO
The Engineer may suspend the progress of the Works or any part of the Works for such time or times and in such manner as he may consider necessary
519 40(1)
misconducts himself
212 16
is incompetent or negligent in the performance of his duties
213 16
fails to conform with any particular provision with regard to safety which may be set out in the Specification
214 16
persists in any conduct which is prejudicial to safety or health
215 16
NO
Engineer does object to and require the Contractor to remove from the Works such persons ?
YES
"Specification" means the specification referred to in the Tender and any modification of it or addition to it as may from time to time be furnished or approved in writing by the Engineer
20 1(1)(f)
Such persons must not be again employed upon the Works without the permission of the Engineer
216 16
Such persons are again employed upon the Works ?
NO
YES
The Employer may, after giving 7 days notice in writing to the Contractor, enter upon the Site and Works and expel the Contractor from the Site and Works:–
1184 63(1)
if the Engineer certifies in writing to the Employer that in his opinion the Contractor is persistently or fundamentally in breach of his obligations under the Contract
1201 63(1)(d)
There is a breach by the Contractor
The Contractor must provide and maintain at his own cost all lights guards fencing warning signs and watching in connection with the Works:–
237 19(1)
when and where necessary for the safety and convenience of the public or others
238 19(1)
The Contractor must provide and maintain at his own cost all lights guards fencing warning signs and watching in connection with the Works:–
243 19(1)
when and where necessary for the protection of the Works
244 19(1)

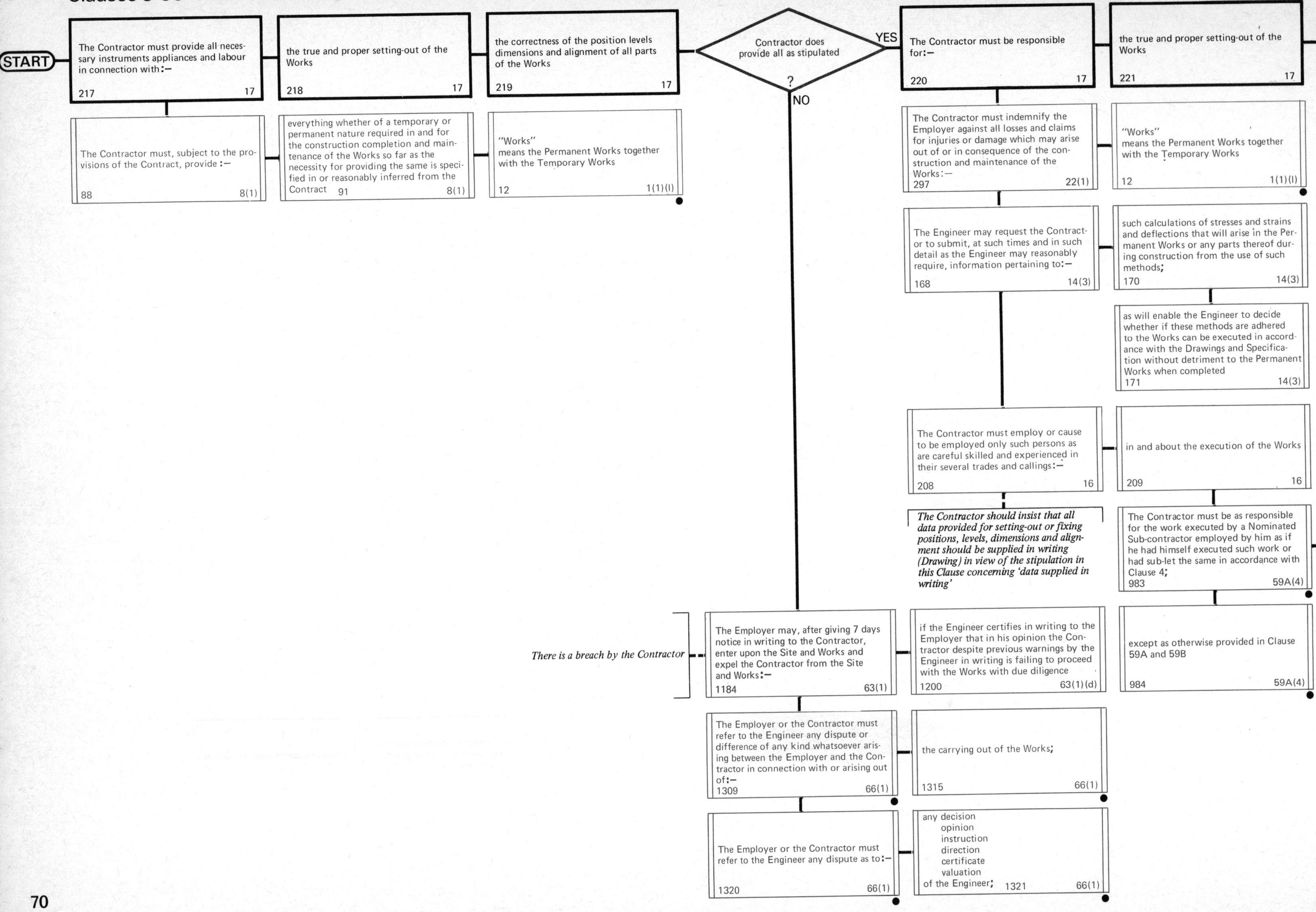
START
The Contractor must provide all necessary instruments appliances and labour in connection with:— 217 17
the true and proper setting-out of the Works 218 17
the correctness of the position levels dimensions and alignment of all parts of the Works 219 17
Contractor does provide all as stipulated ?
YES
NO
The Contractor must be responsible for:— 220 17
the true and proper setting-out of the Works 221 17
The Contractor must, subject to the provisions of the Contract, provide :— 88 8(1)
everything whether of a temporary or permanent nature required in and for the construction completion and maintenance of the Works so far as the necessity for providing the same is specified in or reasonably inferred from the Contract 91 8(1)
"Works" means the Permanent Works together with the Temporary Works 12 1(1)(l)
The Contractor must indemnify the Employer against all losses and claims for injuries or damage which may arise out of or in consequence of the construction and maintenance of the Works:— 297 22(1)
"Works" means the Permanent Works together with the Temporary Works 12 1(1)(l)
The Engineer may request the Contractor to submit, at such times and in such detail as the Engineer may reasonably require, information pertaining to:— 168 14(3)
such calculations of stresses and strains and deflections that will arise in the Permanent Works or any parts thereof during construction from the use of such methods; 170 14(3)
as will enable the Engineer to decide whether if these methods are adhered to the Works can be executed in accordance with the Drawings and Specification without detriment to the Permanent Works when completed 171 14(3)
The Contractor must employ or cause to be employed only such persons as are careful skilled and experienced in their several trades and callings:— 208 16
in and about the execution of the Works 209 16
The Contractor should insist that all data provided for setting-out or fixing positions, levels, dimensions and alignment should be supplied in writing (Drawing) in view of the stipulation in this Clause concerning 'data supplied in writing'
The Contractor must be as responsible for the work executed by a Nominated Sub-contractor employed by him as if he had himself executed such work or had sub-let the same in accordance with Clause 4; 983 59A(4)
except as otherwise provided in Clause 59A and 59B 984 59A(4)
There is a breach by the Contractor
The Employer may, after giving 7 days notice in writing to the Contractor, enter upon the Site and Works and expel the Contractor from the Site and Works:— 1184 63(1)
if the Engineer certifies in writing to the Employer that in his opinion the Contractor despite previous warnings by the Engineer in writing is failing to proceed with the Works with due diligence 1200 63(1)(d)
The Employer or the Contractor must refer to the Engineer any dispute or difference of any kind whatsoever arising between the Employer and the Contractor in connection with or arising out of:— 1309 66(1)
the carrying out of the Works; 1315 66(1)
The Employer or the Contractor must refer to the Engineer any dispute as to:— 1320 66(1)
any decision opinion instruction direction certificate valuation of the Engineer; 1321 66(1)

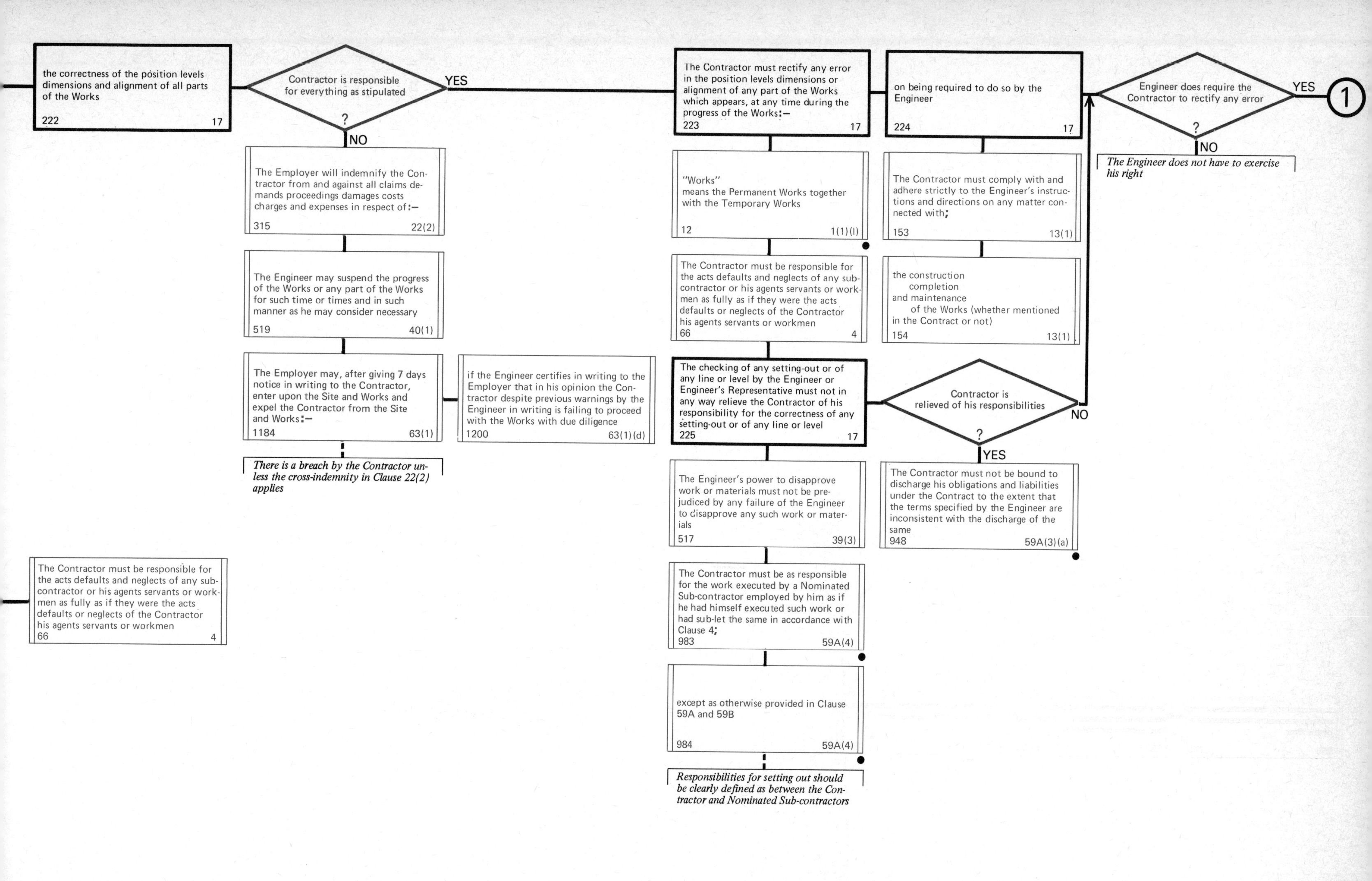

the correctness of the position levels dimensions and alignment of all parts of the Works
222 17
Contractor is responsible for everything as stipulated
?
YES
NO
The Contractor must rectify any error in the position levels dimensions or alignment of any part of the Works which appears, at any time during the progress of the Works:—
223 17
on being required to do so by the Engineer
224 17
Engineer does require the Contractor to rectify any error
?
YES
NO
1
The Engineer does not have to exercise his right
The Employer will indemnify the Contractor from and against all claims demands proceedings damages costs charges and expenses in respect of:—
315 22(2)
"Works" means the Permanent Works together with the Temporary Works
12 1(1)(l)
The Contractor must comply with and adhere strictly to the Engineer's instructions and directions on any matter connected with;
153 13(1)
The Engineer may suspend the progress of the Works or any part of the Works for such time or times and in such manner as he may consider necessary
519 40(1)
The Contractor must be responsible for the acts defaults and neglects of any sub-contractor or his agents servants or workmen as fully as if they were the acts defaults or neglects of the Contractor his agents servants or workmen
66 4
the construction completion and maintenance of the Works (whether mentioned in the Contract or not)
154 13(1)
The Employer may, after giving 7 days notice in writing to the Contractor, enter upon the Site and Works and expel the Contractor from the Site and Works:—
1184 63(1)
if the Engineer certifies in writing to the Employer that in his opinion the Contractor despite previous warnings by the Engineer in writing is failing to proceed with the Works with due diligence
1200 63(1)(d)
The checking of any setting-out or of any line or level by the Engineer or Engineer's Representative must not in any way relieve the Contractor of his responsibility for the correctness of any setting-out or of any line or level
225 17
Contractor is relieved of his responsibilities
?
NO
YES
There is a breach by the Contractor unless the cross-indemnity in Clause 22(2) applies
The Engineer's power to disapprove work or materials must not be prejudiced by any failure of the Engineer to disapprove any such work or materials
517 39(3)
The Contractor must not be bound to discharge his obligations and liabilities under the Contract to the extent that the terms specified by the Engineer are inconsistent with the discharge of the same
948 59A(3)(a)
The Contractor must be responsible for the acts defaults and neglects of any sub-contractor or his agents servants or workmen as fully as if they were the acts defaults or neglects of the Contractor his agents servants or workmen
66 4
The Contractor must be as responsible for the work executed by a Nominated Sub-contractor employed by him as if he had himself executed such work or had sub-let the same in accordance with Clause 4;
983 59A(4)
except as otherwise provided in Clause 59A and 59B
984 59A(4)
Responsibilities for setting out should be clearly defined as between the Contractor and Nominated Sub-contractors

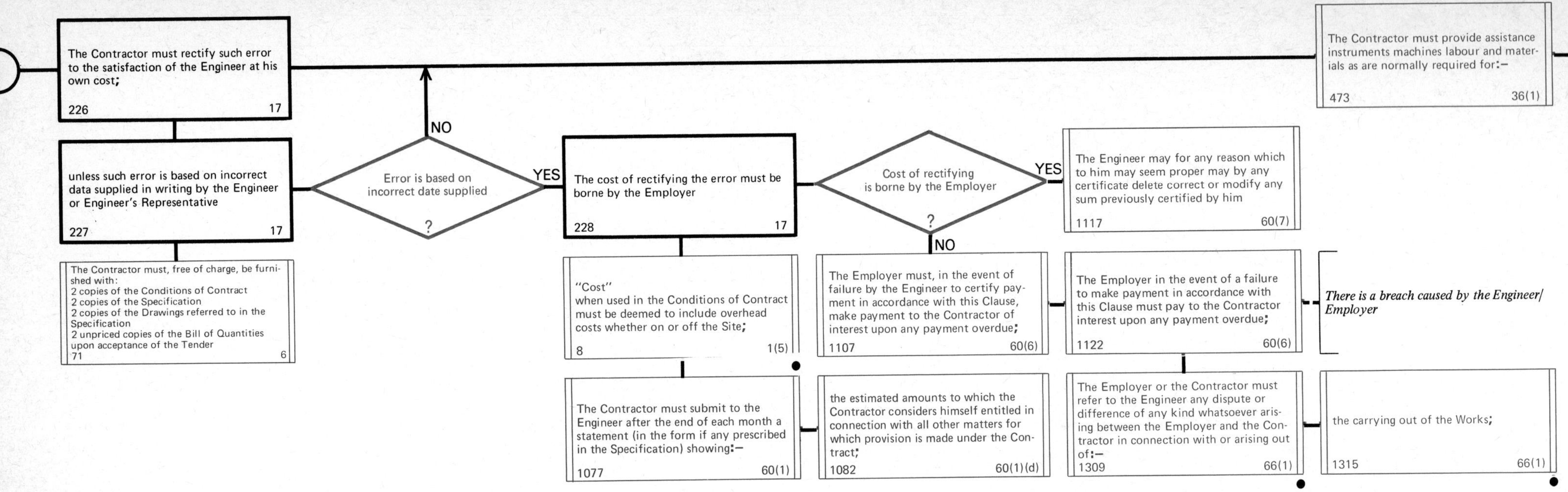
1
The Contractor must rectify such error to the satisfaction of the Engineer at his own cost;
226 17
unless such error is based on incorrect data supplied in writing by the Engineer or Engineer's Representative
227 17
The Contractor must, free of charge, be furnished with:
2 copies of the Conditions of Contract
2 copies of the Specification
2 copies of the Drawings referred to in the Specification
2 unpriced copies of the Bill of Quantities upon acceptance of the Tender
71 6
Error is based on incorrect date supplied
?
NO
YES
The cost of rectifying the error must be borne by the Employer
228 17
"Cost" when used in the Conditions of Contract must be deemed to include overhead costs whether on or off the Site;
8 1(5)
The Contractor must submit to the Engineer after the end of each month a statement (in the form if any prescribed in the Specification) showing:—
1077 60(1)
Cost of rectifying is borne by the Employer
?
YES
NO
The Employer must, in the event of failure by the Engineer to certify payment in accordance with this Clause, make payment to the Contractor of interest upon any payment overdue;
1107 60(6)
the estimated amounts to which the Contractor considers himself entitled in connection with all other matters for which provision is made under the Contract;
1082 60(1)(d)
The Engineer may for any reason which to him may seem proper may by any certificate delete correct or modify any sum previously certified by him
1117 60(7)
The Employer in the event of a failure to make payment in accordance with this Clause must pay to the Contractor interest upon any payment overdue;
1122 60(6)
There is a breach caused by the Engineer/Employer
The Employer or the Contractor must refer to the Engineer any dispute or difference of any kind whatsoever arising between the Employer and the Contractor in connection with or arising out of:—
1309 66(1)
the carrying out of the Works;
1315 66(1)
The Contractor must provide assistance instruments machines labour and materials as are normally required for:—
473 36(1)

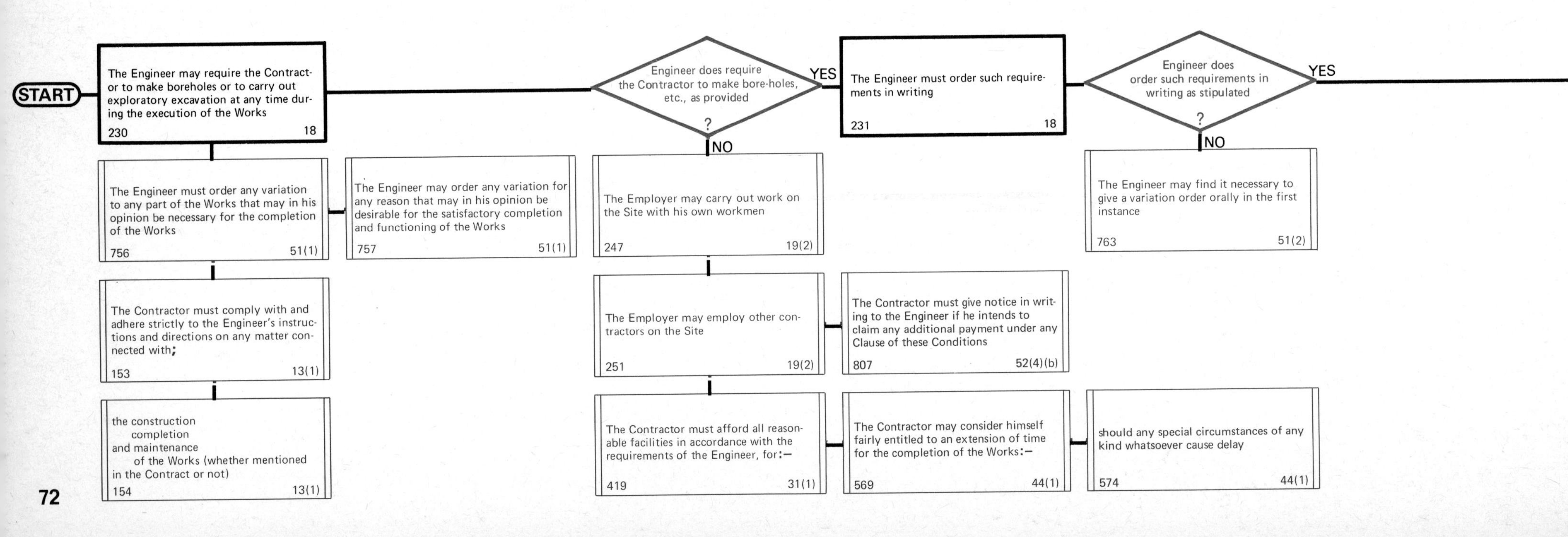
START
The Engineer may require the Contractor to make boreholes or to carry out exploratory excavation at any time during the execution of the Works
230 18
The Engineer must order any variation to any part of the Works that may in his opinion be necessary for the completion of the Works
756 51(1)
The Engineer may order any variation for any reason that may in his opinion be desirable for the satisfactory completion and functioning of the Works
757 51(1)
The Contractor must comply with and adhere strictly to the Engineer's instructions and directions on any matter connected with;
153 13(1)
the construction
completion
and maintenance
of the Works (whether mentioned in the Contract or not)
154 13(1)
Engineer does require the Contractor to make bore-holes, etc., as provided
?
YES
NO
The Employer may carry out work on the Site with his own workmen
247 19(2)
The Employer may employ other contractors on the Site
251 19(2)
The Contractor must afford all reasonable facilities in accordance with the requirements of the Engineer, for:—
419 31(1)
The Engineer must order such requirements in writing
231 18
The Contractor must give notice in writing to the Engineer if he intends to claim any additional payment under any Clause of these Conditions
807 52(4)(b)
The Contractor may consider himself fairly entitled to an extension of time for the completion of the Works:—
569 44(1)
Engineer does order such requirements in writing as stipulated
?
YES
NO
The Engineer may find it necessary to give a variation order orally in the first instance
763 51(2)
should any special circumstances of any kind whatsoever cause delay
574 44(1)

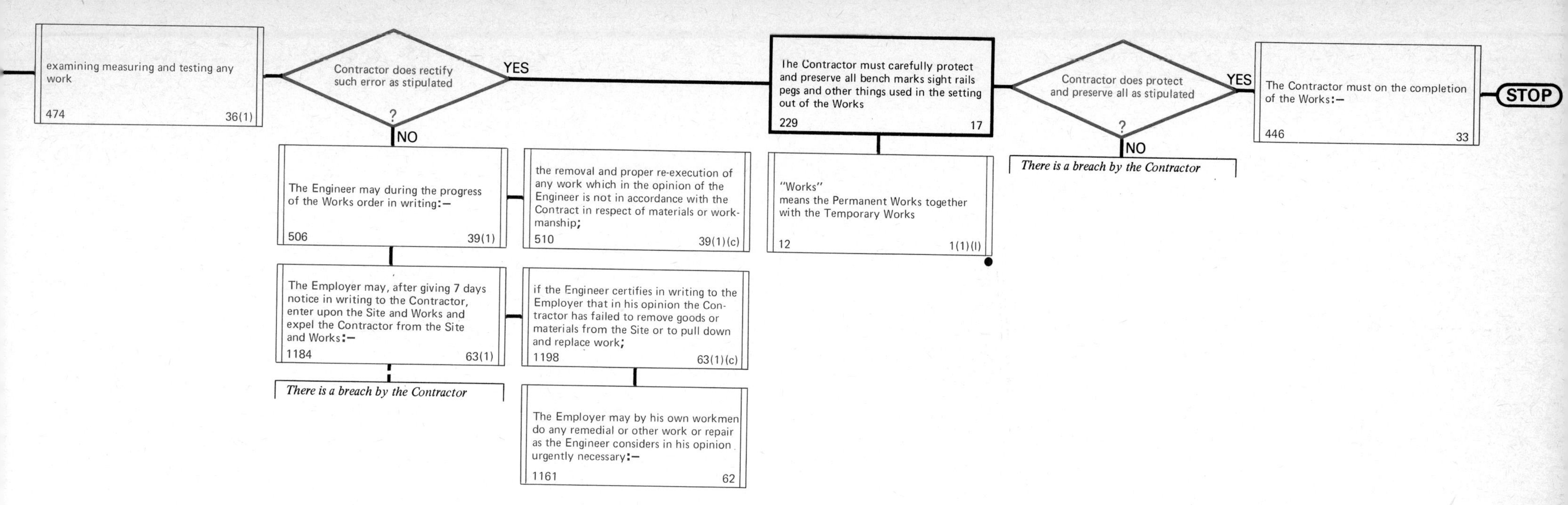

examining measuring and testing any work
474
36(1)
Contractor does rectify such error as stipulated
?
YES
NO
The Contractor must carefully protect and preserve all bench marks sight rails pegs and other things used in the setting out of the Works
229
17
Contractor does protect and preserve all as stipulated
?
YES
NO
The Contractor must on the completion of the Works:—
446
33
STOP
There is a breach by the Contractor
The Engineer may during the progress of the Works order in writing:—
506
39(1)
the removal and proper re-execution of any work which in the opinion of the Engineer is not in accordance with the Contract in respect of materials or workmanship;
510
39(1)(c)
"Works" means the Permanent Works together with the Temporary Works
12
1(1)(l)
The Employer may, after giving 7 days notice in writing to the Contractor, enter upon the Site and Works and expel the Contractor from the Site and Works:—
1184
63(1)
if the Engineer certifies in writing to the Employer that in his opinion the Contractor has failed to remove goods or materials from the Site or to pull down and replace work;
1198
63(1)(c)
There is a breach by the Contractor
The Employer may by his own workmen do any remedial or other work or repair as the Engineer considers in his opinion urgently necessary:—
1161
62

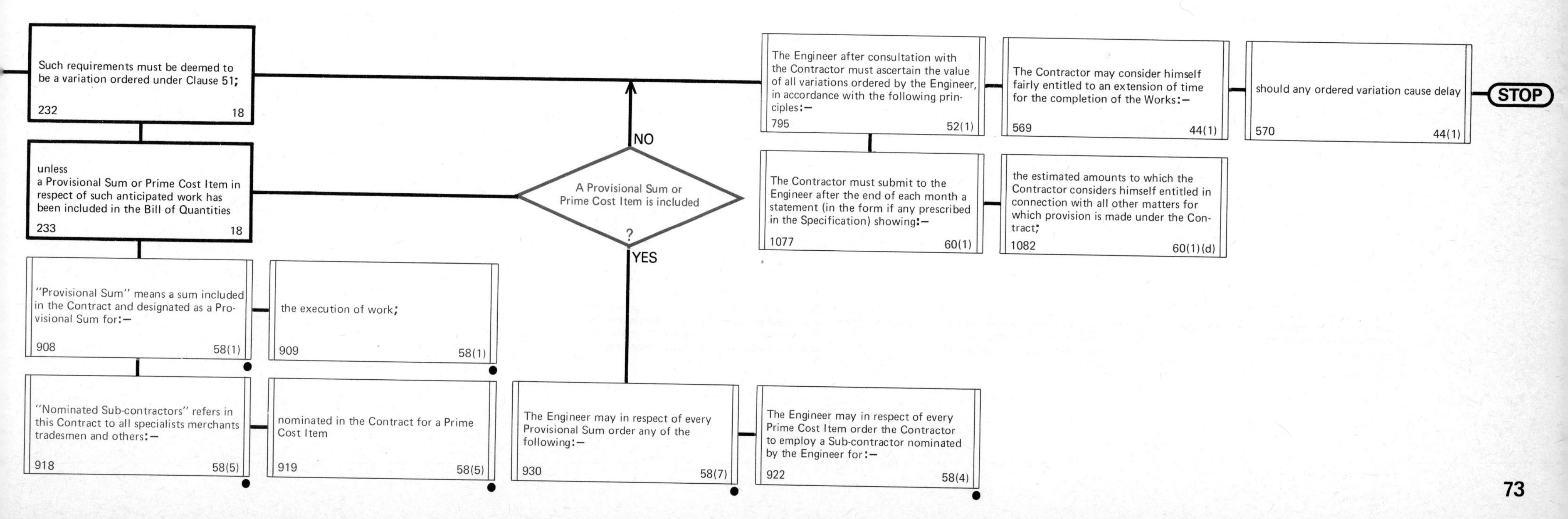

Such requirements must be deemed to be a variation ordered under Clause 51;
232
18
The Engineer after consultation with the Contractor must ascertain the value of all variations ordered by the Engineer, in accordance with the following principles:—
795
52(1)
The Contractor may consider himself fairly entitled to an extension of time for the completion of the Works:—
569
44(1)
should any ordered variation cause delay
570
44(1)
STOP
NO
unless a Provisional Sum or Prime Cost Item in respect of such anticipated work has been included in the Bill of Quantities
233
18
A Provisional Sum or Prime Cost Item is included
?
YES
The Contractor must submit to the Engineer after the end of each month a statement (in the form if any prescribed in the Specification) showing:—
1077
60(1)
the estimated amounts to which the Contractor considers himself entitled in connection with all other matters for which provision is made under the Contract;
1082
60(1)(d)
"Provisional Sum" means a sum included in the Contract and designated as a Provisional Sum for:—
908
58(1)
the execution of work;
909
58(1)
"Nominated Sub-contractors" refers in this Contract to all specialists merchants tradesmen and others:—
918
58(5)
nominated in the Contract for a Prime Cost Item
919
58(5)
The Engineer may in respect of every Provisional Sum order any of the following:—
930
58(7)
The Engineer may in respect of every Prime Cost Item order the Contractor to employ a Sub-contractor nominated by the Engineer for:—
922
58(4)

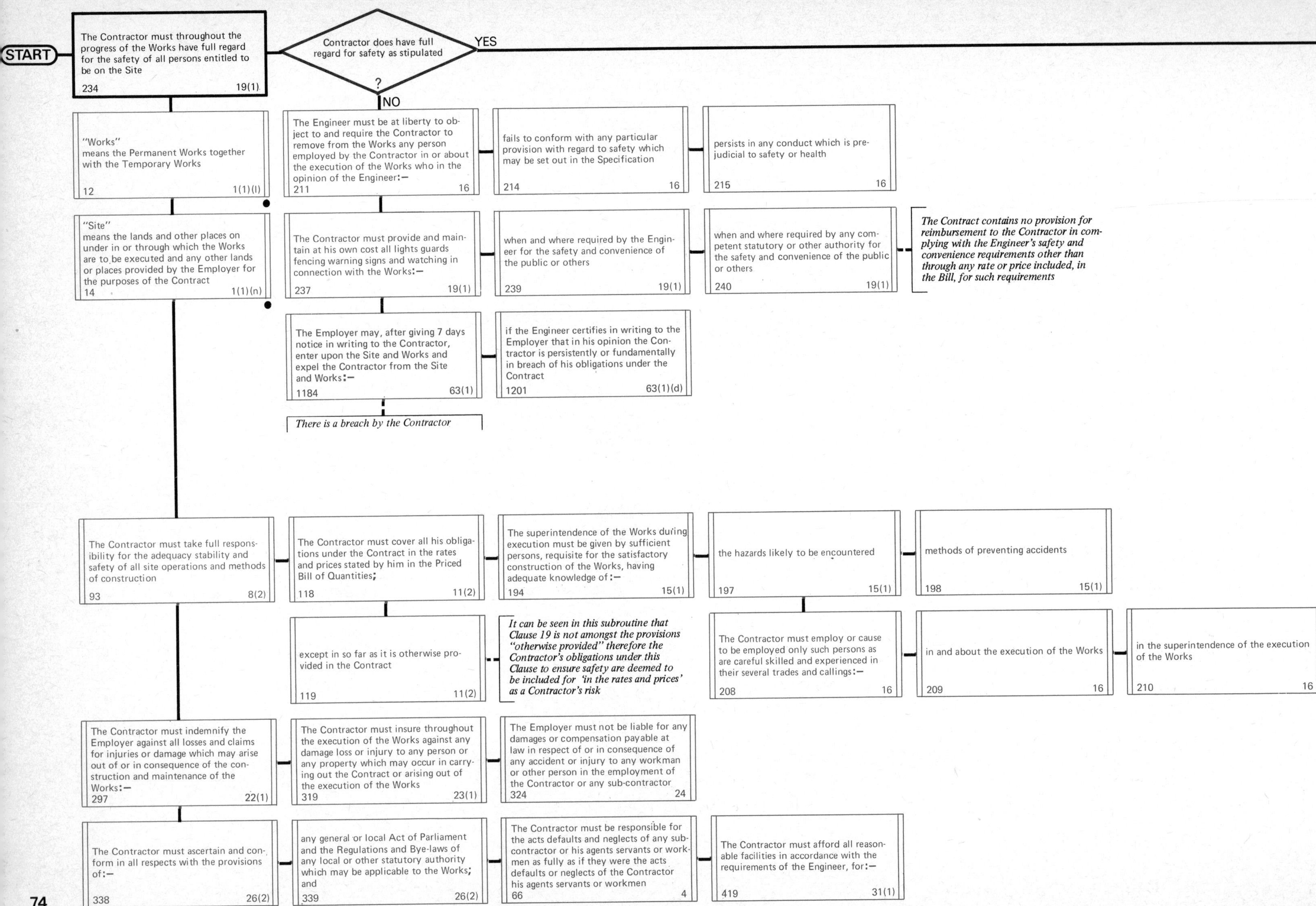
START
The Contractor must throughout the progress of the Works have full regard for the safety of all persons entitled to be on the Site
234 19(1)
Contractor does have full regard for safety as stipulated
?
YES
NO
"Works" means the Permanent Works together with the Temporary Works
12 1(1)(l)
"Site" means the lands and other places on under in or through which the Works are to be executed and any other lands or places provided by the Employer for the purposes of the Contract
14 1(1)(n)
The Engineer must be at liberty to object to and require the Contractor to remove from the Works any person employed by the Contractor in or about the execution of the Works who in the opinion of the Engineer:—
211 16
fails to conform with any particular provision with regard to safety which may be set out in the Specification
214 16
persists in any conduct which is prejudicial to safety or health
215 16
The Contractor must provide and maintain at his own cost all lights guards fencing warning signs and watching in connection with the Works:—
237 19(1)
when and where required by the Engineer for the safety and convenience of the public or others
239 19(1)
when and where required by any competent statutory or other authority for the safety and convenience of the public or others
240 19(1)
The Contract contains no provision for reimbursement to the Contractor in complying with the Engineer's safety and convenience requirements other than through any rate or price included, in the Bill, for such requirements
The Employer may, after giving 7 days notice in writing to the Contractor, enter upon the Site and Works and expel the Contractor from the Site and Works:—
1184 63(1)
if the Engineer certifies in writing to the Employer that in his opinion the Contractor is persistently or fundamentally in breach of his obligations under the Contract
1201 63(1)(d)
There is a breach by the Contractor
The Contractor must take full responsibility for the adequacy stability and safety of all site operations and methods of construction
93 8(2)
The Contractor must cover all his obligations under the Contract in the rates and prices stated by him in the Priced Bill of Quantities;
118 11(2)
The superintendence of the Works during execution must be given by sufficient persons, requisite for the satisfactory construction of the Works, having adequate knowledge of:—
194 15(1)
the hazards likely to be encountered
197 15(1)
methods of preventing accidents
198 15(1)
except in so far as it is otherwise provided in the Contract
119 11(2)
It can be seen in this subroutine that Clause 19 is not amongst the provisions "otherwise provided" therefore the Contractor's obligations under this Clause to ensure safety are deemed to be included for 'in the rates and prices' as a Contractor's risk
The Contractor must employ or cause to be employed only such persons as are careful skilled and experienced in their several trades and callings:—
208 16
in and about the execution of the Works
209 16
in the superintendence of the execution of the Works
210 16
The Contractor must indemnify the Employer against all losses and claims for injuries or damage which may arise out of or in consequence of the construction and maintenance of the Works:—
297 22(1)
The Contractor must insure throughout the execution of the Works against any damage loss or injury to any person or any property which may occur in carrying out the Contract or arising out of the execution of the Works
319 23(1)
The Employer must not be liable for any damages or compensation payable at law in respect of or in consequence of any accident or injury to any workman or other person in the employment of the Contractor or any sub-contractor
324 24
The Contractor must ascertain and conform in all respects with the provisions of:—
338 26(2)
any general or local Act of Parliament and the Regulations and Bye-laws of any local or other statutory authority which may be applicable to the Works; and
339 26(2)
The Contractor must be responsible for the acts defaults and neglects of any sub-contractor or his agents servants or workmen as fully as if they were the acts defaults or neglects of the Contractor his agents servants or workmen
66 4
The Contractor must afford all reasonable facilities in accordance with the requirements of the Engineer, for:—
419 31(1)

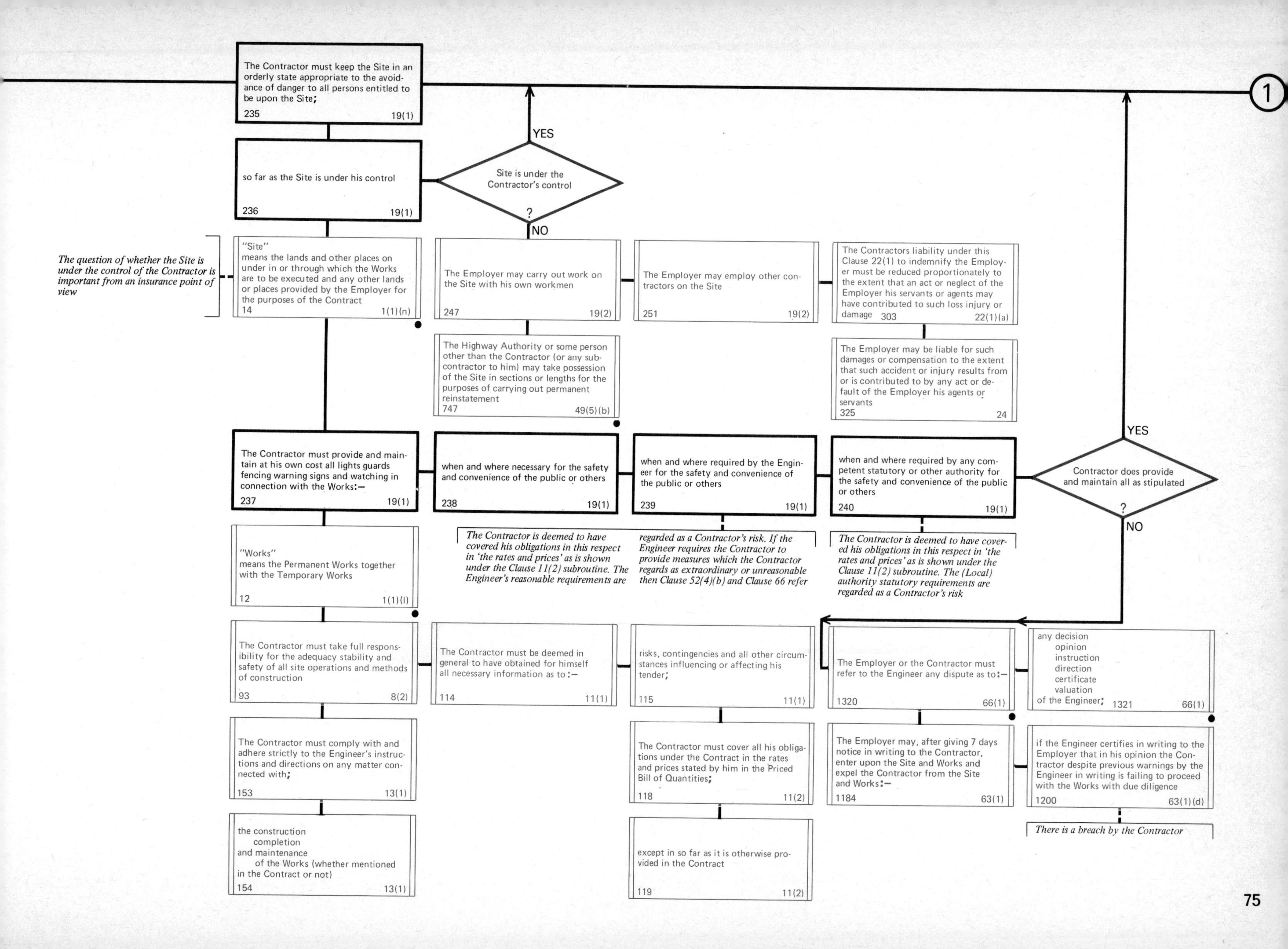
The Contractor must keep the Site in an orderly state appropriate to the avoidance of danger to all persons entitled to be upon the Site;
235 19(1)
so far as the Site is under his control
236 19(1)
Site is under the Contractor's control ?
YES
NO
"Site" means the lands and other places on under in or through which the Works are to be executed and any other lands or places provided by the Employer for the purposes of the Contract
14 1(1)(n)
The question of whether the Site is under the control of the Contractor is important from an insurance point of view
The Employer may carry out work on the Site with his own workmen
247 19(2)
The Employer may employ other contractors on the Site
251 19(2)
The Contractors liability under this Clause 22(1) to indemnify the Employer must be reduced proportionately to the extent that an act or neglect of the Employer his servants or agents may have contributed to such loss injury or damage
303 22(1)(a)
The Highway Authority or some person other than the Contractor (or any subcontractor to him) may take possession of the Site in sections or lengths for the purposes of carrying out permanent reinstatement
747 49(5)(b)
The Employer may be liable for such damages or compensation to the extent that such accident or injury results from or is contributed to by any act or default of the Employer his agents or servants
325 24
The Contractor must provide and maintain at his own cost all lights guards fencing warning signs and watching in connection with the Works:—
237 19(1)
when and where necessary for the safety and convenience of the public or others
238 19(1)
when and where required by the Engineer for the safety and convenience of the public or others
239 19(1)
when and where required by any competent statutory or other authority for the safety and convenience of the public or others
240 19(1)
Contractor does provide and maintain all as stipulated ?
YES
NO
The Contractor is deemed to have covered his obligations in this respect in 'the rates and prices' as is shown under the Clause 11(2) subroutine. The Engineer's reasonable requirements are regarded as a Contractor's risk. If the Engineer requires the Contractor to provide measures which the Contractor regards as extraordinary or unreasonable then Clause 52(4)(b) and Clause 66 refer
The Contractor is deemed to have covered his obligations in this respect in 'the rates and prices' as is shown under the Clause 11(2) subroutine. The (Local) authority statutory requirements are regarded as a Contractor's risk
"Works" means the Permanent Works together with the Temporary Works
12 1(1)(l)
The Contractor must take full responsibility for the adequacy stability and safety of all site operations and methods of construction
93 8(2)
The Contractor must be deemed in general to have obtained for himself all necessary information as to:—
114 11(1)
risks, contingencies and all other circumstances influencing or affecting his tender;
115 11(1)
The Employer or the Contractor must refer to the Engineer any dispute as to:—
1320 66(1)
any decision opinion instruction direction certificate valuation of the Engineer;
1321 66(1)
The Contractor must comply with and adhere strictly to the Engineer's instructions and directions on any matter connected with;
153 13(1)
The Contractor must cover all his obligations under the Contract in the rates and prices stated by him in the Priced Bill of Quantities;
118 11(2)
The Employer may, after giving 7 days notice in writing to the Contractor, enter upon the Site and Works and expel the Contractor from the Site and Works:—
1184 63(1)
if the Engineer certifies in writing to the Employer that in his opinion the Contractor despite previous warnings by the Engineer in writing is failing to proceed with the Works with due diligence
1200 63(1)(d)
There is a breach by the Contractor
the construction completion and maintenance of the Works (whether mentioned in the Contract or not)
154 13(1)
except in so far as it is otherwise provided in the Contract
119 11(2)
1

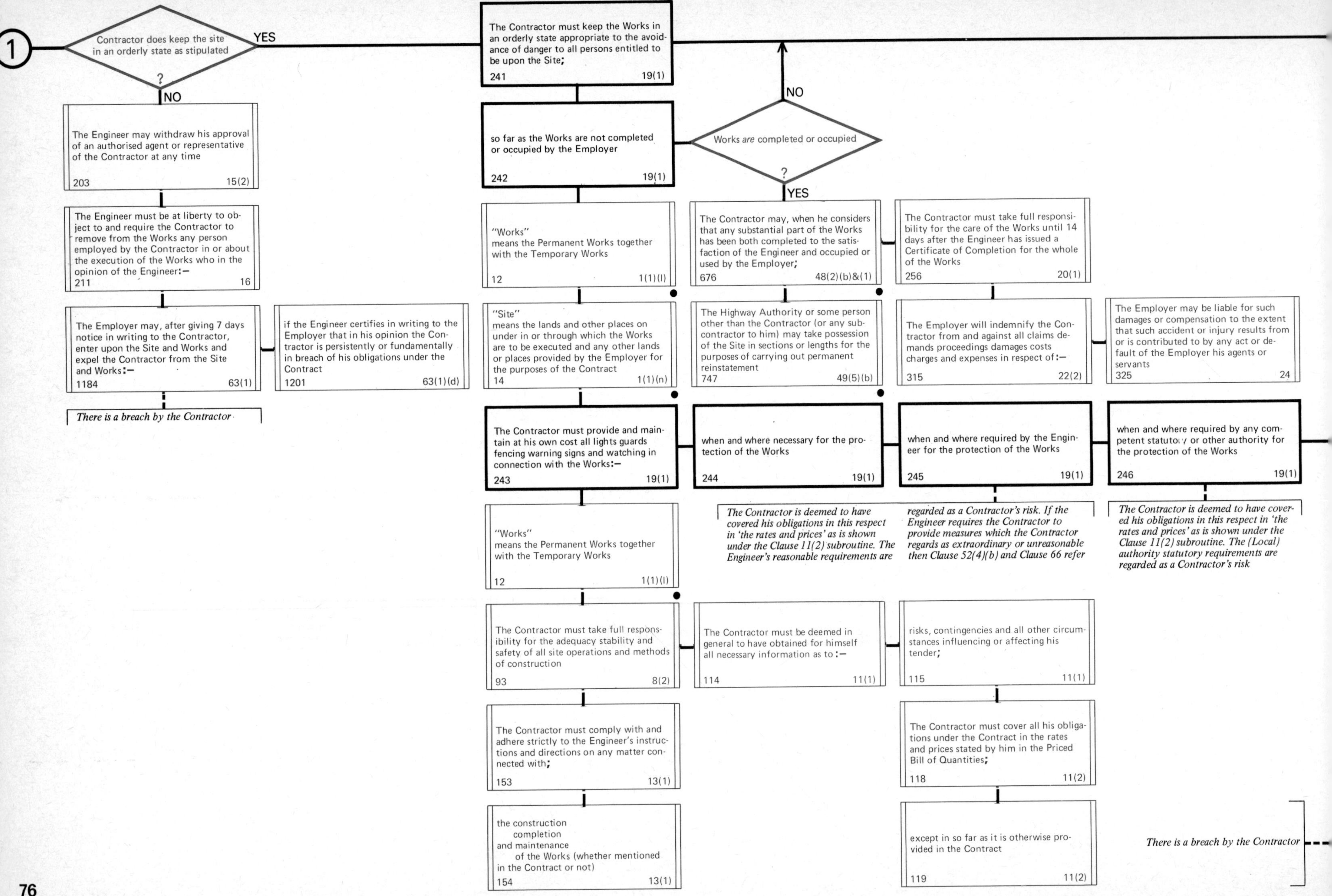

1
Contractor does keep the site in an orderly state as stipulated ?
YES
NO
The Contractor must keep the Works in an orderly state appropriate to the avoidance of danger to all persons entitled to be upon the Site; 241 19(1)
The Engineer may withdraw his approval of an authorised agent or representative of the Contractor at any time 203 15(2)
so far as the Works are not completed or occupied by the Employer 242 19(1)
Works are completed or occupied ?
The Engineer must be at liberty to object to and require the Contractor to remove from the Works any person employed by the Contractor in or about the execution of the Works who in the opinion of the Engineer:— 211 16
"Works" means the Permanent Works together with the Temporary Works 12 1(1)(l)
The Contractor may, when he considers that any substantial part of the Works has been both completed to the satisfaction of the Engineer and occupied or used by the Employer; 676 48(2)(b)&(1)
The Contractor must take full responsibility for the care of the Works until 14 days after the Engineer has issued a Certificate of Completion for the whole of the Works 256 20(1)
The Employer may, after giving 7 days notice in writing to the Contractor, enter upon the Site and Works and expel the Contractor from the Site and Works:— 1184 63(1)
if the Engineer certifies in writing to the Employer that in his opinion the Contractor is persistently or fundamentally in breach of his obligations under the Contract 1201 63(1)(d)
"Site" means the lands and other places on under in or through which the Works are to be executed and any other lands or places provided by the Employer for the purposes of the Contract 14 1(1)(n)
The Highway Authority or some person other than the Contractor (or any subcontractor to him) may take possession of the Site in sections or lengths for the purposes of carrying out permanent reinstatement 747 49(5)(b)
The Employer will indemnify the Contractor from and against all claims demands proceedings damages costs charges and expenses in respect of:— 315 22(2)
The Employer may be liable for such damages or compensation to the extent that such accident or injury results from or is contributed to by any act or default of the Employer his agents or servants 325 24
There is a breach by the Contractor
The Contractor must provide and maintain at his own cost all lights guards fencing warning signs and watching in connection with the Works:— 243 19(1)
when and where necessary for the protection of the Works 244 19(1)
when and where required by the Engineer for the protection of the Works 245 19(1)
when and where required by any competent statutory or other authority for the protection of the Works 246 19(1)
The Contractor is deemed to have covered his obligations in this respect in 'the rates and prices' as is shown under the Clause 11(2) subroutine. The Engineer's reasonable requirements are regarded as a Contractor's risk. If the Engineer requires the Contractor to provide measures which the Contractor regards as extraordinary or unreasonable then Clause 52(4)(b) and Clause 66 refer
The Contractor is deemed to have covered his obligations in this respect in 'the rates and prices' as is shown under the Clause 11(2) subroutine. The (Local) authority statutory requirements are regarded as a Contractor's risk
"Works" means the Permanent Works together with the Temporary Works 12 1(1)(l)
The Contractor must take full responsibility for the adequacy stability and safety of all site operations and methods of construction 93 8(2)
The Contractor must be deemed in general to have obtained for himself all necessary information as to:— 114 11(1)
risks, contingencies and all other circumstances influencing or affecting his tender; 115 11(1)
The Contractor must comply with and adhere strictly to the Engineer's instructions and directions on any matter connected with; 153 13(1)
The Contractor must cover all his obligations under the Contract in the rates and prices stated by him in the Priced Bill of Quantities; 118 11(2)
the construction completion and maintenance of the Works (whether mentioned in the Contract or not) 154 13(1)
except in so far as it is otherwise provided in the Contract 119 11(2)
There is a breach by the Contractor

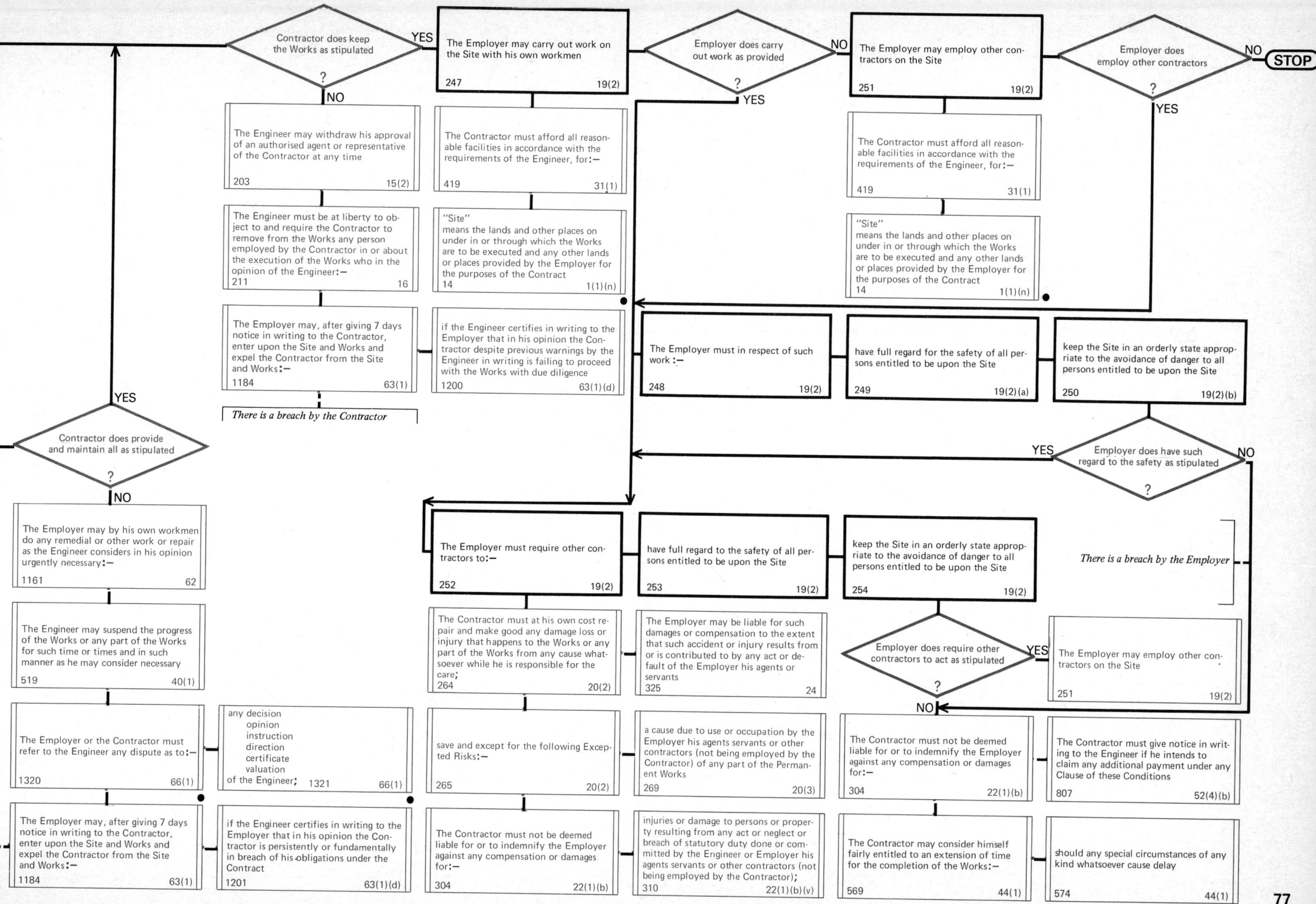

Contractor does keep the Works as stipulated ?
YES
NO
The Employer may carry out work on the Site with his own workmen 247 19(2)
Employer does carry out work as provided ?
NO
YES
The Employer may employ other contractors on the Site 251 19(2)
Employer does employ other contractors ?
NO
STOP
YES
The Engineer may withdraw his approval of an authorised agent or representative of the Contractor at any time 203 15(2)
The Contractor must afford all reasonable facilities in accordance with the requirements of the Engineer, for:— 419 31(1)
The Contractor must afford all reasonable facilities in accordance with the requirements of the Engineer, for:— 419 31(1)
The Engineer must be at liberty to object to and require the Contractor to remove from the Works any person employed by the Contractor in or about the execution of the Works who in the opinion of the Engineer:— 211 16
"Site" means the lands and other places on under in or through which the Works are to be executed and any other lands or places provided by the Employer for the purposes of the Contract 14 1(1)(n)
"Site" means the lands and other places on under in or through which the Works are to be executed and any other lands or places provided by the Employer for the purposes of the Contract 14 1(1)(n)
The Employer may, after giving 7 days notice in writing to the Contractor, enter upon the Site and Works and expel the Contractor from the Site and Works:— 1184 63(1)
if the Engineer certifies in writing to the Employer that in his opinion the Contractor despite previous warnings by the Engineer in writing is failing to proceed with the Works with due diligence 1200 63(1)(d)
The Employer must in respect of such work:— 248 19(2)
have full regard for the safety of all persons entitled to be upon the Site 249 19(2)(a)
keep the Site in an orderly state appropriate to the avoidance of danger to all persons entitled to be upon the Site 250 19(2)(b)
There is a breach by the Contractor
YES
Contractor does provide and maintain all as stipulated ?
NO
YES
Employer does have such regard to the safety as stipulated ?
NO
The Employer may by his own workmen do any remedial or other work or repair as the Engineer considers in his opinion urgently necessary:— 1161 62
The Employer must require other contractors to:— 252 19(2)
have full regard to the safety of all persons entitled to be upon the Site 253 19(2)
keep the Site in an orderly state appropriate to the avoidance of danger to all persons entitled to be upon the Site 254 19(2)
There is a breach by the Employer
The Engineer may suspend the progress of the Works or any part of the Works for such time or times and in such manner as he may consider necessary 519 40(1)
The Contractor must at his own cost repair and make good any damage loss or injury that happens to the Works or any part of the Works from any cause whatsoever while he is responsible for the care; 264 20(2)
The Employer may be liable for such damages or compensation to the extent that such accident or injury results from or is contributed to by any act or default of the Employer his agents or servants 325 24
Employer does require other contractors to act as stipulated ?
YES
NO
The Employer may employ other contractors on the Site 251 19(2)
The Employer or the Contractor must refer to the Engineer any dispute as to:— 1320 66(1)
any decision opinion instruction direction certificate valuation of the Engineer; 1321 66(1)
save and except for the following Excepted Risks:— 265 20(2)
a cause due to use or occupation by the Employer his agents servants or other contractors (not being employed by the Contractor) of any part of the Permanent Works 269 20(3)
The Contractor must not be deemed liable for or to indemnify the Employer against any compensation or damages for:— 304 22(1)(b)
The Contractor must give notice in writing to the Engineer if he intends to claim any additional payment under any Clause of these Conditions 807 52(4)(b)
The Employer may, after giving 7 days notice in writing to the Contractor, enter upon the Site and Works and expel the Contractor from the Site and Works:— 1184 63(1)
if the Engineer certifies in writing to the Employer that in his opinion the Contractor is persistently or fundamentally in breach of his obligations under the Contract 1201 63(1)(d)
The Contractor must not be deemed liable for or to indemnify the Employer against any compensation or damages for:— 304 22(1)(b)
injuries or damage to persons or property resulting from any act or neglect or breach of statutory duty done or committed by the Engineer or Employer his agents servants or other contractors (not being employed by the Contractor); 310 22(1)(b)(v)
The Contractor may consider himself fairly entitled to an extension of time for the completion of the Works:— 569 44(1)
should any special circumstances of any kind whatsoever cause delay 574 44(1)

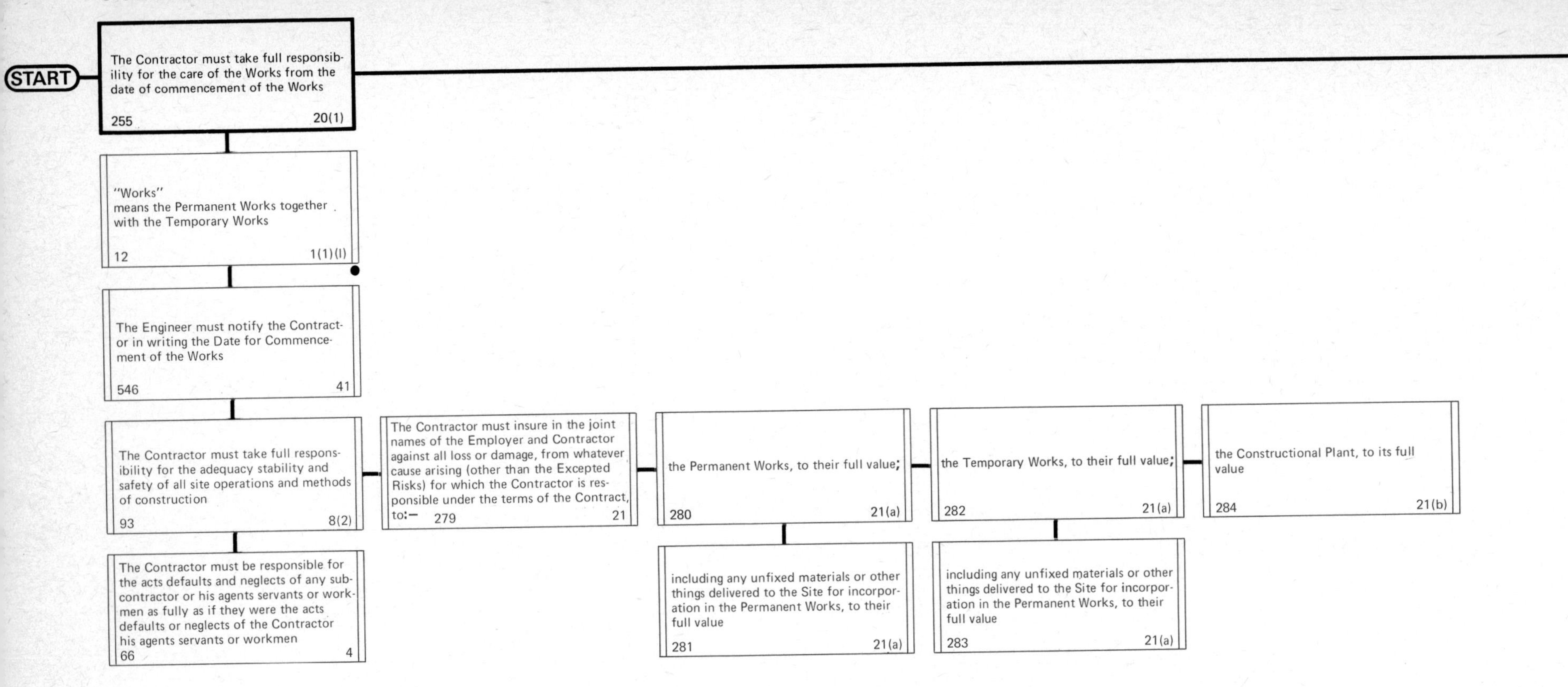
START
The Contractor must take full responsibility for the care of the Works from the date of commencement of the Works
255 20(1)
"Works"
means the Permanent Works together with the Temporary Works
12 1(1)(l)
The Engineer must notify the Contractor in writing the Date for Commencement of the Works
546 41
The Contractor must take full responsibility for the adequacy stability and safety of all site operations and methods of construction
93 8(2)
The Contractor must be responsible for the acts defaults and neglects of any sub-contractor or his agents servants or workmen as fully as if they were the acts defaults or neglects of the Contractor his agents servants or workmen
66 4
The Contractor must insure in the joint names of the Employer and Contractor against all loss or damage, from whatever cause arising (other than the Excepted Risks) for which the Contractor is responsible under the terms of the Contract, to:–
279 21
the Permanent Works, to their full value;
280 21(a)
including any unfixed materials or other things delivered to the Site for incorporation in the Permanent Works, to their full value
281 21(a)
the Temporary Works, to their full value;
282 21(a)
including any unfixed materials or other things delivered to the Site for incorporation in the Permanent Works, to their full value
283 21(a)
the Constructional Plant, to its full value
284 21(b)

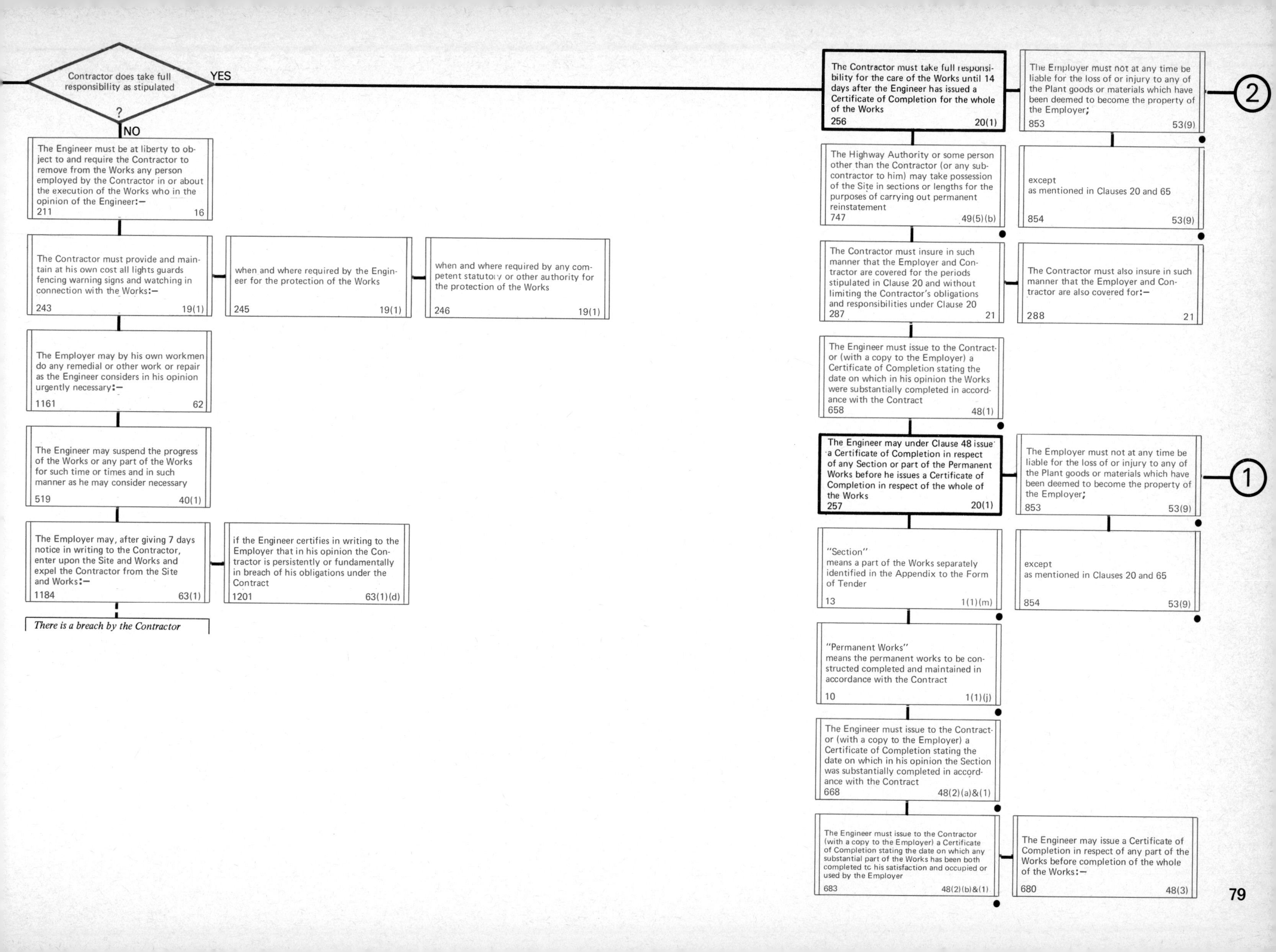
Contractor does take full responsibility as stipulated ?
YES
NO
The Engineer must be at liberty to object to and require the Contractor to remove from the Works any person employed by the Contractor in or about the execution of the Works who in the opinion of the Engineer:— 211 16
The Contractor must provide and maintain at his own cost all lights guards fencing warning signs and watching in connection with the Works:— 243 19(1)
when and where required by the Engineer for the protection of the Works 245 19(1)
when and where required by any competent statutory or other authority for the protection of the Works 246 19(1)
The Employer may by his own workmen do any remedial or other work or repair as the Engineer considers in his opinion urgently necessary:— 1161 62
The Engineer may suspend the progress of the Works or any part of the Works for such time or times and in such manner as he may consider necessary 519 40(1)
The Employer may, after giving 7 days notice in writing to the Contractor, enter upon the Site and Works and expel the Contractor from the Site and Works:— 1184 63(1)
if the Engineer certifies in writing to the Employer that in his opinion the Contractor is persistently or fundamentally in breach of his obligations under the Contract 1201 63(1)(d)
There is a breach by the Contractor
The Contractor must take full responsibility for the care of the Works until 14 days after the Engineer has issued a Certificate of Completion for the whole of the Works 256 20(1)
The Employer must not at any time be liable for the loss of or injury to any of the Plant goods or materials which have been deemed to become the property of the Employer; 853 53(9)
2
The Highway Authority or some person other than the Contractor (or any subcontractor to him) may take possession of the Site in sections or lengths for the purposes of carrying out permanent reinstatement 747 49(5)(b)
except as mentioned in Clauses 20 and 65 854 53(9)
The Contractor must insure in such manner that the Employer and Contractor are covered for the periods stipulated in Clause 20 and without limiting the Contractor's obligations and responsibilities under Clause 20 287 21
The Contractor must also insure in such manner that the Employer and Contractor are also covered for:— 288 21
The Engineer must issue to the Contractor (with a copy to the Employer) a Certificate of Completion stating the date on which in his opinion the Works were substantially completed in accordance with the Contract 658 48(1)
The Engineer may under Clause 48 issue a Certificate of Completion in respect of any Section or part of the Permanent Works before he issues a Certificate of Completion in respect of the whole of the Works 257 20(1)
The Employer must not at any time be liable for the loss of or injury to any of the Plant goods or materials which have been deemed to become the property of the Employer; 853 53(9)
1
"Section" means a part of the Works separately identified in the Appendix to the Form of Tender 13 1(1)(m)
except as mentioned in Clauses 20 and 65 854 53(9)
"Permanent Works" means the permanent works to be constructed completed and maintained in accordance with the Contract 10 1(1)(j)
The Engineer must issue to the Contractor (with a copy to the Employer) a Certificate of Completion stating the date on which in his opinion the Section was substantially completed in accordance with the Contract 668 48(2)(a)&(1)
The Engineer must issue to the Contractor (with a copy to the Employer) a Certificate of Completion stating the date on which any substantial part of the Works has been both completed to his satisfaction and occupied or used by the Employer 683 48(2)(b)&(1)
The Engineer may issue a Certificate of Completion in respect of any part of the Works before completion of the whole of the Works:— 680 48(3)

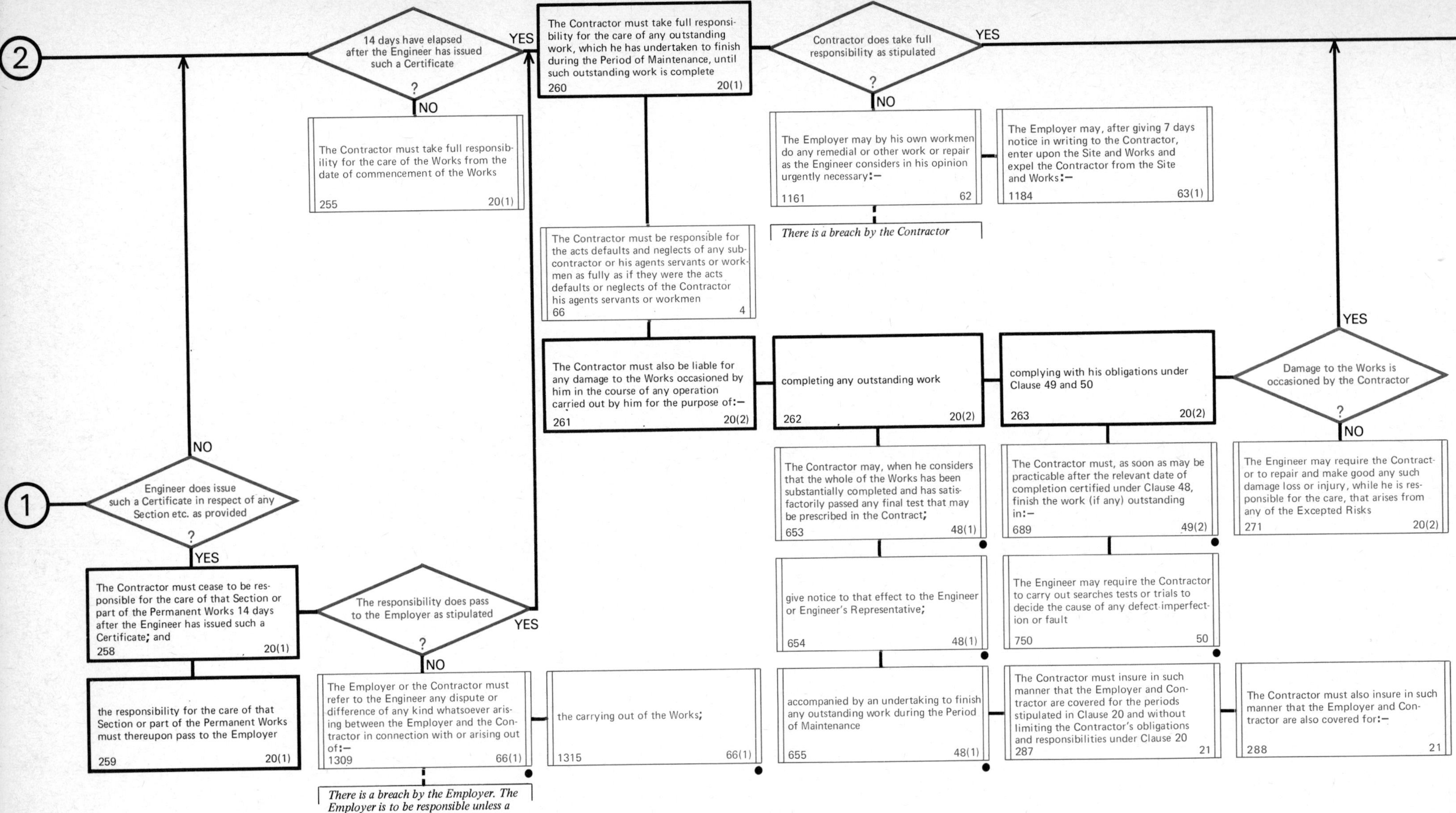
2
1
14 days have elapsed after the Engineer has issued such a Certificate ?
YES
NO
The Contractor must take full responsibility for the care of the Works from the date of commencement of the Works 255 20(1)
The Contractor must take full responsibility for the care of any outstanding work, which he has undertaken to finish during the Period of Maintenance, until such outstanding work is complete 260 20(1)
Contractor does take full responsibility as stipulated ?
YES
NO
The Employer may by his own workmen do any remedial or other work or repair as the Engineer considers in his opinion urgently necessary:— 1161 62
There is a breach by the Contractor
The Employer may, after giving 7 days notice in writing to the Contractor, enter upon the Site and Works and expel the Contractor from the Site and Works:— 1184 63(1)
The Contractor must be responsible for the acts defaults and neglects of any sub-contractor or his agents servants or workmen as fully as if they were the acts defaults or neglects of the Contractor his agents servants or workmen 66 4
The Contractor must also be liable for any damage to the Works occasioned by him in the course of any operation carried out by him for the purpose of:— 261 20(2)
completing any outstanding work 262 20(2)
complying with his obligations under Clause 49 and 50 263 20(2)
Damage to the Works is occasioned by the Contractor ?
YES
NO
The Engineer may require the Contractor to repair and make good any such damage loss or injury, while he is responsible for the care, that arises from any of the Excepted Risks 271 20(2)
The Contractor may, when he considers that the whole of the Works has been substantially completed and has satisfactorily passed any final test that may be prescribed in the Contract; 653 48(1)
The Contractor must, as soon as may be practicable after the relevant date of completion certified under Clause 48, finish the work (if any) outstanding in:— 689 49(2)
Engineer does issue such a Certificate in respect of any Section etc. as provided ?
NO
YES
The Contractor must cease to be responsible for the care of that Section or part of the Permanent Works 14 days after the Engineer has issued such a Certificate; and 258 20(1)
The responsibility does pass to the Employer as stipulated ?
YES
NO
give notice to that effect to the Engineer or Engineer's Representative; 654 48(1)
The Engineer may require the Contractor to carry out searches tests or trials to decide the cause of any defect imperfection or fault 750 50
the responsibility for the care of that Section or part of the Permanent Works must thereupon pass to the Employer 259 20(1)
The Employer or the Contractor must refer to the Engineer any dispute or difference of any kind whatsoever arising between the Employer and the Contractor in connection with or arising out of:— 1309 66(1)
There is a breach by the Employer. The Employer is to be responsible unless a special condition governs
the carrying out of the Works; 1315 66(1)
accompanied by an undertaking to finish any outstanding work during the Period of Maintenance 655 48(1)
The Contractor must insure in such manner that the Employer and Contractor are covered for the periods stipulated in Clause 20 and without limiting the Contractor's obligations and responsibilities under Clause 20 287 21
The Contractor must also insure in such manner that the Employer and Contractor are also covered for:— 288 21

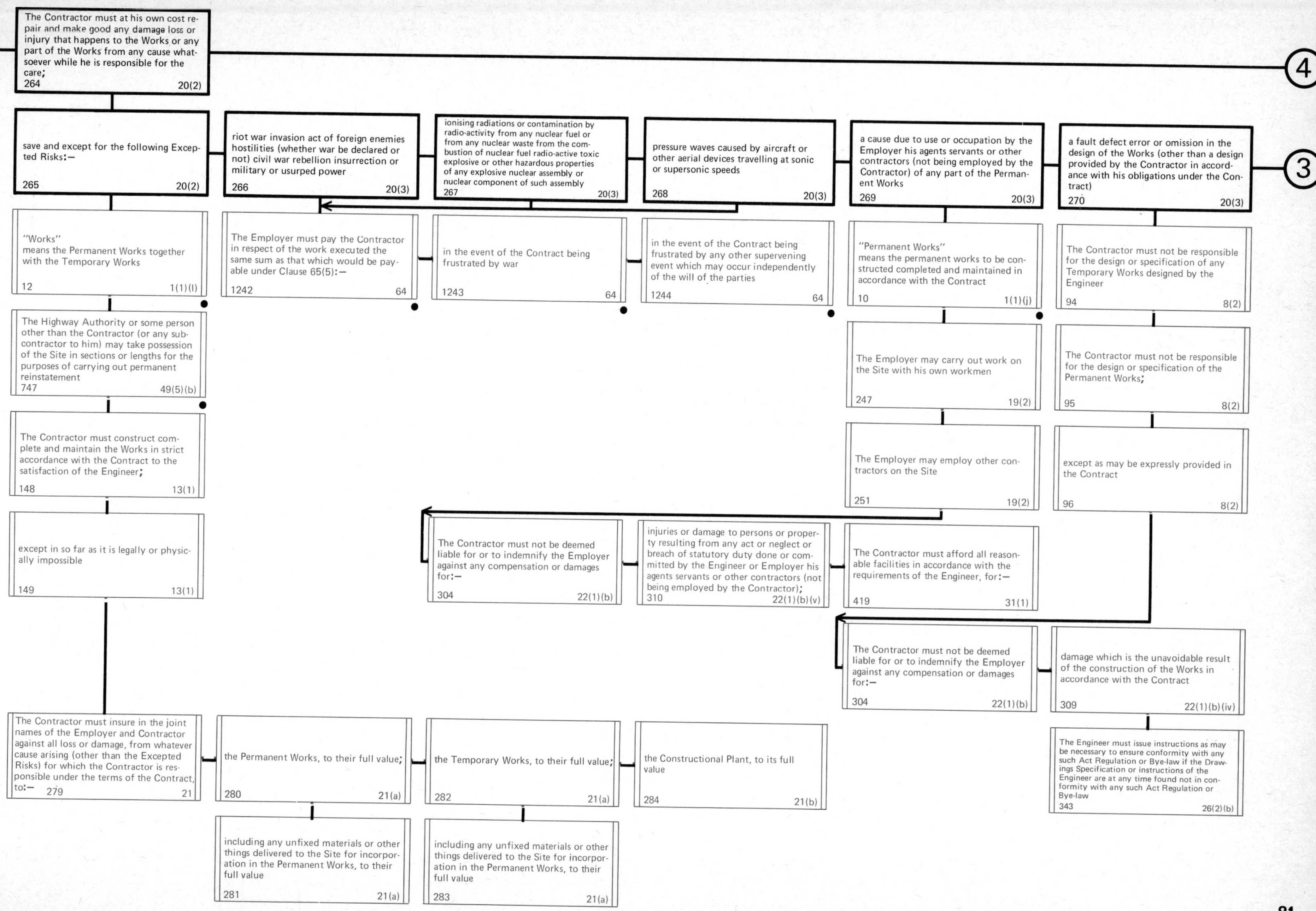

The Contractor must at his own cost repair and make good any damage loss or injury that happens to the Works or any part of the Works from any cause whatsoever while he is responsible for the care; 264 20(2)
4
save and except for the following Excepted Risks:— 265 20(2)
riot war invasion act of foreign enemies hostilities (whether war be declared or not) civil war rebellion insurrection or military or usurped power 266 20(3)
ionising radiations or contamination by radio-activity from any nuclear fuel or from any nuclear waste from the combustion of nuclear fuel radio-active toxic explosive or other hazardous properties of any explosive nuclear assembly or nuclear component of such assembly 267 20(3)
pressure waves caused by aircraft or other aerial devices travelling at sonic or supersonic speeds 268 20(3)
a cause due to use or occupation by the Employer his agents servants or other contractors (not being employed by the Contractor) of any part of the Permanent Works 269 20(3)
a fault defect error or omission in the design of the Works (other than a design provided by the Contractor in accordance with his obligations under the Contract) 270 20(3)
3
"Works" means the Permanent Works together with the Temporary Works 12 1(1)(l)
The Employer must pay the Contractor in respect of the work executed the same sum as that which would be payable under Clause 65(5):— 1242 64
in the event of the Contract being frustrated by war 1243 64
in the event of the Contract being frustrated by any other supervening event which may occur independently of the will of the parties 1244 64
"Permanent Works" means the permanent works to be constructed completed and maintained in accordance with the Contract 10 1(1)(j)
The Contractor must not be responsible for the design or specification of any Temporary Works designed by the Engineer 94 8(2)
The Highway Authority or some person other than the Contractor (or any subcontractor to him) may take possession of the Site in sections or lengths for the purposes of carrying out permanent reinstatement 747 49(5)(b)
The Employer may carry out work on the Site with his own workmen 247 19(2)
The Contractor must not be responsible for the design or specification of the Permanent Works; 95 8(2)
The Contractor must construct complete and maintain the Works in strict accordance with the Contract to the satisfaction of the Engineer; 148 13(1)
The Employer may employ other contractors on the Site 251 19(2)
except as may be expressly provided in the Contract 96 8(2)
except in so far as it is legally or physically impossible 149 13(1)
The Contractor must not be deemed liable for or to indemnify the Employer against any compensation or damages for:— 304 22(1)(b)
injuries or damage to persons or property resulting from any act or neglect or breach of statutory duty done or committed by the Engineer or Employer his agents servants or other contractors (not being employed by the Contractor); 310 22(1)(b)(v)
The Contractor must afford all reasonable facilities in accordance with the requirements of the Engineer, for:— 419 31(1)
The Contractor must not be deemed liable for or to indemnify the Employer against any compensation or damages for:— 304 22(1)(b)
damage which is the unavoidable result of the construction of the Works in accordance with the Contract 309 22(1)(b)(iv)
The Engineer must issue instructions as may be necessary to ensure conformity with any such Act Regulation or Bye-law if the Drawings Specification or instructions of the Engineer are at any time found not in conformity with any such Act Regulation or Bye-law 343 26(2)(b)
The Contractor must insure in the joint names of the Employer and Contractor against all loss or damage, from whatever cause arising (other than the Excepted Risks) for which the Contractor is responsible under the terms of the Contract, to:— 279 21
the Permanent Works, to their full value; 280 21(a)
the Temporary Works, to their full value; 282 21(a)
the Constructional Plant, to its full value 284 21(b)
including any unfixed materials or other things delivered to the Site for incorporation in the Permanent Works, to their full value 281 21(a)
including any unfixed materials or other things delivered to the Site for incorporation in the Permanent Works, to their full value 283 21(a)

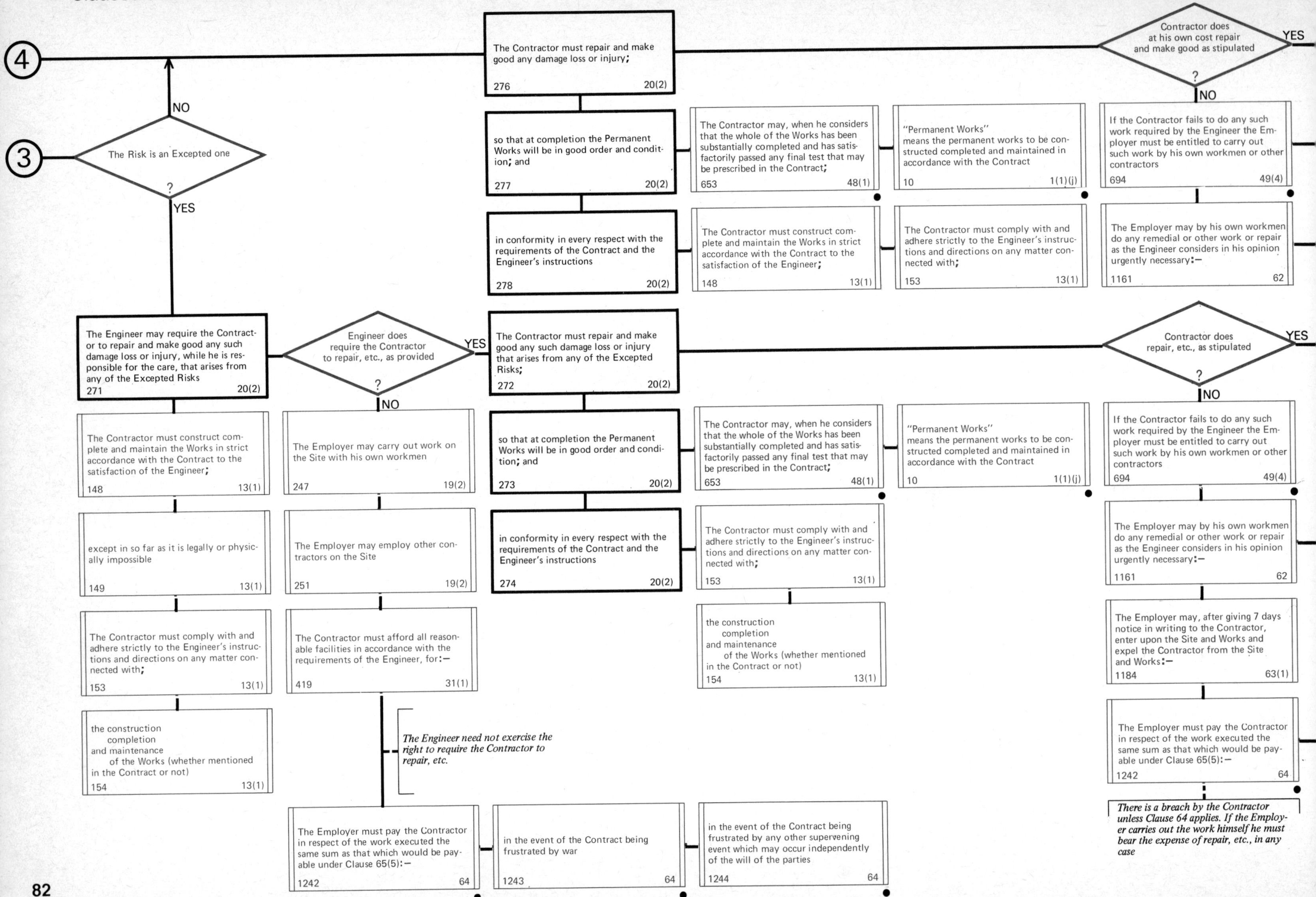
4
The Contractor must repair and make good any damage loss or injury; 276 20(2)
Contractor does at his own cost repair and make good as stipulated ?
YES
NO
so that at completion the Permanent Works will be in good order and condition; and 277 20(2)
The Contractor may, when he considers that the whole of the Works has been substantially completed and has satisfactorily passed any final test that may be prescribed in the Contract; 653 48(1)
"Permanent Works" means the permanent works to be constructed completed and maintained in accordance with the Contract 10 1(1)(j)
If the Contractor fails to do any such work required by the Engineer the Employer must be entitled to carry out such work by his own workmen or other contractors 694 49(4)
3
The Risk is an Excepted one ?
NO
YES
in conformity in every respect with the requirements of the Contract and the Engineer's instructions 278 20(2)
The Contractor must construct complete and maintain the Works in strict accordance with the Contract to the satisfaction of the Engineer; 148 13(1)
The Contractor must comply with and adhere strictly to the Engineer's instructions and directions on any matter connected with; 153 13(1)
The Employer may by his own workmen do any remedial or other work or repair as the Engineer considers in his opinion urgently necessary:— 1161 62
The Engineer may require the Contractor to repair and make good any such damage loss or injury, while he is responsible for the care, that arises from any of the Excepted Risks 271 20(2)
Engineer does require the Contractor to repair, etc., as provided ?
YES
NO
The Contractor must repair and make good any such damage loss or injury that arises from any of the Excepted Risks; 272 20(2)
Contractor does repair, etc., as stipulated ?
YES
NO
The Contractor must construct complete and maintain the Works in strict accordance with the Contract to the satisfaction of the Engineer; 148 13(1)
The Employer may carry out work on the Site with his own workmen 247 19(2)
so that at completion the Permanent Works will be in good order and condition; and 273 20(2)
The Contractor may, when he considers that the whole of the Works has been substantially completed and has satisfactorily passed any final test that may be prescribed in the Contract; 653 48(1)
"Permanent Works" means the permanent works to be constructed completed and maintained in accordance with the Contract 10 1(1)(j)
If the Contractor fails to do any such work required by the Engineer the Employer must be entitled to carry out such work by his own workmen or other contractors 694 49(4)
except in so far as it is legally or physically impossible 149 13(1)
The Employer may employ other contractors on the Site 251 19(2)
in conformity in every respect with the requirements of the Contract and the Engineer's instructions 274 20(2)
The Contractor must comply with and adhere strictly to the Engineer's instructions and directions on any matter connected with; 153 13(1)
The Employer may by his own workmen do any remedial or other work or repair as the Engineer considers in his opinion urgently necessary:— 1161 62
The Contractor must comply with and adhere strictly to the Engineer's instructions and directions on any matter connected with; 153 13(1)
The Contractor must afford all reasonable facilities in accordance with the requirements of the Engineer, for:— 419 31(1)
the construction completion and maintenance of the Works (whether mentioned in the Contract or not) 154 13(1)
The Employer may, after giving 7 days notice in writing to the Contractor, enter upon the Site and Works and expel the Contractor from the Site and Works:— 1184 63(1)
the construction completion and maintenance of the Works (whether mentioned in the Contract or not) 154 13(1)
The Engineer need not exercise the right to require the Contractor to repair, etc.
The Employer must pay the Contractor in respect of the work executed the same sum as that which would be payable under Clause 65(5):— 1242 64
There is a breach by the Contractor unless Clause 64 applies. If the Employer carries out the work himself he must bear the expense of repair, etc., in any case
The Employer must pay the Contractor in respect of the work executed the same sum as that which would be payable under Clause 65(5):— 1242 64
in the event of the Contract being frustrated by war 1243 64
in the event of the Contract being frustrated by any other supervening event which may occur independently of the will of the parties 1244 64

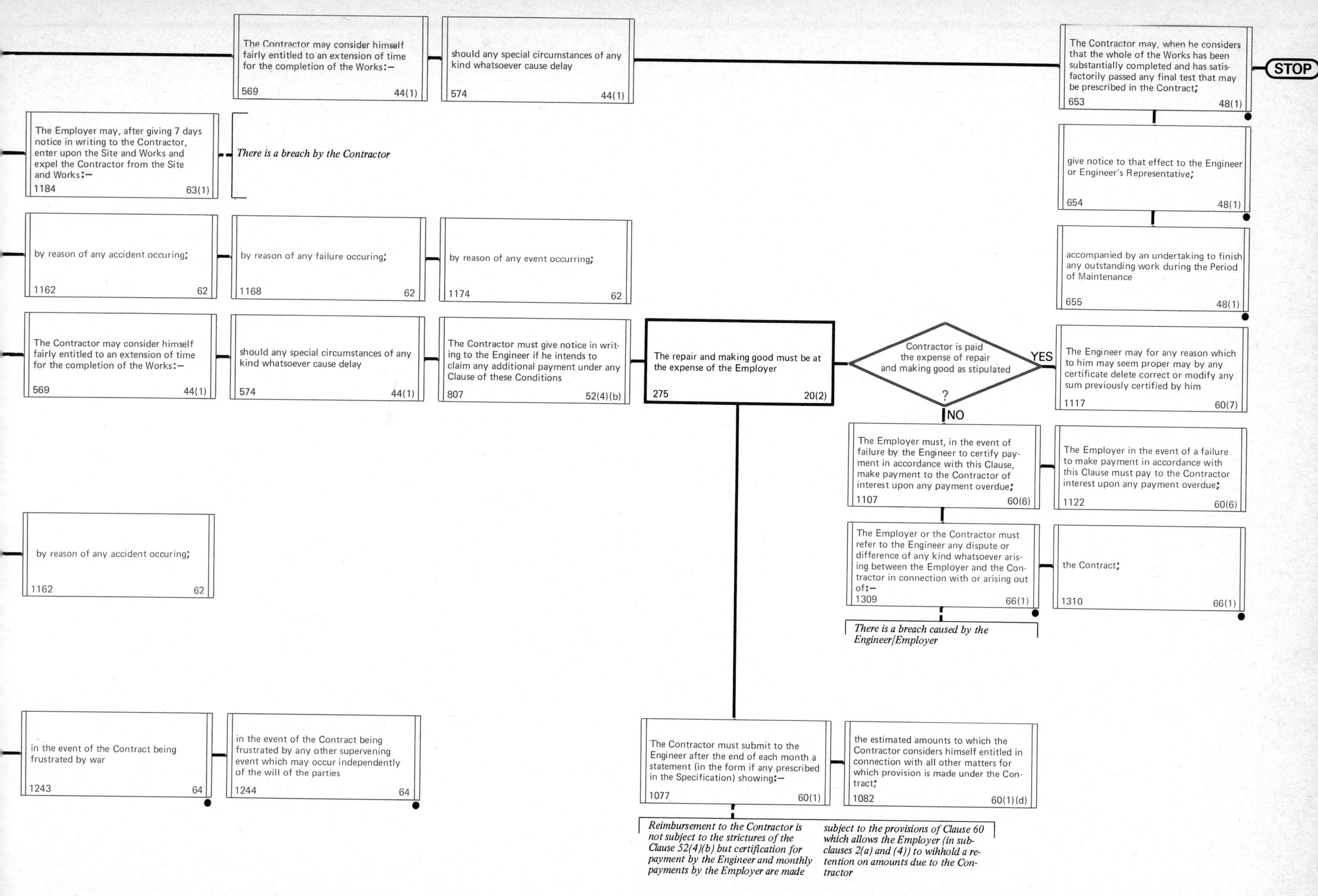
The Contractor may consider himself fairly entitled to an extension of time for the completion of the Works:—
569 44(1)
should any special circumstances of any kind whatsoever cause delay
574 44(1)
The Contractor may, when he considers that the whole of the Works has been substantially completed and has satisfactorily passed any final test that may be prescribed in the Contract;
653 48(1)
STOP
The Employer may, after giving 7 days notice in writing to the Contractor, enter upon the Site and Works and expel the Contractor from the Site and Works:—
1184 63(1)
There is a breach by the Contractor
give notice to that effect to the Engineer or Engineer's Representative;
654 48(1)
by reason of any accident occuring;
1162 62
by reason of any failure occuring;
1168 62
by reason of any event occurring;
1174 62
accompanied by an undertaking to finish any outstanding work during the Period of Maintenance
655 48(1)
The Contractor may consider himself fairly entitled to an extension of time for the completion of the Works:—
569 44(1)
should any special circumstances of any kind whatsoever cause delay
574 44(1)
The Contractor must give notice in writing to the Engineer if he intends to claim any additional payment under any Clause of these Conditions
807 52(4)(b)
The repair and making good must be at the expense of the Employer
275 20(2)
Contractor is paid the expense of repair and making good as stipulated ?
YES
NO
The Engineer may for any reason which to him may seem proper may by any certificate delete correct or modify any sum previously certified by him
1117 60(7)
The Employer must, in the event of failure by the Engineer to certify payment in accordance with this Clause, make payment to the Contractor of interest upon any payment overdue;
1107 60(6)
The Employer in the event of a failure to make payment in accordance with this Clause must pay to the Contractor interest upon any payment overdue;
1122 60(6)
by reason of any accident occuring;
1162 62
The Employer or the Contractor must refer to the Engineer any dispute or difference of any kind whatsoever arising between the Employer and the Contractor in connection with or arising out of:—
1309 66(1)
the Contract;
1310 66(1)
There is a breach caused by the Engineer/Employer
in the event of the Contract being frustrated by war
1243 64
in the event of the Contract being frustrated by any other supervening event which may occur independently of the will of the parties
1244 64
The Contractor must submit to the Engineer after the end of each month a statement (in the form if any prescribed in the Specification) showing:—
1077 60(1)
the estimated amounts to which the Contractor considers himself entitled in connection with all other matters for which provision is made under the Contract;
1082 60(1)(d)
Reimbursement to the Contractor is not subject to the strictures of the Clause 52(4)(b) but certification for payment by the Engineer and monthly payments by the Employer are made subject to the provisions of Clause 60 which allows the Employer (in sub-clauses 2(a) and (4)) to wihhold a retention on amounts due to the Contractor

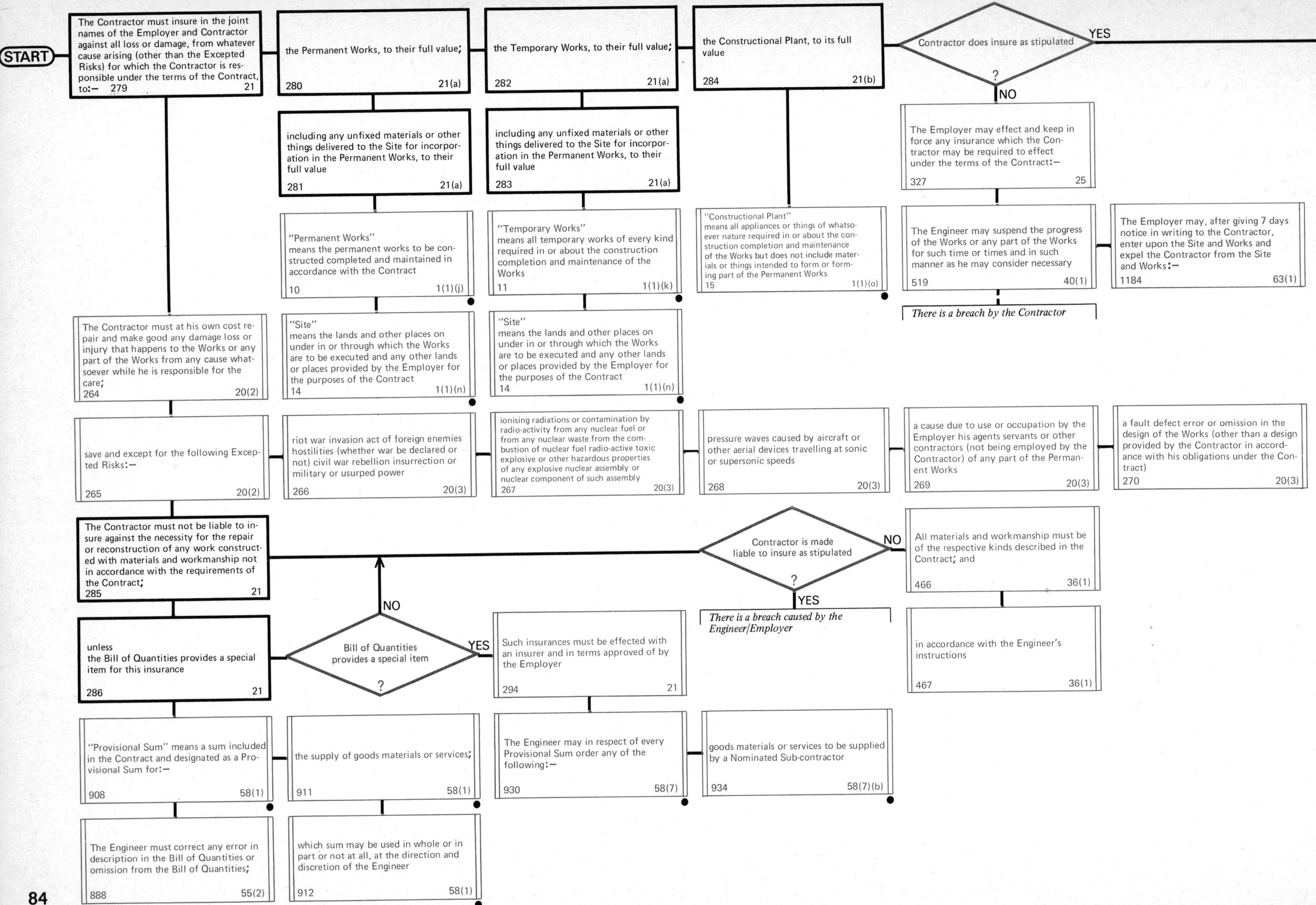
START
The Contractor must insure in the joint names of the Employer and Contractor against all loss or damage, from whatever cause arising (other than the Excepted Risks) for which the Contractor is responsible under the terms of the Contract, to:— 279 21
the Permanent Works, to their full value; 280 21(a)
the Temporary Works, to their full value; 282 21(a)
the Constructional Plant, to its full value 284 21(b)
Contractor does insure as stipulated ?
YES
NO
including any unfixed materials or other things delivered to the Site for incorporation in the Permanent Works, to their full value 281 21(a)
including any unfixed materials or other things delivered to the Site for incorporation in the Permanent Works, to their full value 283 21(a)
The Employer may effect and keep in force any insurance which the Contractor may be required to effect under the terms of the Contract:— 327 25
"Permanent Works" means the permanent works to be constructed completed and maintained in accordance with the Contract 10 1(1)(j)
"Temporary Works" means all temporary works of every kind required in or about the construction completion and maintenance of the Works 11 1(1)(k)
"Constructional Plant" means all appliances or things of whatsoever nature required in or about the construction completion and maintenance of the Works but does not include materials or things intended to form or forming part of the Permanent Works 15 1(1)(o)
The Engineer may suspend the progress of the Works or any part of the Works for such time or times and in such manner as he may consider necessary 519 40(1)
The Employer may, after giving 7 days notice in writing to the Contractor, enter upon the Site and Works and expel the Contractor from the Site and Works:— 1184 63(1)
There is a breach by the Contractor
The Contractor must at his own cost repair and make good any damage loss or injury that happens to the Works or any part of the Works from any cause whatsoever while he is responsible for the care; 264 20(2)
"Site" means the lands and other places on under in or through which the Works are to be executed and any other lands or places provided by the Employer for the purposes of the Contract 14 1(1)(n)
"Site" means the lands and other places on under in or through which the Works are to be executed and any other lands or places provided by the Employer for the purposes of the Contract 14 1(1)(n)
save and except for the following Excepted Risks:— 265 20(2)
riot war invasion act of foreign enemies hostilities (whether war be declared or not) civil war rebellion insurrection or military or usurped power 266 20(3)
ionising radiations or contamination by radio-activity from any nuclear fuel or from any nuclear waste from the combustion of nuclear fuel radio-active toxic explosive or other hazardous properties of any explosive nuclear assembly or nuclear component of such assembly 267 20(3)
pressure waves caused by aircraft or other aerial devices travelling at sonic or supersonic speeds 268 20(3)
a cause due to use or occupation by the Employer his agents servants or other contractors (not being employed by the Contractor) of any part of the Permanent Works 269 20(3)
a fault defect error or omission in the design of the Works (other than a design provided by the Contractor in accordance with his obligations under the Contract) 270 20(3)
The Contractor must not be liable to insure against the necessity for the repair or reconstruction of any work constructed with materials and workmanship not in accordance with the requirements of the Contract; 285 21
Contractor is made liable to insure as stipulated ?
NO
YES
There is a breach caused by the Engineer/Employer
All materials and workmanship must be of the respective kinds described in the Contract; and 466 36(1)
in accordance with the Engineer's instructions 467 36(1)
unless the Bill of Quantities provides a special item for this insurance 286 21
Bill of Quantities provides a special item ?
NO
YES
Such insurances must be effected with an insurer and in terms approved of by the Employer 294 21
"Provisional Sum" means a sum included in the Contract and designated as a Provisional Sum for:— 908 58(1)
the supply of goods materials or services; 911 58(1)
The Engineer may in respect of every Provisional Sum order any of the following:— 930 58(7)
goods materials or services to be supplied by a Nominated Sub-contractor 934 58(7)(b)
The Engineer must correct any error in description in the Bill of Quantities or omission from the Bill of Quantities; 888 55(2)
which sum may be used in whole or in part or not at all, at the direction and discretion of the Engineer 912 58(1)

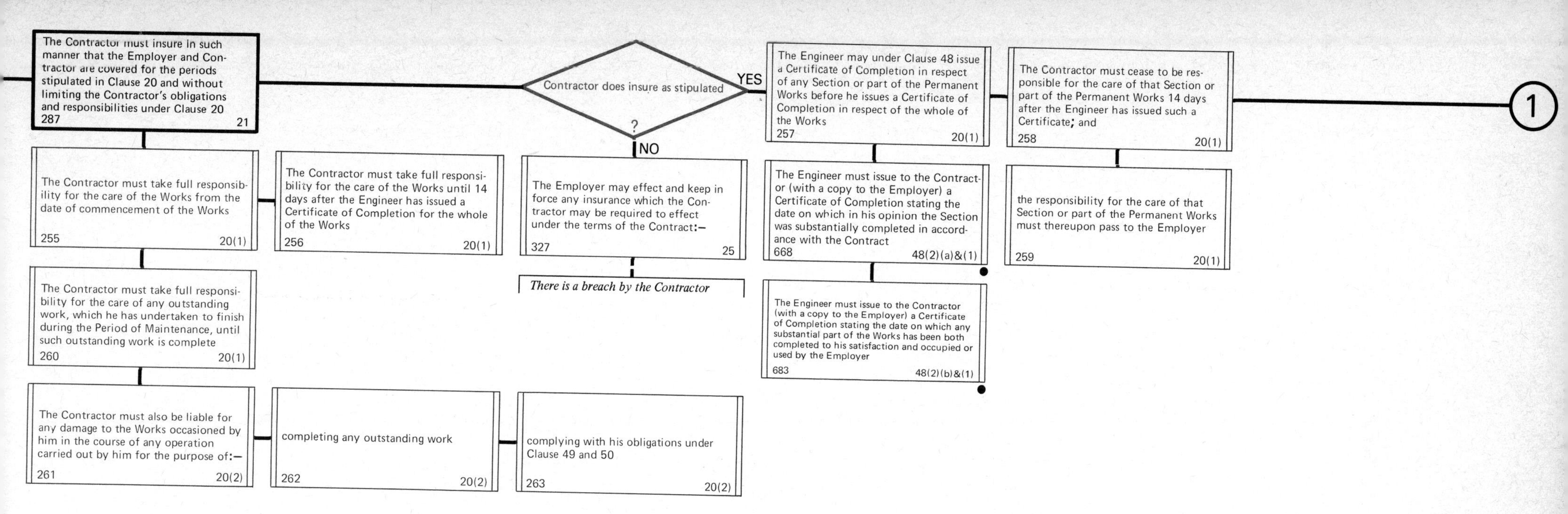

The Contractor must insure in such manner that the Employer and Contractor are covered for the periods stipulated in Clause 20 and without limiting the Contractor's obligations and responsibilities under Clause 20
287
21
Contractor does insure as stipulated
?
YES
NO
The Engineer may under Clause 48 issue a Certificate of Completion in respect of any Section or part of the Permanent Works before he issues a Certificate of Completion in respect of the whole of the Works
257
20(1)
The Contractor must cease to be responsible for the care of that Section or part of the Permanent Works 14 days after the Engineer has issued such a Certificate; and
258
20(1)
1
The Contractor must take full responsibility for the care of the Works from the date of commencement of the Works
255
20(1)
The Contractor must take full responsibility for the care of the Works until 14 days after the Engineer has issued a Certificate of Completion for the whole of the Works
256
20(1)
The Employer may effect and keep in force any insurance which the Contractor may be required to effect under the terms of the Contract:—
327
25
There is a breach by the Contractor
The Engineer must issue to the Contractor (with a copy to the Employer) a Certificate of Completion stating the date on which in his opinion the Section was substantially completed in accordance with the Contract
668
48(2)(a)&(1)
the responsibility for the care of that Section or part of the Permanent Works must thereupon pass to the Employer
259
20(1)
The Contractor must take full responsibility for the care of any outstanding work, which he has undertaken to finish during the Period of Maintenance, until such outstanding work is complete
260
20(1)
The Engineer must issue to the Contractor (with a copy to the Employer) a Certificate of Completion stating the date on which any substantial part of the Works has been both completed to his satisfaction and occupied or used by the Employer
683
48(2)(b)&(1)
The Contractor must also be liable for any damage to the Works occasioned by him in the course of any operation carried out by him for the purpose of:—
261
20(2)
completing any outstanding work
262
20(2)
complying with his obligations under Clause 49 and 50
263
20(2)

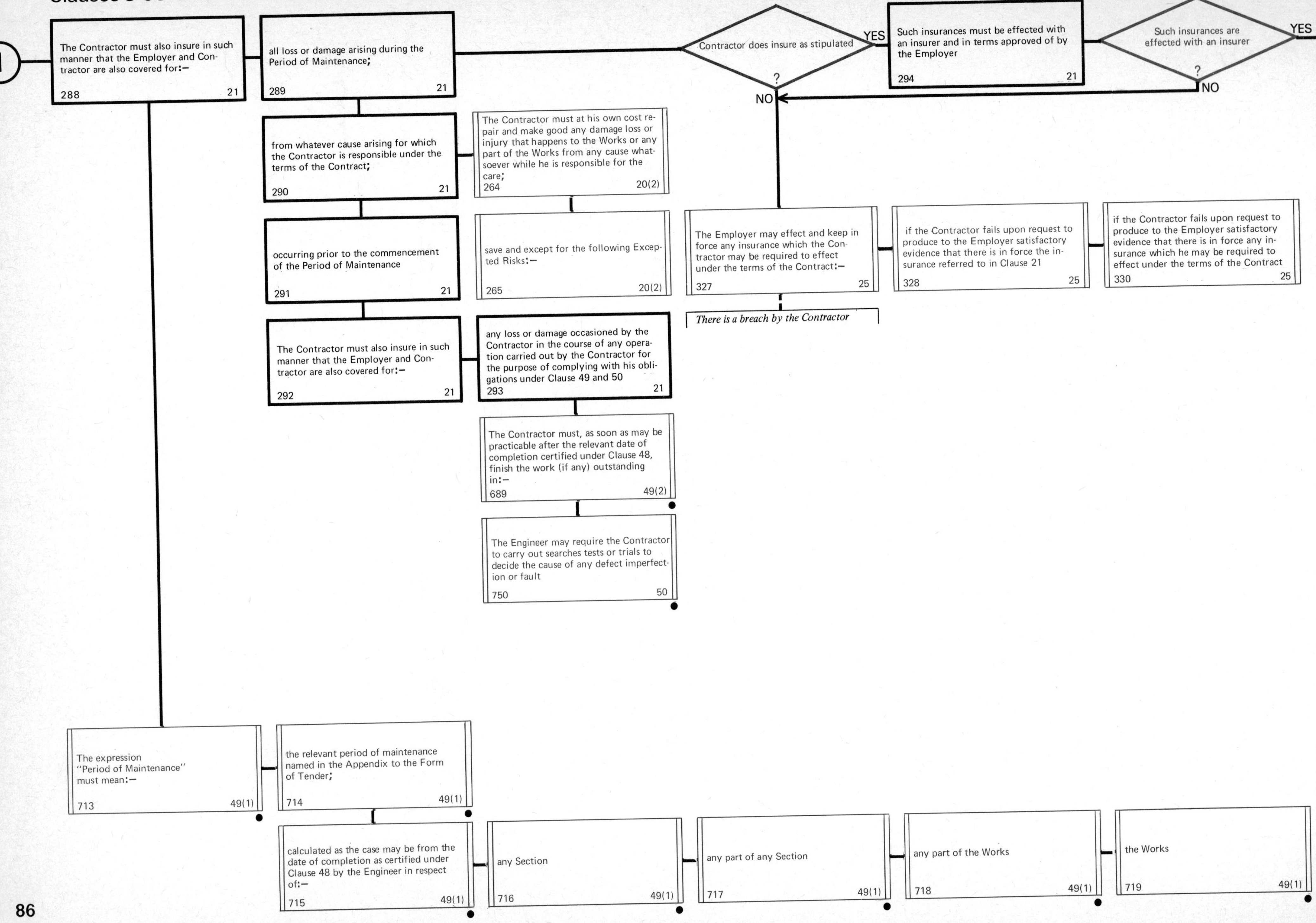
1
The Contractor must also insure in such manner that the Employer and Contractor are also covered for:—
288 21
all loss or damage arising during the Period of Maintenance;
289 21
Contractor does insure as stipulated ?
YES
NO
Such insurances must be effected with an insurer and in terms approved of by the Employer
294 21
Such insurances are effected with an insurer ?
YES
NO
from whatever cause arising for which the Contractor is responsible under the terms of the Contract;
290 21
The Contractor must at his own cost repair and make good any damage loss or injury that happens to the Works or any part of the Works from any cause whatsoever while he is responsible for the care;
264 20(2)
occurring prior to the commencement of the Period of Maintenance
291 21
save and except for the following Excepted Risks:—
265 20(2)
The Employer may effect and keep in force any insurance which the Contractor may be required to effect under the terms of the Contract:—
327 25
if the Contractor fails upon request to produce to the Employer satisfactory evidence that there is in force the insurance referred to in Clause 21
328 25
if the Contractor fails upon request to produce to the Employer satisfactory evidence that there is in force any insurance which he may be required to effect under the terms of the Contract
330 25
There is a breach by the Contractor
The Contractor must also insure in such manner that the Employer and Contractor are also covered for:—
292 21
any loss or damage occasioned by the Contractor in the course of any operation carried out by the Contractor for the purpose of complying with his obligations under Clause 49 and 50
293 21
The Contractor must, as soon as may be practicable after the relevant date of completion certified under Clause 48, finish the work (if any) outstanding in:—
689 49(2)
The Engineer may require the Contractor to carry out searches tests or trials to decide the cause of any defect imperfection or fault
750 50
The expression "Period of Maintenance" must mean:—
713 49(1)
the relevant period of maintenance named in the Appendix to the Form of Tender;
714 49(1)
calculated as the case may be from the date of completion as certified under Clause 48 by the Engineer in respect of:—
715 49(1)
any Section
716 49(1)
any part of any Section
717 49(1)
any part of the Works
718 49(1)
the Works
719 49(1)

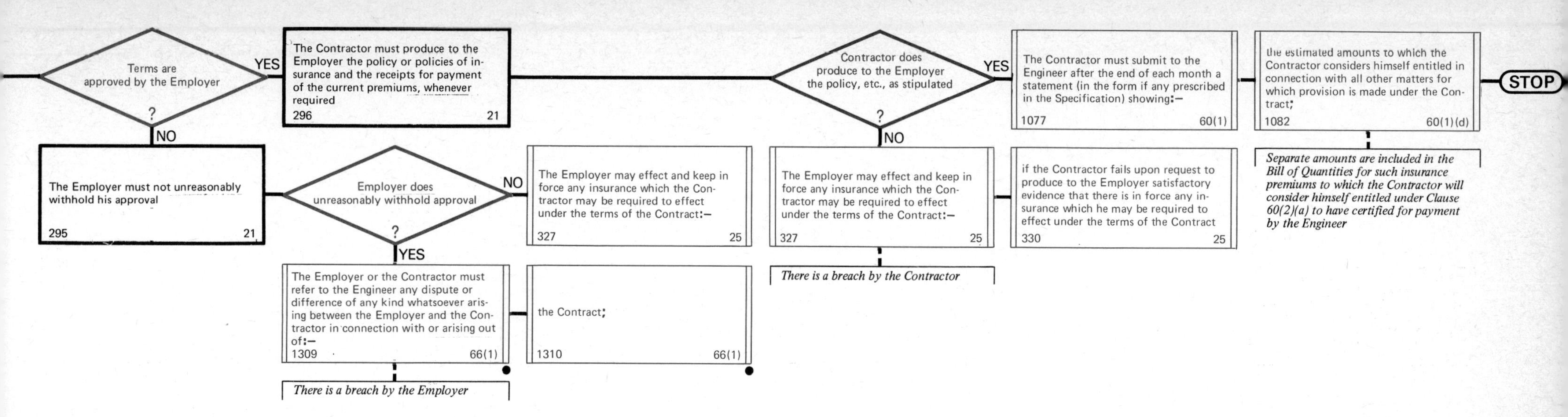
Terms are approved by the Employer
?
YES
NO
The Contractor must produce to the Employer the policy or policies of insurance and the receipts for payment of the current premiums, whenever required
296
21
Contractor does produce to the Employer the policy, etc., as stipulated
?
YES
NO
The Contractor must submit to the Engineer after the end of each month a statement (in the form if any prescribed in the Specification) showing:—
1077
60(1)
the estimated amounts to which the Contractor considers himself entitled in connection with all other matters for which provision is made under the Contract;
1082
60(1)(d)
STOP
The Employer must not unreasonably withhold his approval
295
21
Employer does unreasonably withhold approval
?
NO
YES
The Employer may effect and keep in force any insurance which the Contractor may be required to effect under the terms of the Contract:—
327
25
The Employer may effect and keep in force any insurance which the Contractor may be required to effect under the terms of the Contract:—
327
25
if the Contractor fails upon request to produce to the Employer satisfactory evidence that there is in force any insurance which he may be required to effect under the terms of the Contract
330
25
Separate amounts are included in the Bill of Quantities for such insurance premiums to which the Contractor will consider himself entitled under Clause 60(2)(a) to have certified for payment by the Engineer
The Employer or the Contractor must refer to the Engineer any dispute or difference of any kind whatsoever arising between the Employer and the Contractor in connection with or arising out of:—
1309
66(1)
the Contract;
1310
66(1)
There is a breach by the Contractor
There is a breach by the Employer

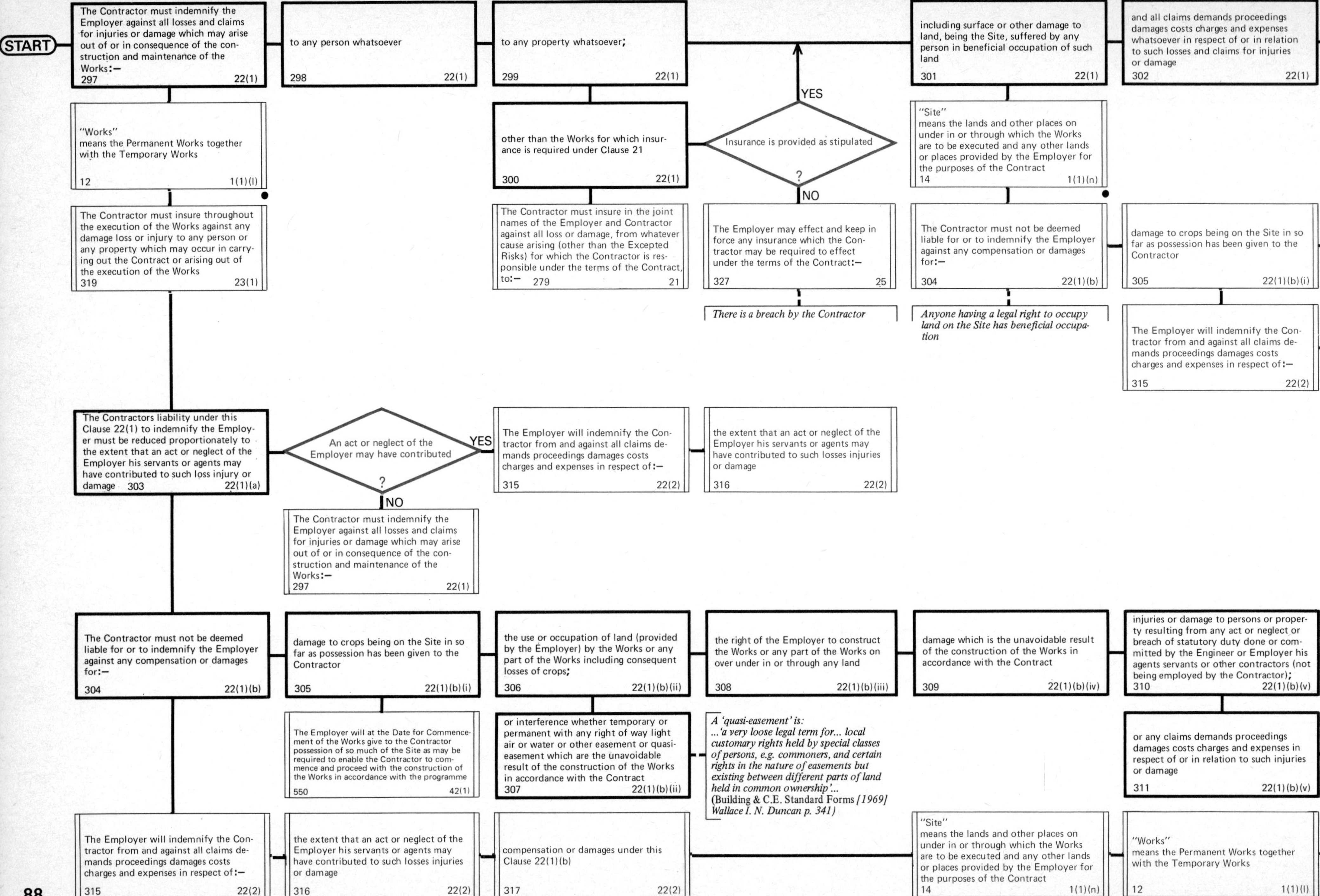
START
The Contractor must indemnify the Employer against all losses and claims for injuries or damage which may arise out of or in consequence of the construction and maintenance of the Works:— 297 22(1)
to any person whatsoever 298 22(1)
to any property whatsoever; 299 22(1)
including surface or other damage to land, being the Site, suffered by any person in beneficial occupation of such land 301 22(1)
and all claims demands proceedings damages costs charges and expenses whatsoever in respect of or in relation to such losses and claims for injuries or damage 302 22(1)
"Works" means the Permanent Works together with the Temporary Works 12 1(1)(l)
other than the Works for which insurance is required under Clause 21 300 22(1)
YES
Insurance is provided as stipulated
?
NO
"Site" means the lands and other places on under in or through which the Works are to be executed and any other lands or places provided by the Employer for the purposes of the Contract 14 1(1)(n)
The Contractor must insure throughout the execution of the Works against any damage loss or injury to any person or any property which may occur in carrying out the Contract or arising out of the execution of the Works 319 23(1)
The Contractor must insure in the joint names of the Employer and Contractor against all loss or damage, from whatever cause arising (other than the Excepted Risks) for which the Contractor is responsible under the terms of the Contract, to:— 279 21
The Employer may effect and keep in force any insurance which the Contractor may be required to effect under the terms of the Contract:— 327 25
The Contractor must not be deemed liable for or to indemnify the Employer against any compensation or damages for:— 304 22(1)(b)
damage to crops being on the Site in so far as possession has been given to the Contractor 305 22(1)(b)(i)
There is a breach by the Contractor
Anyone having a legal right to occupy land on the Site has beneficial occupation
The Employer will indemnify the Contractor from and against all claims demands proceedings damages costs charges and expenses in respect of:— 315 22(2)
The Contractors liability under this Clause 22(1) to indemnify the Employer must be reduced proportionately to the extent that an act or neglect of the Employer his servants or agents may have contributed to such loss injury or damage 303 22(1)(a)
An act or neglect of the Employer may have contributed
YES
?
NO
The Employer will indemnify the Contractor from and against all claims demands proceedings damages costs charges and expenses in respect of:— 315 22(2)
the extent that an act or neglect of the Employer his servants or agents may have contributed to such losses injuries or damage 316 22(2)
The Contractor must indemnify the Employer against all losses and claims for injuries or damage which may arise out of or in consequence of the construction and maintenance of the Works:— 297 22(1)
The Contractor must not be deemed liable for or to indemnify the Employer against any compensation or damages for:— 304 22(1)(b)
damage to crops being on the Site in so far as possession has been given to the Contractor 305 22(1)(b)(i)
the use or occupation of land (provided by the Employer) by the Works or any part of the Works including consequent losses of crops; 306 22(1)(b)(ii)
the right of the Employer to construct the Works or any part of the Works on over under in or through any land 308 22(1)(b)(iii)
damage which is the unavoidable result of the construction of the Works in accordance with the Contract 309 22(1)(b)(iv)
injuries or damage to persons or property resulting from any act or neglect or breach of statutory duty done or committed by the Engineer or Employer his agents servants or other contractors (not being employed by the Contractor); 310 22(1)(b)(v)
The Employer will at the Date for Commencement of the Works give to the Contractor possession of so much of the Site as may be required to enable the Contractor to commence and proceed with the construction of the Works in accordance with the programme 550 42(1)
or interference whether temporary or permanent with any right of way light air or water or other easement or quasi-easement which are the unavoidable result of the construction of the Works in accordance with the Contract 307 22(1)(b)(ii)
A 'quasi-easement' is: ...'a very loose legal term for... local customary rights held by special classes of persons, e.g. commoners, and certain rights in the nature of easements but existing between different parts of land held in common ownership'... (Building & C.E. Standard Forms [1969] Wallace I. N. Duncan p. 341)
or any claims demands proceedings damages costs charges and expenses in respect of or in relation to such injuries or damage 311 22(1)(b)(v)
The Employer will indemnify the Contractor from and against all claims demands proceedings damages costs charges and expenses in respect of:— 315 22(2)
the extent that an act or neglect of the Employer his servants or agents may have contributed to such losses injuries or damage 316 22(2)
compensation or damages under this Clause 22(1)(b) 317 22(2)
"Site" means the lands and other places on under in or through which the Works are to be executed and any other lands or places provided by the Employer for the purposes of the Contract 14 1(1)(n)
"Works" means the Permanent Works together with the Temporary Works 12 1(1)(l)

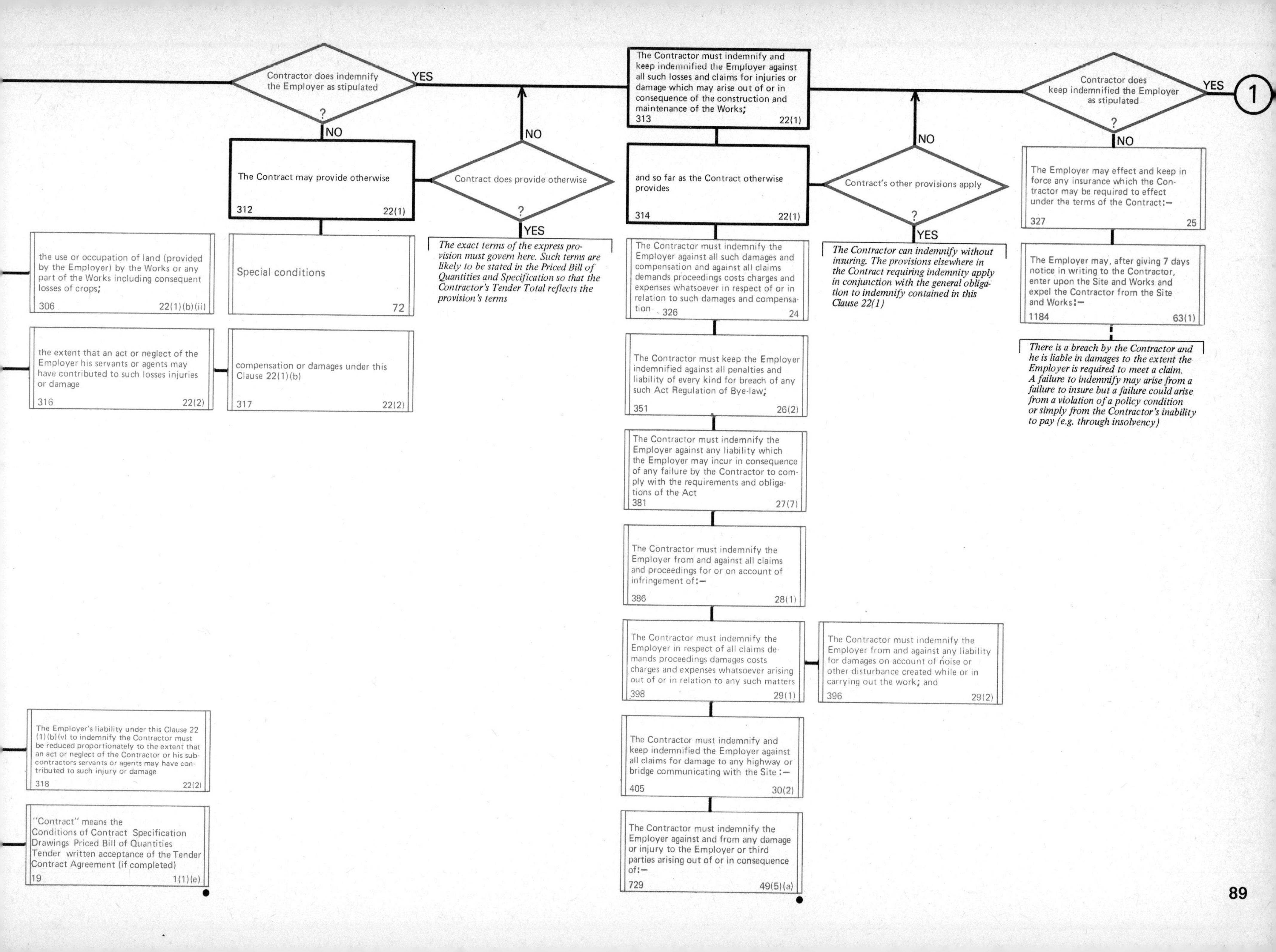

Contractor does indemnify the Employer as stipulated ?
YES
NO
The Contract may provide otherwise 312 22(1)
Contract does provide otherwise ?
NO
YES
The exact terms of the express provision must govern here. Such terms are likely to be stated in the Priced Bill of Quantities and Specification so that the Contractor's Tender Total reflects the provision's terms
The Contractor must indemnify and keep indemnified the Employer against all such losses and claims for injuries or damage which may arise out of or in consequence of the construction and maintenance of the Works; 313 22(1)
and so far as the Contract otherwise provides 314 22(1)
Contract's other provisions apply ?
NO
YES
The Contractor can indemnify without insuring. The provisions elsewhere in the Contract requiring indemnity apply in conjunction with the general obligation to indemnify contained in this Clause 22(1)
Contractor does keep indemnified the Employer as stipulated ?
YES
NO
1
The Employer may effect and keep in force any insurance which the Contractor may be required to effect under the terms of the Contract:— 327 25
The Employer may, after giving 7 days notice in writing to the Contractor, enter upon the Site and Works and expel the Contractor from the Site and Works:— 1184 63(1)
There is a breach by the Contractor and he is liable in damages to the extent the Employer is required to meet a claim. A failure to indemnify may arise from a failure to insure but a failure could arise from a violation of a policy condition or simply from the Contractor's inability to pay (e.g. through insolvency)
the use or occupation of land (provided by the Employer) by the Works or any part of the Works including consequent losses of crops; 306 22(1)(b)(ii)
Special conditions 72
the extent that an act or neglect of the Employer his servants or agents may have contributed to such losses injuries or damage 316 22(2)
compensation or damages under this Clause 22(1)(b) 317 22(2)
The Contractor must indemnify the Employer against all such damages and compensation and against all claims demands proceedings costs charges and expenses whatsoever in respect of or in relation to such damages and compensation 326 24
The Contractor must keep the Employer indemnified against all penalties and liability of every kind for breach of any such Act Regulation of Bye-law; 351 26(2)
The Contractor must indemnify the Employer against any liability which the Employer may incur in consequence of any failure by the Contractor to comply with the requirements and obligations of the Act 381 27(7)
The Contractor must indemnify the Employer from and against all claims and proceedings for or on account of infringement of:— 386 28(1)
The Contractor must indemnify the Employer in respect of all claims demands proceedings damages costs charges and expenses whatsoever arising out of or in relation to any such matters 398 29(1)
The Contractor must indemnify the Employer from and against any liability for damages on account of noise or other disturbance created while or in carrying out the work; and 396 29(2)
The Contractor must indemnify and keep indemnified the Employer against all claims for damage to any highway or bridge communicating with the Site :— 405 30(2)
The Contractor must indemnify the Employer against and from any damage or injury to the Employer or third parties arising out of or in consequence of:— 729 49(5)(a)
The Employer's liability under this Clause 22 (1)(b)(v) to indemnify the Contractor must be reduced proportionately to the extent that an act or neglect of the Contractor or his subcontractors servants or agents may have contributed to such injury or damage 318 22(2)
"Contract" means the Conditions of Contract Specification Drawings Priced Bill of Quantities Tender written acceptance of the Tender Contract Agreement (if completed) 19 1(1)(e)

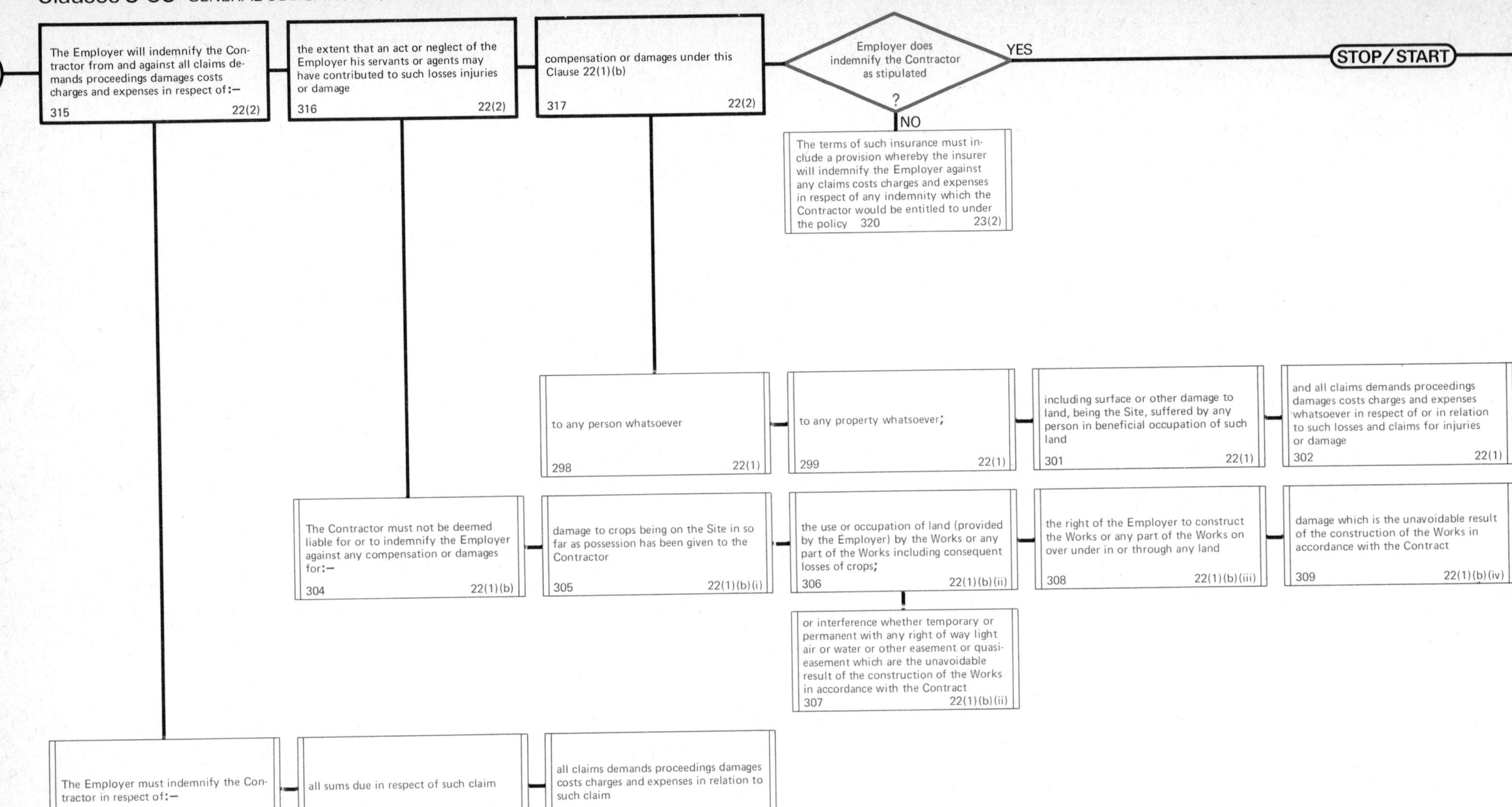
1
The Employer will indemnify the Contractor from and against all claims demands proceedings damages costs charges and expenses in respect of:— 315 22(2)
the extent that an act or neglect of the Employer his servants or agents may have contributed to such losses injuries or damage 316 22(2)
compensation or damages under this Clause 22(1)(b) 317 22(2)
Employer does indemnify the Contractor as stipulated ?
YES
STOP/START
NO
The terms of such insurance must include a provision whereby the insurer will indemnify the Employer against any claims costs charges and expenses in respect of any indemnity which the Contractor would be entitled to under the policy 320 23(2)
to any person whatsoever 298 22(1)
to any property whatsoever; 299 22(1)
including surface or other damage to land, being the Site, suffered by any person in beneficial occupation of such land 301 22(1)
and all claims demands proceedings damages costs charges and expenses whatsoever in respect of or in relation to such losses and claims for injuries or damage 302 22(1)
The Contractor must not be deemed liable for or to indemnify the Employer against any compensation or damages for:— 304 22(1)(b)
damage to crops being on the Site in so far as possession has been given to the Contractor 305 22(1)(b)(i)
the use or occupation of land (provided by the Employer) by the Works or any part of the Works including consequent losses of crops; 306 22(1)(b)(ii)
the right of the Employer to construct the Works or any part of the Works on over under in or through any land 308 22(1)(b)(iii)
damage which is the unavoidable result of the construction of the Works in accordance with the Contract 309 22(1)(b)(iv)
or interference whether temporary or permanent with any right of way light air or water or other easement or quasi-easement which are the unavoidable result of the construction of the Works in accordance with the Contract 307 22(1)(b)(ii)
The Employer must indemnify the Contractor in respect of:— 413 30(3)
all sums due in respect of such claim 414 30(3)
all claims demands proceedings damages costs charges and expenses in relation to such claim 415 30(3)

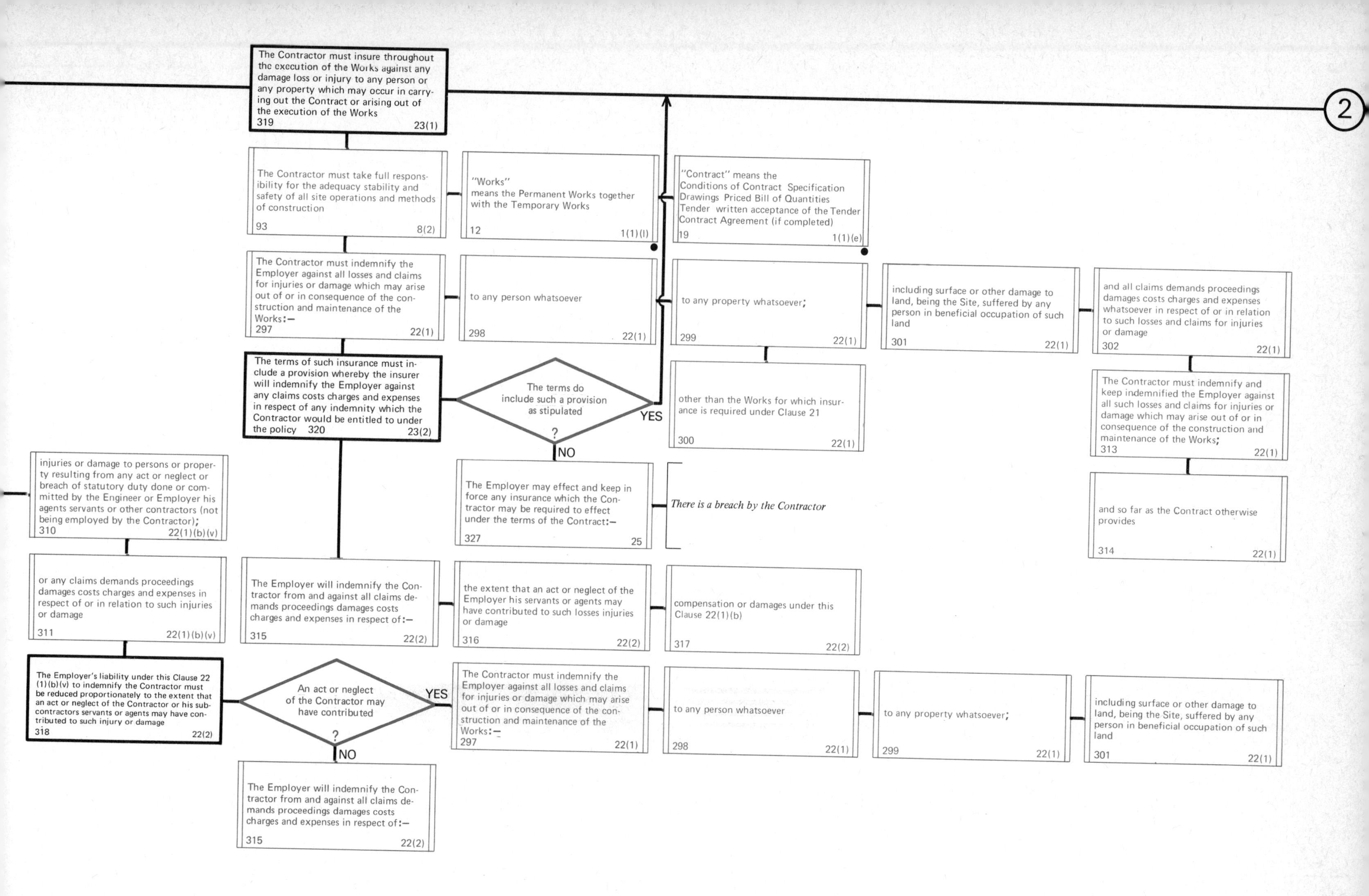

The Contractor must insure throughout the execution of the Works against any damage loss or injury to any person or any property which may occur in carrying out the Contract or arising out of the execution of the Works 319 23(1)
2
The Contractor must take full responsibility for the adequacy stability and safety of all site operations and methods of construction 93 8(2)
"Works" means the Permanent Works together with the Temporary Works 12 1(1)(l)
"Contract" means the Conditions of Contract Specification Drawings Priced Bill of Quantities Tender written acceptance of the Tender Contract Agreement (if completed) 19 1(1)(e)
The Contractor must indemnify the Employer against all losses and claims for injuries or damage which may arise out of or in consequence of the construction and maintenance of the Works:— 297 22(1)
to any person whatsoever 298 22(1)
to any property whatsoever; 299 22(1)
including surface or other damage to land, being the Site, suffered by any person in beneficial occupation of such land 301 22(1)
and all claims demands proceedings damages costs charges and expenses whatsoever in respect of or in relation to such losses and claims for injuries or damage 302 22(1)
The terms of such insurance must include a provision whereby the insurer will indemnify the Employer against any claims costs charges and expenses in respect of any indemnity which the Contractor would be entitled to under the policy 320 23(2)
The terms do include such a provision as stipulated ?
YES
NO
other than the Works for which insurance is required under Clause 21 300 22(1)
The Contractor must indemnify and keep indemnified the Employer against all such losses and claims for injuries or damage which may arise out of or in consequence of the construction and maintenance of the Works; 313 22(1)
injuries or damage to persons or property resulting from any act or neglect or breach of statutory duty done or committed by the Engineer or Employer his agents servants or other contractors (not being employed by the Contractor); 310 22(1)(b)(v)
The Employer may effect and keep in force any insurance which the Contractor may be required to effect under the terms of the Contract:— 327 25
There is a breach by the Contractor
and so far as the Contract otherwise provides 314 22(1)
or any claims demands proceedings damages costs charges and expenses in respect of or in relation to such injuries or damage 311 22(1)(b)(v)
The Employer will indemnify the Contractor from and against all claims demands proceedings damages costs charges and expenses in respect of:— 315 22(2)
the extent that an act or neglect of the Employer his servants or agents may have contributed to such losses injuries or damage 316 22(2)
compensation or damages under this Clause 22(1)(b) 317 22(2)
The Employer's liability under this Clause 22 (1)(b)(v) to indemnify the Contractor must be reduced proportionately to the extent that an act or neglect of the Contractor or his subcontractors servants or agents may have contributed to such injury or damage 318 22(2)
An act or neglect of the Contractor may have contributed ?
YES
NO
The Contractor must indemnify the Employer against all losses and claims for injuries or damage which may arise out of or in consequence of the construction and maintenance of the Works:— 297 22(1)
to any person whatsoever 298 22(1)
to any property whatsoever; 299 22(1)
including surface or other damage to land, being the Site, suffered by any person in beneficial occupation of such land 301 22(1)
The Employer will indemnify the Contractor from and against all claims demands proceedings damages costs charges and expenses in respect of:— 315 22(2)

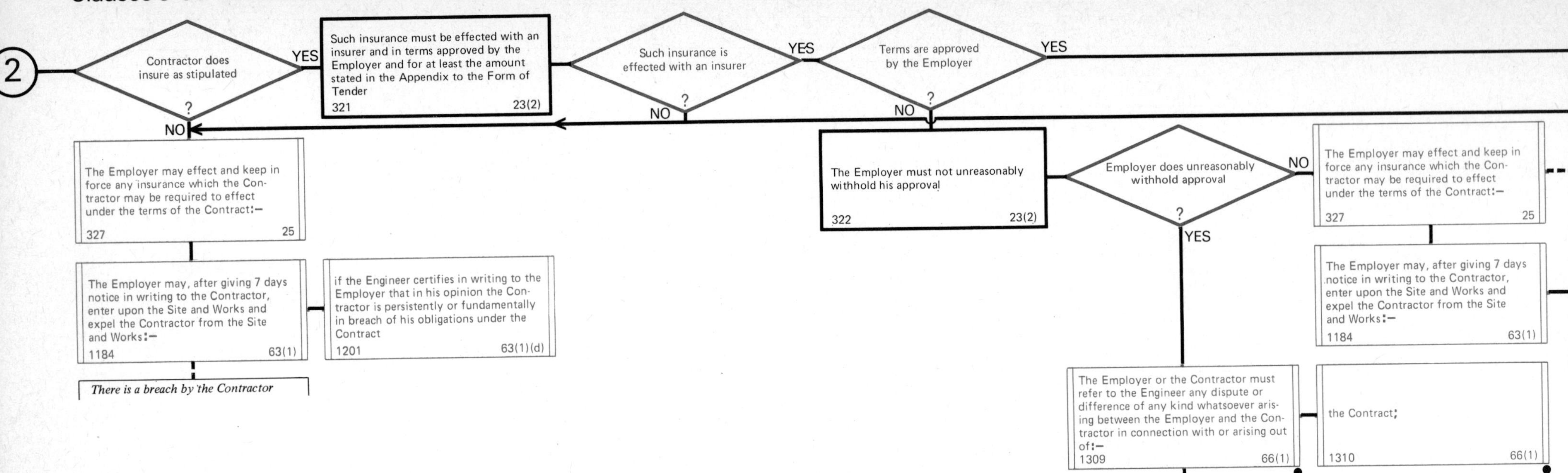
2
Contractor does insure as stipulated ?
YES
NO
Such insurance must be effected with an insurer and in terms approved by the Employer and for at least the amount stated in the Appendix to the Form of Tender
321 23(2)
Such insurance is effected with an insurer ?
YES
NO
Terms are approved by the Employer ?
YES
NO
The Employer may effect and keep in force any insurance which the Contractor may be required to effect under the terms of the Contract:—
327 25
The Employer may, after giving 7 days notice in writing to the Contractor, enter upon the Site and Works and expel the Contractor from the Site and Works:—
1184 63(1)
if the Engineer certifies in writing to the Employer that in his opinion the Contractor is persistently or fundamentally in breach of his obligations under the Contract
1201 63(1)(d)
There is a breach by the Contractor
The Employer must not unreasonably withhold his approval
322 23(2)
Employer does unreasonably withhold approval ?
NO
YES
The Employer may effect and keep in force any insurance which the Contractor may be required to effect under the terms of the Contract:—
327 25
The Employer may, after giving 7 days notice in writing to the Contractor, enter upon the Site and Works and expel the Contractor from the Site and Works:—
1184 63(1)
The Employer or the Contractor must refer to the Engineer any dispute or difference of any kind whatsoever arising between the Employer and the Contractor in connection with or arising out of:—
1309 66(1)
the Contract;
1310 66(1)
There is a breach by the Employer

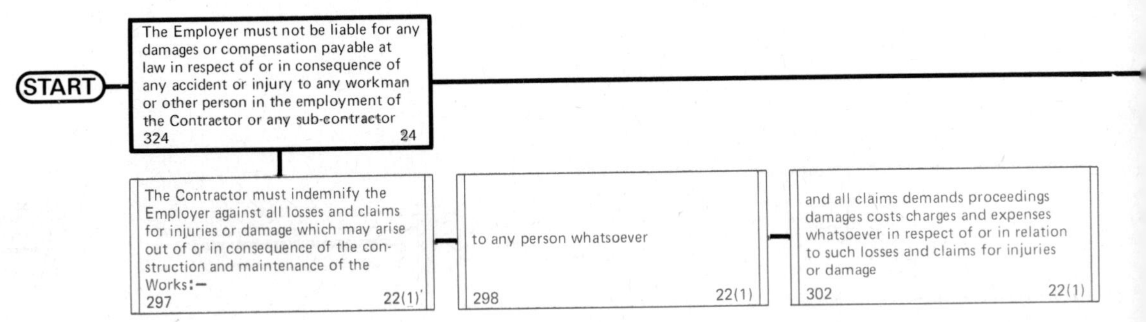
START
The Employer must not be liable for any damages or compensation payable at law in respect of or in consequence of any accident or injury to any workman or other person in the employment of the Contractor or any sub-contractor
324 24
The Contractor must indemnify the Employer against all losses and claims for injuries or damage which may arise out of or in consequence of the construction and maintenance of the Works:—
297 22(1)
to any person whatsoever
298 22(1)
and all claims demands proceedings damages costs charges and expenses whatsoever in respect of or in relation to such losses and claims for injuries or damage
302 22(1)

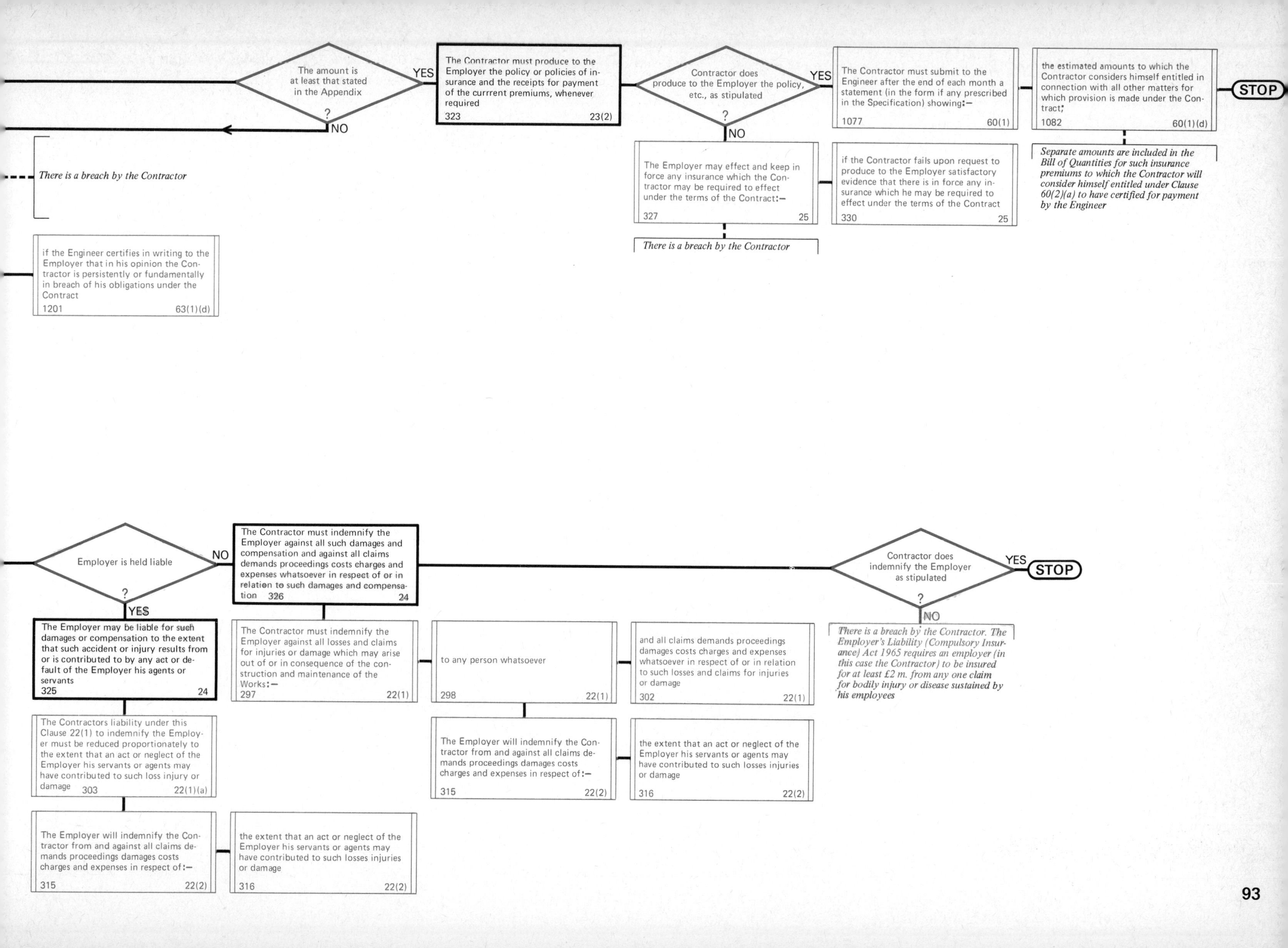
The amount is at least that stated in the Appendix ?
YES
NO
The Contractor must produce to the Employer the policy or policies of insurance and the receipts for payment of the currrent premiums, whenever required
323 23(2)
Contractor does produce to the Employer the policy, etc., as stipulated ?
YES
NO
The Contractor must submit to the Engineer after the end of each month a statement (in the form if any prescribed in the Specification) showing:—
1077 60(1)
the estimated amounts to which the Contractor considers himself entitled in connection with all other matters for which provision is made under the Contract;
1082 60(1)(d)
STOP
Separate amounts are included in the Bill of Quantities for such insurance premiums to which the Contractor will consider himself entitled under Clause 60(2)(a) to have certified for payment by the Engineer
There is a breach by the Contractor
The Employer may effect and keep in force any insurance which the Contractor may be required to effect under the terms of the Contract:—
327 25
if the Contractor fails upon request to produce to the Employer satisfactory evidence that there is in force any insurance which he may be required to effect under the terms of the Contract
330 25
There is a breach by the Contractor
if the Engineer certifies in writing to the Employer that in his opinion the Contractor is persistently or fundamentally in breach of his obligations under the Contract
1201 63(1)(d)
Employer is held liable ?
NO
YES
The Contractor must indemnify the Employer against all such damages and compensation and against all claims demands proceedings costs charges and expenses whatsoever in respect of or in relation to such damages and compensation 326 24
Contractor does indemnify the Employer as stipulated ?
YES
STOP
NO
There is a breach by the Contractor. The Employer's Liability (Compulsory Insurance) Act 1965 requires an employer (in this case the Contractor) to be insured for at least £2 m. from any one claim for bodily injury or disease sustained by his employees
The Employer may be liable for such damages or compensation to the extent that such accident or injury results from or is contributed to by any act or default of the Employer his agents or servants
325 24
The Contractor must indemnify the Employer against all losses and claims for injuries or damage which may arise out of or in consequence of the construction and maintenance of the Works:—
297 22(1)
to any person whatsoever
298 22(1)
and all claims demands proceedings damages costs charges and expenses whatsoever in respect of or in relation to such losses and claims for injuries or damage
302 22(1)
The Contractors liability under this Clause 22(1) to indemnify the Employer must be reduced proportionately to the extent that an act or neglect of the Employer his servants or agents may have contributed to such loss injury or damage 303 22(1)(a)
The Employer will indemnify the Contractor from and against all claims demands proceedings damages costs charges and expenses in respect of:—
315 22(2)
the extent that an act or neglect of the Employer his servants or agents may have contributed to such losses injuries or damage
316 22(2)
The Employer will indemnify the Contractor from and against all claims demands proceedings damages costs charges and expenses in respect of:—
315 22(2)
the extent that an act or neglect of the Employer his servants or agents may have contributed to such losses injuries or damage
316 22(2)

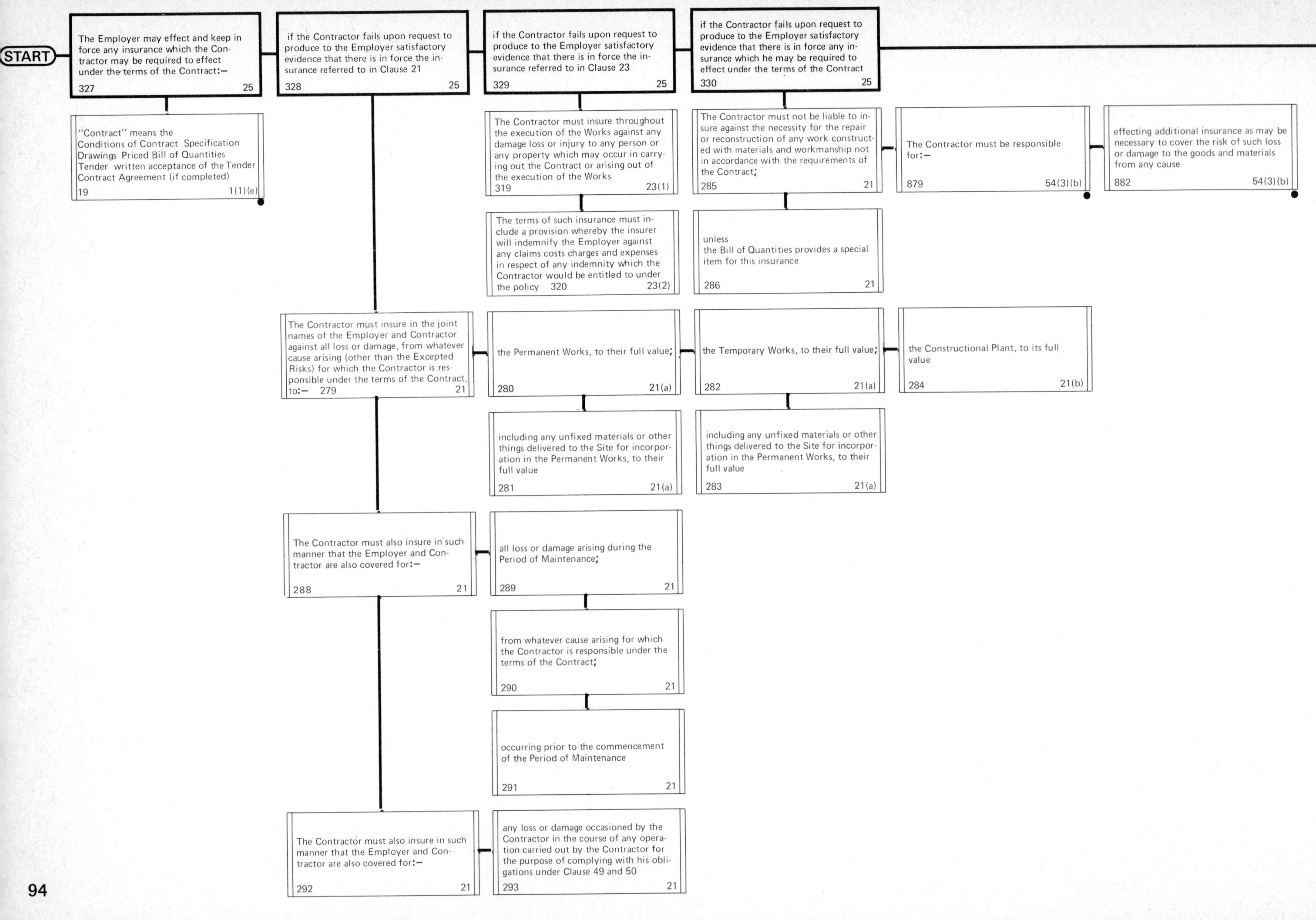
START
The Employer may effect and keep in force any insurance which the Contractor may be required to effect under the terms of the Contract:— 327 25
"Contract" means the Conditions of Contract Specification Drawings Priced Bill of Quantities Tender written acceptance of the Tender Contract Agreement (if completed) 19 1(1)(e)
if the Contractor fails upon request to produce to the Employer satisfactory evidence that there is in force the insurance referred to in Clause 21 328 25
if the Contractor fails upon request to produce to the Employer satisfactory evidence that there is in force the insurance referred to in Clause 23 329 25
if the Contractor fails upon request to produce to the Employer satisfactory evidence that there is in force any insurance which he may be required to effect under the terms of the Contract 330 25
The Contractor must insure throughout the execution of the Works against any damage loss or injury to any person or any property which may occur in carrying out the Contract or arising out of the execution of the Works 319 23(1)
The terms of such insurance must include a provision whereby the insurer will indemnify the Employer against any claims costs charges and expenses in respect of any indemnity which the Contractor would be entitled to under the policy 320 23(2)
The Contractor must not be liable to insure against the necessity for the repair or reconstruction of any work constructed with materials and workmanship not in accordance with the requirements of the Contract; 285 21
The Contractor must be responsible for:— 879 54(3)(b)
effecting additional insurance as may be necessary to cover the risk of such loss or damage to the goods and materials from any cause 882 54(3)(b)
unless the Bill of Quantities provides a special item for this insurance 286 21
The Contractor must insure in the joint names of the Employer and Contractor against all loss or damage, from whatever cause arising (other than the Excepted Risks) for which the Contractor is responsible under the terms of the Contract, to:— 279 21
the Permanent Works, to their full value; 280 21(a)
the Temporary Works, to their full value; 282 21(a)
the Constructional Plant, to its full value 284 21(b)
including any unfixed materials or other things delivered to the Site for incorporation in the Permanent Works, to their full value 281 21(a)
including any unfixed materials or other things delivered to the Site for incorporation in the Permanent Works, to their full value 283 21(a)
The Contractor must also insure in such manner that the Employer and Contractor are also covered for:— 288 21
all loss or damage arising during the Period of Maintenance; 289 21
from whatever cause arising for which the Contractor is responsible under the terms of the Contract; 290 21
occurring prior to the commencement of the Period of Maintenance 291 21
The Contractor must also insure in such manner that the Employer and Contractor are also covered for:— 292 21
any loss or damage occasioned by the Contractor in the course of any operation carried out by the Contractor for the purpose of complying with his obligations under Clause 49 and 50 293 21

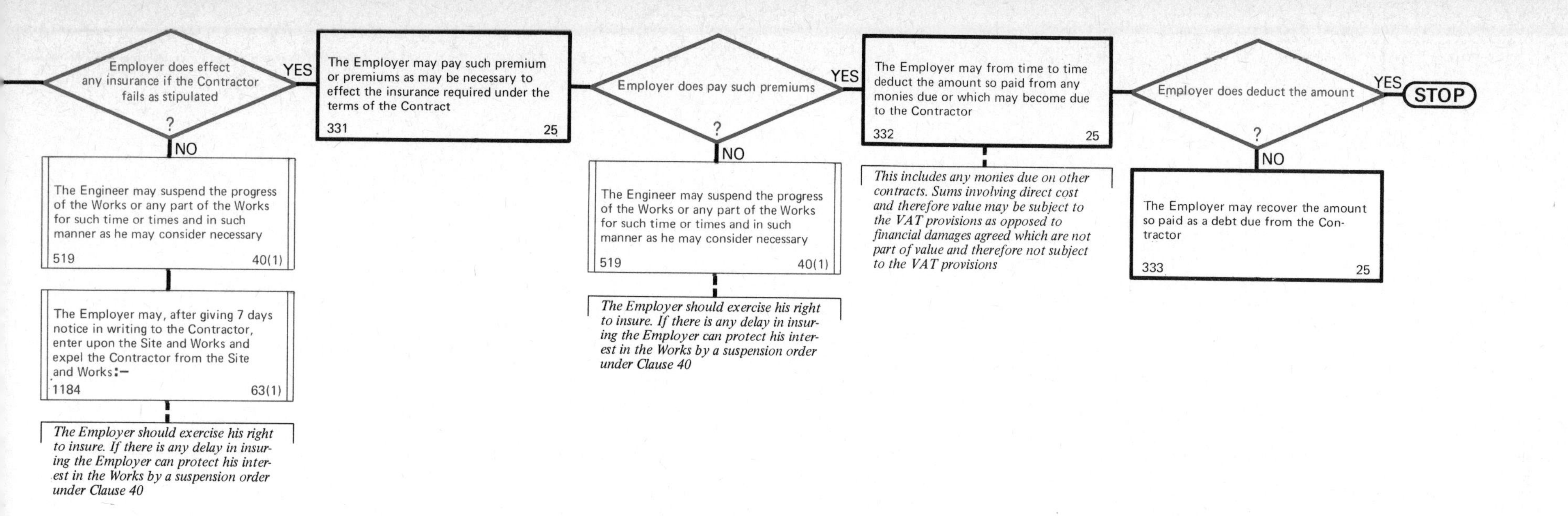
Employer does effect any insurance if the Contractor fails as stipulated
?
YES
NO
The Employer may pay such premium or premiums as may be necessary to effect the insurance required under the terms of the Contract
331 25
Employer does pay such premiums
?
YES
NO
The Employer may from time to time deduct the amount so paid from any monies due or which may become due to the Contractor
332 25
Employer does deduct the amount
?
YES
NO
STOP
The Engineer may suspend the progress of the Works or any part of the Works for such time or times and in such manner as he may consider necessary
519 40(1)
The Employer may, after giving 7 days notice in writing to the Contractor, enter upon the Site and Works and expel the Contractor from the Site and Works:—
1184 63(1)
The Employer should exercise his right to insure. If there is any delay in insuring the Employer can protect his interest in the Works by a suspension order under Clause 40
The Engineer may suspend the progress of the Works or any part of the Works for such time or times and in such manner as he may consider necessary
519 40(1)
The Employer should exercise his right to insure. If there is any delay in insuring the Employer can protect his interest in the Works by a suspension order under Clause 40
This includes any monies due on other contracts. Sums involving direct cost and therefore value may be subject to the VAT provisions as opposed to financial damages agreed which are not part of value and therefore not subject to the VAT provisions
The Employer may recover the amount so paid as a debt due from the Contractor
333 25

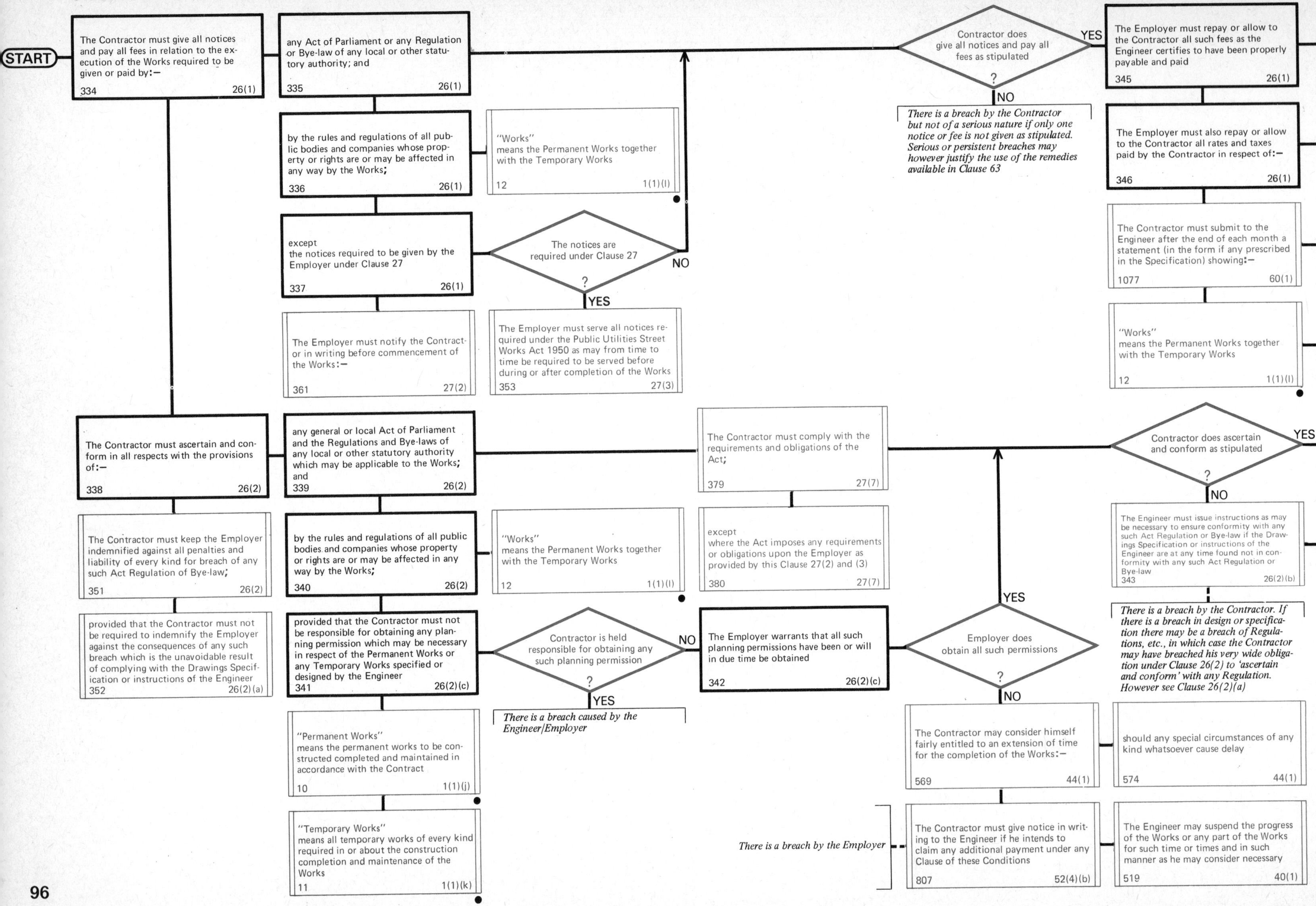
START
The Contractor must give all notices and pay all fees in relation to the execution of the Works required to be given or paid by:— 334 26(1)
any Act of Parliament or any Regulation or Bye-law of any local or other statutory authority; and 335 26(1)
by the rules and regulations of all public bodies and companies whose property or rights are or may be affected in any way by the Works; 336 26(1)
"Works" means the Permanent Works together with the Temporary Works 12 1(1)(l)
except the notices required to be given by the Employer under Clause 27 337 26(1)
The notices are required under Clause 27 ?
NO
YES
The Employer must notify the Contractor in writing before commencement of the Works:— 361 27(2)
The Employer must serve all notices required under the Public Utilities Street Works Act 1950 as may from time to time be required to be served before during or after completion of the Works 353 27(3)
Contractor does give all notices and pay all fees as stipulated ?
YES
NO
There is a breach by the Contractor but not of a serious nature if only one notice or fee is not given as stipulated. Serious or persistent breaches may however justify the use of the remedies available in Clause 63
The Employer must repay or allow to the Contractor all such fees as the Engineer certifies to have been properly payable and paid 345 26(1)
The Employer must also repay or allow to the Contractor all rates and taxes paid by the Contractor in respect of:— 346 26(1)
The Contractor must submit to the Engineer after the end of each month a statement (in the form if any prescribed in the Specification) showing:— 1077 60(1)
"Works" means the Permanent Works together with the Temporary Works 12 1(1)(l)
The Contractor must ascertain and conform in all respects with the provisions of:— 338 26(2)
any general or local Act of Parliament and the Regulations and Bye-laws of any local or other statutory authority which may be applicable to the Works; and 339 26(2)
The Contractor must comply with the requirements and obligations of the Act; 379 27(7)
except where the Act imposes any requirements or obligations upon the Employer as provided by this Clause 27(2) and (3) 380 27(7)
Contractor does ascertain and conform as stipulated ?
YES
NO
The Engineer must issue instructions as may be necessary to ensure conformity with any such Act Regulation or Bye-law if the Drawings Specification or instructions of the Engineer are at any time found not in conformity with any such Act Regulation or Bye-law 343 26(2)(b)
There is a breach by the Contractor. If there is a breach in design or specification there may be a breach of Regulations, etc., in which case the Contractor may have breached his very wide obligation under Clause 26(2) to 'ascertain and conform' with any Regulation. However see Clause 26(2)(a)
The Contractor must keep the Employer indemnified against all penalties and liability of every kind for breach of any such Act Regulation of Bye-law; 351 26(2)
provided that the Contractor must not be required to indemnify the Employer against the consequences of any such breach which is the unavoidable result of complying with the Drawings Specification or instructions of the Engineer 352 26(2)(a)
by the rules and regulations of all public bodies and companies whose property or rights are or may be affected in any way by the Works; 340 26(2)
"Works" means the Permanent Works together with the Temporary Works 12 1(1)(l)
provided that the Contractor must not be responsible for obtaining any planning permission which may be necessary in respect of the Permanent Works or any Temporary Works specified or designed by the Engineer 341 26(2)(c)
Contractor is held responsible for obtaining any such planning permission ?
NO
YES
There is a breach caused by the Engineer/Employer
The Employer warrants that all such planning permissions have been or will in due time be obtained 342 26(2)(c)
Employer does obtain all such permissions ?
YES
NO
"Permanent Works" means the permanent works to be constructed completed and maintained in accordance with the Contract 10 1(1)(j)
"Temporary Works" means all temporary works of every kind required in or about the construction completion and maintenance of the Works 11 1(1)(k)
The Contractor may consider himself fairly entitled to an extension of time for the completion of the Works:— 569 44(1)
should any special circumstances of any kind whatsoever cause delay 574 44(1)
There is a breach by the Employer
The Contractor must give notice in writing to the Engineer if he intends to claim any additional payment under any Clause of these Conditions 807 52(4)(b)
The Engineer may suspend the progress of the Works or any part of the Works for such time or times and in such manner as he may consider necessary 519 40(1)

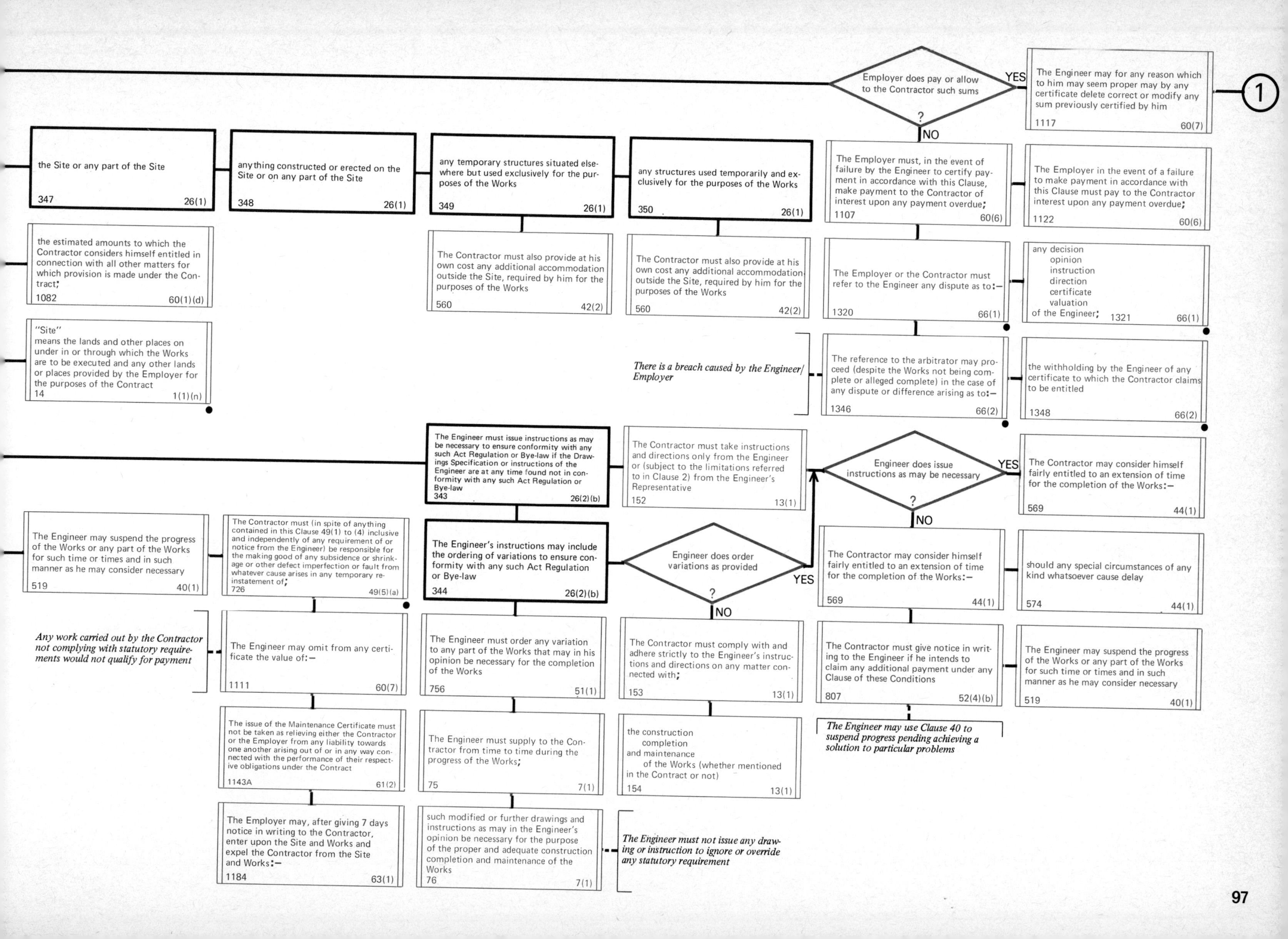

Employer does pay or allow to the Contractor such sums
YES
?
NO
The Engineer may for any reason which to him may seem proper may by any certificate delete correct or modify any sum previously certified by him
1117 60(7)
1
the Site or any part of the Site
347 26(1)
anything constructed or erected on the Site or on any part of the Site
348 26(1)
any temporary structures situated elsewhere but used exclusively for the purposes of the Works
349 26(1)
any structures used temporarily and exclusively for the purposes of the Works
350 26(1)
The Employer must, in the event of failure by the Engineer to certify payment in accordance with this Clause, make payment to the Contractor of interest upon any payment overdue;
1107 60(6)
The Employer in the event of a failure to make payment in accordance with this Clause must pay to the Contractor interest upon any payment overdue;
1122 60(6)
the estimated amounts to which the Contractor considers himself entitled in connection with all other matters for which provision is made under the Contract;
1082 60(1)(d)
The Contractor must also provide at his own cost any additional accommodation outside the Site, required by him for the purposes of the Works
560 42(2)
The Contractor must also provide at his own cost any additional accommodation outside the Site, required by him for the purposes of the Works
560 42(2)
The Employer or the Contractor must refer to the Engineer any dispute as to:—
1320 66(1)
any decision opinion instruction direction certificate valuation of the Engineer;
1321 66(1)
"Site" means the lands and other places on under in or through which the Works are to be executed and any other lands or places provided by the Employer for the purposes of the Contract
14 1(1)(n)
There is a breach caused by the Engineer/Employer
The reference to the arbitrator may proceed (despite the Works not being complete or alleged complete) in the case of any dispute or difference arising as to:—
1346 66(2)
the withholding by the Engineer of any certificate to which the Contractor claims to be entitled
1348 66(2)
The Engineer must issue instructions as may be necessary to ensure conformity with any such Act Regulation or Bye-law if the Drawings Specification or instructions of the Engineer are at any time found not in conformity with any such Act Regulation or Bye-law
343 26(2)(b)
The Contractor must take instructions and directions only from the Engineer or (subject to the limitations referred to in Clause 2) from the Engineer's Representative
152 13(1)
Engineer does issue instructions as may be necessary
YES
?
NO
The Contractor may consider himself fairly entitled to an extension of time for the completion of the Works:—
569 44(1)
The Engineer may suspend the progress of the Works or any part of the Works for such time or times and in such manner as he may consider necessary
519 40(1)
The Contractor must (in spite of anything contained in this Clause 49(1) to (4) inclusive and independently of any requirement of or notice from the Engineer) be responsible for the making good of any subsidence or shrinkage or other defect imperfection or fault from whatever cause arises in any temporary reinstatement of;
726 49(5)(a)
The Engineer's instructions may include the ordering of variations to ensure conformity with any such Act Regulation or Bye-law
344 26(2)(b)
Engineer does order variations as provided
YES
?
NO
The Contractor may consider himself fairly entitled to an extension of time for the completion of the Works:—
569 44(1)
should any special circumstances of any kind whatsoever cause delay
574 44(1)
Any work carried out by the Contractor not complying with statutory requirements would not qualify for payment
The Engineer may omit from any certificate the value of:—
1111 60(7)
The Engineer must order any variation to any part of the Works that may in his opinion be necessary for the completion of the Works
756 51(1)
The Contractor must comply with and adhere strictly to the Engineer's instructions and directions on any matter connected with;
153 13(1)
The Contractor must give notice in writing to the Engineer if he intends to claim any additional payment under any Clause of these Conditions
807 52(4)(b)
The Engineer may suspend the progress of the Works or any part of the Works for such time or times and in such manner as he may consider necessary
519 40(1)
The Engineer may use Clause 40 to suspend progress pending achieving a solution to particular problems
The issue of the Maintenance Certificate must not be taken as relieving either the Contractor or the Employer from any liability towards one another arising out of or in any way connected with the performance of their respective obligations under the Contract
1143A 61(2)
The Engineer must supply to the Contractor from time to time during the progress of the Works;
75 7(1)
the construction completion and maintenance of the Works (whether mentioned in the Contract or not)
154 13(1)
The Employer may, after giving 7 days notice in writing to the Contractor, enter upon the Site and Works and expel the Contractor from the Site and Works:—
1184 63(1)
such modified or further drawings and instructions as may in the Engineer's opinion be necessary for the purpose of the proper and adequate construction completion and maintenance of the Works
76 7(1)
The Engineer must not issue any drawing or instruction to ignore or override any statutory requirement

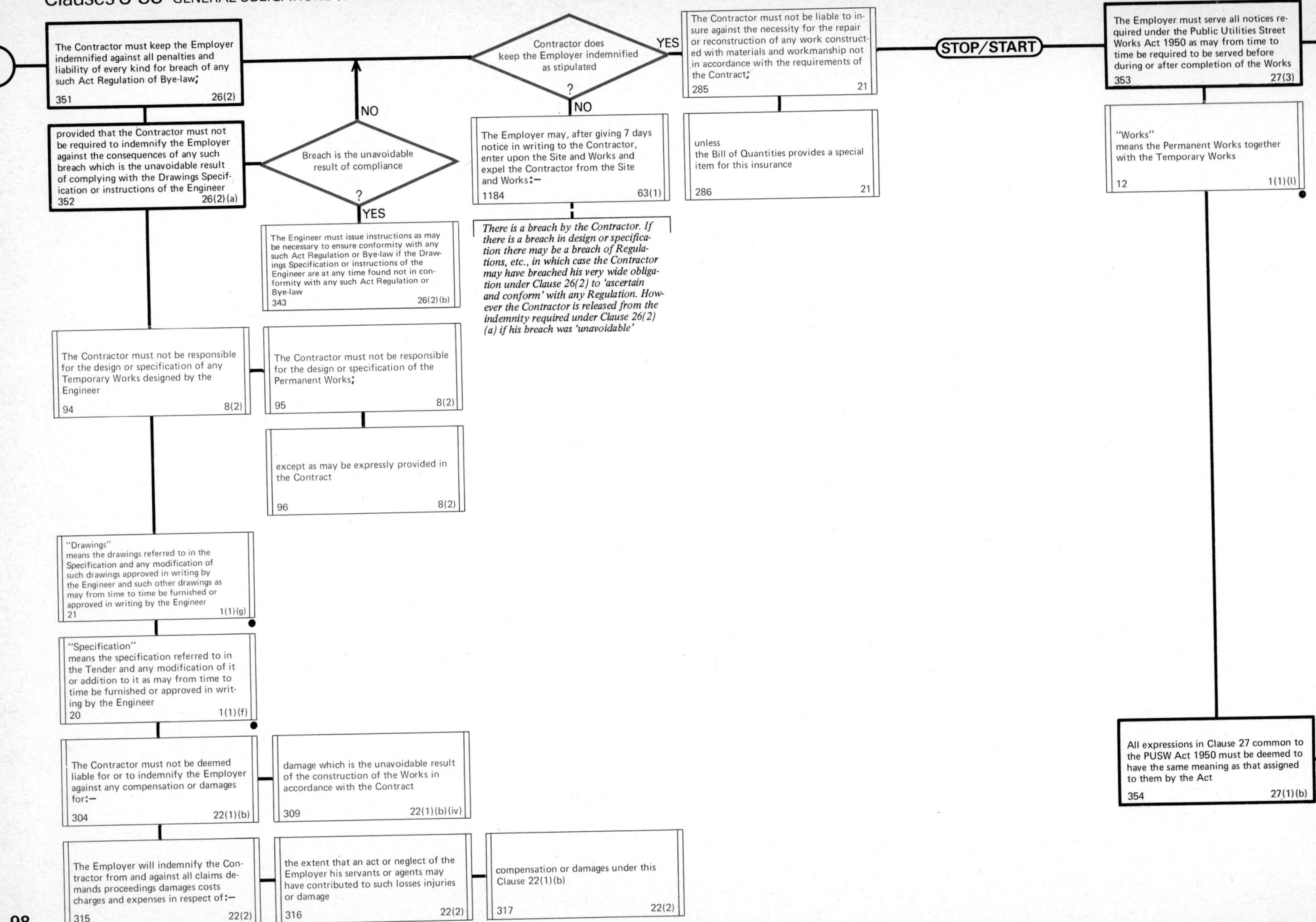
1
The Contractor must keep the Employer indemnified against all penalties and liability of every kind for breach of any such Act Regulation of Bye-law;
351 26(2)
provided that the Contractor must not be required to indemnify the Employer against the consequences of any such breach which is the unavoidable result of complying with the Drawings Specification or instructions of the Engineer
352 26(2)(a)
NO
Breach is the unavoidable result of compliance
?
YES
The Engineer must issue instructions as may be necessary to ensure conformity with any such Act Regulation or Bye-law if the Drawings Specification or instructions of the Engineer are at any time found not in conformity with any such Act Regulation or Bye-law
343 26(2)(b)
Contractor does keep the Employer indemnified as stipulated
?
YES
NO
The Employer may, after giving 7 days notice in writing to the Contractor, enter upon the Site and Works and expel the Contractor from the Site and Works:—
1184 63(1)
There is a breach by the Contractor. If there is a breach in design or specification there may be a breach of Regulations, etc., in which case the Contractor may have breached his very wide obligation under Clause 26(2) to 'ascertain and conform' with any Regulation. However the Contractor is released from the indemnity required under Clause 26(2)(a) if his breach was 'unavoidable'
The Contractor must not be liable to insure against the necessity for the repair or reconstruction of any work constructed with materials and workmanship not in accordance with the requirements of the Contract;
285 21
unless the Bill of Quantities provides a special item for this insurance
286 21
STOP/START
The Employer must serve all notices required under the Public Utilities Street Works Act 1950 as may from time to time be required to be served before during or after completion of the Works
353 27(3)
"Works" means the Permanent Works together with the Temporary Works
12 1(1)(l)
All expressions in Clause 27 common to the PUSW Act 1950 must be deemed to have the same meaning as that assigned to them by the Act
354 27(1)(b)
The Contractor must not be responsible for the design or specification of any Temporary Works designed by the Engineer
94 8(2)
The Contractor must not be responsible for the design or specification of the Permanent Works;
95 8(2)
except as may be expressly provided in the Contract
96 8(2)
"Drawings" means the drawings referred to in the Specification and any modification of such drawings approved in writing by the Engineer and such other drawings as may from time to time be furnished or approved in writing by the Engineer
21 1(1)(g)
"Specification" means the specification referred to in the Tender and any modification of it or addition to it as may from time to time be furnished or approved in writing by the Engineer
20 1(1)(f)
The Contractor must not be deemed liable for or to indemnify the Employer against any compensation or damages for:—
304 22(1)(b)
damage which is the unavoidable result of the construction of the Works in accordance with the Contract
309 22(1)(b)(iv)
The Employer will indemnify the Contractor from and against all claims demands proceedings damages costs charges and expenses in respect of:—
315 22(2)
the extent that an act or neglect of the Employer his servants or agents may have contributed to such losses injuries or damage
316 22(2)
compensation or damages under this Clause 22(1)(b)
317 22(2)

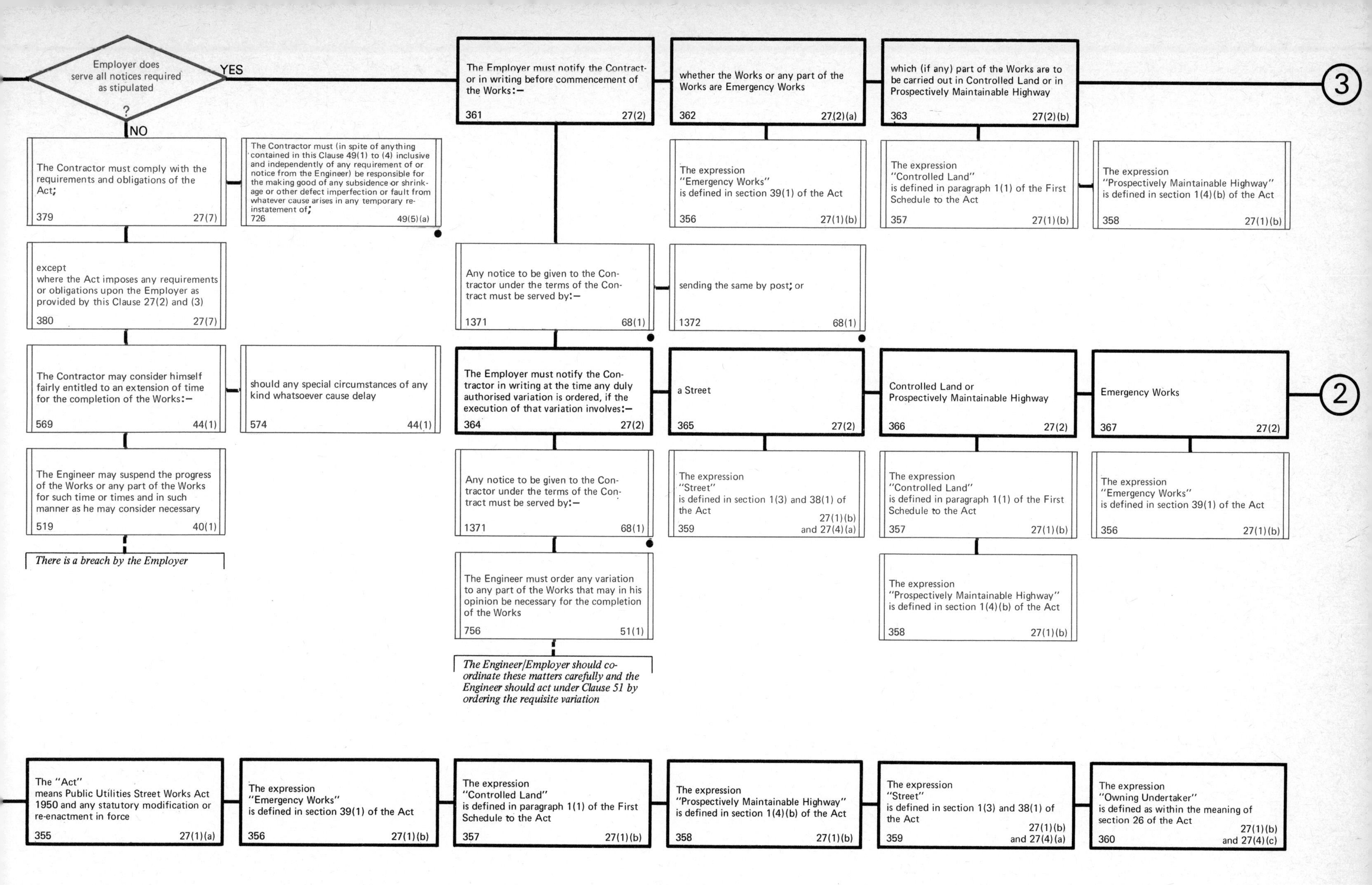
Employer does serve all notices required as stipulated ?
YES
NO
The Employer must notify the Contractor in writing before commencement of the Works:—
361 27(2)
whether the Works or any part of the Works are Emergency Works
362 27(2)(a)
which (if any) part of the Works are to be carried out in Controlled Land or in Prospectively Maintainable Highway
363 27(2)(b)
3
The Contractor must comply with the requirements and obligations of the Act;
379 27(7)
The Contractor must (in spite of anything contained in this Clause 49(1) to (4) inclusive and independently of any requirement of or notice from the Engineer) be responsible for the making good of any subsidence or shrinkage or other defect imperfection or fault from whatever cause arises in any temporary reinstatement of;
726 49(5)(a)
The expression "Emergency Works" is defined in section 39(1) of the Act
356 27(1)(b)
The expression "Controlled Land" is defined in paragraph 1(1) of the First Schedule to the Act
357 27(1)(b)
The expression "Prospectively Maintainable Highway" is defined in section 1(4)(b) of the Act
358 27(1)(b)
except where the Act imposes any requirements or obligations upon the Employer as provided by this Clause 27(2) and (3)
380 27(7)
Any notice to be given to the Contractor under the terms of the Contract must be served by:—
1371 68(1)
sending the same by post; or
1372 68(1)
The Contractor may consider himself fairly entitled to an extension of time for the completion of the Works:—
569 44(1)
should any special circumstances of any kind whatsoever cause delay
574 44(1)
The Employer must notify the Contractor in writing at the time any duly authorised variation is ordered, if the execution of that variation involves:—
364 27(2)
a Street
365 27(2)
Controlled Land or Prospectively Maintainable Highway
366 27(2)
Emergency Works
367 27(2)
2
The Engineer may suspend the progress of the Works or any part of the Works for such time or times and in such manner as he may consider necessary
519 40(1)
Any notice to be given to the Contractor under the terms of the Contract must be served by:—
1371 68(1)
The expression "Street" is defined in section 1(3) and 38(1) of the Act
359 27(1)(b) and 27(4)(a)
The expression "Controlled Land" is defined in paragraph 1(1) of the First Schedule to the Act
357 27(1)(b)
The expression "Emergency Works" is defined in section 39(1) of the Act
356 27(1)(b)
There is a breach by the Employer
The Engineer must order any variation to any part of the Works that may in his opinion be necessary for the completion of the Works
756 51(1)
The expression "Prospectively Maintainable Highway" is defined in section 1(4)(b) of the Act
358 27(1)(b)
The Engineer/Employer should co-ordinate these matters carefully and the Engineer should act under Clause 51 by ordering the requisite variation
The "Act" means Public Utilities Street Works Act 1950 and any statutory modification or re-enactment in force
355 27(1)(a)
The expression "Emergency Works" is defined in section 39(1) of the Act
356 27(1)(b)
The expression "Controlled Land" is defined in paragraph 1(1) of the First Schedule to the Act
357 27(1)(b)
The expression "Prospectively Maintainable Highway" is defined in section 1(4)(b) of the Act
358 27(1)(b)
The expression "Street" is defined in section 1(3) and 38(1) of the Act
359 27(1)(b) and 27(4)(a)
The expression "Owning Undertaker" is defined as within the meaning of section 26 of the Act
360 27(1)(b) and 27(4)(c)

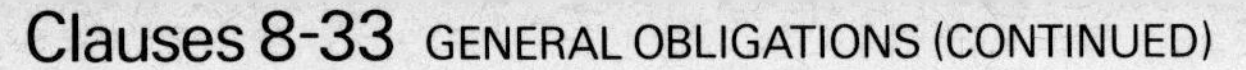

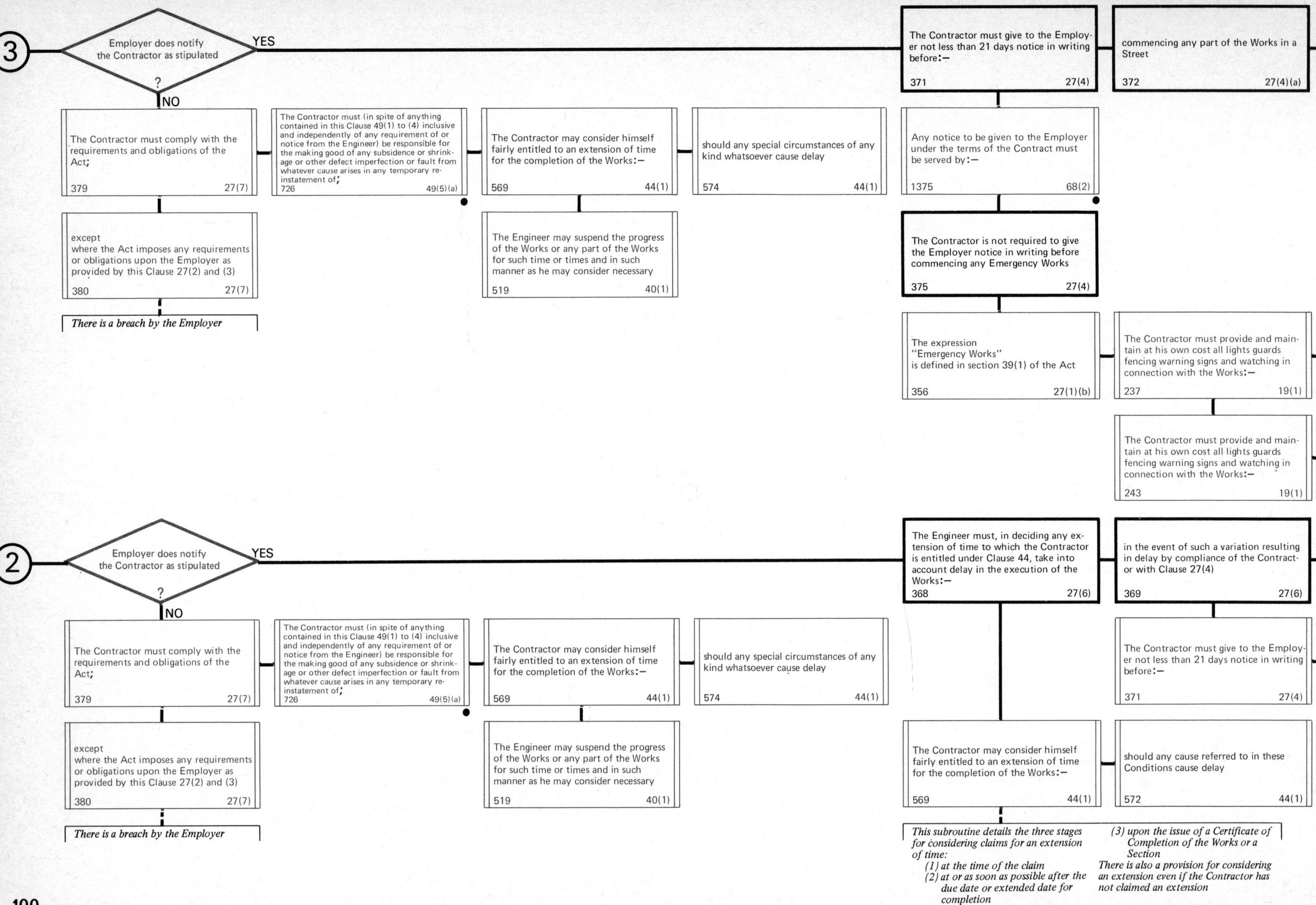
3
Employer does notify the Contractor as stipulated ?
YES
NO
The Contractor must give to the Employer not less than 21 days notice in writing before:— 371 27(4)
commencing any part of the Works in a Street 372 27(4)(a)
The Contractor must comply with the requirements and obligations of the Act; 379 27(7)
The Contractor must (in spite of anything contained in this Clause 49(1) to (4) inclusive and independently of any requirement of or notice from the Engineer) be responsible for the making good of any subsidence or shrinkage or other defect imperfection or fault from whatever cause arises in any temporary reinstatement of; 726 49(5)(a)
The Contractor may consider himself fairly entitled to an extension of time for the completion of the Works:— 569 44(1)
should any special circumstances of any kind whatsoever cause delay 574 44(1)
Any notice to be given to the Employer under the terms of the Contract must be served by:— 1375 68(2)
except where the Act imposes any requirements or obligations upon the Employer as provided by this Clause 27(2) and (3) 380 27(7)
The Engineer may suspend the progress of the Works or any part of the Works for such time or times and in such manner as he may consider necessary 519 40(1)
The Contractor is not required to give the Employer notice in writing before commencing any Emergency Works 375 27(4)
There is a breach by the Employer
The expression "Emergency Works" is defined in section 39(1) of the Act 356 27(1)(b)
The Contractor must provide and maintain at his own cost all lights guards fencing warning signs and watching in connection with the Works:— 237 19(1)
The Contractor must provide and maintain at his own cost all lights guards fencing warning signs and watching in connection with the Works:— 243 19(1)
2
Employer does notify the Contractor as stipulated ?
YES
NO
The Engineer must, in deciding any extension of time to which the Contractor is entitled under Clause 44, take into account delay in the execution of the Works:— 368 27(6)
in the event of such a variation resulting in delay by compliance of the Contractor with Clause 27(4) 369 27(6)
The Contractor must comply with the requirements and obligations of the Act; 379 27(7)
The Contractor must (in spite of anything contained in this Clause 49(1) to (4) inclusive and independently of any requirement of or notice from the Engineer) be responsible for the making good of any subsidence or shrinkage or other defect imperfection or fault from whatever cause arises in any temporary reinstatement of; 726 49(5)(a)
The Contractor may consider himself fairly entitled to an extension of time for the completion of the Works:— 569 44(1)
should any special circumstances of any kind whatsoever cause delay 574 44(1)
The Contractor must give to the Employer not less than 21 days notice in writing before:— 371 27(4)
except where the Act imposes any requirements or obligations upon the Employer as provided by this Clause 27(2) and (3) 380 27(7)
The Engineer may suspend the progress of the Works or any part of the Works for such time or times and in such manner as he may consider necessary 519 40(1)
The Contractor may consider himself fairly entitled to an extension of time for the completion of the Works:— 569 44(1)
should any cause referred to in these Conditions cause delay 572 44(1)
There is a breach by the Employer
This subroutine details the three stages for considering claims for an extension of time:
(1) at the time of the claim
(2) at or as soon as possible after the due date or extended date for completion
(3) upon the issue of a Certificate of Completion of the Works or a Section
There is also a provision for considering an extension even if the Contractor has not claimed an extension

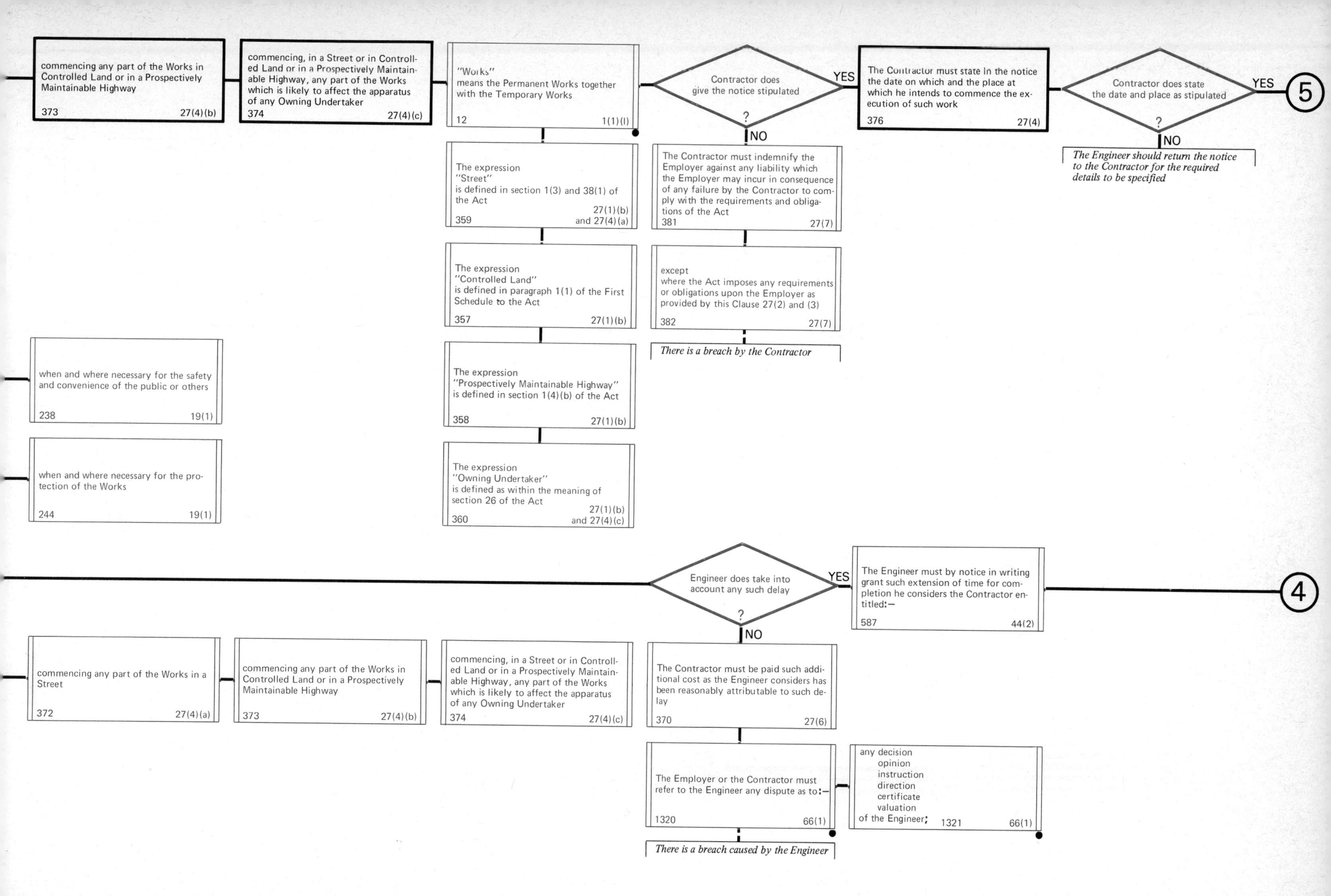
commencing any part of the Works in Controlled Land or in a Prospectively Maintainable Highway
373 27(4)(b)
commencing, in a Street or in Controlled Land or in a Prospectively Maintainable Highway, any part of the Works which is likely to affect the apparatus of any Owning Undertaker
374 27(4)(c)
"Works" means the Permanent Works together with the Temporary Works
12 1(1)(l)
Contractor does give the notice stipulated ?
YES
NO
The Contractor must state in the notice the date on which and the place at which he intends to commence the execution of such work
376 27(4)
Contractor does state the date and place as stipulated ?
YES
NO
5
The Engineer should return the notice to the Contractor for the required details to be specified
The expression "Street" is defined in section 1(3) and 38(1) of the Act
359 27(1)(b) and 27(4)(a)
The Contractor must indemnify the Employer against any liability which the Employer may incur in consequence of any failure by the Contractor to comply with the requirements and obligations of the Act
381 27(7)
The expression "Controlled Land" is defined in paragraph 1(1) of the First Schedule to the Act
357 27(1)(b)
except where the Act imposes any requirements or obligations upon the Employer as provided by this Clause 27(2) and (3)
382 27(7)
There is a breach by the Contractor
when and where necessary for the safety and convenience of the public or others
238 19(1)
The expression "Prospectively Maintainable Highway" is defined in section 1(4)(b) of the Act
358 27(1)(b)
when and where necessary for the protection of the Works
244 19(1)
The expression "Owning Undertaker" is defined as within the meaning of section 26 of the Act
360 27(1)(b) and 27(4)(c)
Engineer does take into account any such delay ?
YES
NO
The Engineer must by notice in writing grant such extension of time for completion he considers the Contractor entitled:—
587 44(2)
4
commencing any part of the Works in a Street
372 27(4)(a)
commencing any part of the Works in Controlled Land or in a Prospectively Maintainable Highway
373 27(4)(b)
commencing, in a Street or in Controlled Land or in a Prospectively Maintainable Highway, any part of the Works which is likely to affect the apparatus of any Owning Undertaker
374 27(4)(c)
The Contractor must be paid such additional cost as the Engineer considers has been reasonably attributable to such delay
370 27(6)
The Employer or the Contractor must refer to the Engineer any dispute as to:—
1320 66(1)
any decision opinion instruction direction certificate valuation of the Engineer;
1321 66(1)
There is a breach caused by the Engineer

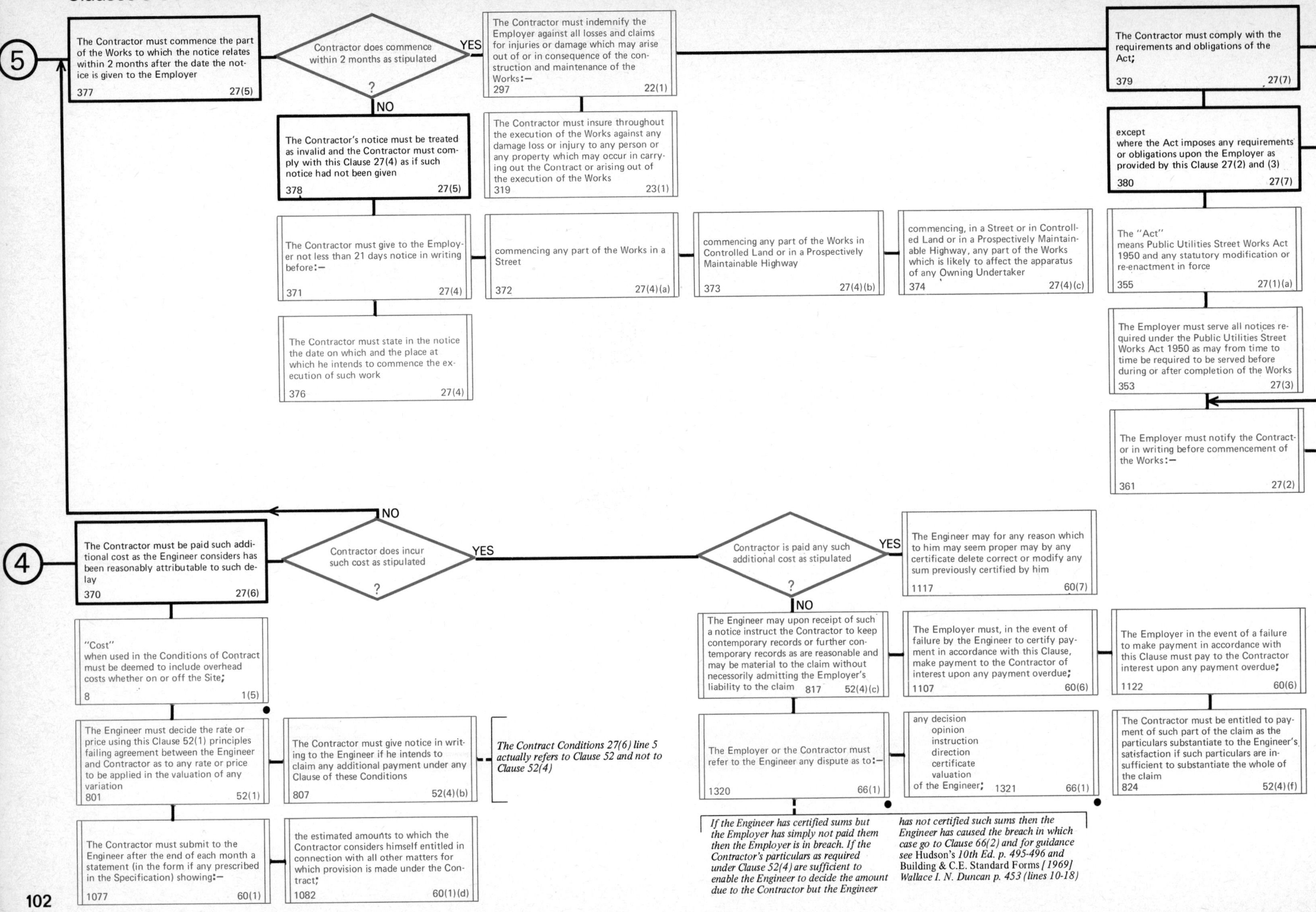

5
The Contractor must commence the part of the Works to which the notice relates within 2 months after the date the notice is given to the Employer 377 27(5)
Contractor does commence within 2 months as stipulated ?
YES
NO
The Contractor must indemnify the Employer against all losses and claims for injuries or damage which may arise out of or in consequence of the construction and maintenance of the Works:— 297 22(1)
The Contractor must insure throughout the execution of the Works against any damage loss or injury to any person or any property which may occur in carrying out the Contract or arising out of the execution of the Works 319 23(1)
The Contractor's notice must be treated as invalid and the Contractor must comply with this Clause 27(4) as if such notice had not been given 378 27(5)
The Contractor must give to the Employer not less than 21 days notice in writing before:— 371 27(4)
commencing any part of the Works in a Street 372 27(4)(a)
commencing any part of the Works in Controlled Land or in a Prospectively Maintainable Highway 373 27(4)(b)
commencing, in a Street or in Controlled Land or in a Prospectively Maintainable Highway, any part of the Works which is likely to affect the apparatus of any Owning Undertaker 374 27(4)(c)
The Contractor must state in the notice the date on which and the place at which he intends to commence the execution of such work 376 27(4)
The Contractor must comply with the requirements and obligations of the Act; 379 27(7)
except where the Act imposes any requirements or obligations upon the Employer as provided by this Clause 27(2) and (3) 380 27(7)
The "Act" means Public Utilities Street Works Act 1950 and any statutory modification or re-enactment in force 355 27(1)(a)
The Employer must serve all notices required under the Public Utilities Street Works Act 1950 as may from time to time be required to be served before during or after completion of the Works 353 27(3)
The Employer must notify the Contractor in writing before commencement of the Works:— 361 27(2)
4
The Contractor must be paid such additional cost as the Engineer considers has been reasonably attributable to such delay 370 27(6)
Contractor does incur such cost as stipulated ?
NO
YES
Contractor is paid any such additional cost as stipulated ?
YES
NO
The Engineer may for any reason which to him may seem proper may by any certificate delete correct or modify any sum previously certified by him 1117 60(7)
"Cost" when used in the Conditions of Contract must be deemed to include overhead costs whether on or off the Site; 8 1(5)
The Engineer must decide the rate or price using this Clause 52(1) principles failing agreement between the Engineer and Contractor as to any rate or price to be applied in the valuation of any variation 801 52(1)
The Contractor must give notice in writing to the Engineer if he intends to claim any additional payment under any Clause of these Conditions 807 52(4)(b)
The Contract Conditions 27(6) line 5 actually refers to Clause 52 and not to Clause 52(4)
The Contractor must submit to the Engineer after the end of each month a statement (in the form if any prescribed in the Specification) showing:— 1077 60(1)
the estimated amounts to which the Contractor considers himself entitled in connection with all other matters for which provision is made under the Contract; 1082 60(1)(d)
The Engineer may upon receipt of such a notice instruct the Contractor to keep contemporary records or further contemporary records as are reasonable and may be material to the claim without necessorily admitting the Employer's liability to the claim 817 52(4)(c)
The Employer must, in the event of failure by the Engineer to certify payment in accordance with this Clause, make payment to the Contractor of interest upon any payment overdue; 1107 60(6)
The Employer in the event of a failure to make payment in accordance with this Clause must pay to the Contractor interest upon any payment overdue; 1122 60(6)
The Employer or the Contractor must refer to the Engineer any dispute as to:— 1320 66(1)
any decision opinion instruction direction certificate valuation of the Engineer; 1321 66(1)
The Contractor must be entitled to payment of such part of the claim as the particulars substantiate to the Engineer's satisfaction if such particulars are insufficient to substantiate the whole of the claim 824 52(4)(f)
If the Engineer has certified sums but the Employer has simply not paid them then the Employer is in breach. If the Contractor's particulars as required under Clause 52(4) are sufficient to enable the Engineer to decide the amount due to the Contractor but the Engineer has not certified such sums then the Engineer has caused the breach in which case go to Clause 66(2) and for guidance see Hudson's 10th Ed. p. 495-496 and Building & C.E. Standard Forms [1969] Wallace I. N. Duncan p. 453 (lines 10-18)

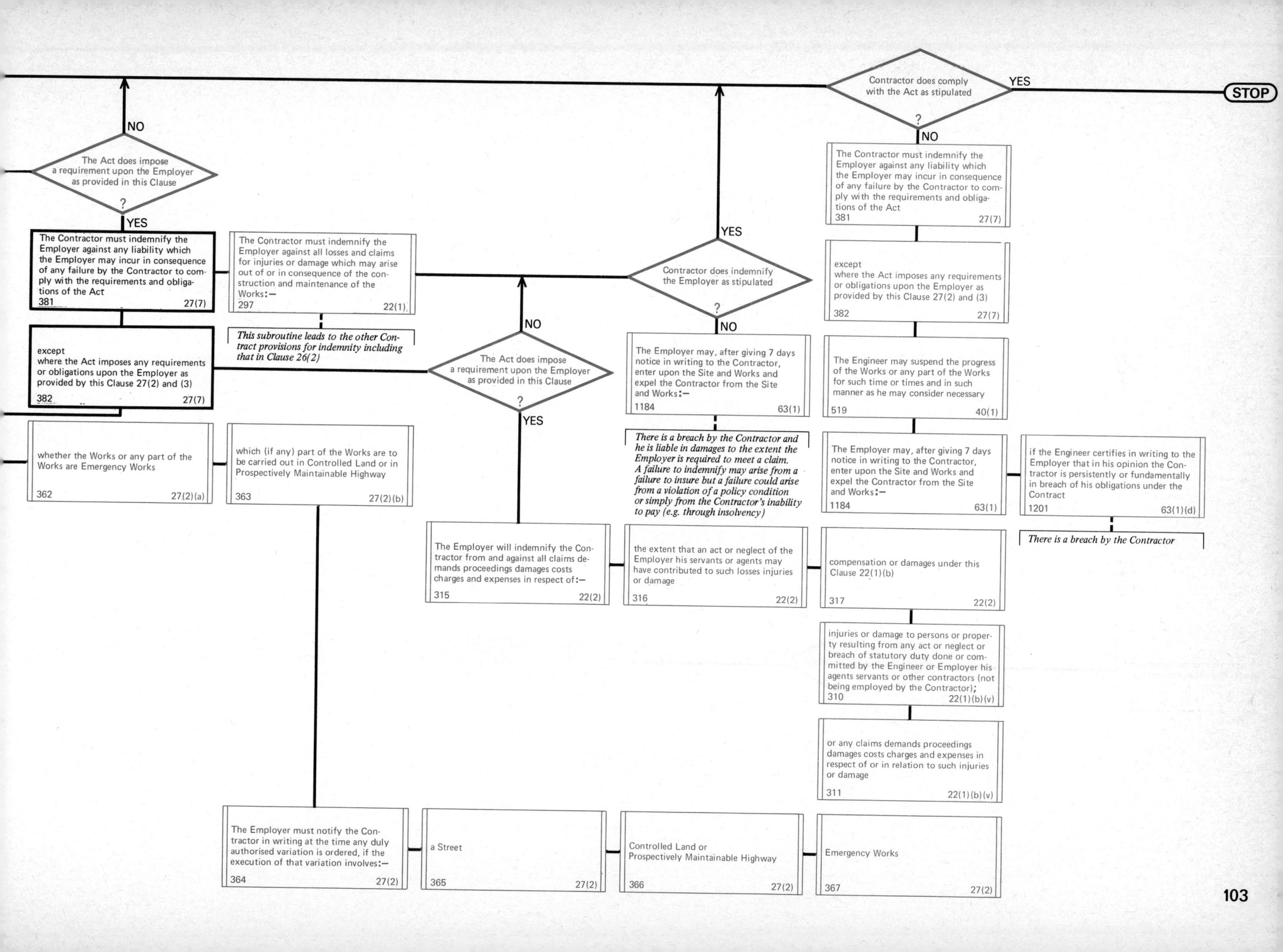
Contractor does comply with the Act as stipulated ?
YES
STOP
NO
The Contractor must indemnify the Employer against any liability which the Employer may incur in consequence of any failure by the Contractor to comply with the requirements and obligations of the Act 381 27(7)
except where the Act imposes any requirements or obligations upon the Employer as provided by this Clause 27(2) and (3) 382 27(7)
The Engineer may suspend the progress of the Works or any part of the Works for such time or times and in such manner as he may consider necessary 519 40(1)
The Employer may, after giving 7 days notice in writing to the Contractor, enter upon the Site and Works and expel the Contractor from the Site and Works:— 1184 63(1)
if the Engineer certifies in writing to the Employer that in his opinion the Contractor is persistently or fundamentally in breach of his obligations under the Contract 1201 63(1)(d)
There is a breach by the Contractor
compensation or damages under this Clause 22(1)(b) 317 22(2)
injuries or damage to persons or property resulting from any act or neglect or breach of statutory duty done or committed by the Engineer or Employer his agents servants or other contractors (not being employed by the Contractor); 310 22(1)(b)(v)
or any claims demands proceedings damages costs charges and expenses in respect of or in relation to such injuries or damage 311 22(1)(b)(v)
NO
The Act does impose a requirement upon the Employer as provided in this Clause ?
YES
The Contractor must indemnify the Employer against any liability which the Employer may incur in consequence of any failure by the Contractor to comply with the requirements and obligations of the Act 381 27(7)
except where the Act imposes any requirements or obligations upon the Employer as provided by this Clause 27(2) and (3) 382 27(7)
whether the Works or any part of the Works are Emergency Works 362 27(2)(a)
which (if any) part of the Works are to be carried out in Controlled Land or in Prospectively Maintainable Highway 363 27(2)(b)
The Contractor must indemnify the Employer against all losses and claims for injuries or damage which may arise out of or in consequence of the construction and maintenance of the Works:— 297 22(1)
This subroutine leads to the other Contract provisions for indemnity including that in Clause 26(2)
NO
The Act does impose a requirement upon the Employer as provided in this Clause ?
YES
The Employer will indemnify the Contractor from and against all claims demands proceedings damages costs charges and expenses in respect of:— 315 22(2)
YES
Contractor does indemnify the Employer as stipulated ?
NO
The Employer may, after giving 7 days notice in writing to the Contractor, enter upon the Site and Works and expel the Contractor from the Site and Works:— 1184 63(1)
There is a breach by the Contractor and he is liable in damages to the extent the Employer is required to meet a claim. A failure to indemnify may arise from a failure to insure but a failure could arise from a violation of a policy condition or simply from the Contractor's inability to pay (e.g. through insolvency)
the extent that an act or neglect of the Employer his servants or agents may have contributed to such losses injuries or damage 316 22(2)
The Employer must notify the Contractor in writing at the time any duly authorised variation is ordered, if the execution of that variation involves:— 364 27(2)
a Street 365 27(2)
Controlled Land or Prospectively Maintainable Highway 366 27(2)
Emergency Works 367 27(2)

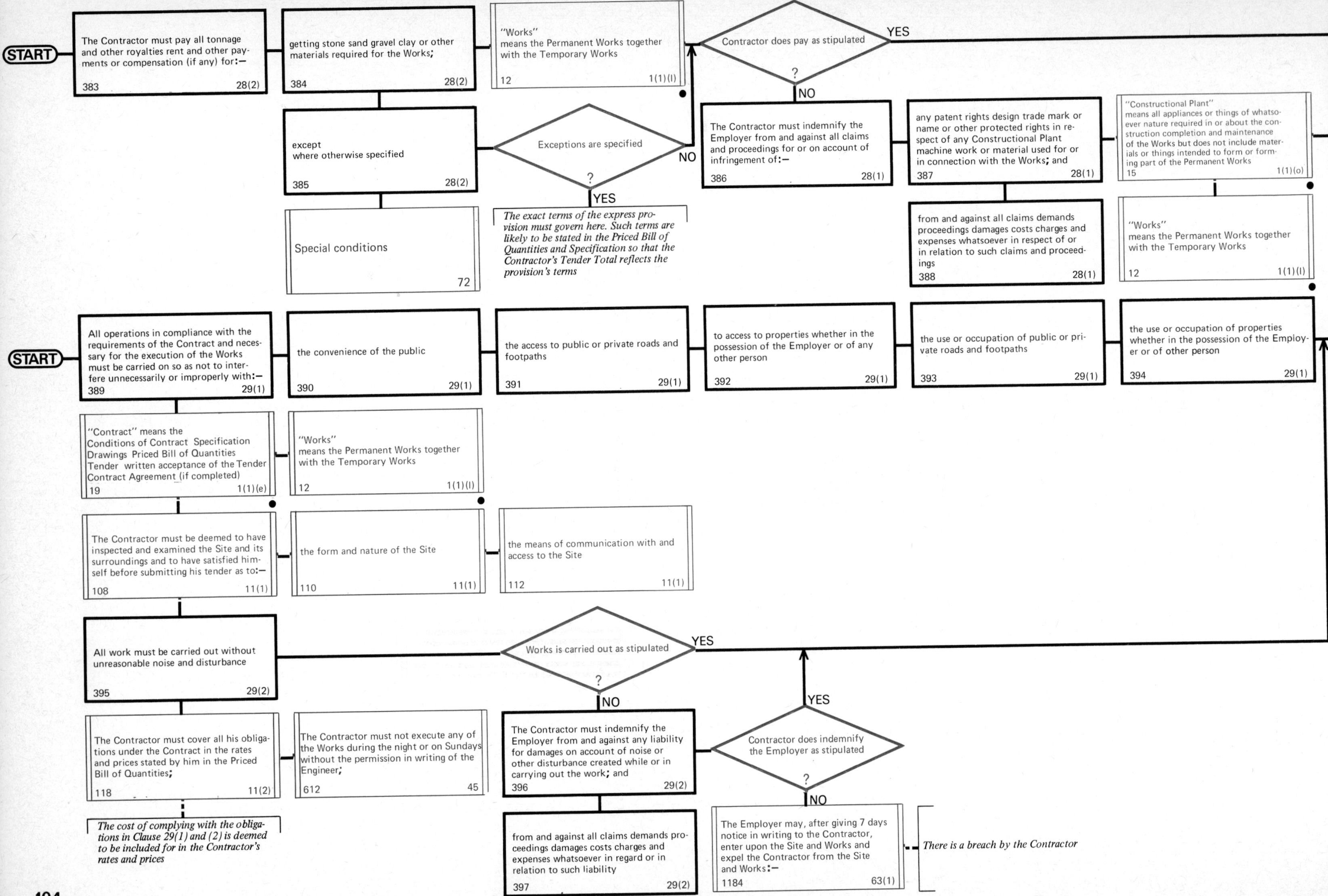
START
The Contractor must pay all tonnage and other royalties rent and other payments or compensation (if any) for:—
383 28(2)
getting stone sand gravel clay or other materials required for the Works;
384 28(2)
"Works" means the Permanent Works together with the Temporary Works
12 1(1)(l)
Contractor does pay as stipulated ?
YES
NO
except where otherwise specified
385 28(2)
Exceptions are specified ?
NO
YES
Special conditions
72
The exact terms of the express provision must govern here. Such terms are likely to be stated in the Priced Bill of Quantities and Specification so that the Contractor's Tender Total reflects the provision's terms
The Contractor must indemnify the Employer from and against all claims and proceedings for or on account of infringement of:—
386 28(1)
any patent rights design trade mark or name or other protected rights in respect of any Constructional Plant machine work or material used for or in connection with the Works; and
387 28(1)
"Constructional Plant" means all appliances or things of whatsoever nature required in or about the construction completion and maintenance of the Works but does not include materials or things intended to form or forming part of the Permanent Works
15 1(1)(o)
from and against all claims demands proceedings damages costs charges and expenses whatsoever in respect of or in relation to such claims and proceedings
388 28(1)
"Works" means the Permanent Works together with the Temporary Works
12 1(1)(l)
START
All operations in compliance with the requirements of the Contract and necessary for the execution of the Works must be carried on so as not to interfere unnecessarily or improperly with:—
389 29(1)
the convenience of the public
390 29(1)
the access to public or private roads and footpaths
391 29(1)
to access to properties whether in the possession of the Employer or of any other person
392 29(1)
the use or occupation of public or private roads and footpaths
393 29(1)
the use or occupation of properties whether in the possession of the Employer or of other person
394 29(1)
"Contract" means the Conditions of Contract Specification Drawings Priced Bill of Quantities Tender written acceptance of the Tender Contract Agreement (if completed)
19 1(1)(e)
"Works" means the Permanent Works together with the Temporary Works
12 1(1)(l)
The Contractor must be deemed to have inspected and examined the Site and its surroundings and to have satisfied himself before submitting his tender as to:—
108 11(1)
the form and nature of the Site
110 11(1)
the means of communication with and access to the Site
112 11(1)
All work must be carried out without unreasonable noise and disturbance
395 29(2)
Works is carried out as stipulated ?
YES
NO
The Contractor must cover all his obligations under the Contract in the rates and prices stated by him in the Priced Bill of Quantities;
118 11(2)
The Contractor must not execute any of the Works during the night or on Sundays without the permission in writing of the Engineer;
612 45
The Contractor must indemnify the Employer from and against any liability for damages on account of noise or other disturbance created while or in carrying out the work; and
396 29(2)
Contractor does indemnify the Employer as stipulated ?
YES
NO
The cost of complying with the obligations in Clause 29(1) and (2) is deemed to be included for in the Contractor's rates and prices
from and against all claims demands proceedings damages costs charges and expenses whatsoever in regard or in relation to such liability
397 29(2)
The Employer may, after giving 7 days notice in writing to the Contractor, enter upon the Site and Works and expel the Contractor from the Site and Works:—
1184 63(1)
There is a breach by the Contractor

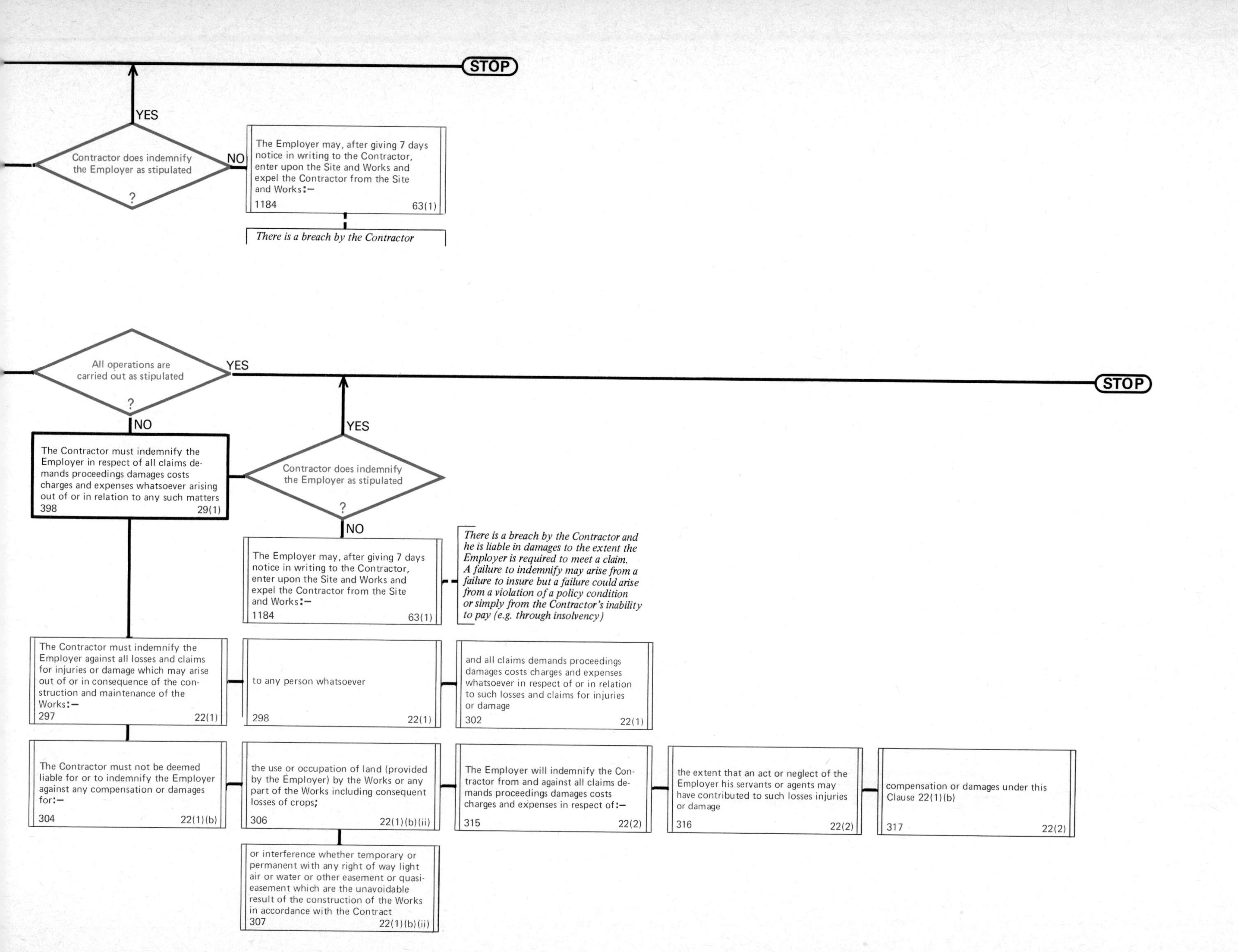

STOP
YES
Contractor does indemnify the Employer as stipulated ?
NO
The Employer may, after giving 7 days notice in writing to the Contractor, enter upon the Site and Works and expel the Contractor from the Site and Works:— 1184 63(1)
There is a breach by the Contractor
All operations are carried out as stipulated ?
YES
STOP
NO
YES
The Contractor must indemnify the Employer in respect of all claims demands proceedings damages costs charges and expenses whatsoever arising out of or in relation to any such matters 398 29(1)
Contractor does indemnify the Employer as stipulated ?
NO
The Employer may, after giving 7 days notice in writing to the Contractor, enter upon the Site and Works and expel the Contractor from the Site and Works:— 1184 63(1)
There is a breach by the Contractor and he is liable in damages to the extent the Employer is required to meet a claim. A failure to indemnify may arise from a failure to insure but a failure could arise from a violation of a policy condition or simply from the Contractor's inability to pay (e.g. through insolvency)
The Contractor must indemnify the Employer against all losses and claims for injuries or damage which may arise out of or in consequence of the construction and maintenance of the Works:— 297 22(1)
to any person whatsoever 298 22(1)
and all claims demands proceedings damages costs charges and expenses whatsoever in respect of or in relation to such losses and claims for injuries or damage 302 22(1)
The Contractor must not be deemed liable for or to indemnify the Employer against any compensation or damages for:— 304 22(1)(b)
the use or occupation of land (provided by the Employer) by the Works or any part of the Works including consequent losses of crops; 306 22(1)(b)(ii)
The Employer will indemnify the Contractor from and against all claims demands proceedings damages costs charges and expenses in respect of:— 315 22(2)
the extent that an act or neglect of the Employer his servants or agents may have contributed to such losses injuries or damage 316 22(2)
compensation or damages under this Clause 22(1)(b) 317 22(2)
or interference whether temporary or permanent with any right of way light air or water or other easement or quasi-easement which are the unavoidable result of the construction of the Works in accordance with the Contract 307 22(1)(b)(ii)

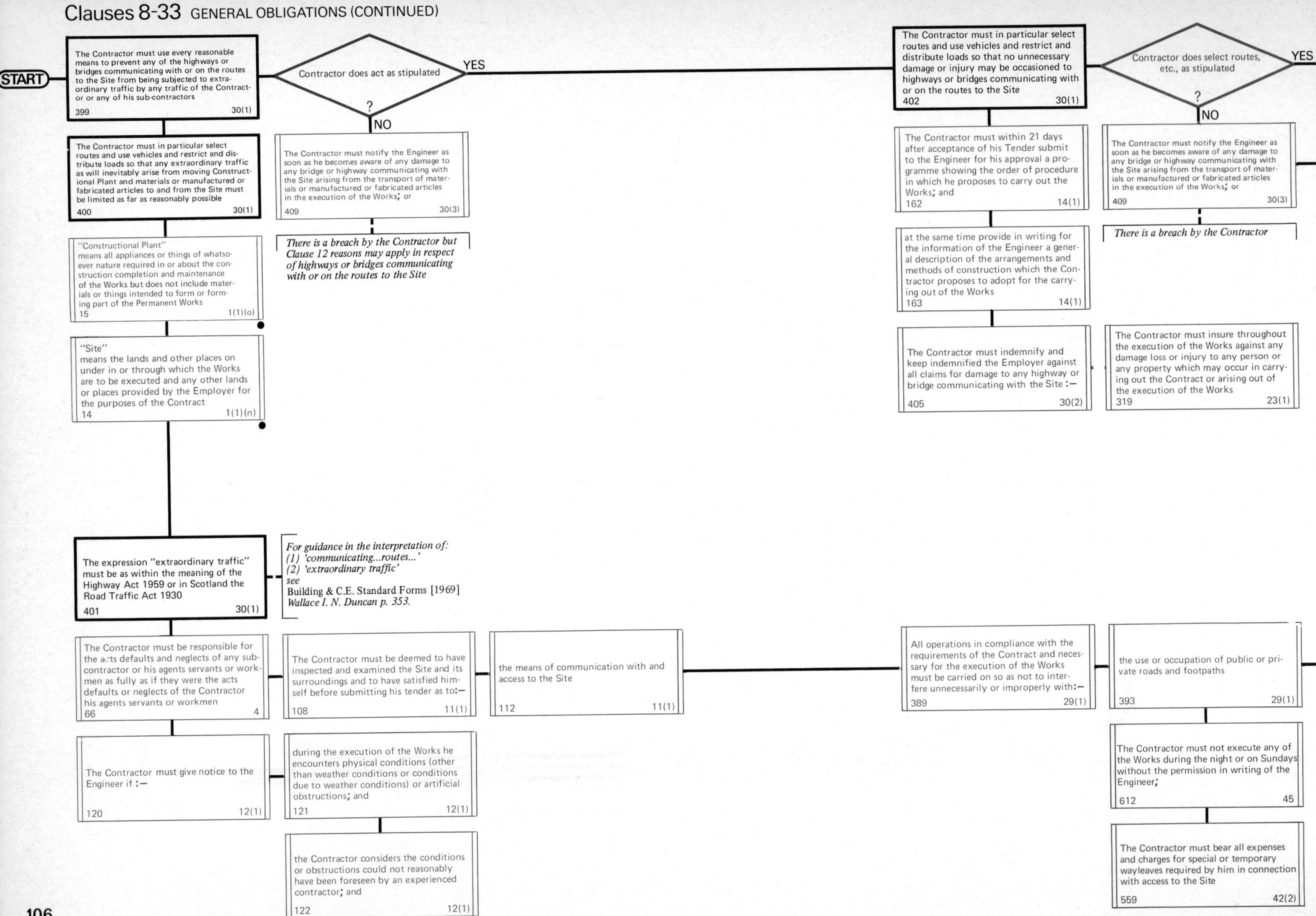
START
The Contractor must use every reasonable means to prevent any of the highways or bridges communicating with or on the routes to the Site from being subjected to extraordinary traffic by any traffic of the Contractor or any of his sub-contractors 399 30(1)
Contractor does act as stipulated ?
YES
NO
The Contractor must notify the Engineer as soon as he becomes aware of any damage to any bridge or highway communicating with the Site arising from the transport of materials or manufactured or fabricated articles in the execution of the Works; or 409 30(3)
There is a breach by the Contractor but Clause 12 reasons may apply in respect of highways or bridges communicating with or on the routes to the Site
The Contractor must in particular select routes and use vehicles and restrict and distribute loads so that any extraordinary traffic as will inevitably arise from moving Constructional Plant and materials or manufactured or fabricated articles to and from the Site must be limited as far as reasonably possible 400 30(1)
"Constructional Plant" means all appliances or things of whatsoever nature required in or about the construction completion and maintenance of the Works but does not include materials or things intended to form or forming part of the Permanent Works 15 1(1)(o)
"Site" means the lands and other places on under in or through which the Works are to be executed and any other lands or places provided by the Employer for the purposes of the Contract 14 1(1)(n)
The expression "extraordinary traffic" must be as within the meaning of the Highway Act 1959 or in Scotland the Road Traffic Act 1930 401 30(1)
For guidance in the interpretation of: (1) 'communicating...routes...' (2) 'extraordinary traffic' see Building & C.E. Standard Forms [1969] Wallace I. N. Duncan p. 353.
The Contractor must be responsible for the acts defaults and neglects of any sub-contractor or his agents servants or workmen as fully as if they were the acts defaults or neglects of the Contractor his agents servants or workmen 66 4
The Contractor must be deemed to have inspected and examined the Site and its surroundings and to have satisfied himself before submitting his tender as to:— 108 11(1)
the means of communication with and access to the Site 112 11(1)
The Contractor must give notice to the Engineer if :— 120 12(1)
during the execution of the Works he encounters physical conditions (other than weather conditions or conditions due to weather conditions) or artificial obstructions; and 121 12(1)
the Contractor considers the conditions or obstructions could not reasonably have been foreseen by an experienced contractor; and 122 12(1)
The Contractor must in particular select routes and use vehicles and restrict and distribute loads so that no unnecessary damage or injury may be occasioned to highways or bridges communicating with or on the routes to the Site 402 30(1)
Contractor does select routes, etc., as stipulated ?
YES
NO
The Contractor must within 21 days after acceptance of his Tender submit to the Engineer for his approval a programme showing the order of procedure in which he proposes to carry out the Works; and 162 14(1)
at the same time provide in writing for the information of the Engineer a general description of the arrangements and methods of construction which the Contractor proposes to adopt for the carrying out of the Works 163 14(1)
The Contractor must indemnify and keep indemnified the Employer against all claims for damage to any highway or bridge communicating with the Site :— 405 30(2)
The Contractor must notify the Engineer as soon as he becomes aware of any damage to any bridge or highway communicating with the Site arising from the transport of materials or manufactured or fabricated articles in the execution of the Works; or 409 30(3)
There is a breach by the Contractor
The Contractor must insure throughout the execution of the Works against any damage loss or injury to any person or any property which may occur in carrying out the Contract or arising out of the execution of the Works 319 23(1)
All operations in compliance with the requirements of the Contract and necessary for the execution of the Works must be carried on so as not to interfere unnecessarily or improperly with:— 389 29(1)
the use or occupation of public or private roads and footpaths 393 29(1)
The Contractor must not execute any of the Works during the night or on Sundays without the permission in writing of the Engineer; 612 45
The Contractor must bear all expenses and charges for special or temporary wayleaves required by him in connection with access to the Site 559 42(2)

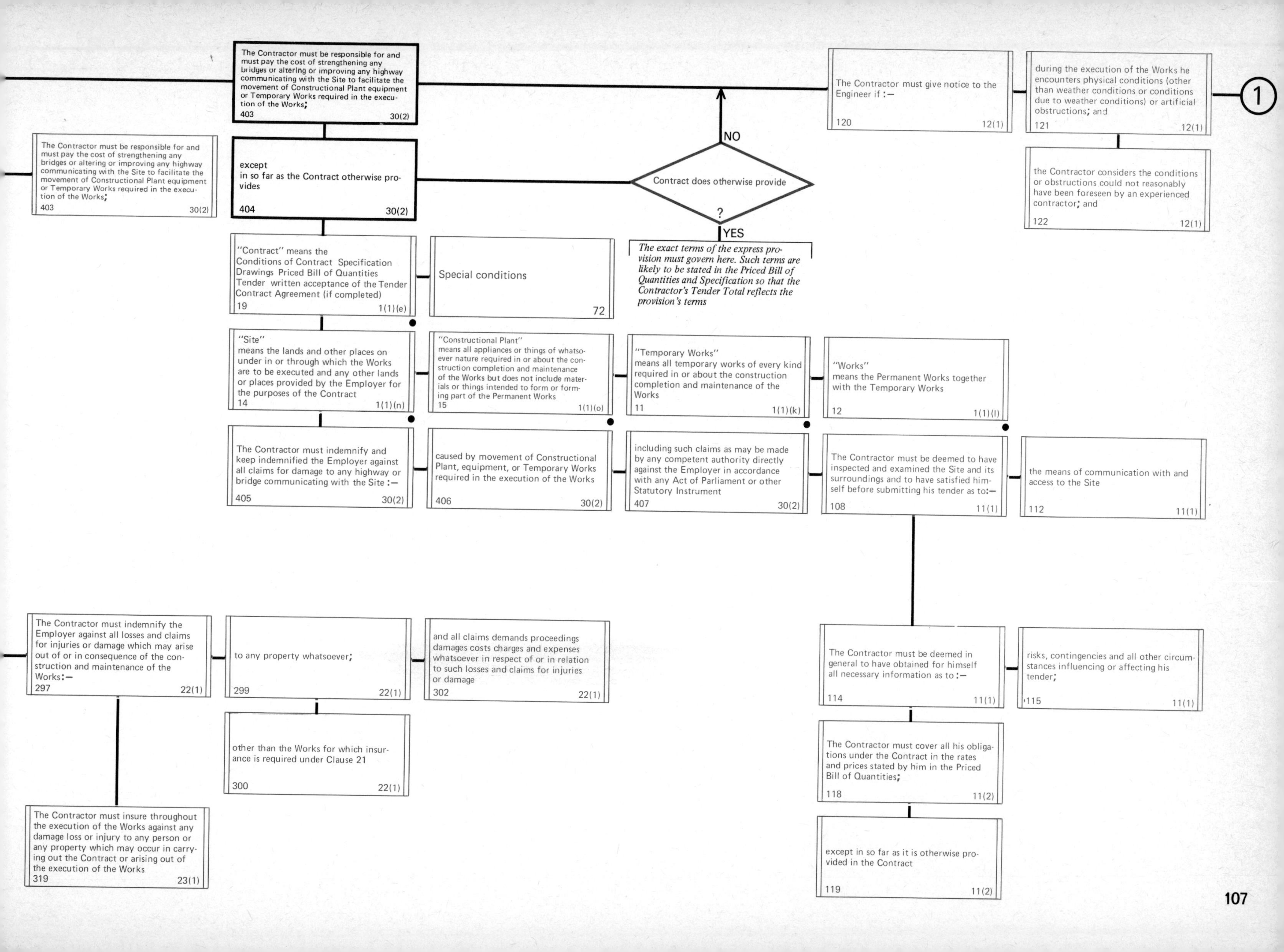
The Contractor must be responsible for and must pay the cost of strengthening any bridges or altering or improving any highway communicating with the Site to facilitate the movement of Constructional Plant equipment or Temporary Works required in the execution of the Works;
403 30(2)
The Contractor must be responsible for and must pay the cost of strengthening any bridges or altering or improving any highway communicating with the Site to facilitate the movement of Constructional Plant equipment or Temporary Works required in the execution of the Works;
403 30(2)
except in so far as the Contract otherwise provides
404 30(2)
Contract does otherwise provide ?
NO
YES
The exact terms of the express provision must govern here. Such terms are likely to be stated in the Priced Bill of Quantities and Specification so that the Contractor's Tender Total reflects the provision's terms
The Contractor must give notice to the Engineer if :—
120 12(1)
during the execution of the Works he encounters physical conditions (other than weather conditions or conditions due to weather conditions) or artificial obstructions; and
121 12(1)
1
the Contractor considers the conditions or obstructions could not reasonably have been foreseen by an experienced contractor; and
122 12(1)
"Contract" means the Conditions of Contract Specification Drawings Priced Bill of Quantities Tender written acceptance of the Tender Contract Agreement (if completed)
19 1(1)(e)
Special conditions
72
"Site" means the lands and other places on under in or through which the Works are to be executed and any other lands or places provided by the Employer for the purposes of the Contract
14 1(1)(n)
"Constructional Plant" means all appliances or things of whatsoever nature required in or about the construction completion and maintenance of the Works but does not include materials or things intended to form or forming part of the Permanent Works
15 1(1)(o)
"Temporary Works" means all temporary works of every kind required in or about the construction completion and maintenance of the Works
11 1(1)(k)
"Works" means the Permanent Works together with the Temporary Works
12 1(1)(l)
The Contractor must indemnify and keep indemnified the Employer against all claims for damage to any highway or bridge communicating with the Site :—
405 30(2)
caused by movement of Constructional Plant, equipment, or Temporary Works required in the execution of the Works
406 30(2)
including such claims as may be made by any competent authority directly against the Employer in accordance with any Act of Parliament or other Statutory Instrument
407 30(2)
The Contractor must be deemed to have inspected and examined the Site and its surroundings and to have satisfied himself before submitting his tender as to:—
108 11(1)
the means of communication with and access to the Site
112 11(1)
The Contractor must indemnify the Employer against all losses and claims for injuries or damage which may arise out of or in consequence of the construction and maintenance of the Works:—
297 22(1)
to any property whatsoever;
299 22(1)
and all claims demands proceedings damages costs charges and expenses whatsoever in respect of or in relation to such losses and claims for injuries or damage
302 22(1)
other than the Works for which insurance is required under Clause 21
300 22(1)
The Contractor must insure throughout the execution of the Works against any damage loss or injury to any person or any property which may occur in carrying out the Contract or arising out of the execution of the Works
319 23(1)
The Contractor must be deemed in general to have obtained for himself all necessary information as to :—
114 11(1)
risks, contingencies and all other circumstances influencing or affecting his tender;
115 11(1)
The Contractor must cover all his obligations under the Contract in the rates and prices stated by him in the Priced Bill of Quantities;
118 11(2)
except in so far as it is otherwise provided in the Contract
119 11(2)

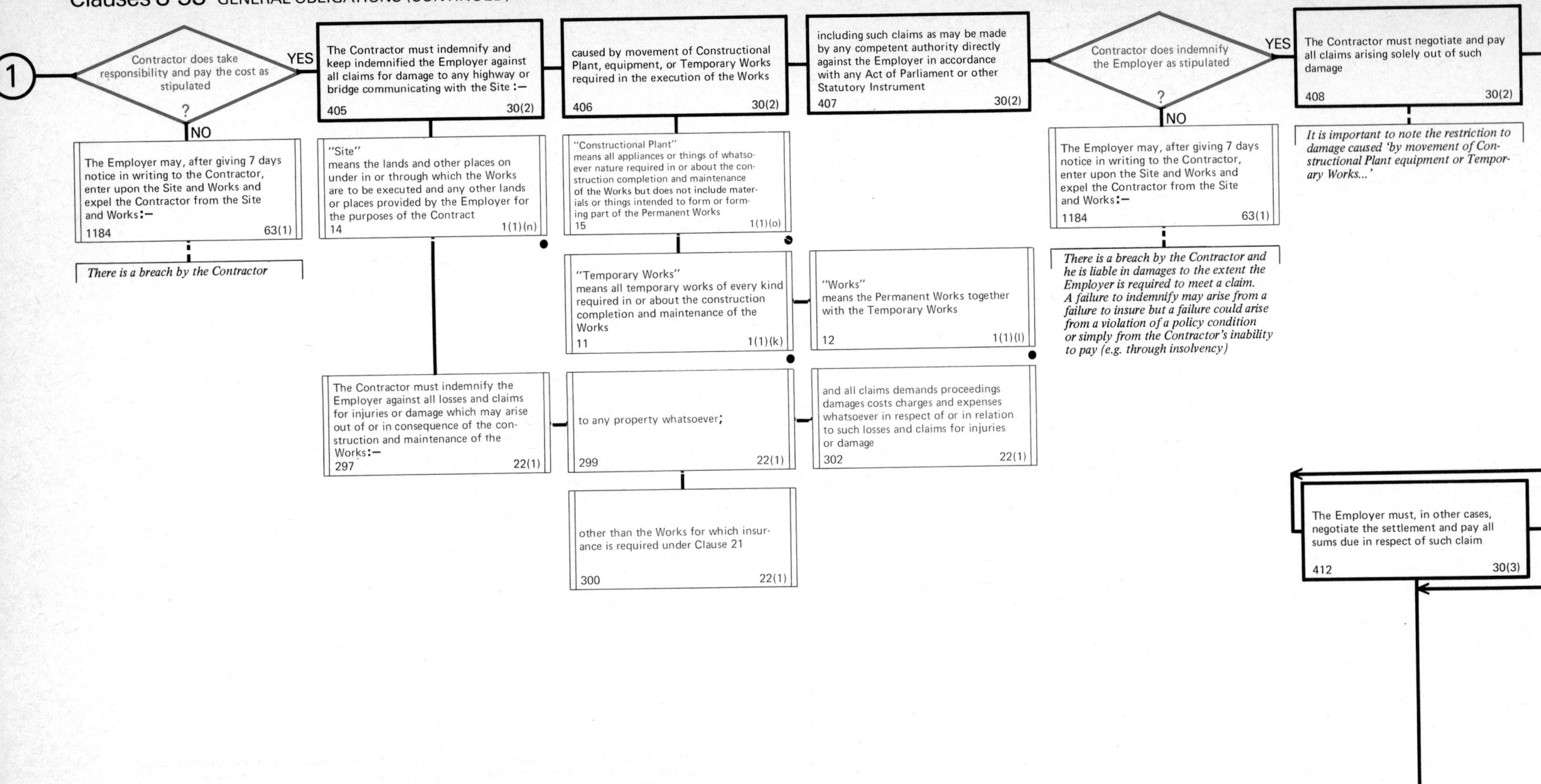
1
Contractor does take responsibility and pay the cost as stipulated ?
YES
NO
The Contractor must indemnify and keep indemnified the Employer against all claims for damage to any highway or bridge communicating with the Site :—
405 30(2)
caused by movement of Constructional Plant, equipment, or Temporary Works required in the execution of the Works
406 30(2)
including such claims as may be made by any competent authority directly against the Employer in accordance with any Act of Parliament or other Statutory Instrument
407 30(2)
Contractor does indemnify the Employer as stipulated ?
YES
NO
The Contractor must negotiate and pay all claims arising solely out of such damage
408 30(2)
The Employer may, after giving 7 days notice in writing to the Contractor, enter upon the Site and Works and expel the Contractor from the Site and Works:—
1184 63(1)
There is a breach by the Contractor
"Site" means the lands and other places on under in or through which the Works are to be executed and any other lands or places provided by the Employer for the purposes of the Contract
14 1(1)(n)
"Constructional Plant" means all appliances or things of whatsoever nature required in or about the construction completion and maintenance of the Works but does not include materials or things intended to form or forming part of the Permanent Works
15 1(1)(o)
The Employer may, after giving 7 days notice in writing to the Contractor, enter upon the Site and Works and expel the Contractor from the Site and Works:—
1184 63(1)
It is important to note the restriction to damage caused 'by movement of Constructional Plant equipment or Temporary Works...'
There is a breach by the Contractor and he is liable in damages to the extent the Employer is required to meet a claim. A failure to indemnify may arise from a failure to insure but a failure could arise from a violation of a policy condition or simply from the Contractor's inability to pay (e.g. through insolvency)
"Temporary Works" means all temporary works of every kind required in or about the construction completion and maintenance of the Works
11 1(1)(k)
"Works" means the Permanent Works together with the Temporary Works
12 1(1)(l)
The Contractor must indemnify the Employer against all losses and claims for injuries or damage which may arise out of or in consequence of the construction and maintenance of the Works:—
297 22(1)
to any property whatsoever;
299 22(1)
and all claims demands proceedings damages costs charges and expenses whatsoever in respect of or in relation to such losses and claims for injuries or damage
302 22(1)
other than the Works for which insurance is required under Clause 21
300 22(1)
The Employer must, in other cases, negotiate the settlement and pay all sums due in respect of such claim
412 30(3)
The Employer must indemnify the Contractor in respect of:—
413 30(3)

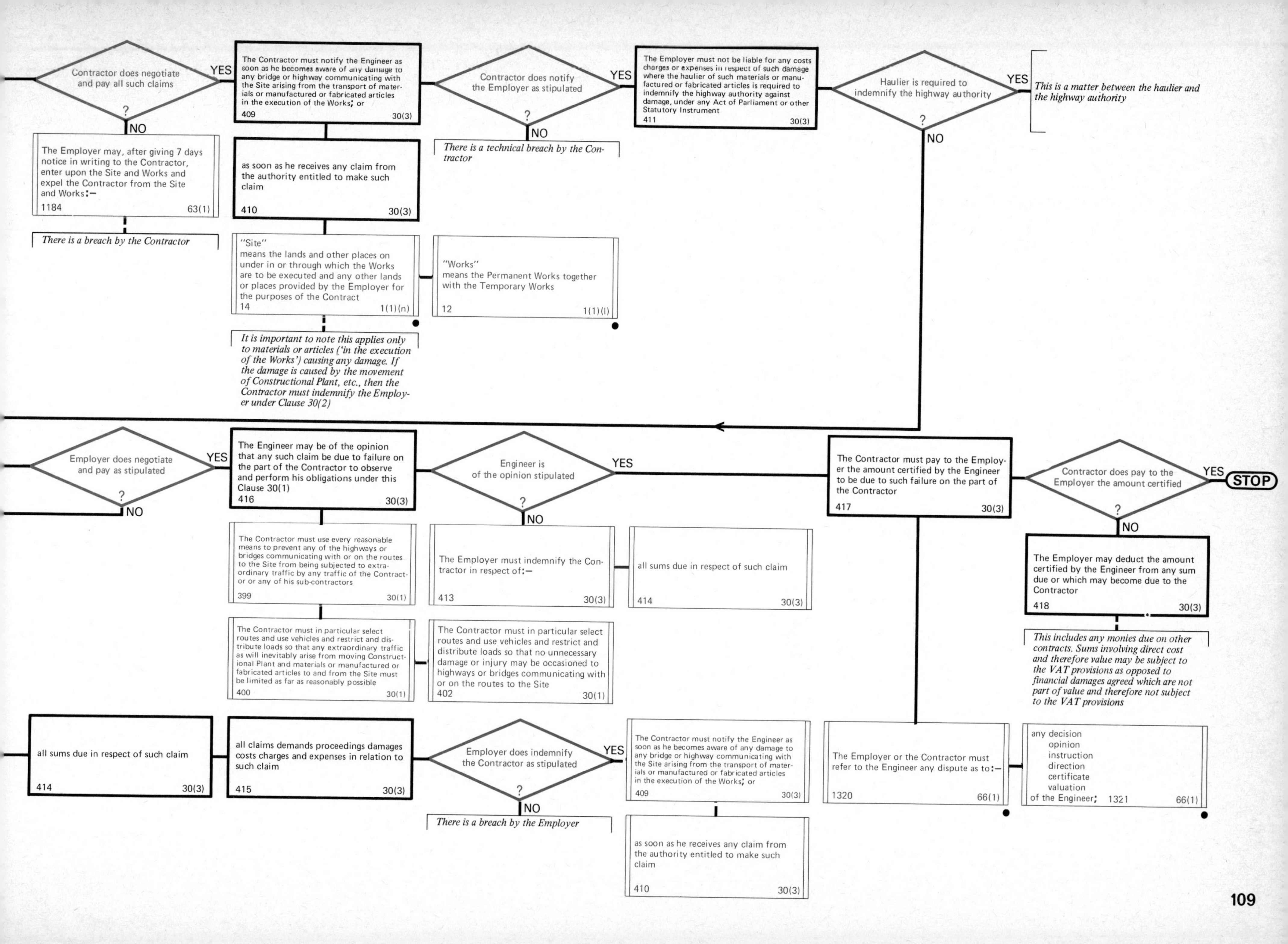

Contractor does negotiate and pay all such claims ?
YES
NO
The Contractor must notify the Engineer as soon as he becomes aware of any damage to any bridge or highway communicating with the Site arising from the transport of materials or manufactured or fabricated articles in the execution of the Works; or
409 30(3)
Contractor does notify the Employer as stipulated ?
YES
NO
The Employer must not be liable for any costs charges or expenses in respect of such damage where the haulier of such materials or manufactured or fabricated articles is required to indemnify the highway authority against damage, under any Act of Parliament or other Statutory Instrument
411 30(3)
Haulier is required to indemnify the highway authority ?
YES
NO
This is a matter between the haulier and the highway authority
The Employer may, after giving 7 days notice in writing to the Contractor, enter upon the Site and Works and expel the Contractor from the Site and Works:—
1184 63(1)
There is a breach by the Contractor
as soon as he receives any claim from the authority entitled to make such claim
410 30(3)
There is a technical breach by the Contractor
"Site" means the lands and other places on under in or through which the Works are to be executed and any other lands or places provided by the Employer for the purposes of the Contract
14 1(1)(n)
"Works" means the Permanent Works together with the Temporary Works
12 1(1)(l)
It is important to note this applies only to materials or articles ('in the execution of the Works') causing any damage. If the damage is caused by the movement of Constructional Plant, etc., then the Contractor must indemnify the Employer under Clause 30(2)
Employer does negotiate and pay as stipulated ?
YES
NO
The Engineer may be of the opinion that any such claim be due to failure on the part of the Contractor to observe and perform his obligations under this Clause 30(1)
416 30(3)
Engineer is of the opinion stipulated ?
YES
NO
The Contractor must pay to the Employer the amount certified by the Engineer to be due to such failure on the part of the Contractor
417 30(3)
Contractor does pay to the Employer the amount certified ?
YES
STOP
NO
The Contractor must use every reasonable means to prevent any of the highways or bridges communicating with or on the routes to the Site from being subjected to extraordinary traffic by any traffic of the Contractor or any of his sub-contractors
399 30(1)
The Employer must indemnify the Contractor in respect of:—
413 30(3)
all sums due in respect of such claim
414 30(3)
The Employer may deduct the amount certified by the Engineer from any sum due or which may become due to the Contractor
418 30(3)
This includes any monies due on other contracts. Sums involving direct cost and therefore value may be subject to the VAT provisions as opposed to financial damages agreed which are not part of value and therefore not subject to the VAT provisions
The Contractor must in particular select routes and use vehicles and restrict and distribute loads so that any extraordinary traffic as will inevitably arise from moving Constructional Plant and materials or manufactured or fabricated articles to and from the Site must be limited as far as reasonably possible
400 30(1)
The Contractor must in particular select routes and use vehicles and restrict and distribute loads so that no unnecessary damage or injury may be occasioned to highways or bridges communicating with or on the routes to the Site
402 30(1)
all sums due in respect of such claim
414 30(3)
all claims demands proceedings damages costs charges and expenses in relation to such claim
415 30(3)
Employer does indemnify the Contractor as stipulated ?
YES
NO
There is a breach by the Employer
The Contractor must notify the Engineer as soon as he becomes aware of any damage to any bridge or highway communicating with the Site arising from the transport of materials or manufactured or fabricated articles in the execution of the Works; or
409 30(3)
as soon as he receives any claim from the authority entitled to make such claim
410 30(3)
The Employer or the Contractor must refer to the Engineer any dispute as to:—
1320 66(1)
any decision opinion instruction direction certificate valuation of the Engineer; 1321 66(1)

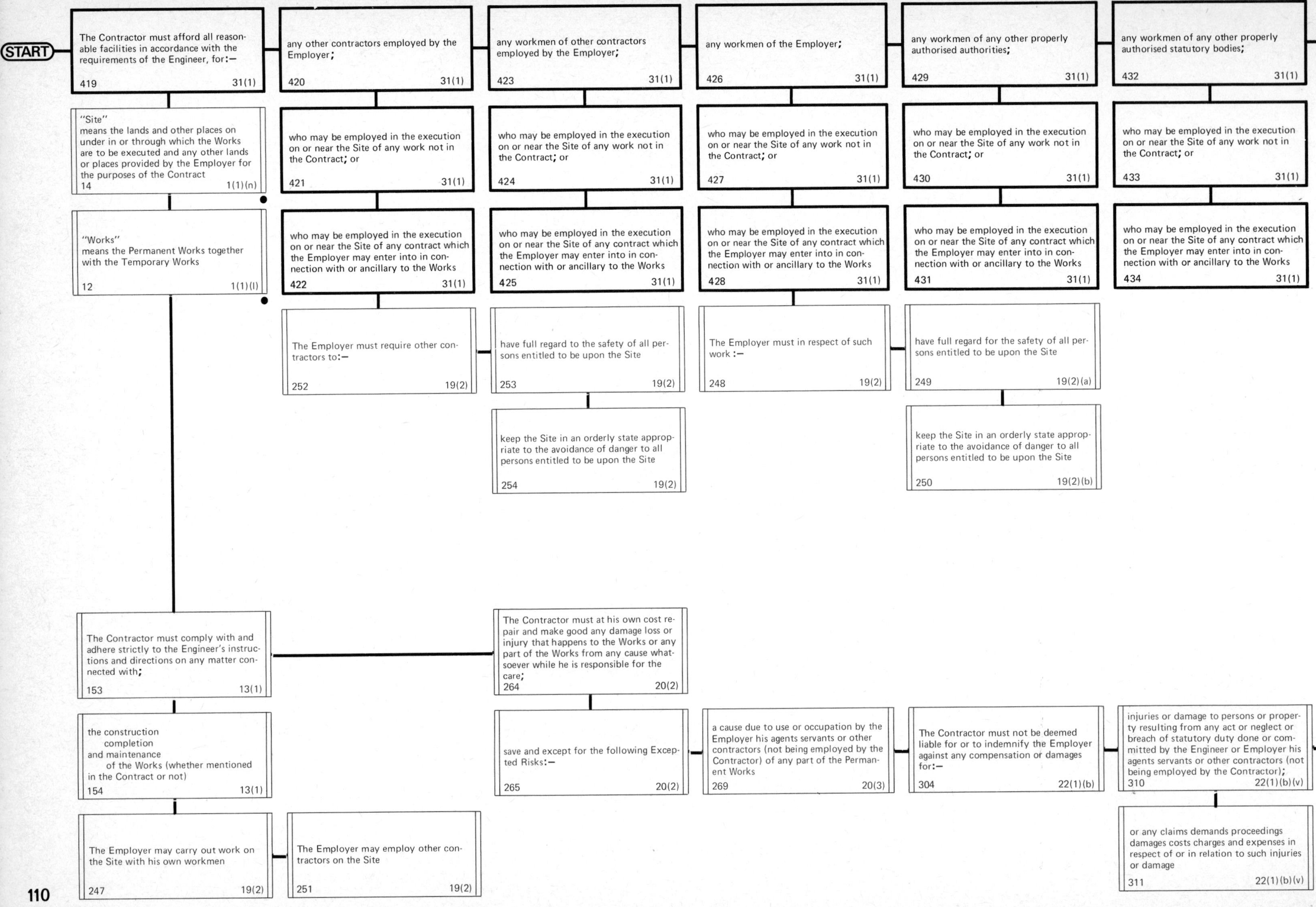
START
The Contractor must afford all reasonable facilities in accordance with the requirements of the Engineer, for:— 419 31(1)
any other contractors employed by the Employer; 420 31(1)
any workmen of other contractors employed by the Employer; 423 31(1)
any workmen of the Employer; 426 31(1)
any workmen of any other properly authorised authorities; 429 31(1)
any workmen of any other properly authorised statutory bodies; 432 31(1)
"Site" means the lands and other places on under in or through which the Works are to be executed and any other lands or places provided by the Employer for the purposes of the Contract 14 1(1)(n)
who may be employed in the execution on or near the Site of any work not in the Contract; or 421 31(1)
who may be employed in the execution on or near the Site of any work not in the Contract; or 424 31(1)
who may be employed in the execution on or near the Site of any work not in the Contract; or 427 31(1)
who may be employed in the execution on or near the Site of any work not in the Contract; or 430 31(1)
who may be employed in the execution on or near the Site of any work not in the Contract; or 433 31(1)
"Works" means the Permanent Works together with the Temporary Works 12 1(1)(l)
who may be employed in the execution on or near the Site of any contract which the Employer may enter into in connection with or ancillary to the Works 422 31(1)
who may be employed in the execution on or near the Site of any contract which the Employer may enter into in connection with or ancillary to the Works 425 31(1)
who may be employed in the execution on or near the Site of any contract which the Employer may enter into in connection with or ancillary to the Works 428 31(1)
who may be employed in the execution on or near the Site of any contract which the Employer may enter into in connection with or ancillary to the Works 431 31(1)
who may be employed in the execution on or near the Site of any contract which the Employer may enter into in connection with or ancillary to the Works 434 31(1)
The Employer must require other contractors to:— 252 19(2)
have full regard to the safety of all persons entitled to be upon the Site 253 19(2)
The Employer must in respect of such work :— 248 19(2)
have full regard for the safety of all persons entitled to be upon the Site 249 19(2)(a)
keep the Site in an orderly state appropriate to the avoidance of danger to all persons entitled to be upon the Site 254 19(2)
keep the Site in an orderly state appropriate to the avoidance of danger to all persons entitled to be upon the Site 250 19(2)(b)
The Contractor must comply with and adhere strictly to the Engineer's instructions and directions on any matter connected with; 153 13(1)
The Contractor must at his own cost repair and make good any damage loss or injury that happens to the Works or any part of the Works from any cause whatsoever while he is responsible for the care; 264 20(2)
the construction completion and maintenance of the Works (whether mentioned in the Contract or not) 154 13(1)
save and except for the following Excepted Risks:— 265 20(2)
a cause due to use or occupation by the Employer his agents servants or other contractors (not being employed by the Contractor) of any part of the Permanent Works 269 20(3)
The Contractor must not be deemed liable for or to indemnify the Employer against any compensation or damages for:— 304 22(1)(b)
injuries or damage to persons or property resulting from any act or neglect or breach of statutory duty done or committed by the Engineer or Employer his agents servants or other contractors (not being employed by the Contractor); 310 22(1)(b)(v)
or any claims demands proceedings damages costs charges and expenses in respect of or in relation to such injuries or damage 311 22(1)(b)(v)
The Employer may carry out work on the Site with his own workmen 247 19(2)
The Employer may employ other contractors on the Site 251 19(2)

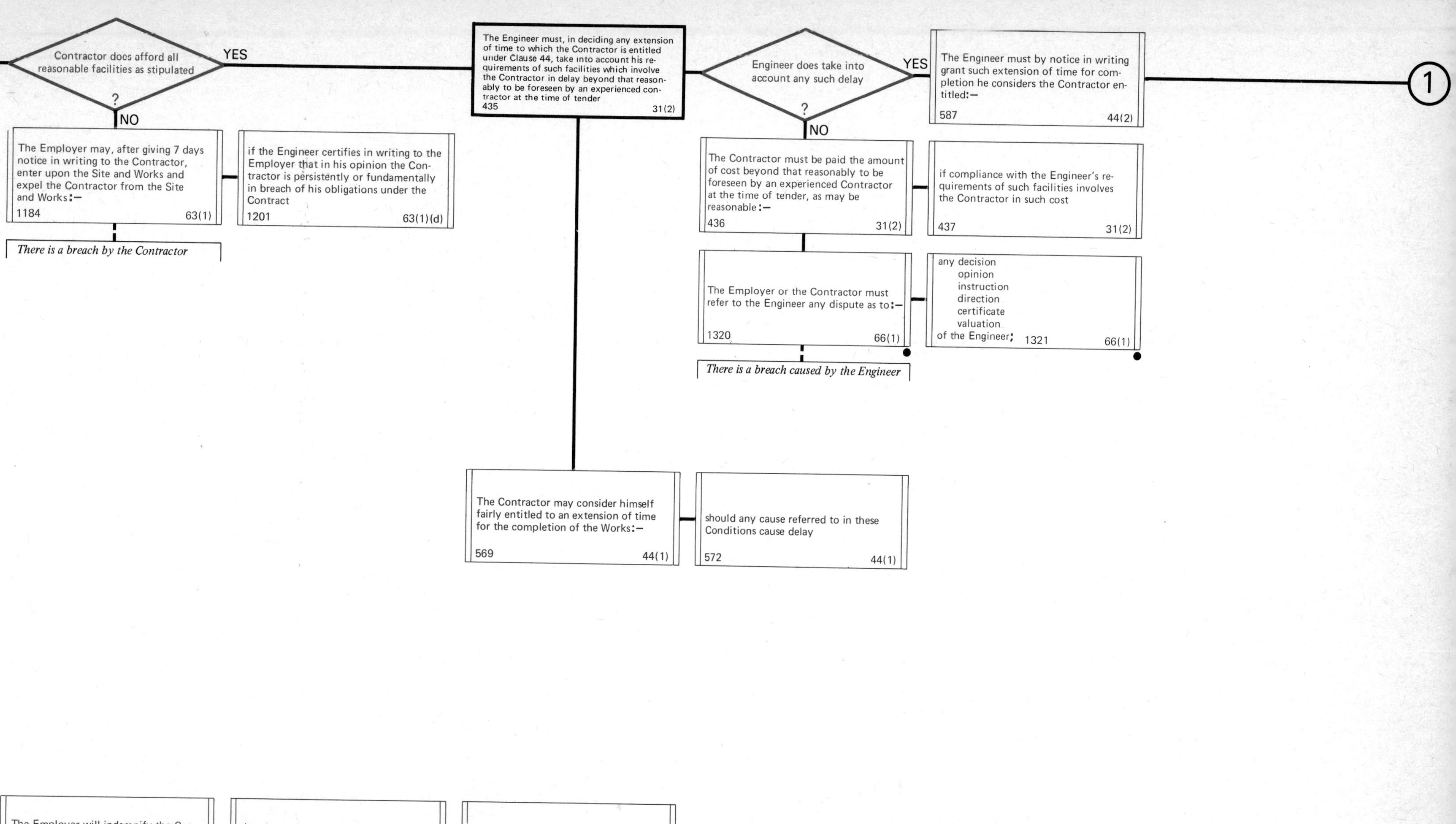
Contractor does afford all reasonable facilities as stipulated
?
YES
NO
The Engineer must, in deciding any extension of time to which the Contractor is entitled under Clause 44, take into account his requirements of such facilities which involve the Contractor in delay beyond that reasonably to be foreseen by an experienced contractor at the time of tender
435 31(2)
Engineer does take into account any such delay
?
YES
NO
The Engineer must by notice in writing grant such extension of time for completion he considers the Contractor entitled:—
587 44(2)
1
The Employer may, after giving 7 days notice in writing to the Contractor, enter upon the Site and Works and expel the Contractor from the Site and Works:—
1184 63(1)
if the Engineer certifies in writing to the Employer that in his opinion the Contractor is persistently or fundamentally in breach of his obligations under the Contract
1201 63(1)(d)
There is a breach by the Contractor
The Contractor must be paid the amount of cost beyond that reasonably to be foreseen by an experienced Contractor at the time of tender, as may be reasonable:—
436 31(2)
if compliance with the Engineer's requirements of such facilities involves the Contractor in such cost
437 31(2)
The Employer or the Contractor must refer to the Engineer any dispute as to:—
1320 66(1)
any decision opinion instruction direction certificate valuation of the Engineer;
1321 66(1)
There is a breach caused by the Engineer
The Contractor may consider himself fairly entitled to an extension of time for the completion of the Works:—
569 44(1)
should any cause referred to in these Conditions cause delay
572 44(1)
The Employer will indemnify the Contractor from and against all claims demands proceedings damages costs charges and expenses in respect of:—
315 22(2)
the extent that an act or neglect of the Employer his servants or agents may have contributed to such losses injuries or damage
316 22(2)
compensation or damages under this Clause 22(1)(b)
317 22(2)

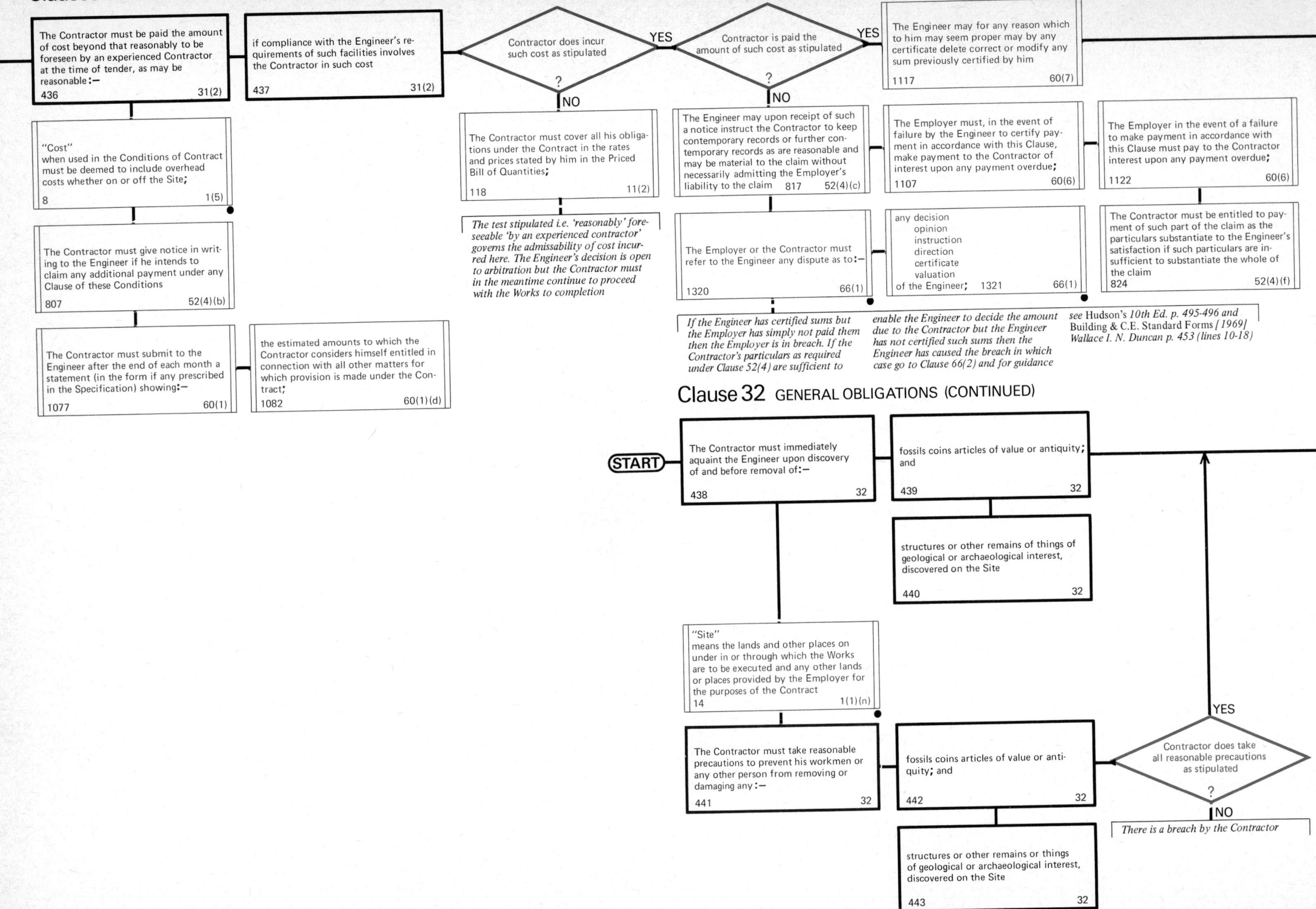

Clause 32 GENERAL OBLIGATIONS (CONTINUED)

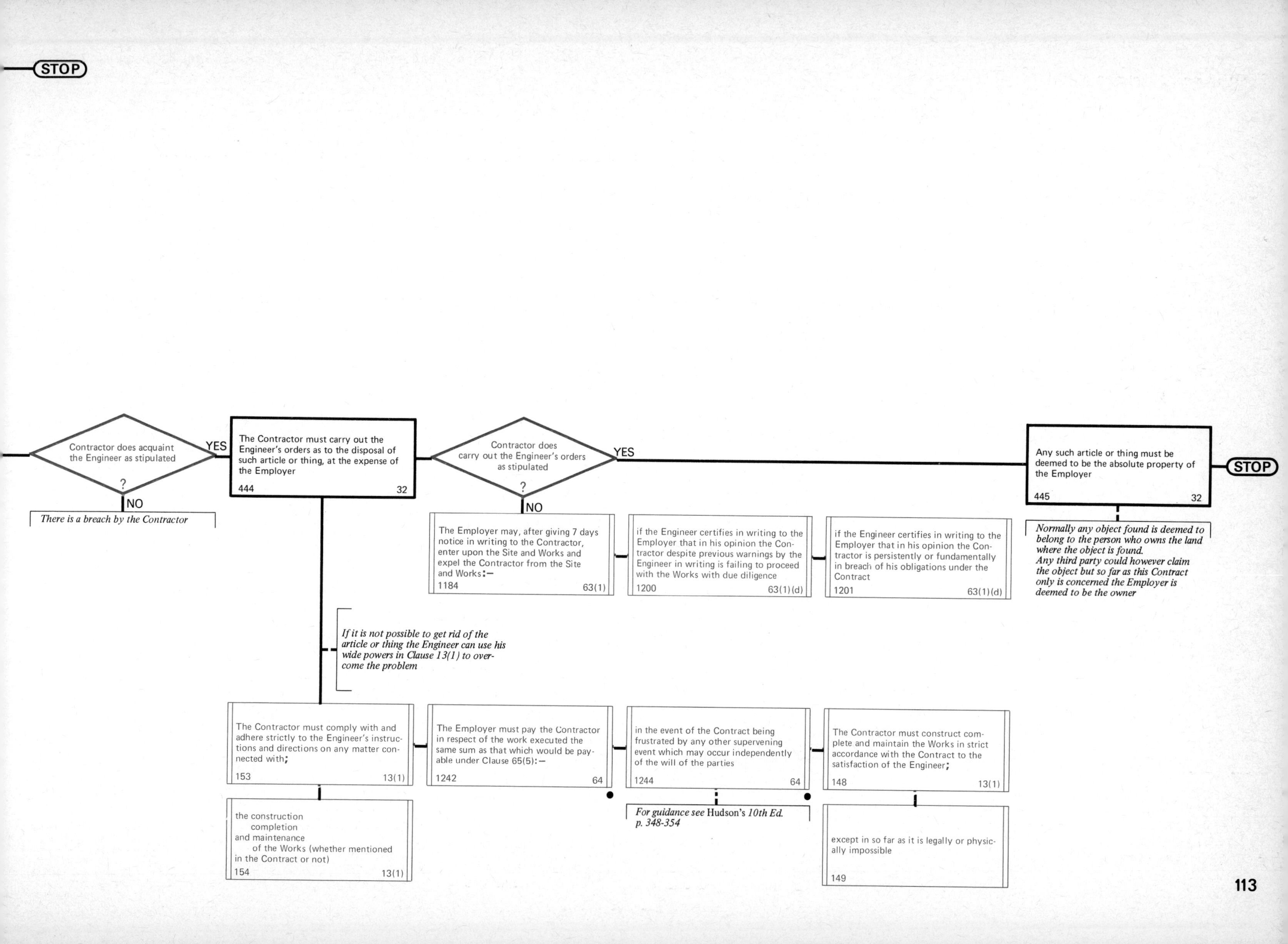
STOP
Contractor does acquaint the Engineer as stipulated ?
YES
NO
There is a breach by the Contractor
The Contractor must carry out the Engineer's orders as to the disposal of such article or thing, at the expense of the Employer
444 32
Contractor does carry out the Engineer's orders as stipulated ?
YES
NO
Any such article or thing must be deemed to be the absolute property of the Employer
445 32
STOP
Normally any object found is deemed to belong to the person who owns the land where the object is found.
Any third party could however claim the object but so far as this Contract only is concerned the Employer is deemed to be the owner
The Employer may, after giving 7 days notice in writing to the Contractor, enter upon the Site and Works and expel the Contractor from the Site and Works:—
1184 63(1)
if the Engineer certifies in writing to the Employer that in his opinion the Contractor despite previous warnings by the Engineer in writing is failing to proceed with the Works with due diligence
1200 63(1)(d)
if the Engineer certifies in writing to the Employer that in his opinion the Contractor is persistently or fundamentally in breach of his obligations under the Contract
1201 63(1)(d)
If it is not possible to get rid of the article or thing the Engineer can use his wide powers in Clause 13(1) to overcome the problem
The Contractor must comply with and adhere strictly to the Engineer's instructions and directions on any matter connected with;
153 13(1)
the construction
completion
and maintenance
of the Works (whether mentioned in the Contract or not)
154 13(1)
The Employer must pay the Contractor in respect of the work executed the same sum as that which would be payable under Clause 65(5):—
1242 64
in the event of the Contract being frustrated by any other supervening event which may occur independently of the will of the parties
1244 64
For guidance see Hudson's 10th Ed. p. 348-354
The Contractor must construct complete and maintain the Works in strict accordance with the Contract to the satisfaction of the Engineer;
148 13(1)
except in so far as it is legally or physically impossible
149

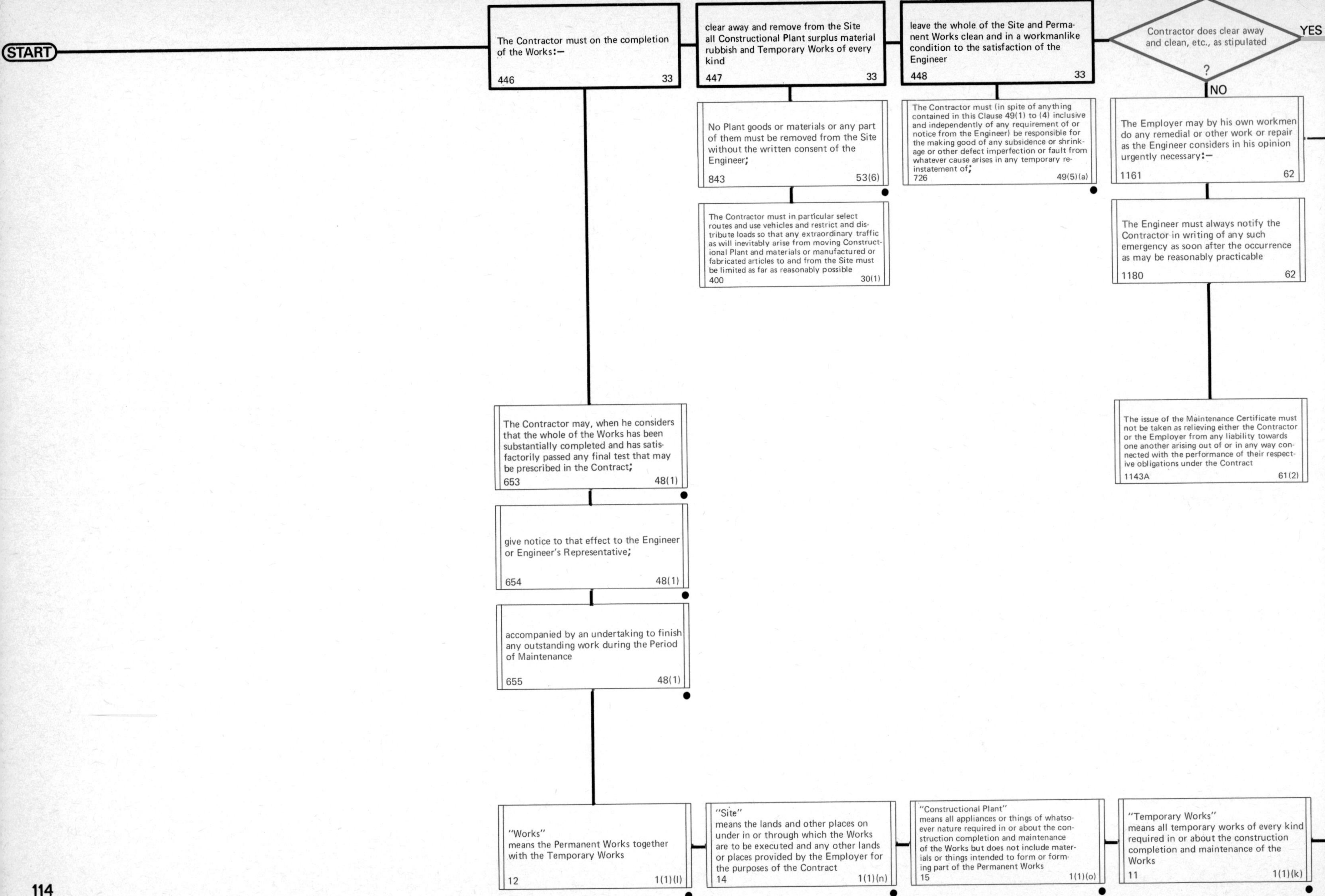
START
The Contractor must on the completion of the Works:—
446 33
clear away and remove from the Site all Constructional Plant surplus material rubbish and Temporary Works of every kind
447 33
leave the whole of the Site and Permanent Works clean and in a workmanlike condition to the satisfaction of the Engineer
448 33
Contractor does clear away and clean, etc., as stipulated ?
YES
NO
No Plant goods or materials or any part of them must be removed from the Site without the written consent of the Engineer;
843 53(6)
The Contractor must in particular select routes and use vehicles and restrict and distribute loads so that any extraordinary traffic as will inevitably arise from moving Constructional Plant and materials or manufactured or fabricated articles to and from the Site must be limited as far as reasonably possible
400 30(1)
The Contractor must (in spite of anything contained in this Clause 49(1) to (4) inclusive and independently of any requirement of or notice from the Engineer) be responsible for the making good of any subsidence or shrinkage or other defect imperfection or fault from whatever cause arises in any temporary reinstatement of;
726 49(5)(a)
The Employer may by his own workmen do any remedial or other work or repair as the Engineer considers in his opinion urgently necessary:—
1161 62
The Engineer must always notify the Contractor in writing of any such emergency as soon after the occurrence as may be reasonably practicable
1180 62
The issue of the Maintenance Certificate must not be taken as relieving either the Contractor or the Employer from any liability towards one another arising out of or in any way connected with the performance of their respective obligations under the Contract
1143A 61(2)
The Contractor may, when he considers that the whole of the Works has been substantially completed and has satisfactorily passed any final test that may be prescribed in the Contract;
653 48(1)
give notice to that effect to the Engineer or Engineer's Representative;
654 48(1)
accompanied by an undertaking to finish any outstanding work during the Period of Maintenance
655 48(1)
"Works" means the Permanent Works together with the Temporary Works
12 1(1)(l)
"Site" means the lands and other places on under in or through which the Works are to be executed and any other lands or places provided by the Employer for the purposes of the Contract
14 1(1)(n)
"Constructional Plant" means all appliances or things of whatsoever nature required in or about the construction completion and maintenance of the Works but does not include materials or things intended to form or forming part of the Permanent Works
15 1(1)(o)
"Temporary Works" means all temporary works of every kind required in or about the construction completion and maintenance of the Works
11 1(1)(k)

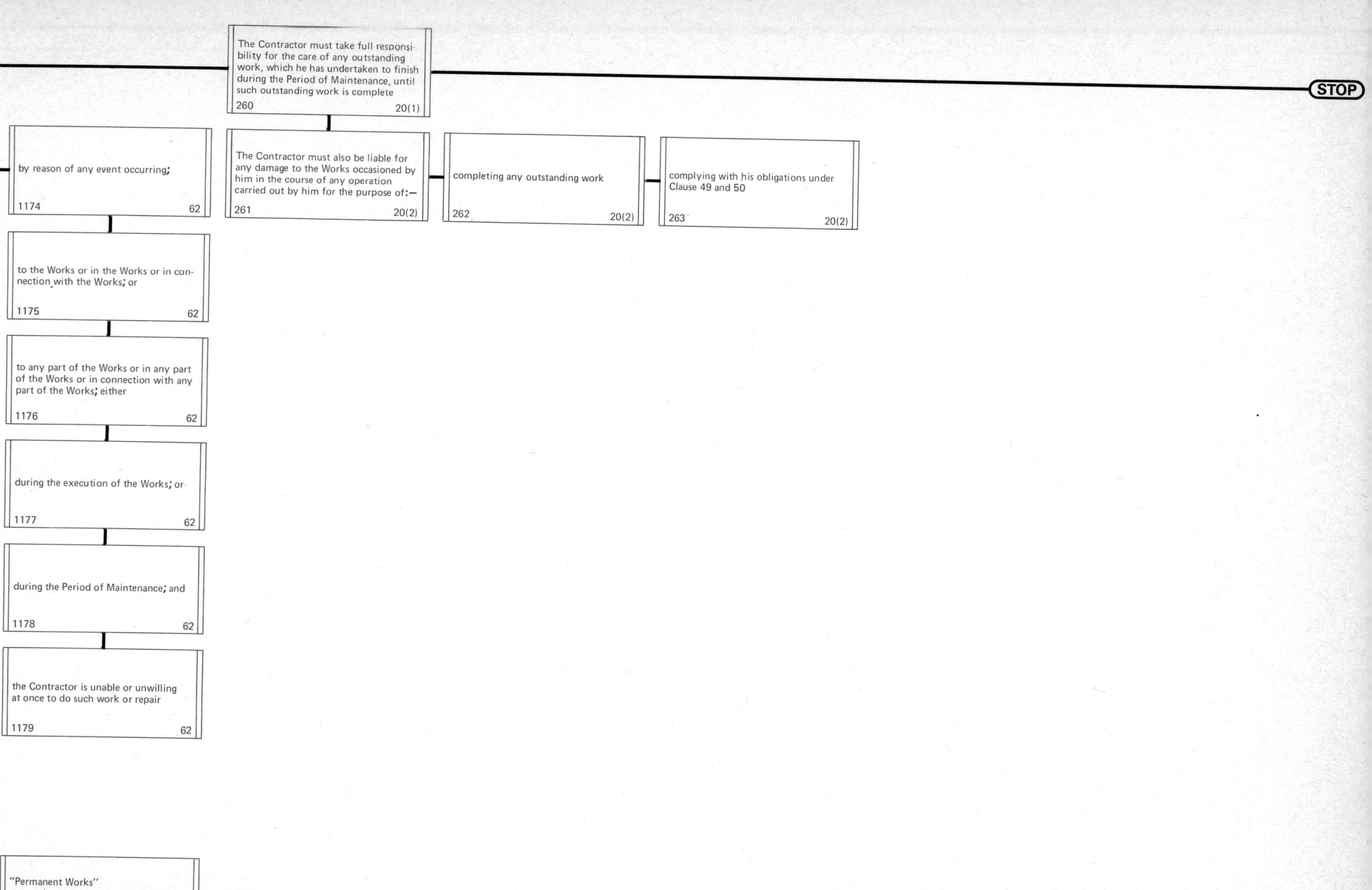
The Contractor must take full responsibility for the care of any outstanding work, which he has undertaken to finish during the Period of Maintenance, until such outstanding work is complete
260
20(1)
STOP
by reason of any event occurring;
1174
62
The Contractor must also be liable for any damage to the Works occasioned by him in the course of any operation carried out by him for the purpose of:—
261
20(2)
completing any outstanding work
262
20(2)
complying with his obligations under Clause 49 and 50
263
20(2)
to the Works or in the Works or in connection with the Works; or
1175
62
to any part of the Works or in any part of the Works or in connection with any part of the Works; either
1176
62
during the execution of the Works; or
1177
62
during the Period of Maintenance; and
1178
62
the Contractor is unable or unwilling at once to do such work or repair
1179
62
"Permanent Works"
means the permanent works to be constructed completed and maintained in accordance with the Contract
10
1(1)(j)

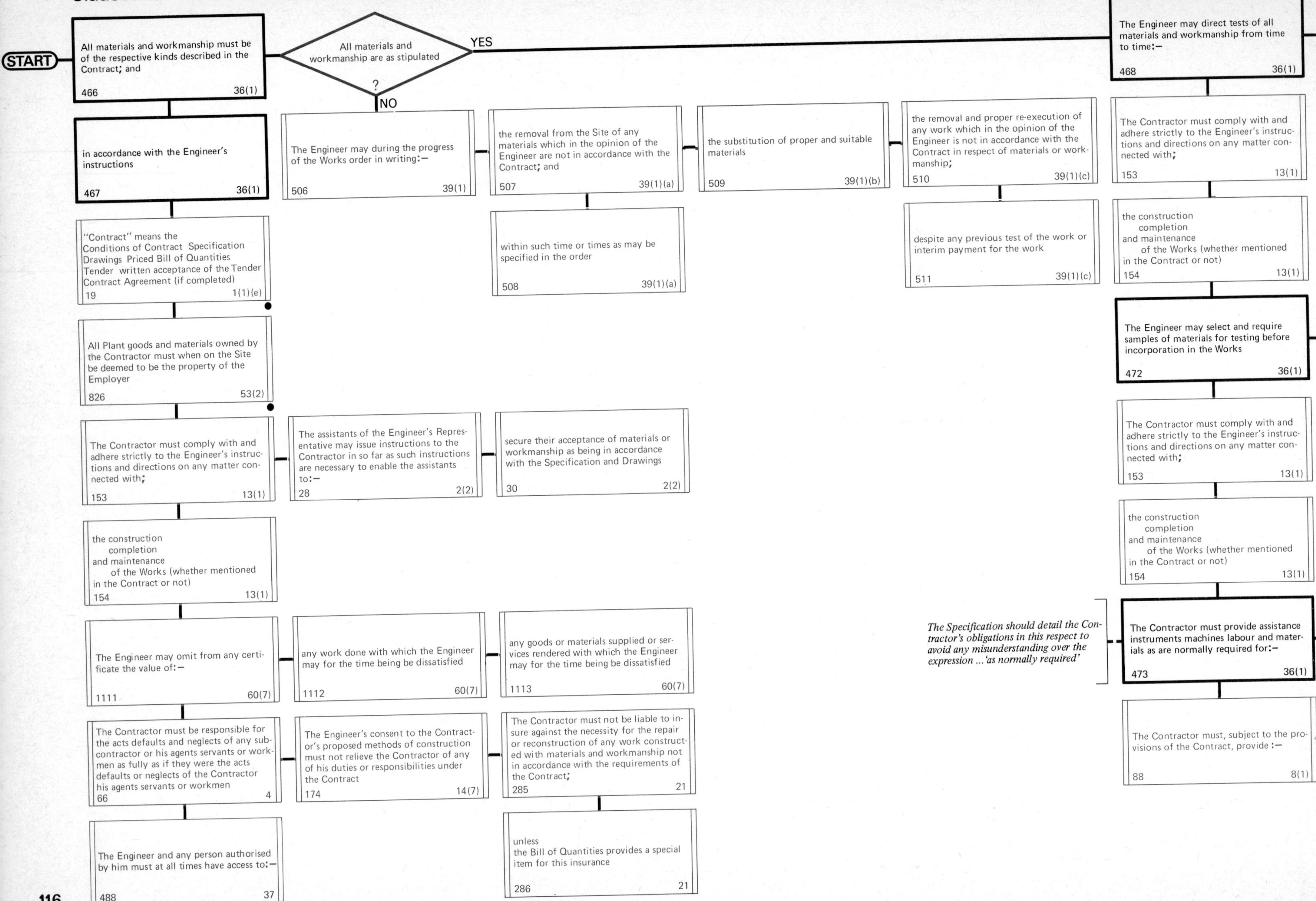
START
All materials and workmanship must be of the respective kinds described in the Contract; and 466 36(1)
All materials and workmanship are as stipulated ?
YES
NO
The Engineer may direct tests of all materials and workmanship from time to time:— 468 36(1)
in accordance with the Engineer's instructions 467 36(1)
The Engineer may during the progress of the Works order in writing:— 506 39(1)
the removal from the Site of any materials which in the opinion of the Engineer are not in accordance with the Contract; and 507 39(1)(a)
the substitution of proper and suitable materials 509 39(1)(b)
the removal and proper re-execution of any work which in the opinion of the Engineer is not in accordance with the Contract in respect of materials or workmanship; 510 39(1)(c)
The Contractor must comply with and adhere strictly to the Engineer's instructions and directions on any matter connected with; 153 13(1)
"Contract" means the Conditions of Contract Specification Drawings Priced Bill of Quantities Tender written acceptance of the Tender Contract Agreement (if completed) 19 1(1)(e)
within such time or times as may be specified in the order 508 39(1)(a)
despite any previous test of the work or interim payment for the work 511 39(1)(c)
the construction completion and maintenance of the Works (whether mentioned in the Contract or not) 154 13(1)
All Plant goods and materials owned by the Contractor must when on the Site be deemed to be the property of the Employer 826 53(2)
The Engineer may select and require samples of materials for testing before incorporation in the Works 472 36(1)
The Contractor must comply with and adhere strictly to the Engineer's instructions and directions on any matter connected with; 153 13(1)
The assistants of the Engineer's Representative may issue instructions to the Contractor in so far as such instructions are necessary to enable the assistants to:— 28 2(2)
secure their acceptance of materials or workmanship as being in accordance with the Specification and Drawings 30 2(2)
The Contractor must comply with and adhere strictly to the Engineer's instructions and directions on any matter connected with; 153 13(1)
the construction completion and maintenance of the Works (whether mentioned in the Contract or not) 154 13(1)
the construction completion and maintenance of the Works (whether mentioned in the Contract or not) 154 13(1)
The Engineer may omit from any certificate the value of:— 1111 60(7)
any work done with which the Engineer may for the time being be dissatisfied 1112 60(7)
any goods or materials supplied or services rendered with which the Engineer may for the time being be dissatisfied 1113 60(7)
The Specification should detail the Contractor's obligations in this respect to avoid any misunderstanding over the expression ...'as normally required'
The Contractor must provide assistance instruments machines labour and materials as are normally required for:— 473 36(1)
The Contractor must be responsible for the acts defaults and neglects of any subcontractor or his agents servants or workmen as fully as if they were the acts defaults or neglects of the Contractor his agents servants or workmen 66 4
The Engineer's consent to the Contractor's proposed methods of construction must not relieve the Contractor of any of his duties or responsibilities under the Contract 174 14(7)
The Contractor must not be liable to insure against the necessity for the repair or reconstruction of any work constructed with materials and workmanship not in accordance with the requirements of the Contract; 285 21
The Contractor must, subject to the provisions of the Contract, provide :— 88 8(1)
The Engineer and any person authorised by him must at all times have access to:— 488 37
unless the Bill of Quantities provides a special item for this insurance 286 21

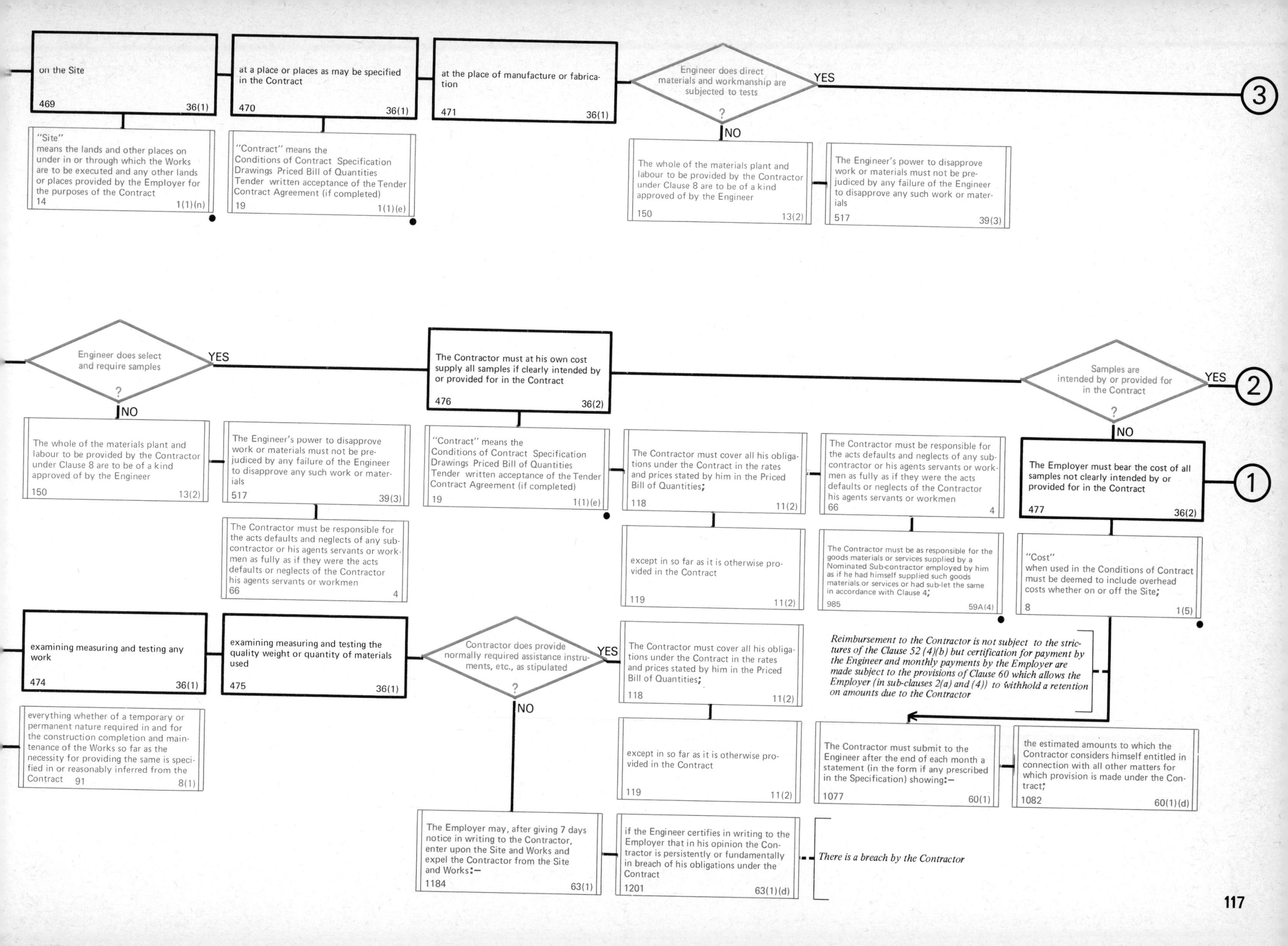

on the Site
469 36(1)
at a place or places as may be specified in the Contract
470 36(1)
at the place of manufacture or fabrication
471 36(1)
Engineer does direct materials and workmanship are subjected to tests ?
YES
NO
3
"Site" means the lands and other places on under in or through which the Works are to be executed and any other lands or places provided by the Employer for the purposes of the Contract
14 1(1)(n)
"Contract" means the Conditions of Contract Specification Drawings Priced Bill of Quantities Tender written acceptance of the Tender Contract Agreement (if completed)
19 1(1)(e)
The whole of the materials plant and labour to be provided by the Contractor under Clause 8 are to be of a kind approved of by the Engineer
150 13(2)
The Engineer's power to disapprove work or materials must not be prejudiced by any failure of the Engineer to disapprove any such work or materials
517 39(3)
Engineer does select and require samples ?
YES
NO
The Contractor must at his own cost supply all samples if clearly intended by or provided for in the Contract
476 36(2)
Samples are intended by or provided for in the Contract ?
YES
2
NO
The whole of the materials plant and labour to be provided by the Contractor under Clause 8 are to be of a kind approved of by the Engineer
150 13(2)
The Engineer's power to disapprove work or materials must not be prejudiced by any failure of the Engineer to disapprove any such work or materials
517 39(3)
The Contractor must be responsible for the acts defaults and neglects of any sub-contractor or his agents servants or workmen as fully as if they were the acts defaults or neglects of the Contractor his agents servants or workmen
66 4
"Contract" means the Conditions of Contract Specification Drawings Priced Bill of Quantities Tender written acceptance of the Tender Contract Agreement (if completed)
19 1(1)(e)
The Contractor must cover all his obligations under the Contract in the rates and prices stated by him in the Priced Bill of Quantities;
118 11(2)
except in so far as it is otherwise provided in the Contract
119 11(2)
The Contractor must be responsible for the acts defaults and neglects of any sub-contractor or his agents servants or workmen as fully as if they were the acts defaults or neglects of the Contractor his agents servants or workmen
66 4
The Contractor must be as responsible for the goods materials or services supplied by a Nominated Sub-contractor employed by him as if he had himself supplied such goods materials or services or had sub-let the same in accordance with Clause 4;
985 59A(4)
The Employer must bear the cost of all samples not clearly intended by or provided for in the Contract
477 36(2)
1
"Cost" when used in the Conditions of Contract must be deemed to include overhead costs whether on or off the Site;
8 1(5)
examining measuring and testing any work
474 36(1)
examining measuring and testing the quality weight or quantity of materials used
475 36(1)
Contractor does provide normally required assistance instruments, etc., as stipulated ?
YES
NO
everything whether of a temporary or permanent nature required in and for the construction completion and maintenance of the Works so far as the necessity for providing the same is specified in or reasonably inferred from the Contract
91 8(1)
The Contractor must cover all his obligations under the Contract in the rates and prices stated by him in the Priced Bill of Quantities;
118 11(2)
except in so far as it is otherwise provided in the Contract
119 11(2)
Reimbursement to the Contractor is not subject to the strictures of the Clause 52 (4)(b) but certification for payment by the Engineer and monthly payments by the Employer are made subject to the provisions of Clause 60 which allows the Employer (in sub-clauses 2(a) and (4)) to withhold a retention on amounts due to the Contractor
The Contractor must submit to the Engineer after the end of each month a statement (in the form if any prescribed in the Specification) showing:—
1077 60(1)
the estimated amounts to which the Contractor considers himself entitled in connection with all other matters for which provision is made under the Contract;
1082 60(1)(d)
The Employer may, after giving 7 days notice in writing to the Contractor, enter upon the Site and Works and expel the Contractor from the Site and Works:—
1184 63(1)
if the Engineer certifies in writing to the Employer that in his opinion the Contractor is persistently or fundamentally in breach of his obligations under the Contract
1201 63(1)(d)
There is a breach by the Contractor

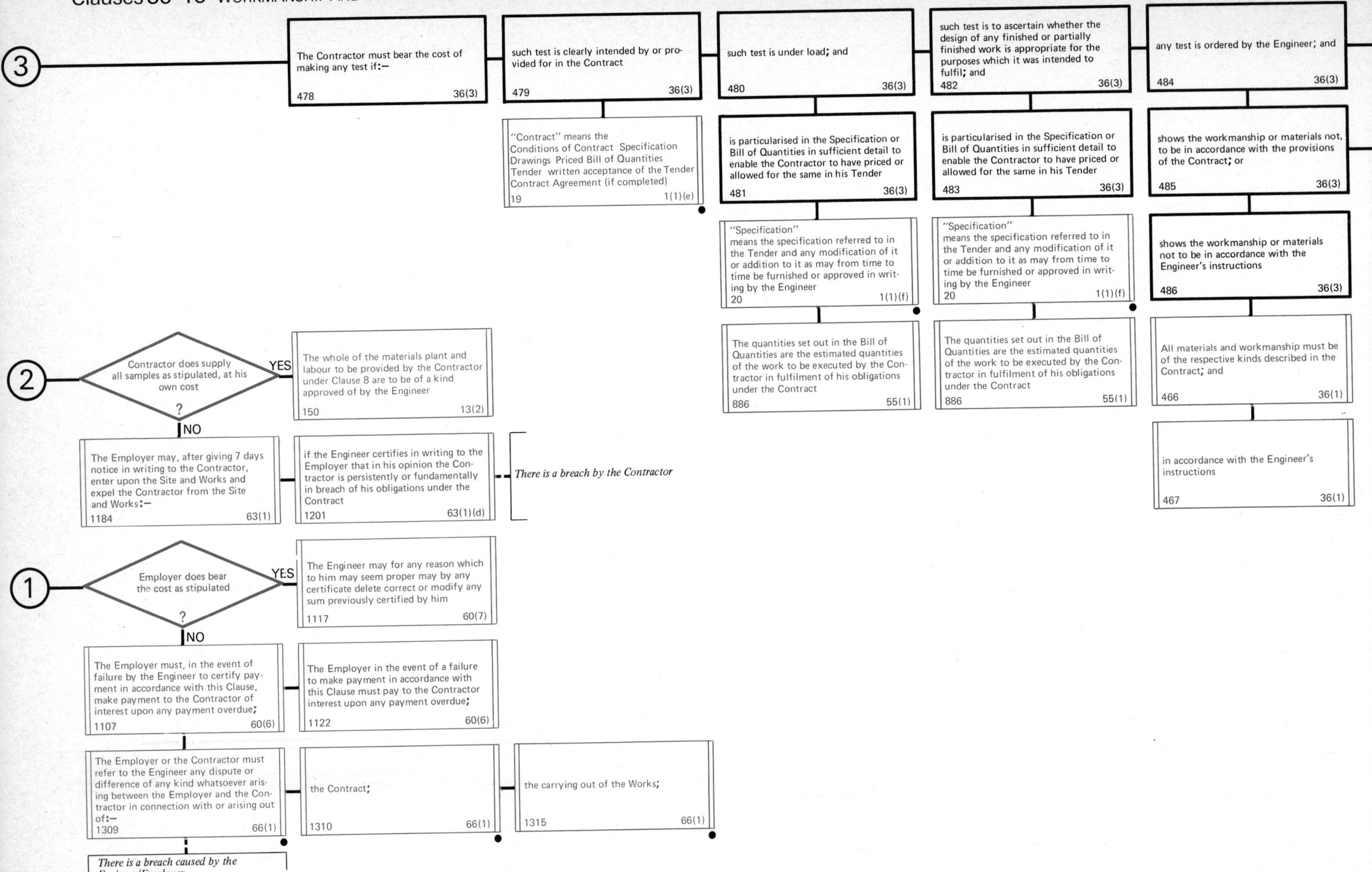
3
The Contractor must bear the cost of making any test if:—
478 36(3)
such test is clearly intended by or provided for in the Contract
479 36(3)
"Contract" means the Conditions of Contract Specification Drawings Priced Bill of Quantities Tender written acceptance of the Tender Contract Agreement (if completed)
19 1(1)(e)
such test is under load; and
480 36(3)
is particularised in the Specification or Bill of Quantities in sufficient detail to enable the Contractor to have priced or allowed for the same in his Tender
481 36(3)
"Specification" means the specification referred to in the Tender and any modification of it or addition to it as may from time to time be furnished or approved in writing by the Engineer
20 1(1)(f)
The quantities set out in the Bill of Quantities are the estimated quantities of the work to be executed by the Contractor in fulfilment of his obligations under the Contract
886 55(1)
such test is to ascertain whether the design of any finished or partially finished work is appropriate for the purposes which it was intended to fulfil; and
482 36(3)
is particularised in the Specification or Bill of Quantities in sufficient detail to enable the Contractor to have priced or allowed for the same in his Tender
483 36(3)
"Specification" means the specification referred to in the Tender and any modification of it or addition to it as may from time to time be furnished or approved in writing by the Engineer
20 1(1)(f)
The quantities set out in the Bill of Quantities are the estimated quantities of the work to be executed by the Contractor in fulfilment of his obligations under the Contract
886 55(1)
any test is ordered by the Engineer; and
484 36(3)
shows the workmanship or materials not to be in accordance with the provisions of the Contract; or
485 36(3)
shows the workmanship or materials not to be in accordance with the Engineer's instructions
486 36(3)
All materials and workmanship must be of the respective kinds described in the Contract; and
466 36(1)
in accordance with the Engineer's instructions
467 36(1)
2
Contractor does supply all samples as stipulated, at his own cost ?
YES
The whole of the materials plant and labour to be provided by the Contractor under Clause 8 are to be of a kind approved of by the Engineer
150 13(2)
NO
The Employer may, after giving 7 days notice in writing to the Contractor, enter upon the Site and Works and expel the Contractor from the Site and Works:—
1184 63(1)
if the Engineer certifies in writing to the Employer that in his opinion the Contractor is persistently or fundamentally in breach of his obligations under the Contract
1201 63(1)(d)
There is a breach by the Contractor
1
Employer does bear the cost as stipulated ?
YES
The Engineer may for any reason which to him may seem proper may by any certificate delete correct or modify any sum previously certified by him
1117 60(7)
NO
The Employer must, in the event of failure by the Engineer to certify payment in accordance with this Clause, make payment to the Contractor of interest upon any payment overdue;
1107 60(6)
The Employer in the event of a failure to make payment in accordance with this Clause must pay to the Contractor interest upon any payment overdue;
1122 60(6)
The Employer or the Contractor must refer to the Engineer any dispute or difference of any kind whatsoever arising between the Employer and the Contractor in connection with or arising out of:—
1309 66(1)
the Contract;
1310 66(1)
the carrying out of the Works;
1315 66(1)
There is a breach caused by the Engineer/Employer

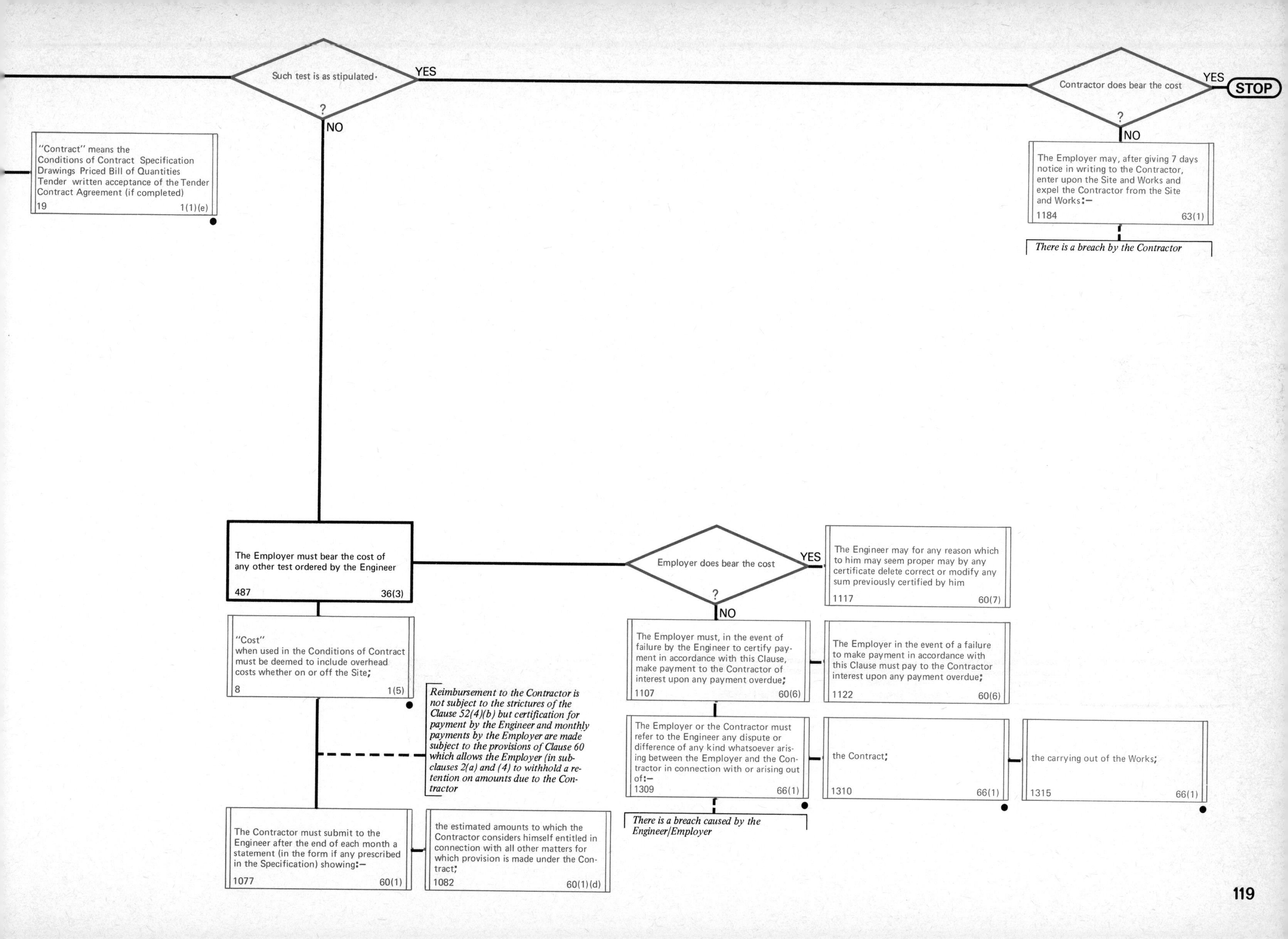
Such test is as stipulated ?
YES
NO
Contractor does bear the cost ?
YES
STOP
NO
"Contract" means the Conditions of Contract Specification Drawings Priced Bill of Quantities Tender written acceptance of the Tender Contract Agreement (if completed)
19 1(1)(e)
The Employer may, after giving 7 days notice in writing to the Contractor, enter upon the Site and Works and expel the Contractor from the Site and Works:—
1184 63(1)
There is a breach by the Contractor
The Employer must bear the cost of any other test ordered by the Engineer
487 36(3)
Employer does bear the cost ?
YES
NO
The Engineer may for any reason which to him may seem proper may by any certificate delete correct or modify any sum previously certified by him
1117 60(7)
"Cost" when used in the Conditions of Contract must be deemed to include overhead costs whether on or off the Site;
8 1(5)
The Employer must, in the event of failure by the Engineer to certify payment in accordance with this Clause, make payment to the Contractor of interest upon any payment overdue;
1107 60(6)
The Employer in the event of a failure to make payment in accordance with this Clause must pay to the Contractor interest upon any payment overdue;
1122 60(6)
Reimbursement to the Contractor is not subject to the strictures of the Clause 52(4)(b) but certification for payment by the Engineer and monthly payments by the Employer are made subject to the provisions of Clause 60 which allows the Employer (in sub-clauses 2(a) and (4) to withhold a retention on amounts due to the Contractor
The Employer or the Contractor must refer to the Engineer any dispute or difference of any kind whatsoever arising between the Employer and the Contractor in connection with or arising out of:—
1309 66(1)
the Contract;
1310 66(1)
the carrying out of the Works;
1315 66(1)
There is a breach caused by the Engineer/Employer
The Contractor must submit to the Engineer after the end of each month a statement (in the form if any prescribed in the Specification) showing:—
1077 60(1)
the estimated amounts to which the Contractor considers himself entitled in connection with all other matters for which provision is made under the Contract;
1082 60(1)(d)

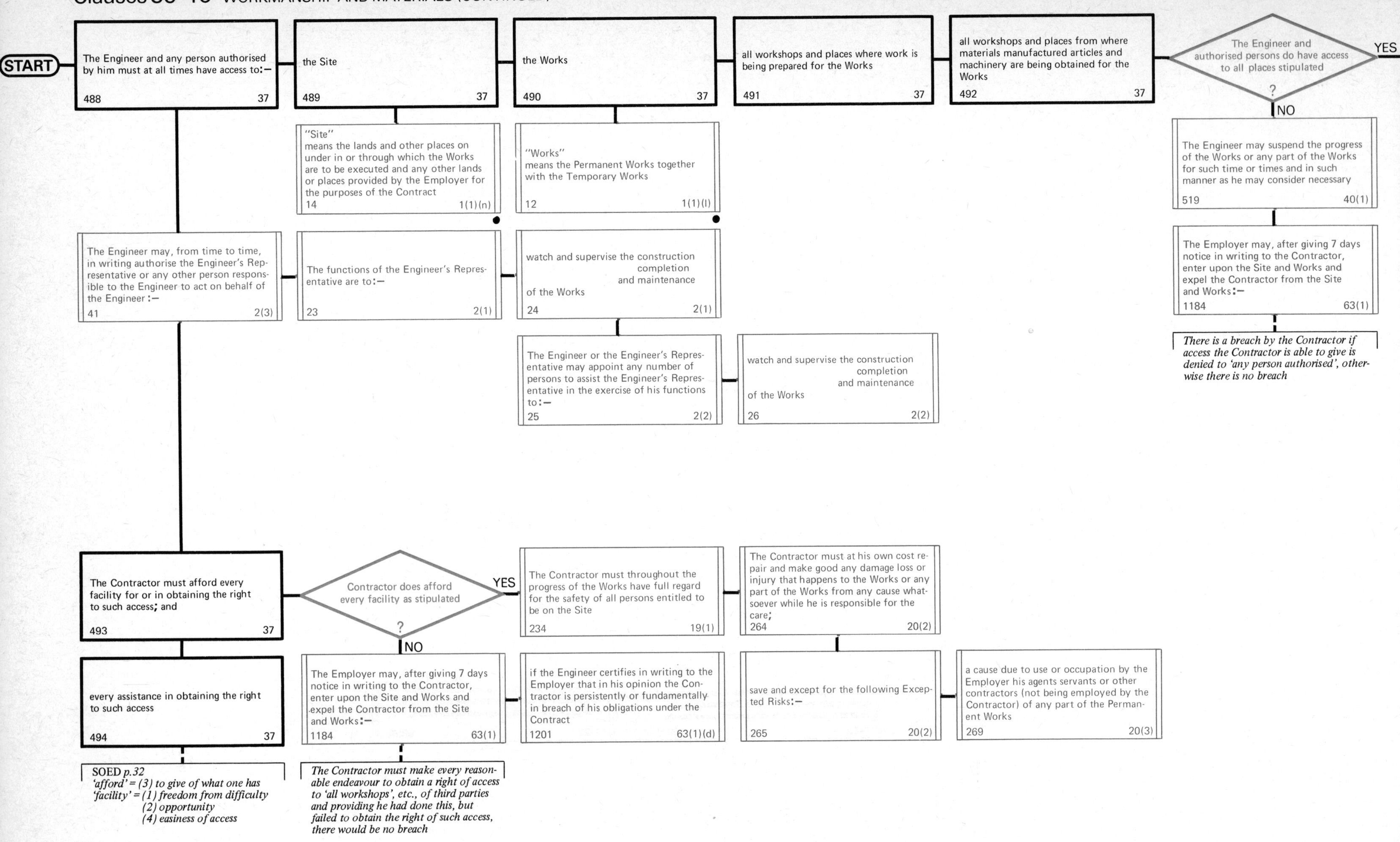
START
The Engineer and any person authorised by him must at all times have access to:— 488 37
the Site 489 37
the Works 490 37
all workshops and places where work is being prepared for the Works 491 37
all workshops and places from where materials manufactured articles and machinery are being obtained for the Works 492 37
The Engineer and authorised persons do have access to all places stipulated ?
YES
NO
"Site" means the lands and other places on under in or through which the Works are to be executed and any other lands or places provided by the Employer for the purposes of the Contract 14 1(1)(n)
"Works" means the Permanent Works together with the Temporary Works 12 1(1)(l)
The Engineer may suspend the progress of the Works or any part of the Works for such time or times and in such manner as he may consider necessary 519 40(1)
The Engineer may, from time to time, in writing authorise the Engineer's Representative or any other person responsible to the Engineer to act on behalf of the Engineer:— 41 2(3)
The functions of the Engineer's Representative are to:— 23 2(1)
watch and supervise the construction completion and maintenance of the Works 24 2(1)
The Employer may, after giving 7 days notice in writing to the Contractor, enter upon the Site and Works and expel the Contractor from the Site and Works:— 1184 63(1)
There is a breach by the Contractor if access the Contractor is able to give is denied to 'any person authorised', otherwise there is no breach
The Engineer or the Engineer's Representative may appoint any number of persons to assist the Engineer's Representative in the exercise of his functions to:— 25 2(2)
watch and supervise the construction completion and maintenance of the Works 26 2(2)
The Contractor must afford every facility for or in obtaining the right to such access; and 493 37
Contractor does afford every facility as stipulated ?
YES
NO
The Contractor must throughout the progress of the Works have full regard for the safety of all persons entitled to be on the Site 234 19(1)
The Contractor must at his own cost repair and make good any damage loss or injury that happens to the Works or any part of the Works from any cause whatsoever while he is responsible for the care; 264 20(2)
every assistance in obtaining the right to such access 494 37
The Employer may, after giving 7 days notice in writing to the Contractor, enter upon the Site and Works and expel the Contractor from the Site and Works:— 1184 63(1)
if the Engineer certifies in writing to the Employer that in his opinion the Contractor is persistently or fundamentally in breach of his obligations under the Contract 1201 63(1)(d)
save and except for the following Excepted Risks:— 265 20(2)
a cause due to use or occupation by the Employer his agents servants or other contractors (not being employed by the Contractor) of any part of the Permanent Works 269 20(3)
SOED p.32
'afford' = (3) to give of what one has
'facility' = (1) freedom from difficulty
(2) opportunity
(4) easiness of access
The Contractor must make every reasonable endeavour to obtain a right of access to 'all workshops', etc., of third parties and providing he had done this, but failed to obtain the right of such access, there would be no breach

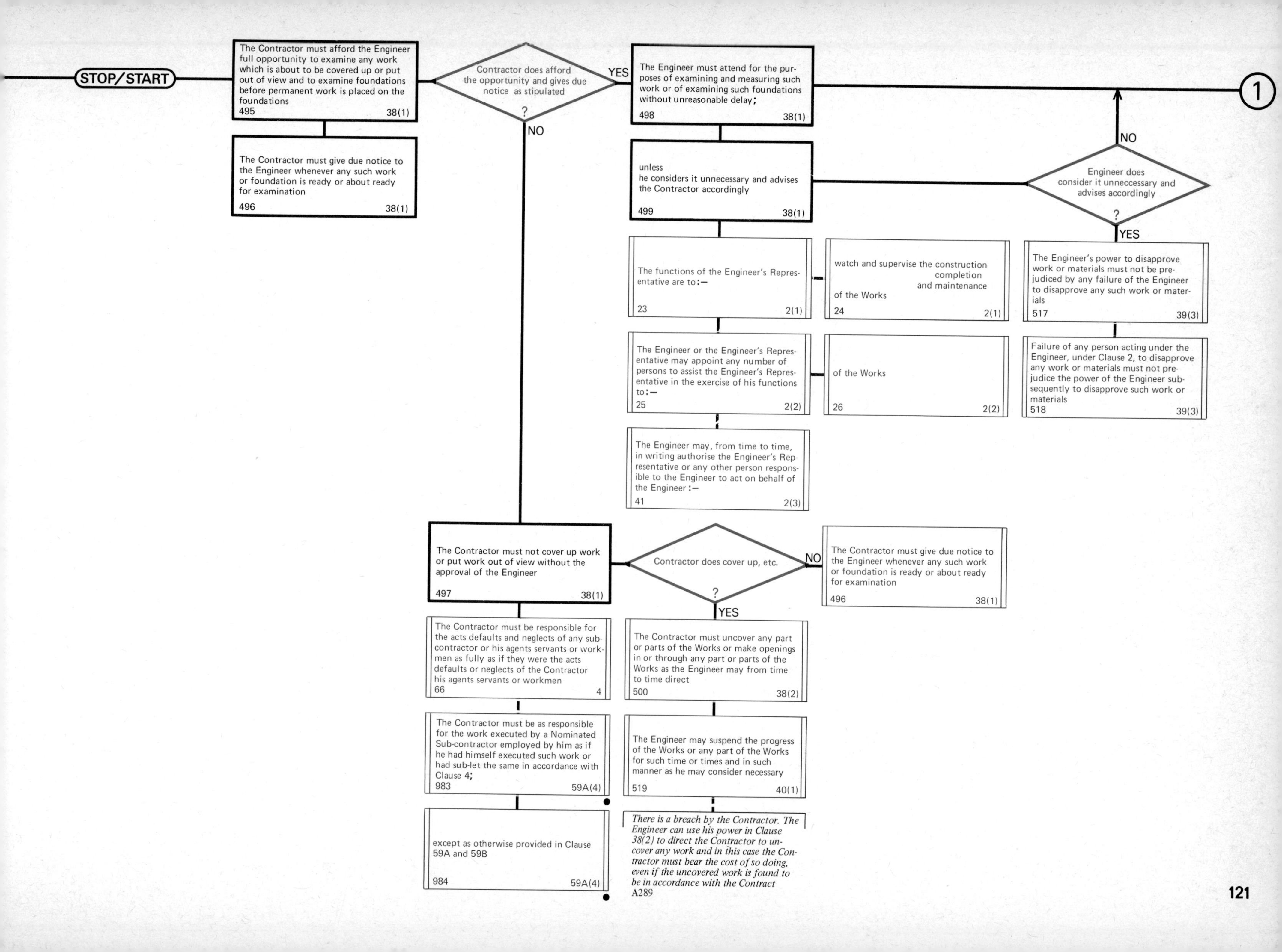
STOP/START
The Contractor must afford the Engineer full opportunity to examine any work which is about to be covered up or put out of view and to examine foundations before permanent work is placed on the foundations
495 38(1)
The Contractor must give due notice to the Engineer whenever any such work or foundation is ready or about ready for examination
496 38(1)
Contractor does afford the opportunity and gives due notice as stipulated ?
YES
NO
The Engineer must attend for the purposes of examining and measuring such work or of examining such foundations without unreasonable delay;
498 38(1)
1
unless he considers it unnecessary and advises the Contractor accordingly
499 38(1)
Engineer does consider it unneccessary and advises accordingly ?
NO
YES
The functions of the Engineer's Representative are to:—
23 2(1)
watch and supervise the construction completion and maintenance of the Works
24 2(1)
The Engineer's power to disapprove work or materials must not be prejudiced by any failure of the Engineer to disapprove any such work or materials
517 39(3)
The Engineer or the Engineer's Representative may appoint any number of persons to assist the Engineer's Representative in the exercise of his functions to:—
25 2(2)
of the Works
26 2(2)
Failure of any person acting under the Engineer, under Clause 2, to disapprove any work or materials must not prejudice the power of the Engineer subsequently to disapprove such work or materials
518 39(3)
The Engineer may, from time to time, in writing authorise the Engineer's Representative or any other person responsible to the Engineer to act on behalf of the Engineer :—
41 2(3)
The Contractor must not cover up work or put work out of view without the approval of the Engineer
497 38(1)
Contractor does cover up, etc. ?
NO
YES
The Contractor must give due notice to the Engineer whenever any such work or foundation is ready or about ready for examination
496 38(1)
The Contractor must be responsible for the acts defaults and neglects of any sub-contractor or his agents servants or workmen as fully as if they were the acts defaults or neglects of the Contractor his agents servants or workmen
66 4
The Contractor must uncover any part or parts of the Works or make openings in or through any part or parts of the Works as the Engineer may from time to time direct
500 38(2)
The Contractor must be as responsible for the work executed by a Nominated Sub-contractor employed by him as if he had himself executed such work or had sub-let the same in accordance with Clause 4;
983 59A(4)
The Engineer may suspend the progress of the Works or any part of the Works for such time or times and in such manner as he may consider necessary
519 40(1)
except as otherwise provided in Clause 59A and 59B
984 59A(4)
There is a breach by the Contractor. The Engineer can use his power in Clause 38(2) to direct the Contractor to uncover any work and in this case the Contractor must bear the cost of so doing, even if the uncovered work is found to be in accordance with the Contract
A289

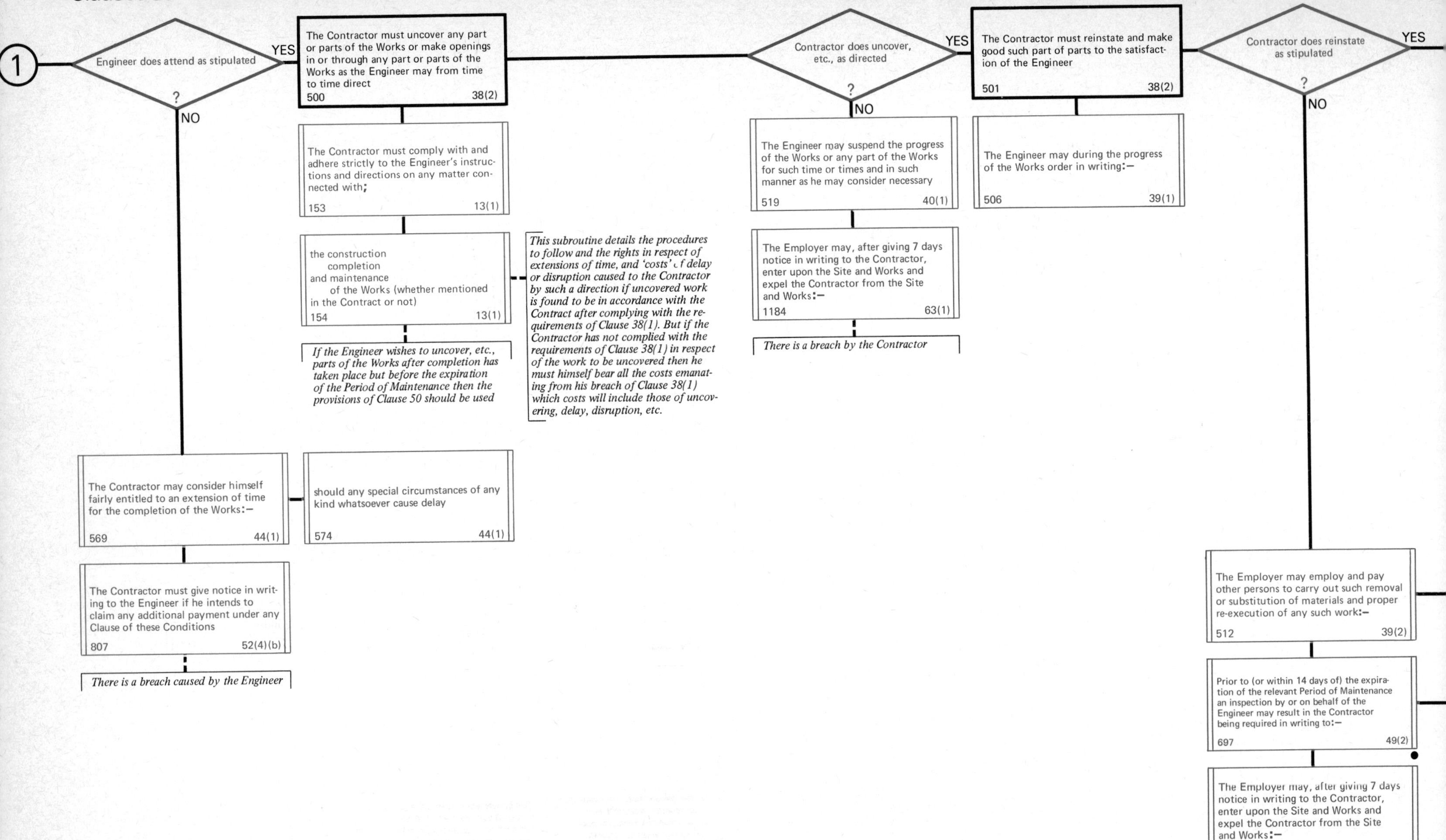
1
Engineer does attend as stipulated
?
YES
NO
The Contractor must uncover any part or parts of the Works or make openings in or through any part or parts of the Works as the Engineer may from time to time direct
500 38(2)
The Contractor must comply with and adhere strictly to the Engineer's instructions and directions on any matter connected with;
153 13(1)
the construction completion and maintenance of the Works (whether mentioned in the Contract or not)
154 13(1)
If the Engineer wishes to uncover, etc., parts of the Works after completion has taken place but before the expiration of the Period of Maintenance then the provisions of Clause 50 should be used
This subroutine details the procedures to follow and the rights in respect of extensions of time, and 'costs' of delay or disruption caused to the Contractor by such a direction if uncovered work is found to be in accordance with the Contract after complying with the requirements of Clause 38(1). But if the Contractor has not complied with the requirements of Clause 38(1) in respect of the work to be uncovered then he must himself bear all the costs emanating from his breach of Clause 38(1) which costs will include those of uncovering, delay, disruption, etc.
The Contractor may consider himself fairly entitled to an extension of time for the completion of the Works:—
569 44(1)
should any special circumstances of any kind whatsoever cause delay
574 44(1)
The Contractor must give notice in writing to the Engineer if he intends to claim any additional payment under any Clause of these Conditions
807 52(4)(b)
There is a breach caused by the Engineer
Contractor does uncover, etc., as directed
?
YES
NO
The Engineer may suspend the progress of the Works or any part of the Works for such time or times and in such manner as he may consider necessary
519 40(1)
The Employer may, after giving 7 days notice in writing to the Contractor, enter upon the Site and Works and expel the Contractor from the Site and Works:—
1184 63(1)
There is a breach by the Contractor
The Contractor must reinstate and make good such part of parts to the satisfaction of the Engineer
501 38(2)
The Engineer may during the progress of the Works order in writing:—
506 39(1)
Contractor does reinstate as stipulated
?
YES
NO
The Employer may employ and pay other persons to carry out such removal or substitution of materials and proper re-execution of any such work:—
512 39(2)
Prior to (or within 14 days of) the expiration of the relevant Period of Maintenance an inspection by or on behalf of the Engineer may result in the Contractor being required in writing to:—
697 49(2)
The Employer may, after giving 7 days notice in writing to the Contractor, enter upon the Site and Works and expel the Contractor from the Site and Works:—
1184 63(1)
There is a breach by the Contractor

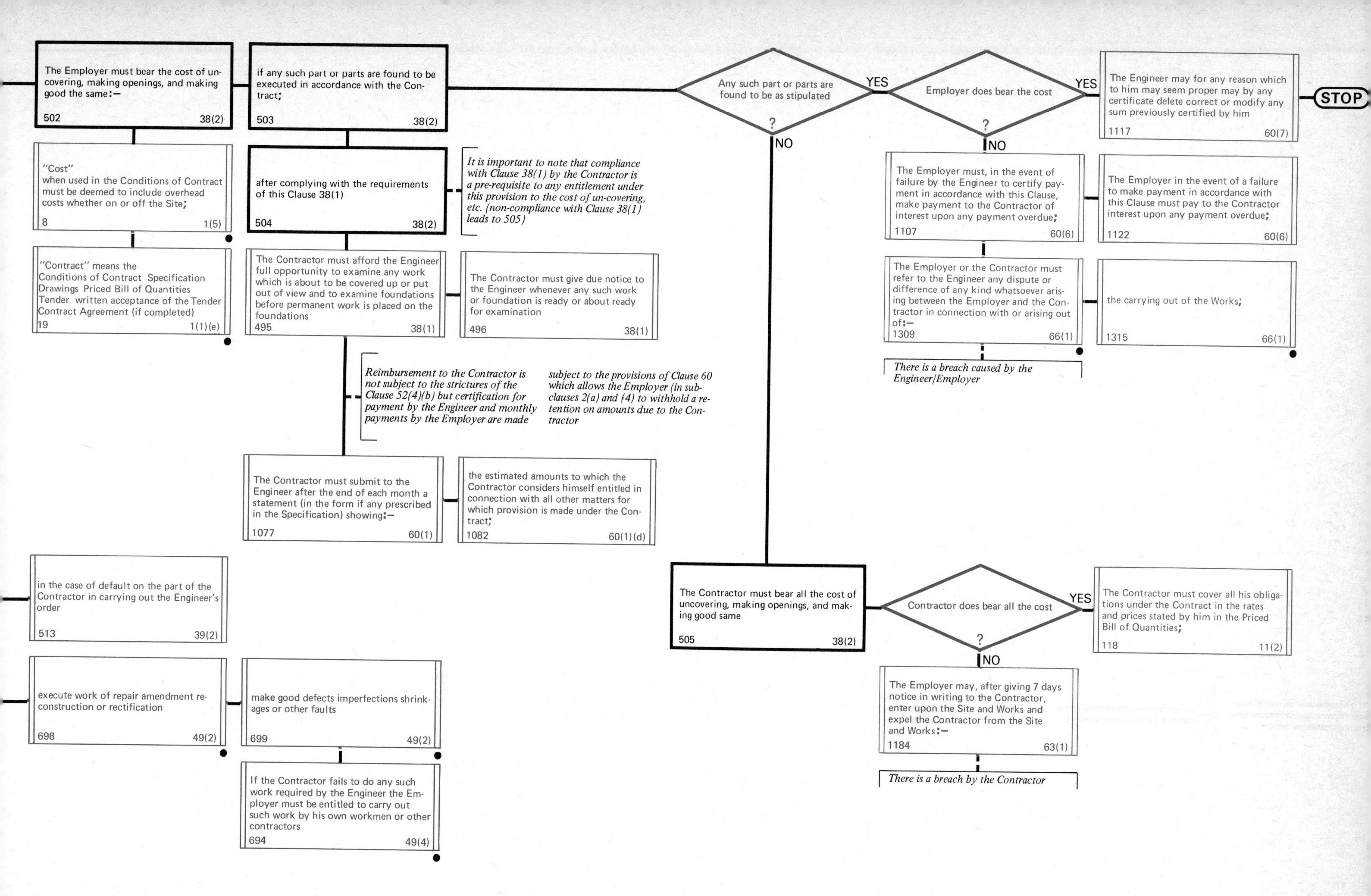
The Employer must bear the cost of uncovering, making openings, and making good the same:— 502 38(2)
if any such part or parts are found to be executed in accordance with the Contract; 503 38(2)
Any such part or parts are found to be as stipulated ?
YES
NO
Employer does bear the cost ?
YES
NO
The Engineer may for any reason which to him may seem proper may by any certificate delete correct or modify any sum previously certified by him 1117 60(7)
STOP
"Cost" when used in the Conditions of Contract must be deemed to include overhead costs whether on or off the Site; 8 1(5)
after complying with the requirements of this Clause 38(1) 504 38(2)
It is important to note that compliance with Clause 38(1) by the Contractor is a pre-requisite to any entitlement under this provision to the cost of un-covering, etc. (non-compliance with Clause 38(1) leads to 505)
The Employer must, in the event of failure by the Engineer to certify payment in accordance with this Clause, make payment to the Contractor of interest upon any payment overdue; 1107 60(6)
The Employer in the event of a failure to make payment in accordance with this Clause must pay to the Contractor interest upon any payment overdue; 1122 60(6)
"Contract" means the Conditions of Contract Specification Drawings Priced Bill of Quantities Tender written acceptance of the Tender Contract Agreement (if completed) 19 1(1)(e)
The Contractor must afford the Engineer full opportunity to examine any work which is about to be covered up or put out of view and to examine foundations before permanent work is placed on the foundations 495 38(1)
The Contractor must give due notice to the Engineer whenever any such work or foundation is ready or about ready for examination 496 38(1)
The Employer or the Contractor must refer to the Engineer any dispute or difference of any kind whatsoever arising between the Employer and the Contractor in connection with or arising out of:— 1309 66(1)
the carrying out of the Works; 1315 66(1)
There is a breach caused by the Engineer/Employer
Reimbursement to the Contractor is not subject to the strictures of the Clause 52(4)(b) but certification for payment by the Engineer and monthly payments by the Employer are made subject to the provisions of Clause 60 which allows the Employer (in sub-clauses 2(a) and (4) to withhold a retention on amounts due to the Contractor
The Contractor must submit to the Engineer after the end of each month a statement (in the form if any prescribed in the Specification) showing:— 1077 60(1)
the estimated amounts to which the Contractor considers himself entitled in connection with all other matters for which provision is made under the Contract; 1082 60(1)(d)
in the case of default on the part of the Contractor in carrying out the Engineer's order 513 39(2)
The Contractor must bear all the cost of uncovering, making openings, and making good same 505 38(2)
Contractor does bear all the cost ?
YES
NO
The Contractor must cover all his obligations under the Contract in the rates and prices stated by him in the Priced Bill of Quantities; 118 11(2)
execute work of repair amendment reconstruction or rectification 698 49(2)
make good defects imperfections shrinkages or other faults 699 49(2)
The Employer may, after giving 7 days notice in writing to the Contractor, enter upon the Site and Works and expel the Contractor from the Site and Works:— 1184 63(1)
If the Contractor fails to do any such work required by the Engineer the Employer must be entitled to carry out such work by his own workmen or other contractors 694 49(4)
There is a breach by the Contractor

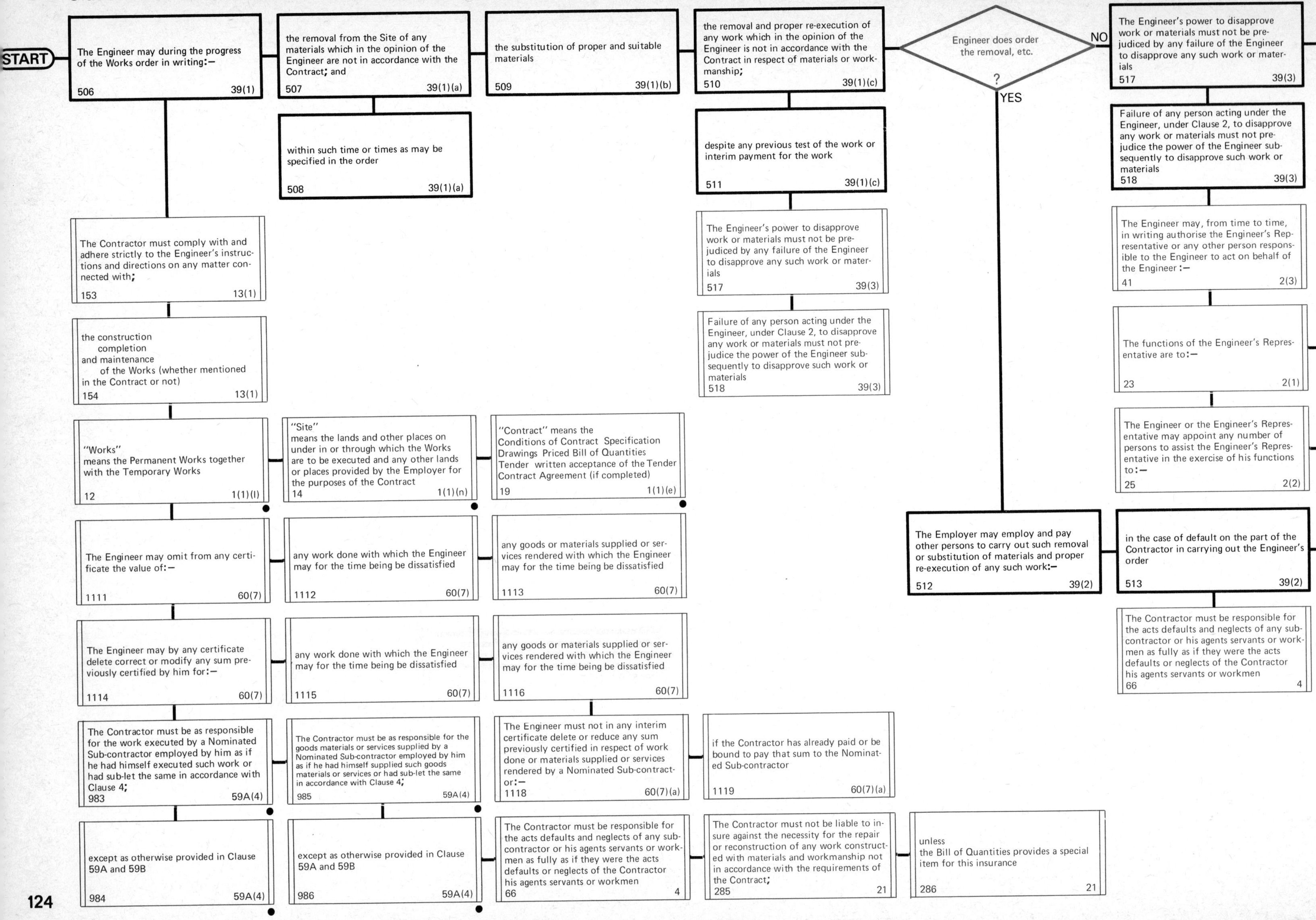
START
The Engineer may during the progress of the Works order in writing:— 506 39(1)
the removal from the Site of any materials which in the opinion of the Engineer are not in accordance with the Contract; and 507 39(1)(a)
the substitution of proper and suitable materials 509 39(1)(b)
the removal and proper re-execution of any work which in the opinion of the Engineer is not in accordance with the Contract in respect of materials or workmanship; 510 39(1)(c)
Engineer does order the removal, etc. ?
NO
YES
The Engineer's power to disapprove work or materials must not be prejudiced by any failure of the Engineer to disapprove any such work or materials 517 39(3)
within such time or times as may be specified in the order 508 39(1)(a)
despite any previous test of the work or interim payment for the work 511 39(1)(c)
Failure of any person acting under the Engineer, under Clause 2, to disapprove any work or materials must not prejudice the power of the Engineer subsequently to disapprove such work or materials 518 39(3)
The Contractor must comply with and adhere strictly to the Engineer's instructions and directions on any matter connected with; 153 13(1)
The Engineer's power to disapprove work or materials must not be prejudiced by any failure of the Engineer to disapprove any such work or materials 517 39(3)
The Engineer may, from time to time, in writing authorise the Engineer's Representative or any other person responsible to the Engineer to act on behalf of the Engineer:— 41 2(3)
the construction completion and maintenance of the Works (whether mentioned in the Contract or not) 154 13(1)
Failure of any person acting under the Engineer, under Clause 2, to disapprove any work or materials must not prejudice the power of the Engineer subsequently to disapprove such work or materials 518 39(3)
The functions of the Engineer's Representative are to:— 23 2(1)
"Works" means the Permanent Works together with the Temporary Works 12 1(1)(l)
"Site" means the lands and other places on under in or through which the Works are to be executed and any other lands or places provided by the Employer for the purposes of the Contract 14 1(1)(n)
"Contract" means the Conditions of Contract Specification Drawings Priced Bill of Quantities Tender written acceptance of the Tender Contract Agreement (if completed) 19 1(1)(e)
The Engineer or the Engineer's Representative may appoint any number of persons to assist the Engineer's Representative in the exercise of his functions to:— 25 2(2)
The Engineer may omit from any certificate the value of:— 1111 60(7)
any work done with which the Engineer may for the time being be dissatisfied 1112 60(7)
any goods or materials supplied or services rendered with which the Engineer may for the time being be dissatisfied 1113 60(7)
The Employer may employ and pay other persons to carry out such removal or substitution of materials and proper re-execution of any such work:— 512 39(2)
in the case of default on the part of the Contractor in carrying out the Engineer's order 513 39(2)
The Engineer may by any certificate delete correct or modify any sum previously certified by him for:— 1114 60(7)
any work done with which the Engineer may for the time being be dissatisfied 1115 60(7)
any goods or materials supplied or services rendered with which the Engineer may for the time being be dissatisfied 1116 60(7)
The Contractor must be responsible for the acts defaults and neglects of any sub-contractor or his agents servants or workmen as fully as if they were the acts defaults or neglects of the Contractor his agents servants or workmen 66 4
The Contractor must be as responsible for the work executed by a Nominated Sub-contractor employed by him as if he had himself executed such work or had sub-let the same in accordance with Clause 4; 983 59A(4)
The Contractor must be as responsible for the goods materials or services supplied by a Nominated Sub-contractor employed by him as if he had himself supplied such goods materials or services or had sub-let the same in accordance with Clause 4; 985 59A(4)
The Engineer must not in any interim certificate delete or reduce any sum previously certified in respect of work done or materials supplied or services rendered by a Nominated Sub-contractor:— 1118 60(7)(a)
if the Contractor has already paid or be bound to pay that sum to the Nominated Sub-contractor 1119 60(7)(a)
except as otherwise provided in Clause 59A and 59B 984 59A(4)
except as otherwise provided in Clause 59A and 59B 986 59A(4)
The Contractor must be responsible for the acts defaults and neglects of any sub-contractor or his agents servants or workmen as fully as if they were the acts defaults or neglects of the Contractor his agents servants or workmen 66 4
The Contractor must not be liable to insure against the necessity for the repair or reconstruction of any work constructed with materials and workmanship not in accordance with the requirements of the Contract; 285 21
unless the Bill of Quantities provides a special item for this insurance 286 21

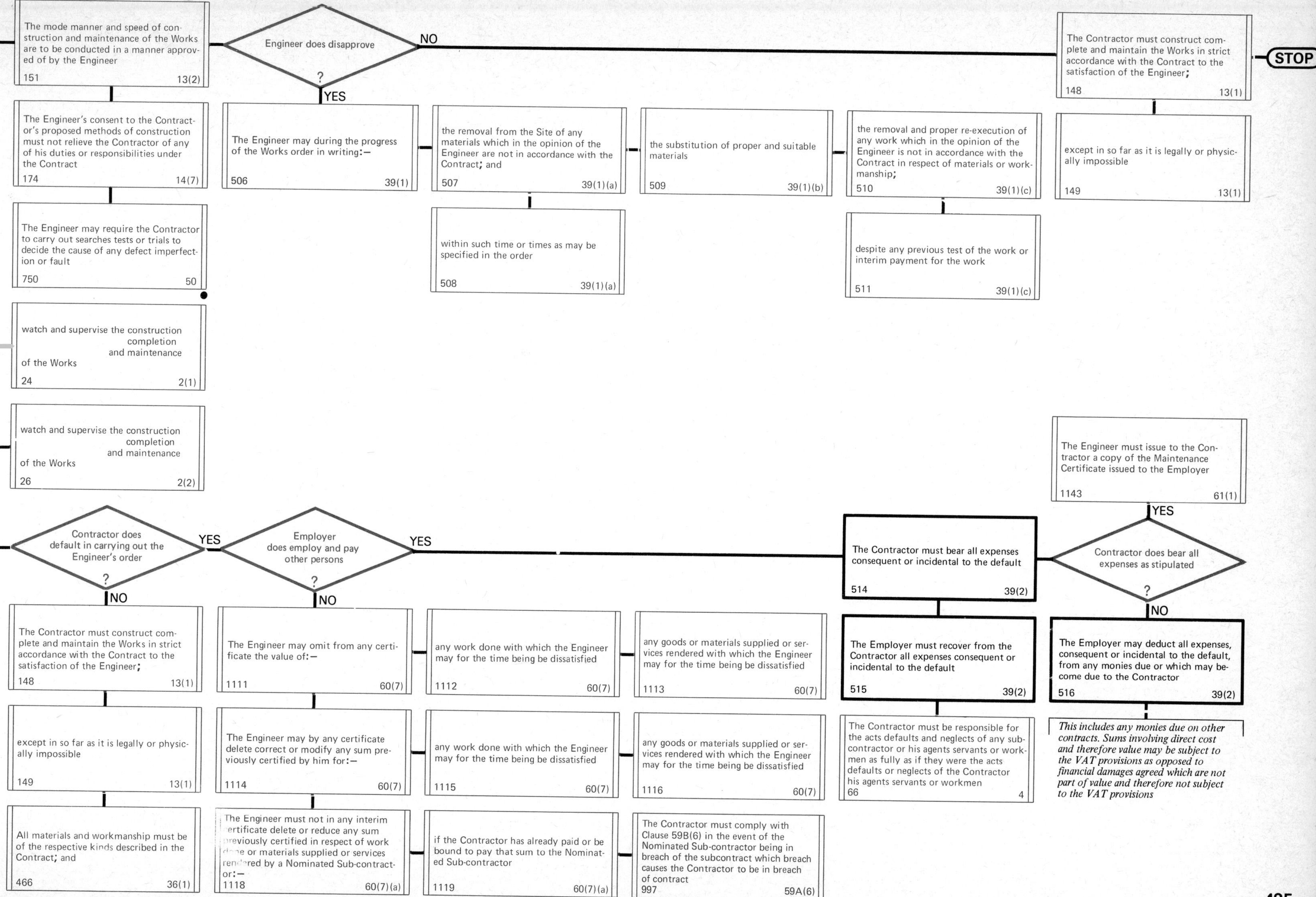

The mode manner and speed of construction and maintenance of the Works are to be conducted in a manner approved of by the Engineer
151 13(2)
Engineer does disapprove ?
NO
YES
The Contractor must construct complete and maintain the Works in strict accordance with the Contract to the satisfaction of the Engineer;
148 13(1)
STOP
The Engineer's consent to the Contractor's proposed methods of construction must not relieve the Contractor of any of his duties or responsibilities under the Contract
174 14(7)
The Engineer may during the progress of the Works order in writing:–
506 39(1)
the removal from the Site of any materials which in the opinion of the Engineer are not in accordance with the Contract; and
507 39(1)(a)
the substitution of proper and suitable materials
509 39(1)(b)
the removal and proper re-execution of any work which in the opinion of the Engineer is not in accordance with the Contract in respect of materials or workmanship;
510 39(1)(c)
except in so far as it is legally or physically impossible
149 13(1)
The Engineer may require the Contractor to carry out searches tests or trials to decide the cause of any defect imperfection or fault
750 50
within such time or times as may be specified in the order
508 39(1)(a)
despite any previous test of the work or interim payment for the work
511 39(1)(c)
watch and supervise the construction completion and maintenance of the Works
24 2(1)
watch and supervise the construction completion and maintenance of the Works
26 2(2)
The Engineer must issue to the Contractor a copy of the Maintenance Certificate issued to the Employer
1143 61(1)
YES
Contractor does default in carrying out the Engineer's order ?
YES
Employer does employ and pay other persons ?
YES
The Contractor must bear all expenses consequent or incidental to the default
514 39(2)
Contractor does bear all expenses as stipulated ?
NO
NO
NO
The Contractor must construct complete and maintain the Works in strict accordance with the Contract to the satisfaction of the Engineer;
148 13(1)
The Engineer may omit from any certificate the value of:–
1111 60(7)
any work done with which the Engineer may for the time being be dissatisfied
1112 60(7)
any goods or materials supplied or services rendered with which the Engineer may for the time being be dissatisfied
1113 60(7)
The Employer must recover from the Contractor all expenses consequent or incidental to the default
515 39(2)
The Employer may deduct all expenses, consequent or incidental to the default, from any monies due or which may become due to the Contractor
516 39(2)
except in so far as it is legally or physically impossible
149 13(1)
The Engineer may by any certificate delete correct or modify any sum previously certified by him for:–
1114 60(7)
any work done with which the Engineer may for the time being be dissatisfied
1115 60(7)
any goods or materials supplied or services rendered with which the Engineer may for the time being be dissatisfied
1116 60(7)
The Contractor must be responsible for the acts defaults and neglects of any sub-contractor or his agents servants or workmen as fully as if they were the acts defaults or neglects of the Contractor his agents servants or workmen
66 4
This includes any monies due on other contracts. Sums involving direct cost and therefore value may be subject to the VAT provisions as opposed to financial damages agreed which are not part of value and therefore not subject to the VAT provisions
All materials and workmanship must be of the respective kinds described in the Contract; and
466 36(1)
The Engineer must not in any interim certificate delete or reduce any sum previously certified in respect of work done or materials supplied or services rendered by a Nominated Sub-contractor:–
1118 60(7)(a)
if the Contractor has already paid or be bound to pay that sum to the Nominated Sub-contractor
1119 60(7)(a)
The Contractor must comply with Clause 59B(6) in the event of the Nominated Sub-contractor being in breach of the subcontract which breach causes the Contractor to be in breach of contract
997 59A(6)

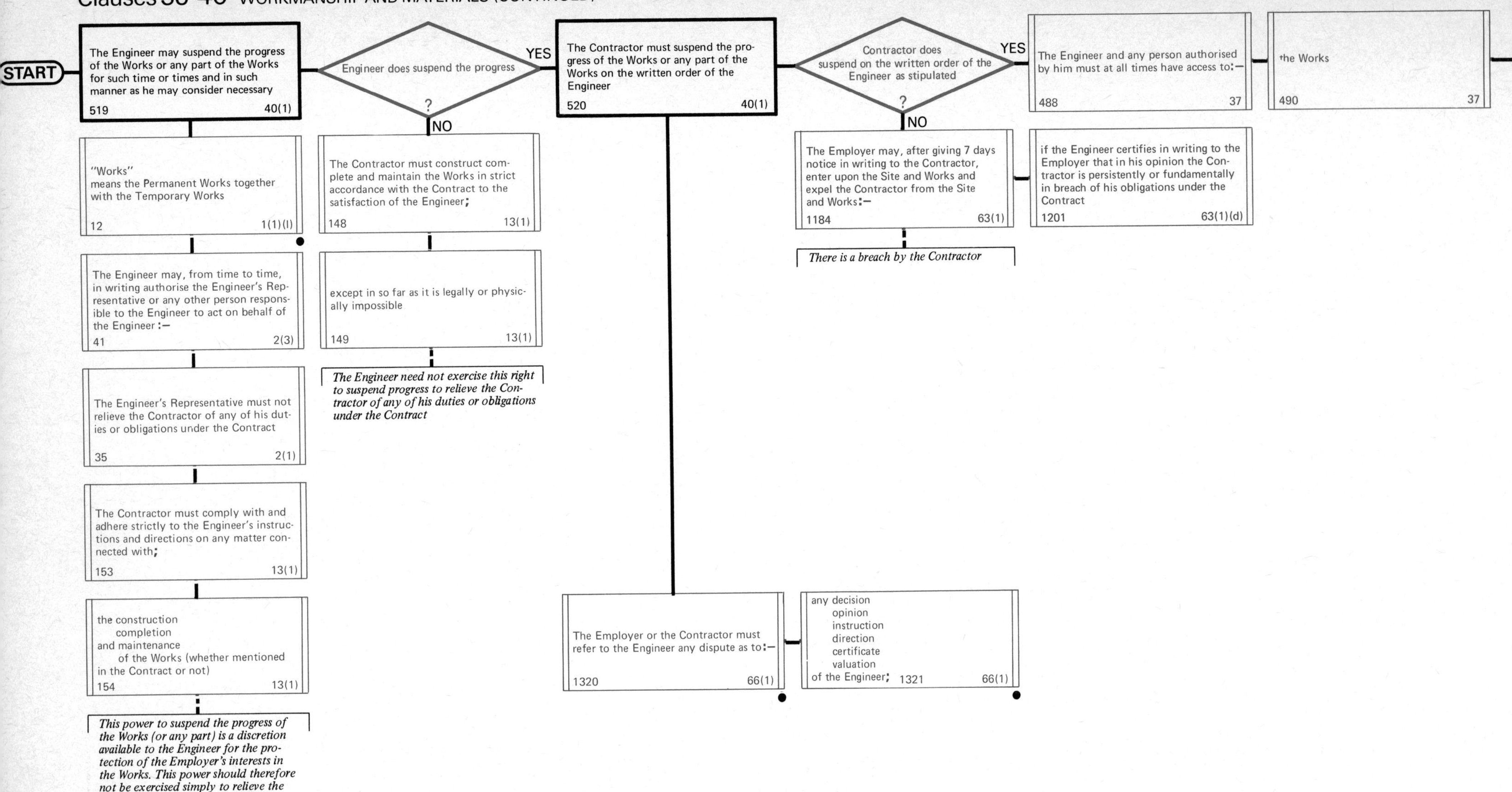
START
The Engineer may suspend the progress of the Works or any part of the Works for such time or times and in such manner as he may consider necessary
519 40(1)
Engineer does suspend the progress ?
YES
NO
The Contractor must suspend the progress of the Works or any part of the Works on the written order of the Engineer
520 40(1)
Contractor does suspend on the written order of the Engineer as stipulated ?
YES
NO
The Engineer and any person authorised by him must at all times have access to:—
488 37
the Works
490 37
"Works" means the Permanent Works together with the Temporary Works
12 1(1)(l)
The Contractor must construct complete and maintain the Works in strict accordance with the Contract to the satisfaction of the Engineer;
148 13(1)
The Employer may, after giving 7 days notice in writing to the Contractor, enter upon the Site and Works and expel the Contractor from the Site and Works:—
1184 63(1)
if the Engineer certifies in writing to the Employer that in his opinion the Contractor is persistently or fundamentally in breach of his obligations under the Contract
1201 63(1)(d)
There is a breach by the Contractor
The Engineer may, from time to time, in writing authorise the Engineer's Representative or any other person responsible to the Engineer to act on behalf of the Engineer:—
41 2(3)
except in so far as it is legally or physically impossible
149 13(1)
The Engineer need not exercise this right to suspend progress to relieve the Contractor of any of his duties or obligations under the Contract
The Engineer's Representative must not relieve the Contractor of any of his duties or obligations under the Contract
35 2(1)
The Contractor must comply with and adhere strictly to the Engineer's instructions and directions on any matter connected with;
153 13(1)
the construction
completion
and maintenance
of the Works (whether mentioned in the Contract or not)
154 13(1)
This power to suspend the progress of the Works (or any part) is a discretion available to the Engineer for the protection of the Employer's interests in the Works. This power should therefore not be exercised simply to relieve the Contractor (or his sub-contractors) from any of the duties of obligations under the Contract unless any such non-compliance threatens the safety of the Works or the proper execution of the work
The Employer or the Contractor must refer to the Engineer any dispute as to:—
1320 66(1)
any decision
opinion
instruction
direction
certificate
valuation
of the Engineer; 1321 66(1)

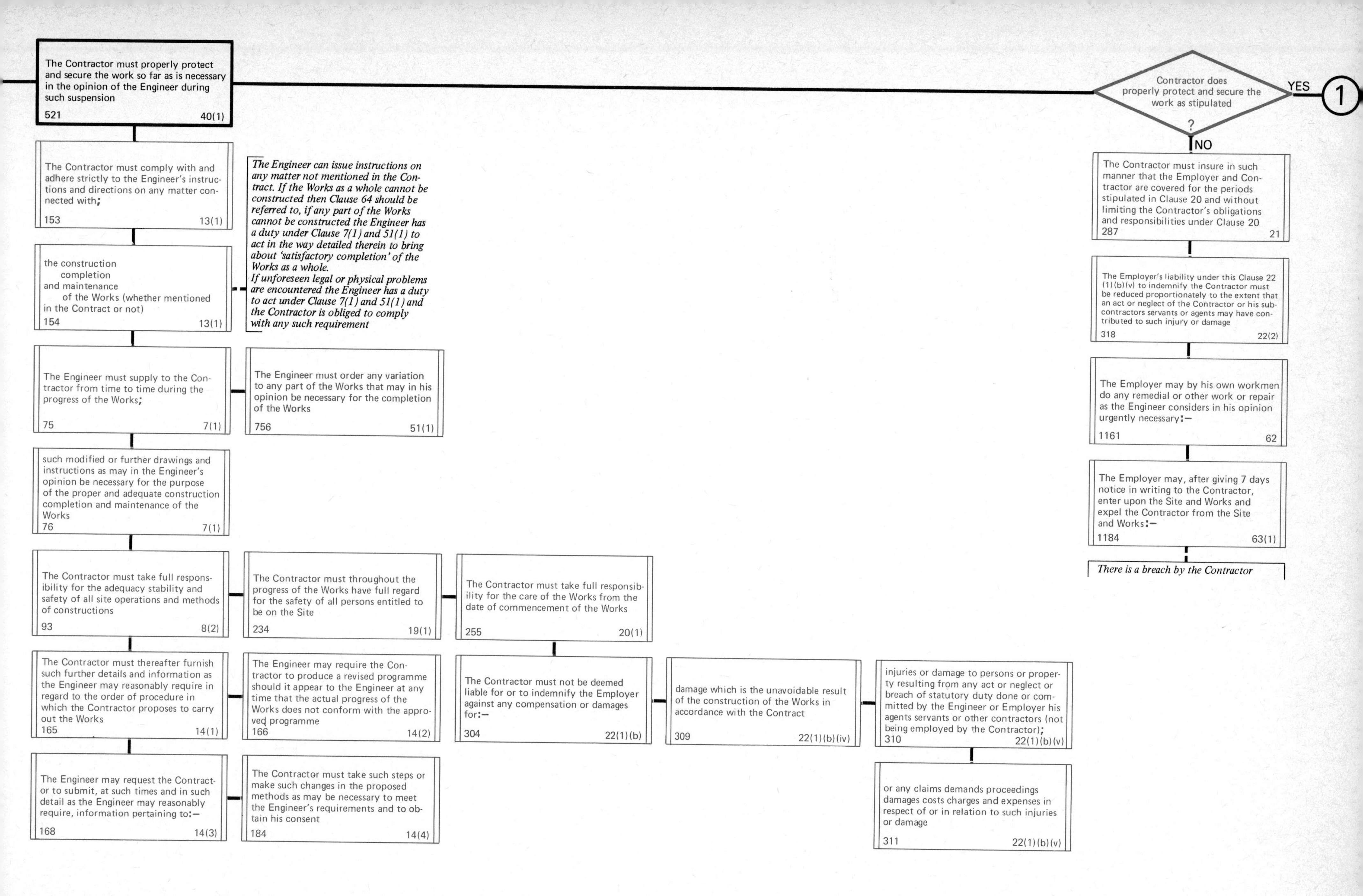
The Contractor must properly protect and secure the work so far as is necessary in the opinion of the Engineer during such suspension
521 40(1)
Contractor does properly protect and secure the work as stipulated ?
YES
1
NO
The Contractor must comply with and adhere strictly to the Engineer's instructions and directions on any matter connected with;
153 13(1)
The Engineer can issue instructions on any matter not mentioned in the Contract. If the Works as a whole cannot be constructed then Clause 64 should be referred to, if any part of the Works cannot be constructed the Engineer has a duty under Clause 7(1) and 51(1) to act in the way detailed therein to bring about 'satisfactory completion' of the Works as a whole. If unforeseen legal or physical problems are encountered the Engineer has a duty to act under Clause 7(1) and 51(1) and the Contractor is obliged to comply with any such requirement
the construction completion and maintenance of the Works (whether mentioned in the Contract or not)
154 13(1)
The Engineer must supply to the Contractor from time to time during the progress of the Works;
75 7(1)
The Engineer must order any variation to any part of the Works that may in his opinion be necessary for the completion of the Works
756 51(1)
such modified or further drawings and instructions as may in the Engineer's opinion be necessary for the purpose of the proper and adequate construction completion and maintenance of the Works
76 7(1)
The Contractor must take full responsibility for the adequacy stability and safety of all site operations and methods of constructions
93 8(2)
The Contractor must throughout the progress of the Works have full regard for the safety of all persons entitled to be on the Site
234 19(1)
The Contractor must take full responsibility for the care of the Works from the date of commencement of the Works
255 20(1)
The Contractor must thereafter furnish such further details and information as the Engineer may reasonably require in regard to the order of procedure in which the Contractor proposes to carry out the Works
165 14(1)
The Engineer may require the Contractor to produce a revised programme should it appear to the Engineer at any time that the actual progress of the Works does not conform with the approved programme
166 14(2)
The Contractor must not be deemed liable for or to indemnify the Employer against any compensation or damages for:—
304 22(1)(b)
damage which is the unavoidable result of the construction of the Works in accordance with the Contract
309 22(1)(b)(iv)
injuries or damage to persons or property resulting from any act or neglect or breach of statutory duty done or committed by the Engineer or Employer his agents servants or other contractors (not being employed by the Contractor);
310 22(1)(b)(v)
or any claims demands proceedings damages costs charges and expenses in respect of or in relation to such injuries or damage
311 22(1)(b)(v)
The Engineer may request the Contractor to submit, at such times and in such detail as the Engineer may reasonably require, information pertaining to:—
168 14(3)
The Contractor must take such steps or make such changes in the proposed methods as may be necessary to meet the Engineer's requirements and to obtain his consent
184 14(4)
The Contractor must insure in such manner that the Employer and Contractor are covered for the periods stipulated in Clause 20 and without limiting the Contractor's obligations and responsibilities under Clause 20
287 21
The Employer's liability under this Clause 22 (1)(b)(v) to indemnify the Contractor must be reduced proportionately to the extent that an act or neglect of the Contractor or his subcontractors servants or agents may have contributed to such injury or damage
318 22(2)
The Employer may by his own workmen do any remedial or other work or repair as the Engineer considers in his opinion urgently necessary:—
1161 62
The Employer may, after giving 7 days notice in writing to the Contractor, enter upon the Site and Works and expel the Contractor from the Site and Works:—
1184 63(1)
There is a breach by the Contractor

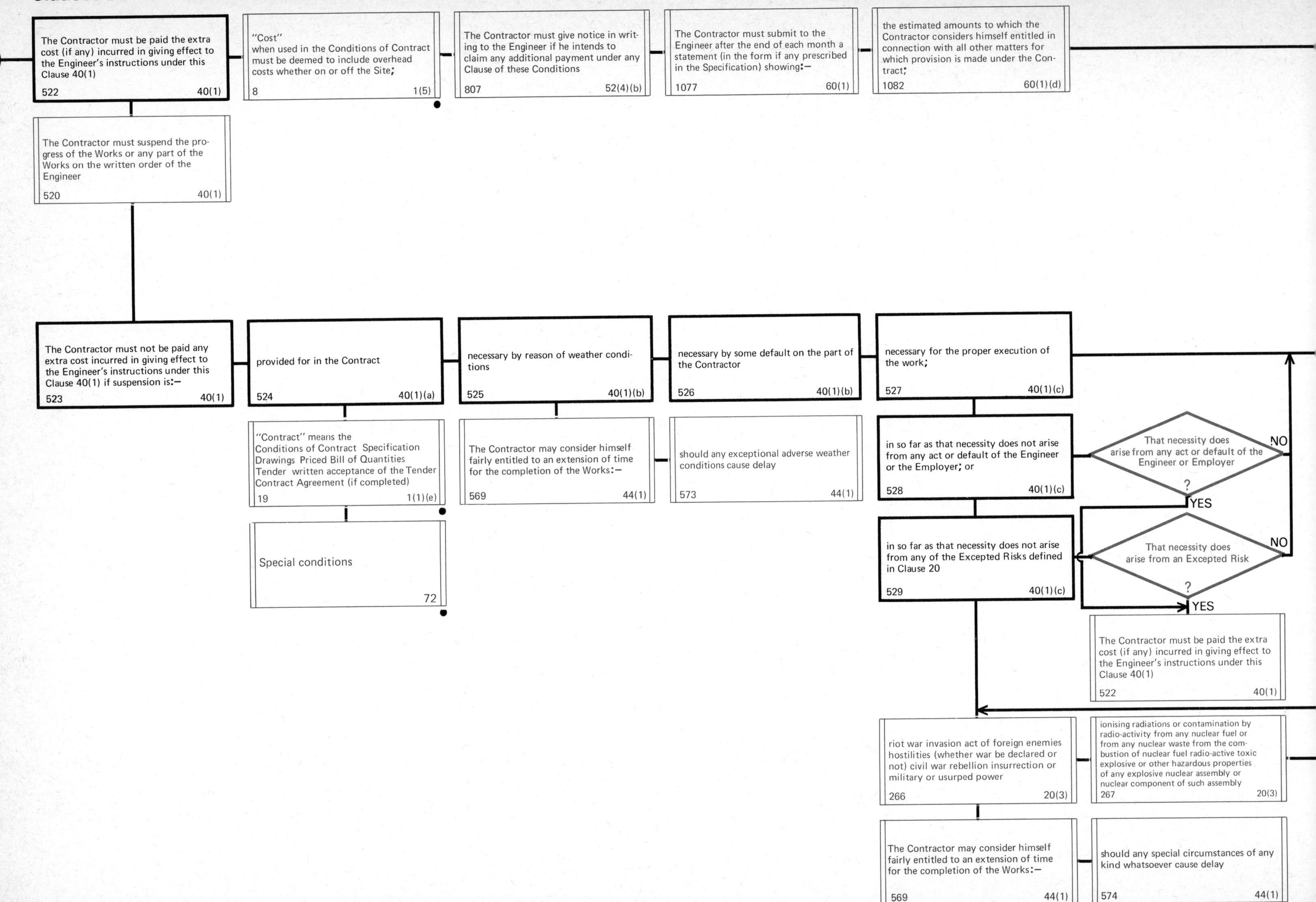
1
The Contractor must be paid the extra cost (if any) incurred in giving effect to the Engineer's instructions under this Clause 40(1)
522 40(1)
"Cost" when used in the Conditions of Contract must be deemed to include overhead costs whether on or off the Site;
8 1(5)
The Contractor must give notice in writing to the Engineer if he intends to claim any additional payment under any Clause of these Conditions
807 52(4)(b)
The Contractor must submit to the Engineer after the end of each month a statement (in the form if any prescribed in the Specification) showing:—
1077 60(1)
the estimated amounts to which the Contractor considers himself entitled in connection with all other matters for which provision is made under the Contract;
1082 60(1)(d)
The Contractor must suspend the progress of the Works or any part of the Works on the written order of the Engineer
520 40(1)
The Contractor must not be paid any extra cost incurred in giving effect to the Engineer's instructions under this Clause 40(1) if suspension is:—
523 40(1)
provided for in the Contract
524 40(1)(a)
necessary by reason of weather conditions
525 40(1)(b)
necessary by some default on the part of the Contractor
526 40(1)(b)
necessary for the proper execution of the work;
527 40(1)(c)
"Contract" means the Conditions of Contract Specification Drawings Priced Bill of Quantities Tender written acceptance of the Tender Contract Agreement (if completed)
19 1(1)(e)
Special conditions
72
The Contractor may consider himself fairly entitled to an extension of time for the completion of the Works:—
569 44(1)
should any exceptional adverse weather conditions cause delay
573 44(1)
in so far as that necessity does not arise from any act or default of the Engineer or the Employer; or
528 40(1)(c)
That necessity does arise from any act or default of the Engineer or Employer ?
NO
YES
in so far as that necessity does not arise from any of the Excepted Risks defined in Clause 20
529 40(1)(c)
That necessity does arise from an Excepted Risk ?
NO
YES
The Contractor must be paid the extra cost (if any) incurred in giving effect to the Engineer's instructions under this Clause 40(1)
522 40(1)
riot war invasion act of foreign enemies hostilities (whether war be declared or not) civil war rebellion insurrection or military or usurped power
266 20(3)
ionising radiations or contamination by radio-activity from any nuclear fuel or from any nuclear waste from the combustion of nuclear fuel radio-active toxic explosive or other hazardous properties of any explosive nuclear assembly or nuclear component of such assembly
267 20(3)
The Contractor may consider himself fairly entitled to an extension of time for the completion of the Works:—
569 44(1)
should any special circumstances of any kind whatsoever cause delay
574 44(1)

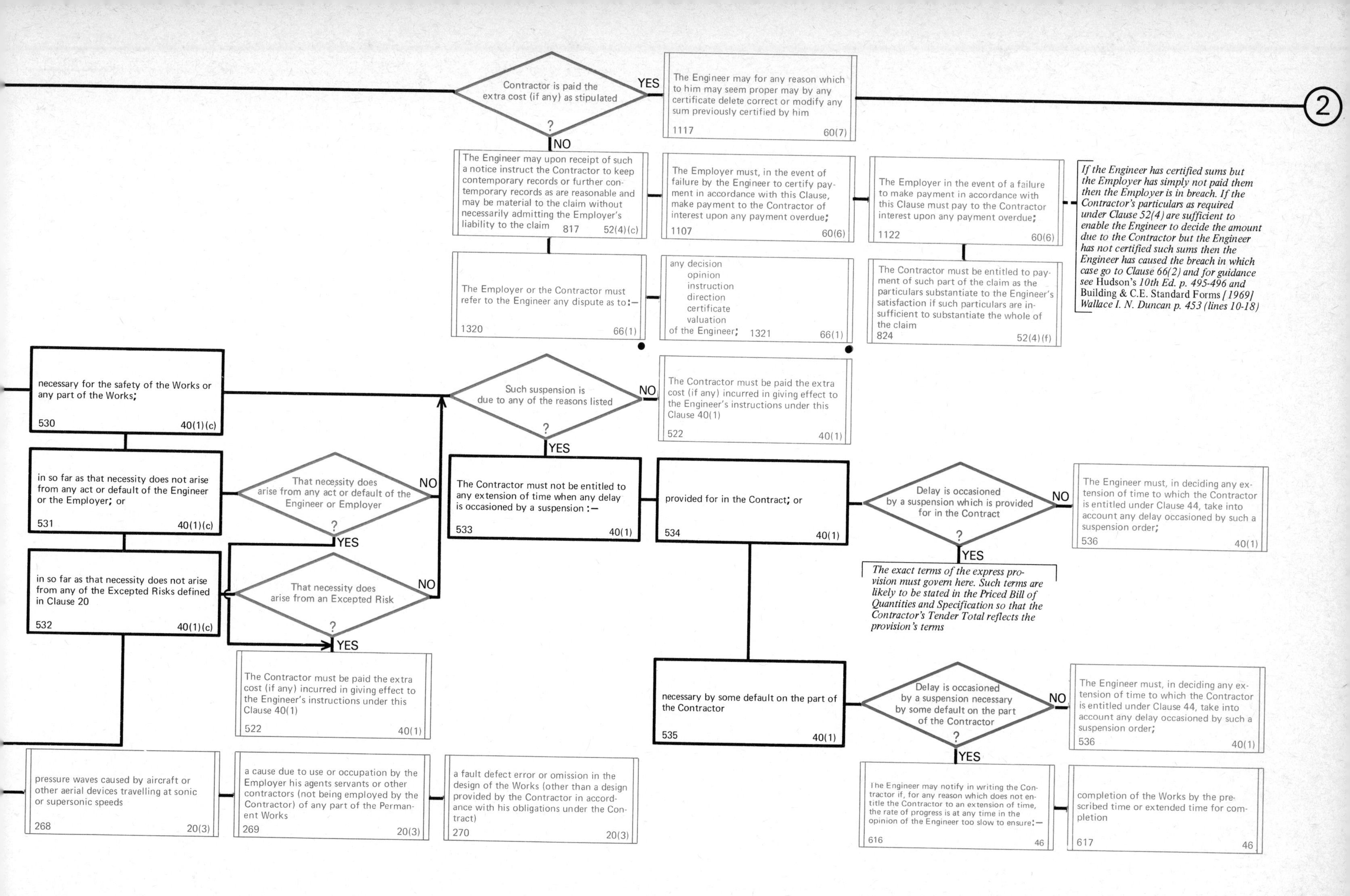

Contractor is paid the extra cost (if any) as stipulated ?
YES
NO
The Engineer may for any reason which to him may seem proper may by any certificate delete correct or modify any sum previously certified by him 1117 60(7)
2
The Engineer may upon receipt of such a notice instruct the Contractor to keep contemporary records or further contemporary records as are reasonable and may be material to the claim without necessarily admitting the Employer's liability to the claim 817 52(4)(c)
The Employer must, in the event of failure by the Engineer to certify payment in accordance with this Clause, make payment to the Contractor of interest upon any payment overdue; 1107 60(6)
The Employer in the event of a failure to make payment in accordance with this Clause must pay to the Contractor interest upon any payment overdue; 1122 60(6)
If the Engineer has certified sums but the Employer has simply not paid them then the Employer is in breach. If the Contractor's particulars as required under Clause 52(4) are sufficient to enable the Engineer to decide the amount due to the Contractor but the Engineer has not certified such sums then the Engineer has caused the breach in which case go to Clause 66(2) and for guidance see Hudson's 10th Ed. p. 495-496 and Building & C.E. Standard Forms [1969] Wallace I. N. Duncan p. 453 (lines 10-18)
The Employer or the Contractor must refer to the Engineer any dispute as to:— 1320 66(1)
any decision opinion instruction direction certificate valuation of the Engineer; 1321 66(1)
The Contractor must be entitled to payment of such part of the claim as the particulars substantiate to the Engineer's satisfaction if such particulars are insufficient to substantiate the whole of the claim 824 52(4)(f)
necessary for the safety of the Works or any part of the Works; 530 40(1)(c)
Such suspension is due to any of the reasons listed ?
NO
YES
The Contractor must be paid the extra cost (if any) incurred in giving effect to the Engineer's instructions under this Clause 40(1) 522 40(1)
in so far as that necessity does not arise from any act or default of the Engineer or the Employer; or 531 40(1)(c)
That necessity does arise from any act or default of the Engineer or Employer ?
NO
YES
The Contractor must not be entitled to any extension of time when any delay is occasioned by a suspension:— 533 40(1)
provided for in the Contract; or 534 40(1)
Delay is occasioned by a suspension which is provided for in the Contract ?
NO
YES
The Engineer must, in deciding any extension of time to which the Contractor is entitled under Clause 44, take into account any delay occasioned by such a suspension order; 536 40(1)
in so far as that necessity does not arise from any of the Excepted Risks defined in Clause 20 532 40(1)(c)
That necessity does arise from an Excepted Risk ?
NO
YES
The exact terms of the express provision must govern here. Such terms are likely to be stated in the Priced Bill of Quantities and Specification so that the Contractor's Tender Total reflects the provision's terms
The Contractor must be paid the extra cost (if any) incurred in giving effect to the Engineer's instructions under this Clause 40(1) 522 40(1)
necessary by some default on the part of the Contractor 535 40(1)
Delay is occasioned by a suspension necessary by some default on the part of the Contractor ?
NO
YES
The Engineer must, in deciding any extension of time to which the Contractor is entitled under Clause 44, take into account any delay occasioned by such a suspension order; 536 40(1)
pressure waves caused by aircraft or other aerial devices travelling at sonic or supersonic speeds 268 20(3)
a cause due to use or occupation by the Employer his agents servants or other contractors (not being employed by the Contractor) of any part of the Permanent Works 269 20(3)
a fault defect error or omission in the design of the Works (other than a design provided by the Contractor in accordance with his obligations under the Contract) 270 20(3)
The Engineer may notify in writing the Contractor if, for any reason which does not entitle the Contractor to an extension of time, the rate of progress is at any time in the opinion of the Engineer too slow to ensure:— 616 46
completion of the Works by the prescribed time or extended time for completion 617 46

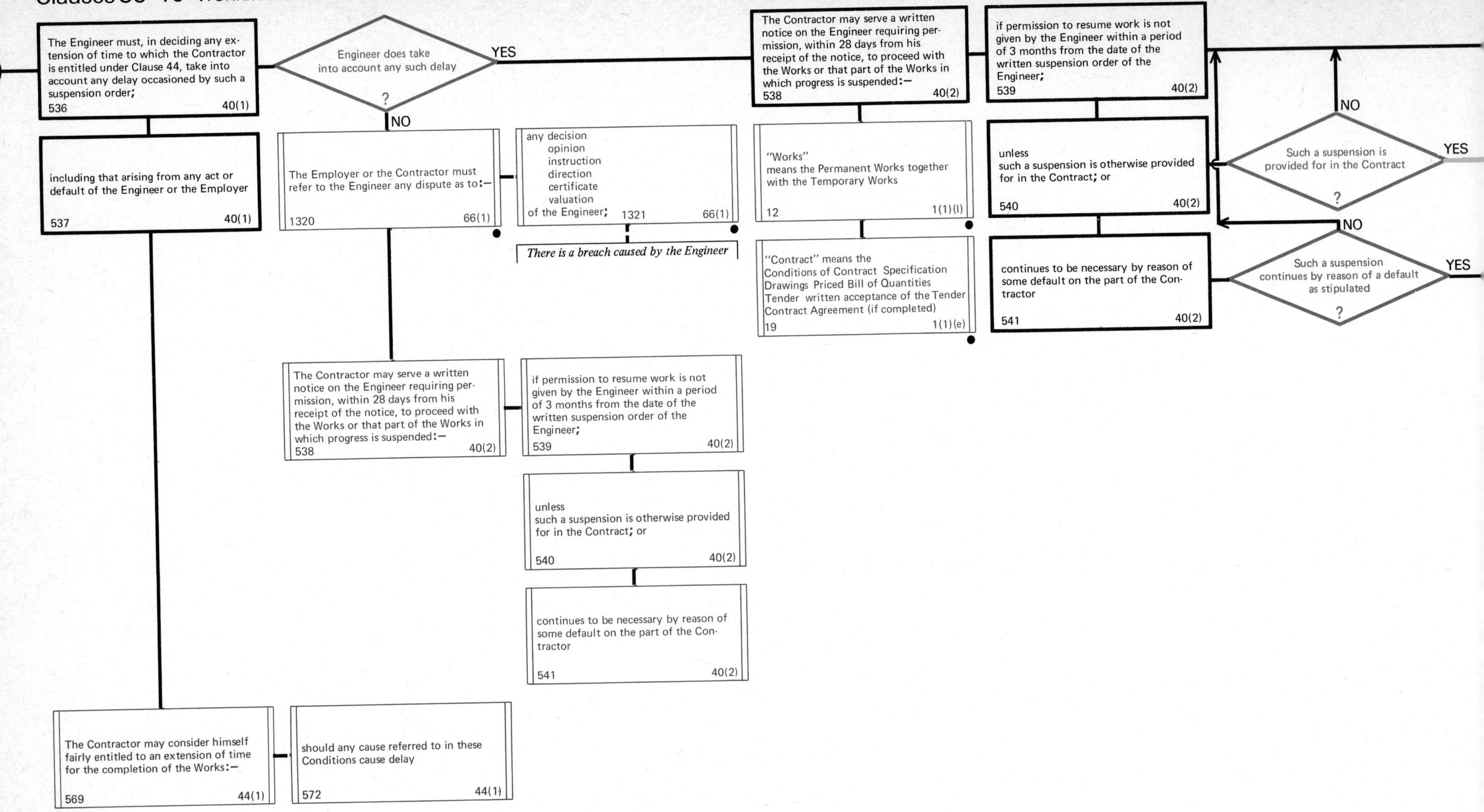
2
The Engineer must, in deciding any extension of time to which the Contractor is entitled under Clause 44, take into account any delay occasioned by such a suspension order;
536 40(1)
Engineer does take into account any such delay ?
YES
NO
The Contractor may serve a written notice on the Engineer requiring permission, within 28 days from his receipt of the notice, to proceed with the Works or that part of the Works in which progress is suspended:—
538 40(2)
if permission to resume work is not given by the Engineer within a period of 3 months from the date of the written suspension order of the Engineer;
539 40(2)
including that arising from any act or default of the Engineer or the Employer
537 40(1)
The Employer or the Contractor must refer to the Engineer any dispute as to:—
1320 66(1)
any decision
opinion
instruction
direction
certificate
valuation
of the Engineer;
1321 66(1)
There is a breach caused by the Engineer
"Works"
means the Permanent Works together with the Temporary Works
12 1(1)(l)
"Contract" means the Conditions of Contract Specification Drawings Priced Bill of Quantities Tender written acceptance of the Tender Contract Agreement (if completed)
19 1(1)(e)
unless
such a suspension is otherwise provided for in the Contract; or
540 40(2)
continues to be necessary by reason of some default on the part of the Contractor
541 40(2)
Such a suspension is provided for in the Contract ?
YES
NO
Such a suspension continues by reason of a default as stipulated ?
YES
NO
The Contractor may serve a written notice on the Engineer requiring permission, within 28 days from his receipt of the notice, to proceed with the Works or that part of the Works in which progress is suspended:—
538 40(2)
if permission to resume work is not given by the Engineer within a period of 3 months from the date of the written suspension order of the Engineer;
539 40(2)
unless
such a suspension is otherwise provided for in the Contract; or
540 40(2)
continues to be necessary by reason of some default on the part of the Contractor
541 40(2)
The Contractor may consider himself fairly entitled to an extension of time for the completion of the Works:—
569 44(1)
should any cause referred to in these Conditions cause delay
572 44(1)

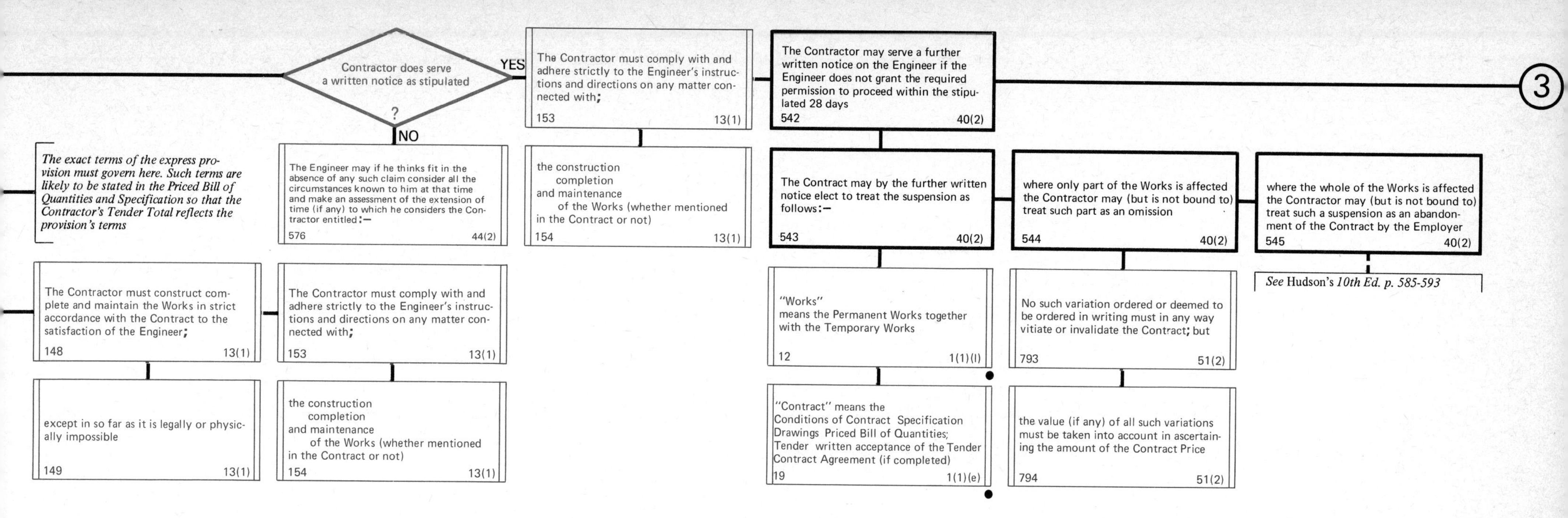
Contractor does serve a written notice as stipulated ?
YES
NO
The Contractor must comply with and adhere strictly to the Engineer's instructions and directions on any matter connected with; 153 13(1)
The Contractor may serve a further written notice on the Engineer if the Engineer does not grant the required permission to proceed within the stipulated 28 days 542 40(2)
3
The exact terms of the express provision must govern here. Such terms are likely to be stated in the Priced Bill of Quantities and Specification so that the Contractor's Tender Total reflects the provision's terms
The Engineer may if he thinks fit in the absence of any such claim consider all the circumstances known to him at that time and make an assessment of the extension of time (if any) to which he considers the Contractor entitled:— 576 44(2)
the construction completion and maintenance of the Works (whether mentioned in the Contract or not) 154 13(1)
The Contract may by the further written notice elect to treat the suspension as follows:— 543 40(2)
where only part of the Works is affected the Contractor may (but is not bound to) treat such part as an omission 544 40(2)
where the whole of the Works is affected the Contractor may (but is not bound to) treat such a suspension as an abandonment of the Contract by the Employer 545 40(2)
See Hudson's 10th Ed. p. 585-593
The Contractor must construct complete and maintain the Works in strict accordance with the Contract to the satisfaction of the Engineer; 148 13(1)
The Contractor must comply with and adhere strictly to the Engineer's instructions and directions on any matter connected with; 153 13(1)
"Works" means the Permanent Works together with the Temporary Works 12 1(1)(l)
No such variation ordered or deemed to be ordered in writing must in any way vitiate or invalidate the Contract; but 793 51(2)
except in so far as it is legally or physically impossible 149 13(1)
the construction completion and maintenance of the Works (whether mentioned in the Contract or not) 154 13(1)
"Contract" means the Conditions of Contract Specification Drawings Priced Bill of Quantities; Tender written acceptance of the Tender Contract Agreement (if completed) 19 1(1)(e)
the value (if any) of all such variations must be taken into account in ascertaining the amount of the Contract Price 794 51(2)

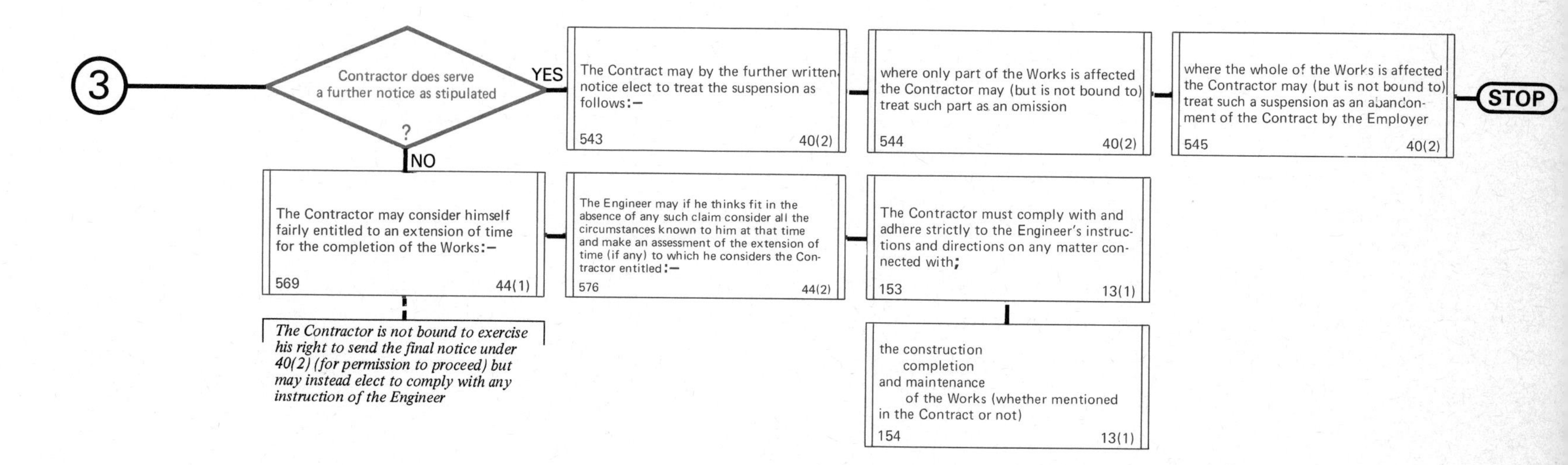
3
Contractor does serve a further notice as stipulated ?
YES
NO
The Contract may by the further written notice elect to treat the suspension as follows:— 543 40(2)
where only part of the Works is affected the Contractor may (but is not bound to) treat such part as an omission 544 40(2)
where the whole of the Works is affected the Contractor may (but is not bound to) treat such a suspension as an abandonment of the Contract by the Employer 545 40(2)
STOP
The Contractor may consider himself fairly entitled to an extension of time for the completion of the Works:— 569 44(1)
The Engineer may if he thinks fit in the absence of any such claim consider all the circumstances known to him at that time and make an assessment of the extension of time (if any) to which he considers the Contractor entitled:— 576 44(2)
The Contractor must comply with and adhere strictly to the Engineer's instructions and directions on any matter connected with; 153 13(1)
The Contractor is not bound to exercise his right to send the final notice under 40(2) (for permission to proceed) but may instead elect to comply with any instruction of the Engineer
the construction completion and maintenance of the Works (whether mentioned in the Contract or not) 154 13(1)

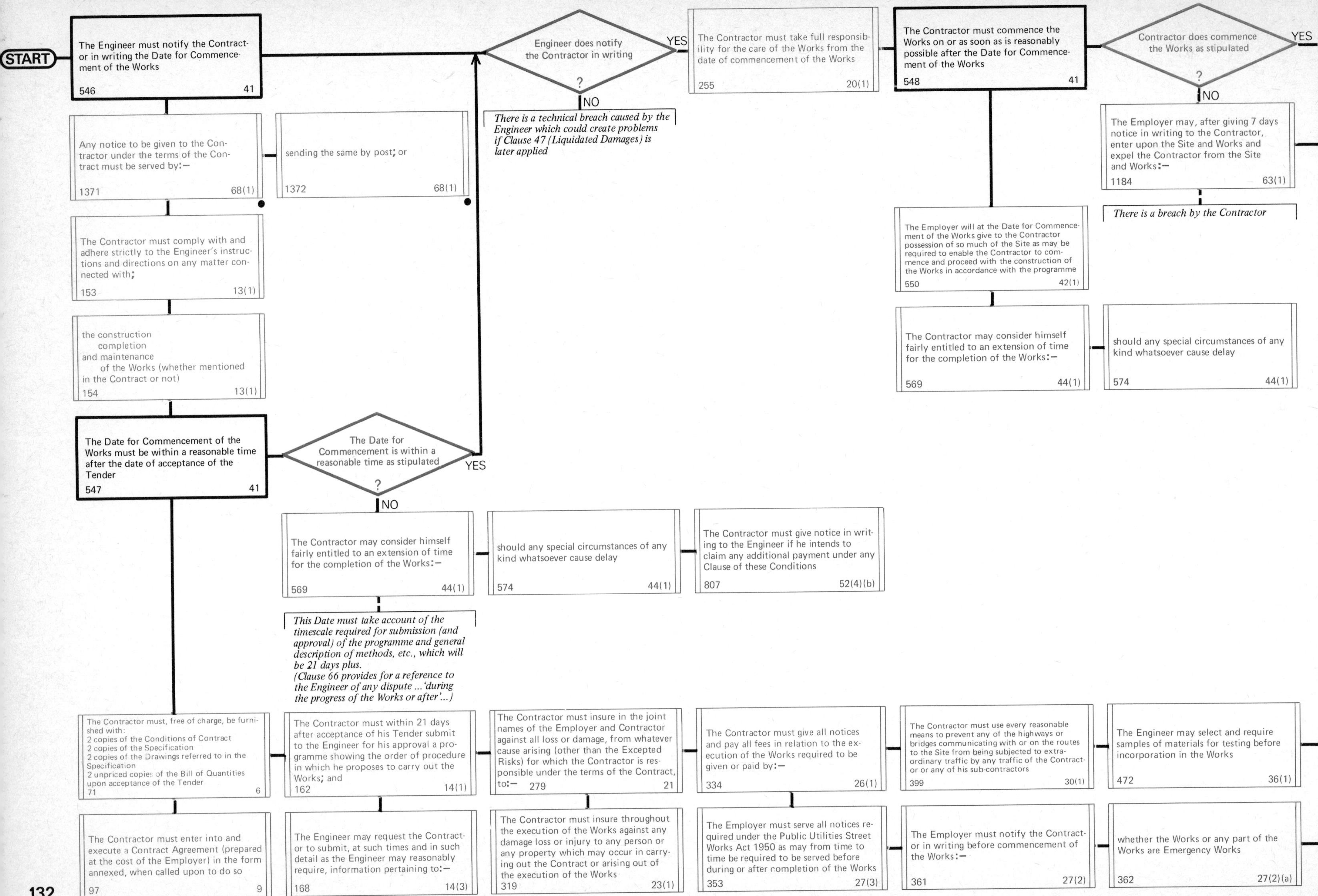
START
The Engineer must notify the Contractor in writing the Date for Commencement of the Works
546 41
Engineer does notify the Contractor in writing
?
YES
NO
There is a technical breach caused by the Engineer which could create problems if Clause 47 (Liquidated Damages) is later applied
The Contractor must take full responsibility for the care of the Works from the date of commencement of the Works
255 20(1)
The Contractor must commence the Works on or as soon as is reasonably possible after the Date for Commencement of the Works
548 41
Contractor does commence the Works as stipulated
?
YES
NO
The Employer may, after giving 7 days notice in writing to the Contractor, enter upon the Site and Works and expel the Contractor from the Site and Works:—
1184 63(1)
There is a breach by the Contractor
Any notice to be given to the Contractor under the terms of the Contract must be served by:—
1371 68(1)
sending the same by post; or
1372 68(1)
The Contractor must comply with and adhere strictly to the Engineer's instructions and directions on any matter connected with;
153 13(1)
the construction completion and maintenance of the Works (whether mentioned in the Contract or not)
154 13(1)
The Employer will at the Date for Commencement of the Works give to the Contractor possession of so much of the Site as may be required to enable the Contractor to commence and proceed with the construction of the Works in accordance with the programme
550 42(1)
The Contractor may consider himself fairly entitled to an extension of time for the completion of the Works:—
569 44(1)
should any special circumstances of any kind whatsoever cause delay
574 44(1)
The Date for Commencement of the Works must be within a reasonable time after the date of acceptance of the Tender
547 41
The Date for Commencement is within a reasonable time as stipulated
?
YES
NO
The Contractor may consider himself fairly entitled to an extension of time for the completion of the Works:—
569 44(1)
should any special circumstances of any kind whatsoever cause delay
574 44(1)
The Contractor must give notice in writing to the Engineer if he intends to claim any additional payment under any Clause of these Conditions
807 52(4)(b)
This Date must take account of the timescale required for submission (and approval) of the programme and general description of methods, etc., which will be 21 days plus.
(Clause 66 provides for a reference to the Engineer of any dispute ... 'during the progress of the Works or after'...)
The Contractor must, free of charge, be furnished with:
2 copies of the Conditions of Contract
2 copies of the Specification
2 copies of the Drawings referred to in the Specification
2 unpriced copies of the Bill of Quantities upon acceptance of the Tender
71 6
The Contractor must within 21 days after acceptance of his Tender submit to the Engineer for his approval a programme showing the order of procedure in which he proposes to carry out the Works; and
162 14(1)
The Contractor must insure in the joint names of the Employer and Contractor against all loss or damage, from whatever cause arising (other than the Excepted Risks) for which the Contractor is responsible under the terms of the Contract, to:—
279 21
The Contractor must give all notices and pay all fees in relation to the execution of the Works required to be given or paid by:—
334 26(1)
The Contractor must use every reasonable means to prevent any of the highways or bridges communicating with or on the routes to the Site from being subjected to extraordinary traffic by any traffic of the Contractor or any of his sub-contractors
399 30(1)
The Engineer may select and require samples of materials for testing before incorporation in the Works
472 36(1)
The Contractor must enter into and execute a Contract Agreement (prepared at the cost of the Employer) in the form annexed, when called upon to do so
97 9
The Engineer may request the Contractor to submit, at such times and in such detail as the Engineer may reasonably require, information pertaining to:—
168 14(3)
The Contractor must insure throughout the execution of the Works against any damage loss or injury to any person or any property which may occur in carrying out the Contract or arising out of the execution of the Works
319 23(1)
The Employer must serve all notices required under the Public Utilities Street Works Act 1950 as may from time to time be required to be served before during or after completion of the Works
353 27(3)
The Employer must notify the Contractor in writing before commencement of the Works:—
361 27(2)
whether the Works or any part of the Works are Emergency Works
362 27(2)(a)

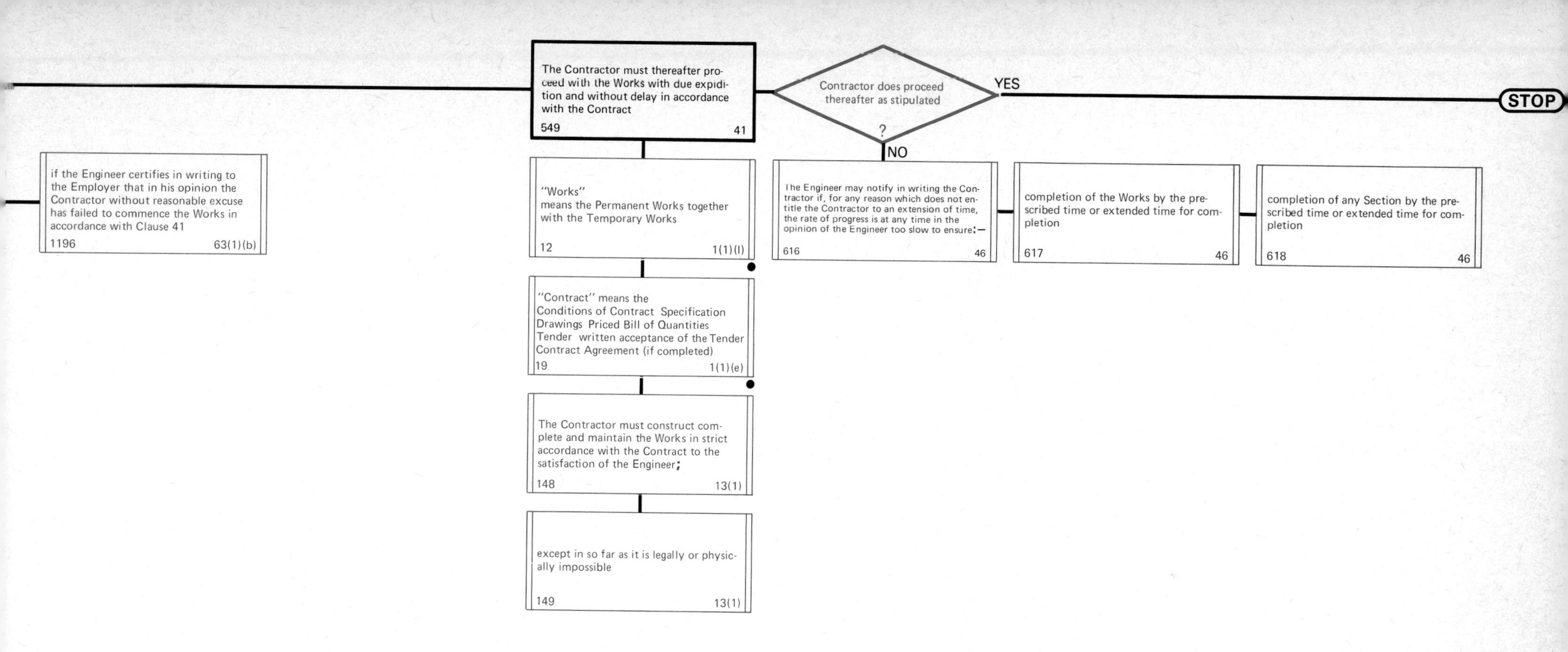

The Contractor must thereafter proceed with the Works with due expidition and without delay in accordance with the Contract
549 41
Contractor does proceed thereafter as stipulated
?
YES
STOP
NO
if the Engineer certifies in writing to the Employer that in his opinion the Contractor without reasonable excuse has failed to commence the Works in accordance with Clause 41
1196 63(1)(b)
"Works" means the Permanent Works together with the Temporary Works
12 1(1)(l)
"Contract" means the Conditions of Contract Specification Drawings Priced Bill of Quantities Tender written acceptance of the Tender Contract Agreement (if completed)
19 1(1)(e)
The Contractor must construct complete and maintain the Works in strict accordance with the Contract to the satisfaction of the Engineer;
148 13(1)
except in so far as it is legally or physically impossible
149 13(1)
The Engineer may notify in writing the Contractor if, for any reason which does not entitle the Contractor to an extension of time, the rate of progress is at any time in the opinion of the Engineer too slow to ensure:—
616 46
completion of the Works by the prescribed time or extended time for completion
617 46
completion of any Section by the prescribed time or extended time for completion
618 46

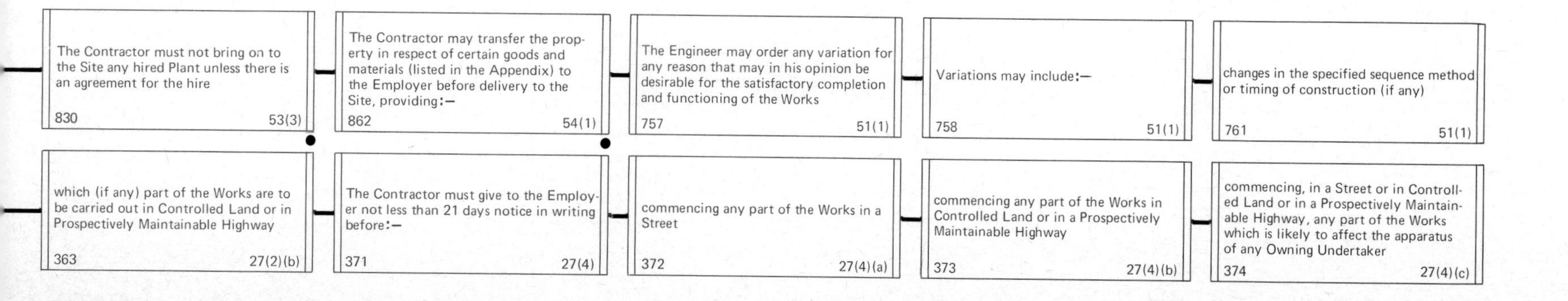

The Contractor must not bring on to the Site any hired Plant unless there is an agreement for the hire
830 53(3)
The Contractor may transfer the property in respect of certain goods and materials (listed in the Appendix) to the Employer before delivery to the Site, providing:—
862 54(1)
The Engineer may order any variation for any reason that may in his opinion be desirable for the satisfactory completion and functioning of the Works
757 51(1)
Variations may include:—
758 51(1)
changes in the specified sequence method or timing of construction (if any)
761 51(1)
which (if any) part of the Works are to be carried out in Controlled Land or in Prospectively Maintainable Highway
363 27(2)(b)
The Contractor must give to the Employer not less than 21 days notice in writing before:—
371 27(4)
commencing any part of the Works in a Street
372 27(4)(a)
commencing any part of the Works in Controlled Land or in a Prospectively Maintainable Highway
373 27(4)(b)
commencing, in a Street or in Controlled Land or in a Prospectively Maintainable Highway, any part of the Works which is likely to affect the apparatus of any Owning Undertaker
374 27(4)(c)

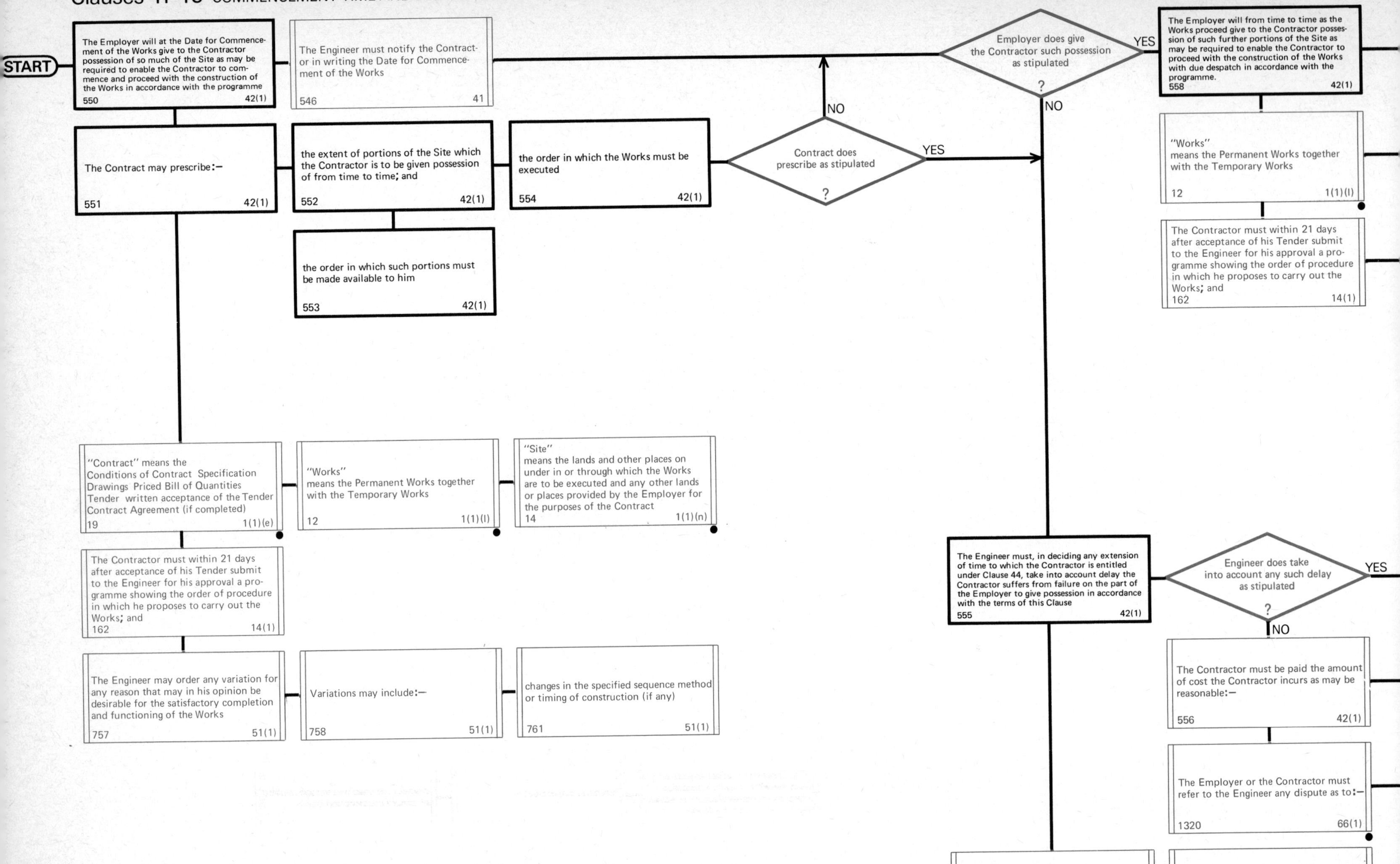
START
The Employer will at the Date for Commencement of the Works give to the Contractor possession of so much of the Site as may be required to enable the Contractor to commence and proceed with the construction of the Works in accordance with the programme 550 42(1)
The Engineer must notify the Contractor in writing the Date for Commencement of the Works 546 41
Employer does give the Contractor such possession as stipulated ?
YES
NO
The Employer will from time to time as the Works proceed give to the Contractor possession of such further portions of the Site as may be required to enable the Contractor to proceed with the construction of the Works with due despatch in accordance with the programme. 558 42(1)
The Contract may prescribe:— 551 42(1)
the extent of portions of the Site which the Contractor is to be given possession of from time to time; and 552 42(1)
the order in which the Works must be executed 554 42(1)
Contract does prescribe as stipulated ?
YES
NO
"Works" means the Permanent Works together with the Temporary Works 12 1(1)(l)
The Contractor must within 21 days after acceptance of his Tender submit to the Engineer for his approval a programme showing the order of procedure in which he proposes to carry out the Works; and 162 14(1)
the order in which such portions must be made available to him 553 42(1)
"Contract" means the Conditions of Contract Specification Drawings Priced Bill of Quantities Tender written acceptance of the Tender Contract Agreement (if completed) 19 1(1)(e)
"Works" means the Permanent Works together with the Temporary Works 12 1(1)(l)
"Site" means the lands and other places on under in or through which the Works are to be executed and any other lands or places provided by the Employer for the purposes of the Contract 14 1(1)(n)
The Engineer must, in deciding any extension of time to which the Contractor is entitled under Clause 44, take into account delay the Contractor suffers from failure on the part of the Employer to give possession in accordance with the terms of this Clause 555 42(1)
Engineer does take into account any such delay as stipulated ?
YES
NO
The Contractor must within 21 days after acceptance of his Tender submit to the Engineer for his approval a programme showing the order of procedure in which he proposes to carry out the Works; and 162 14(1)
The Contractor must be paid the amount of cost the Contractor incurs as may be reasonable:— 556 42(1)
The Engineer may order any variation for any reason that may in his opinion be desirable for the satisfactory completion and functioning of the Works 757 51(1)
Variations may include:— 758 51(1)
changes in the specified sequence method or timing of construction (if any) 761 51(1)
The Employer or the Contractor must refer to the Engineer any dispute as to:— 1320 66(1)
The Contractor may consider himself fairly entitled to an extension of time for the completion of the Works:— 569 44(1)
should any cause referred to in these Conditions cause delay 572 44(1)

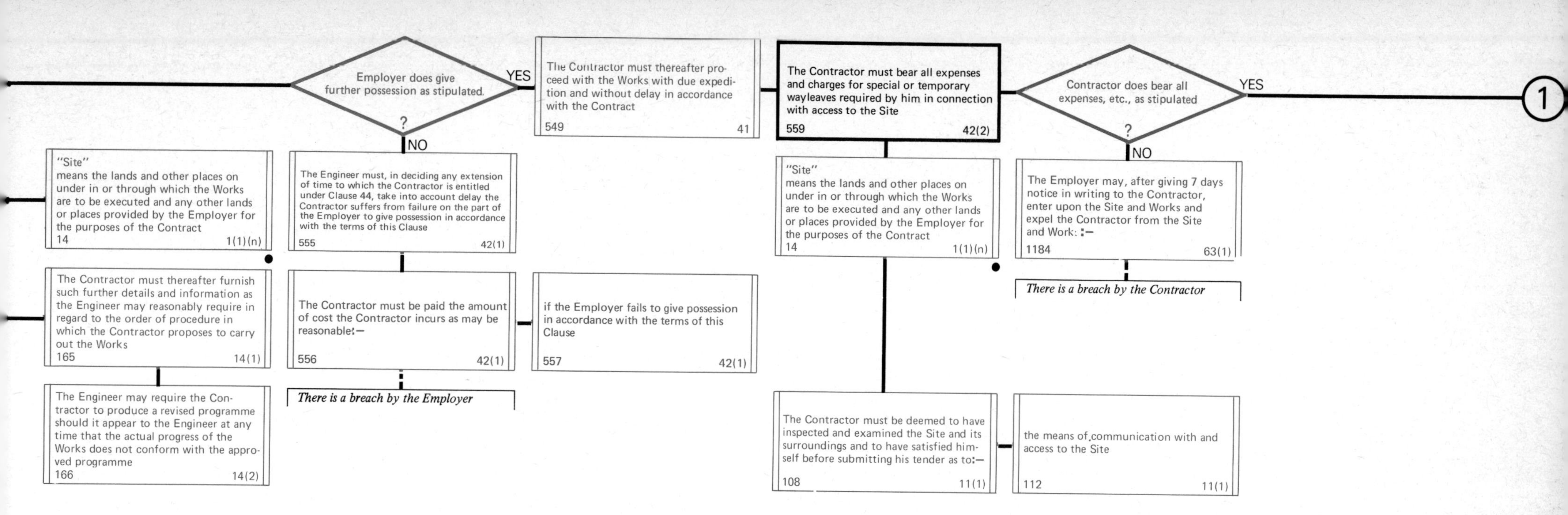
Employer does give further possession as stipulated.
YES
?
NO
The Contractor must thereafter proceed with the Works with due expedition and without delay in accordance with the Contract
549 41
The Contractor must bear all expenses and charges for special or temporary wayleaves required by him in connection with access to the Site
559 42(2)
Contractor does bear all expenses, etc., as stipulated
YES
?
NO
1
"Site" means the lands and other places on under in or through which the Works are to be executed and any other lands or places provided by the Employer for the purposes of the Contract
14 1(1)(n)
The Engineer must, in deciding any extension of time to which the Contractor is entitled under Clause 44, take into account delay the Contractor suffers from failure on the part of the Employer to give possession in accordance with the terms of this Clause
555 42(1)
"Site" means the lands and other places on under in or through which the Works are to be executed and any other lands or places provided by the Employer for the purposes of the Contract
14 1(1)(n)
The Employer may, after giving 7 days notice in writing to the Contractor, enter upon the Site and Works and expel the Contractor from the Site and Work:—
1184 63(1)
There is a breach by the Contractor
The Contractor must thereafter furnish such further details and information as the Engineer may reasonably require in regard to the order of procedure in which the Contractor proposes to carry out the Works
165 14(1)
The Contractor must be paid the amount of cost the Contractor incurs as may be reasonable:—
556 42(1)
if the Employer fails to give possession in accordance with the terms of this Clause
557 42(1)
There is a breach by the Employer
The Engineer may require the Contractor to produce a revised programme should it appear to the Engineer at any time that the actual progress of the Works does not conform with the approved programme
166 14(2)
The Contractor must be deemed to have inspected and examined the Site and its surroundings and to have satisfied himself before submitting his tender as to:—
108 11(1)
the means of communication with and access to the Site
112 11(1)

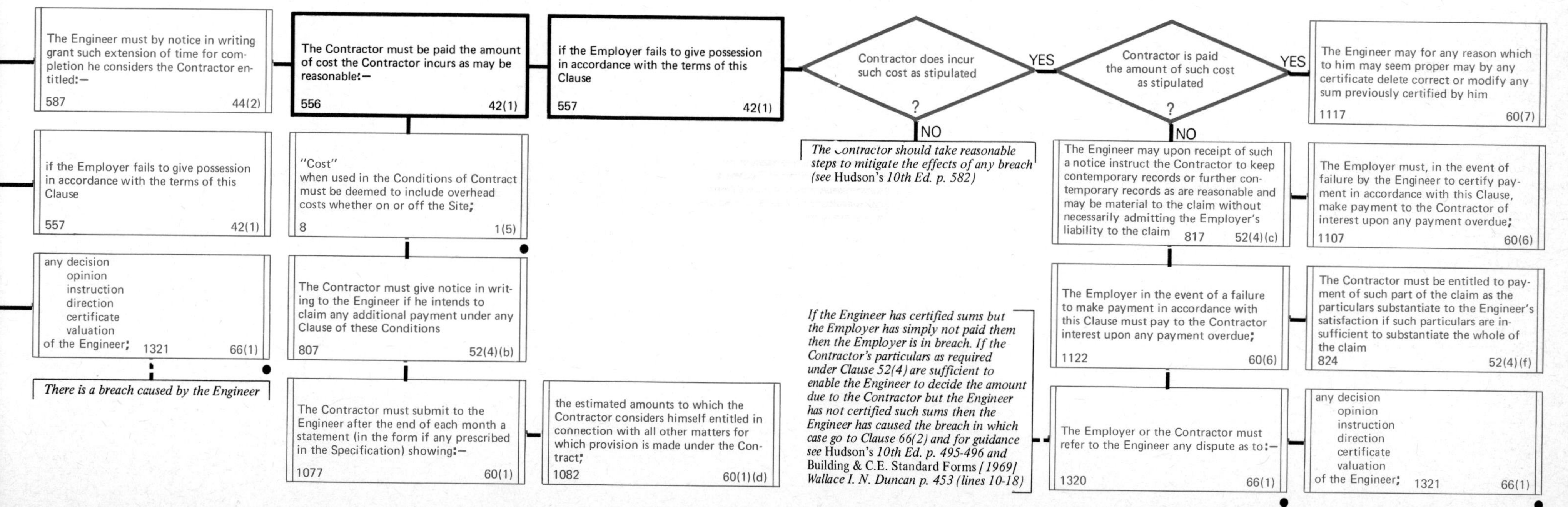
The Engineer must by notice in writing grant such extension of time for completion he considers the Contractor entitled:—
587 44(2)
The Contractor must be paid the amount of cost the Contractor incurs as may be reasonable:—
556 42(1)
if the Employer fails to give possession in accordance with the terms of this Clause
557 42(1)
Contractor does incur such cost as stipulated
YES
?
NO
Contractor is paid the amount of such cost as stipulated
YES
?
NO
The Engineer may for any reason which to him may seem proper may by any certificate delete correct or modify any sum previously certified by him
1117 60(7)
if the Employer fails to give possession in accordance with the terms of this Clause
557 42(1)
"Cost" when used in the Conditions of Contract must be deemed to include overhead costs whether on or off the Site;
8 1(5)
The Contractor should take reasonable steps to mitigate the effects of any breach (see Hudson's 10th Ed. p. 582)
The Engineer may upon receipt of such a notice instruct the Contractor to keep contemporary records or further contemporary records as are reasonable and may be material to the claim without necessarily admitting the Employer's liability to the claim
817 52(4)(c)
The Employer must, in the event of failure by the Engineer to certify payment in accordance with this Clause, make payment to the Contractor of interest upon any payment overdue;
1107 60(6)
any decision opinion instruction direction certificate valuation of the Engineer;
1321 66(1)
There is a breach caused by the Engineer
The Contractor must give notice in writing to the Engineer if he intends to claim any additional payment under any Clause of these Conditions
807 52(4)(b)
If the Engineer has certified sums but the Employer has simply not paid them then the Employer is in breach. If the Contractor's particulars as required under Clause 52(4) are sufficient to enable the Engineer to decide the amount due to the Contractor but the Engineer has not certified such sums then the Engineer has caused the breach in which case go to Clause 66(2) and for guidance see Hudson's 10th Ed. p. 495-496 and Building & C.E. Standard Forms [1969] Wallace I. N. Duncan p. 453 (lines 10-18)
The Employer in the event of a failure to make payment in accordance with this Clause must pay to the Contractor interest upon any payment overdue;
1122 60(6)
The Contractor must be entitled to payment of such part of the claim as the particulars substantiate to the Engineer's satisfaction if such particulars are insufficient to substantiate the whole of the claim
824 52(4)(f)
The Contractor must submit to the Engineer after the end of each month a statement (in the form if any prescribed in the Specification) showing:—
1077 60(1)
the estimated amounts to which the Contractor considers himself entitled in connection with all other matters for which provision is made under the Contract;
1082 60(1)(d)
The Employer or the Contractor must refer to the Engineer any dispute as to:—
1320 66(1)
any decision opinion instruction direction certificate valuation of the Engineer;
1321 66(1)

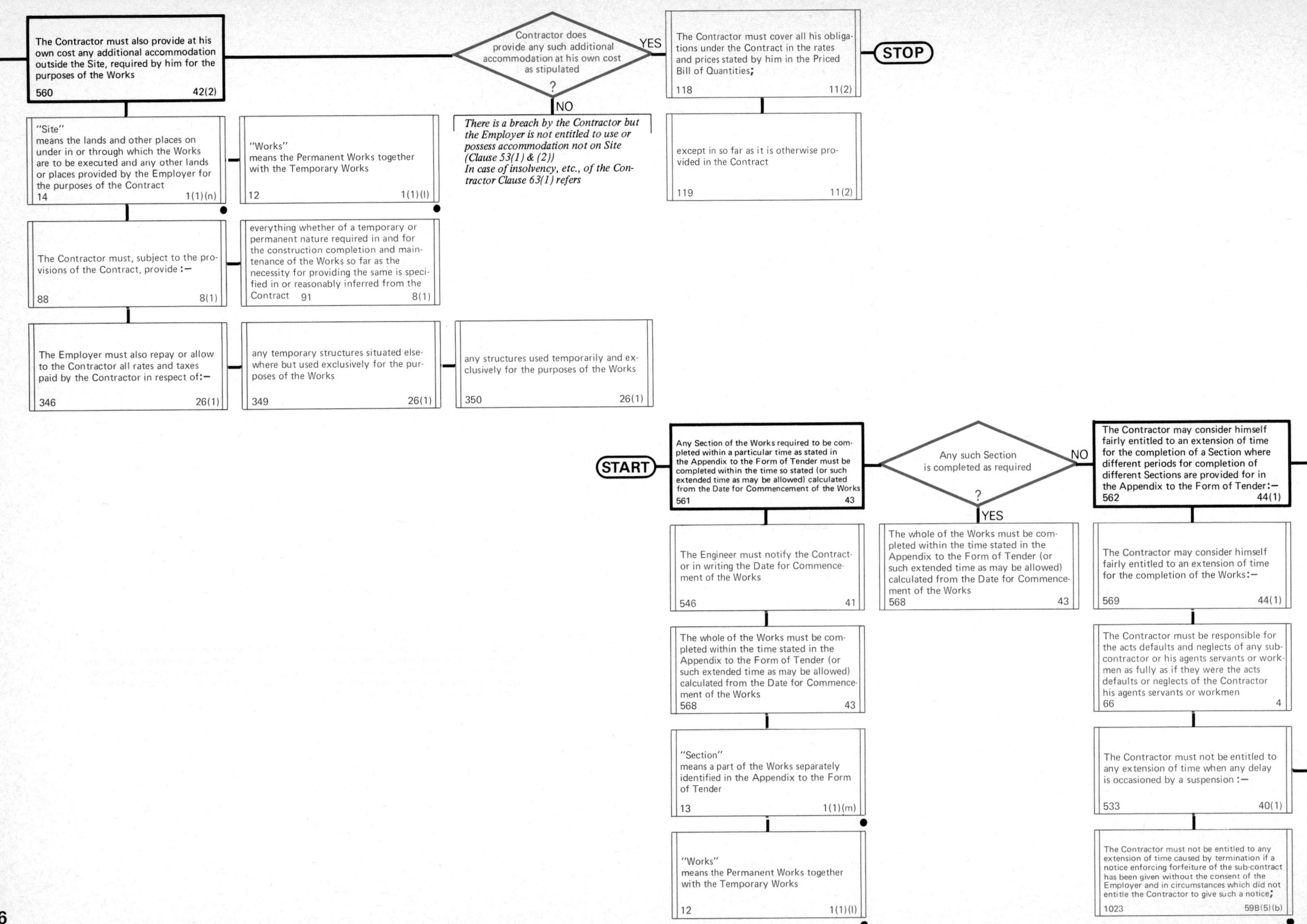
1
The Contractor must also provide at his own cost any additional accommodation outside the Site, required by him for the purposes of the Works 560 42(2)
Contractor does provide any such additional accommodation at his own cost as stipulated ?
YES
The Contractor must cover all his obligations under the Contract in the rates and prices stated by him in the Priced Bill of Quantities; 118 11(2)
STOP
except in so far as it is otherwise provided in the Contract 119 11(2)
NO
There is a breach by the Contractor but the Employer is not entitled to use or possess accommodation not on Site (Clause 53(1) & (2)) In case of insolvency, etc., of the Contractor Clause 63(1) refers
"Site" means the lands and other places on under in or through which the Works are to be executed and any other lands or places provided by the Employer for the purposes of the Contract 14 1(1)(n)
"Works" means the Permanent Works together with the Temporary Works 12 1(1)(l)
The Contractor must, subject to the provisions of the Contract, provide :— 88 8(1)
everything whether of a temporary or permanent nature required in and for the construction completion and maintenance of the Works so far as the necessity for providing the same is specified in or reasonably inferred from the Contract 91 8(1)
The Employer must also repay or allow to the Contractor all rates and taxes paid by the Contractor in respect of:— 346 26(1)
any temporary structures situated elsewhere but used exclusively for the purposes of the Works 349 26(1)
any structures used temporarily and exclusively for the purposes of the Works 350 26(1)
START
Any Section of the Works required to be completed within a particular time as stated in the Appendix to the Form of Tender must be completed within the time so stated (or such extended time as may be allowed) calculated from the Date for Commencement of the Works 561 43
Any such Section is completed as required ?
NO
The Contractor may consider himself fairly entitled to an extension of time for the completion of a Section where different periods for completion of different Sections are provided for in the Appendix to the Form of Tender:— 562 44(1)
YES
The whole of the Works must be completed within the time stated in the Appendix to the Form of Tender (or such extended time as may be allowed) calculated from the Date for Commencement of the Works 568 43
The Engineer must notify the Contractor in writing the Date for Commencement of the Works 546 41
The whole of the Works must be completed within the time stated in the Appendix to the Form of Tender (or such extended time as may be allowed) calculated from the Date for Commencement of the Works 568 43
"Section" means a part of the Works separately identified in the Appendix to the Form of Tender 13 1(1)(m)
"Works" means the Permanent Works together with the Temporary Works 12 1(1)(l)
The Contractor may consider himself fairly entitled to an extension of time for the completion of the Works:— 569 44(1)
The Contractor must be responsible for the acts defaults and neglects of any sub-contractor or his agents servants or workmen as fully as if they were the acts defaults or neglects of the Contractor his agents servants or workmen 66 4
The Contractor must not be entitled to any extension of time when any delay is occasioned by a suspension :— 533 40(1)
The Contractor must not be entitled to any extension of time caused by termination if a notice enforcing forfeiture of the sub-contract has been given without the consent of the Employer and in circumstances which did not entitle the Contractor to give such a notice; 1023 59B(5)(b)

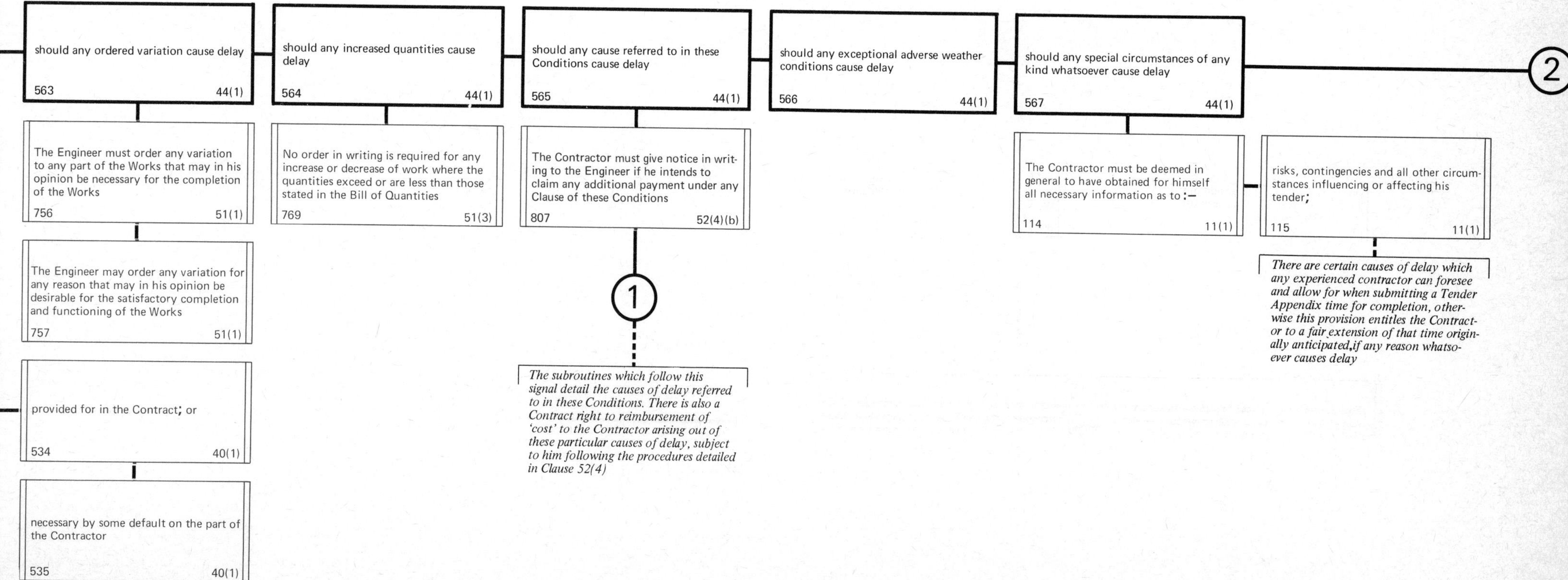
should any ordered variation cause delay
563 44(1)
should any increased quantities cause delay
564 44(1)
should any cause referred to in these Conditions cause delay
565 44(1)
should any exceptional adverse weather conditions cause delay
566 44(1)
should any special circumstances of any kind whatsoever cause delay
567 44(1)
2
The Engineer must order any variation to any part of the Works that may in his opinion be necessary for the completion of the Works
756 51(1)
No order in writing is required for any increase or decrease of work where the quantities exceed or are less than those stated in the Bill of Quantities
769 51(3)
The Contractor must give notice in writing to the Engineer if he intends to claim any additional payment under any Clause of these Conditions
807 52(4)(b)
The Contractor must be deemed in general to have obtained for himself all necessary information as to:—
114 11(1)
risks, contingencies and all other circumstances influencing or affecting his tender;
115 11(1)
There are certain causes of delay which any experienced contractor can foresee and allow for when submitting a Tender Appendix time for completion, otherwise this provision entitles the Contractor to a fair extension of that time originally anticipated, if any reason whatsoever causes delay
The Engineer may order any variation for any reason that may in his opinion be desirable for the satisfactory completion and functioning of the Works
757 51(1)
1
The subroutines which follow this signal detail the causes of delay referred to in these Conditions. There is also a Contract right to reimbursement of 'cost' to the Contractor arising out of these particular causes of delay, subject to him following the procedures detailed in Clause 52(4)
provided for in the Contract; or
534 40(1)
necessary by some default on the part of the Contractor
535 40(1)

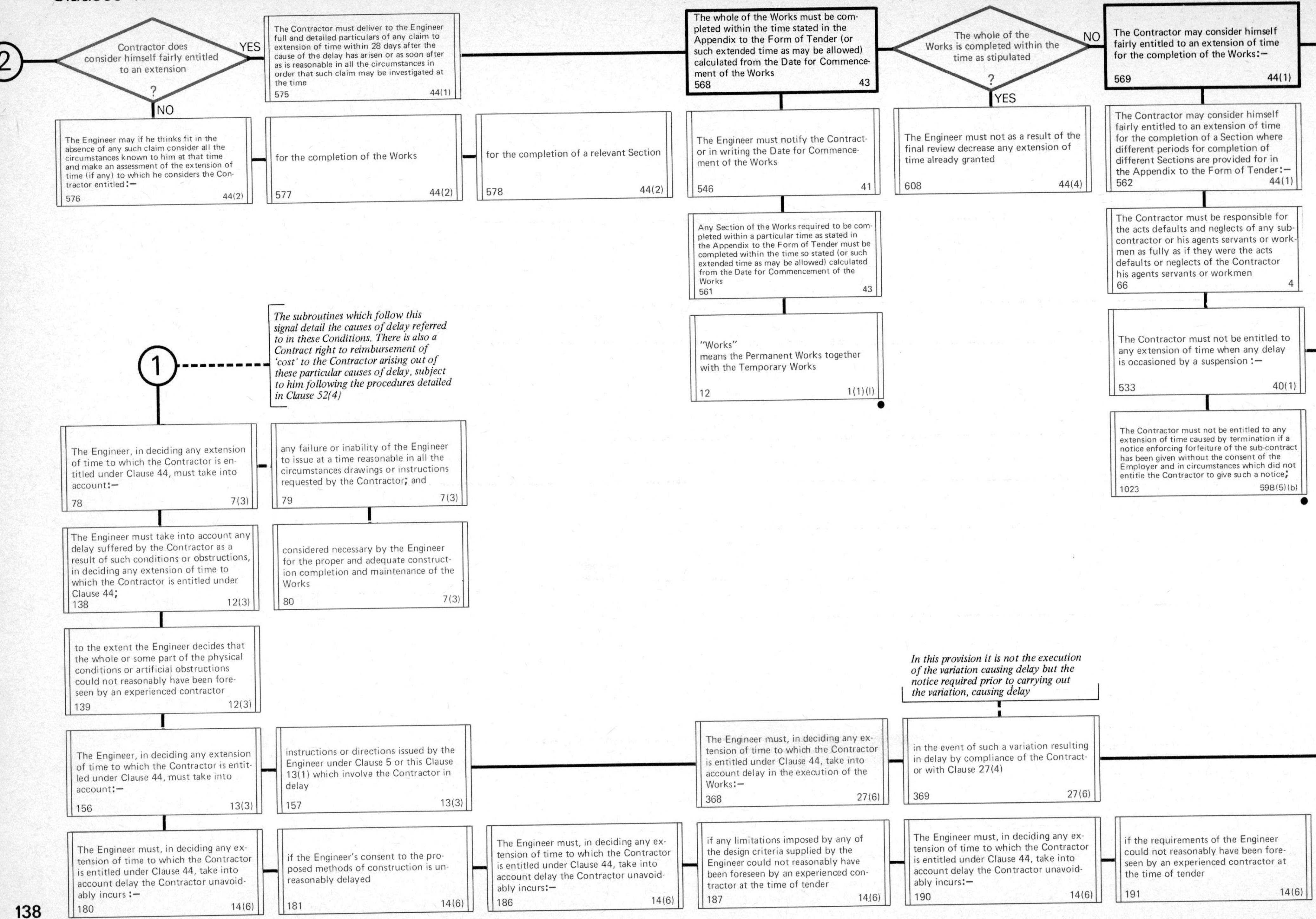

2
Contractor does consider himself fairly entitled to an extension ?
YES
NO
The Contractor must deliver to the Engineer full and detailed particulars of any claim to extension of time within 28 days after the cause of the delay has arisen or as soon after as is reasonable in all the circumstances in order that such claim may be investigated at the time 575 44(1)
The Engineer may if he thinks fit in the absence of any such claim consider all the circumstances known to him at that time and make an assessment of the extension of time (if any) to which he considers the Contractor entitled:— 576 44(2)
for the completion of the Works 577 44(2)
for the completion of a relevant Section 578 44(2)
The whole of the Works must be completed within the time stated in the Appendix to the Form of Tender (or such extended time as may be allowed) calculated from the Date for Commencement of the Works 568 43
The Engineer must notify the Contractor in writing the Date for Commencement of the Works 546 41
Any Section of the Works required to be completed within a particular time as stated in the Appendix to the Form of Tender must be completed within the time so stated (or such extended time as may be allowed) calculated from the Date for Commencement of the Works 561 43
"Works" means the Permanent Works together with the Temporary Works 12 1(1)(l)
The whole of the Works is completed within the time as stipulated ?
NO
YES
The Engineer must not as a result of the final review decrease any extension of time already granted 608 44(4)
The Contractor may consider himself fairly entitled to an extension of time for the completion of the Works:— 569 44(1)
The Contractor may consider himself fairly entitled to an extension of time for the completion of a Section where different periods for completion of different Sections are provided for in the Appendix to the Form of Tender:— 562 44(1)
The Contractor must be responsible for the acts defaults and neglects of any sub-contractor or his agents servants or workmen as fully as if they were the acts defaults or neglects of the Contractor his agents servants or workmen 66 4
The Contractor must not be entitled to any extension of time when any delay is occasioned by a suspension :— 533 40(1)
The Contractor must not be entitled to any extension of time caused by termination if a notice enforcing forfeiture of the sub-contract has been given without the consent of the Employer and in circumstances which did not entitle the Contractor to give such a notice; 1023 59B(5)(b)
1
The subroutines which follow this signal detail the causes of delay referred to in these Conditions. There is also a Contract right to reimbursement of 'cost' to the Contractor arising out of these particular causes of delay, subject to him following the procedures detailed in Clause 52(4)
The Engineer, in deciding any extension of time to which the Contractor is entitled under Clause 44, must take into account:— 78 7(3)
any failure or inability of the Engineer to issue at a time reasonable in all the circumstances drawings or instructions requested by the Contractor; and 79 7(3)
considered necessary by the Engineer for the proper and adequate construction completion and maintenance of the Works 80 7(3)
The Engineer must take into account any delay suffered by the Contractor as a result of such conditions or obstructions, in deciding any extension of time to which the Contractor is entitled under Clause 44; 138 12(3)
to the extent the Engineer decides that the whole or some part of the physical conditions or artificial obstructions could not reasonably have been foreseen by an experienced contractor 139 12(3)
The Engineer, in deciding any extension of time to which the Contractor is entitled under Clause 44, must take into account:— 156 13(3)
instructions or directions issued by the Engineer under Clause 5 or this Clause 13(1) which involve the Contractor in delay 157 13(3)
The Engineer must, in deciding any extension of time to which the Contractor is entitled under Clause 44, take into account delay in the execution of the Works:— 368 27(6)
In this provision it is not the execution of the variation causing delay but the notice required prior to carrying out the variation, causing delay
in the event of such a variation resulting in delay by compliance of the Contractor with Clause 27(4) 369 27(6)
The Engineer must, in deciding any extension of time to which the Contractor is entitled under Clause 44, take into account delay the Contractor unavoidably incurs :— 180 14(6)
if the Engineer's consent to the proposed methods of construction is unreasonably delayed 181 14(6)
The Engineer must, in deciding any extension of time to which the Contractor is entitled under Clause 44, take into account delay the Contractor unavoidably incurs:— 186 14(6)
if any limitations imposed by any of the design criteria supplied by the Engineer could not reasonably have been foreseen by an experienced contractor at the time of tender 187 14(6)
The Engineer must, in deciding any extension of time to which the Contractor is entitled under Clause 44, take into account delay the Contractor unavoidably incurs:— 190 14(6)
if the requirements of the Engineer could not reasonably have been foreseen by an experienced contractor at the time of tender 191 14(6)

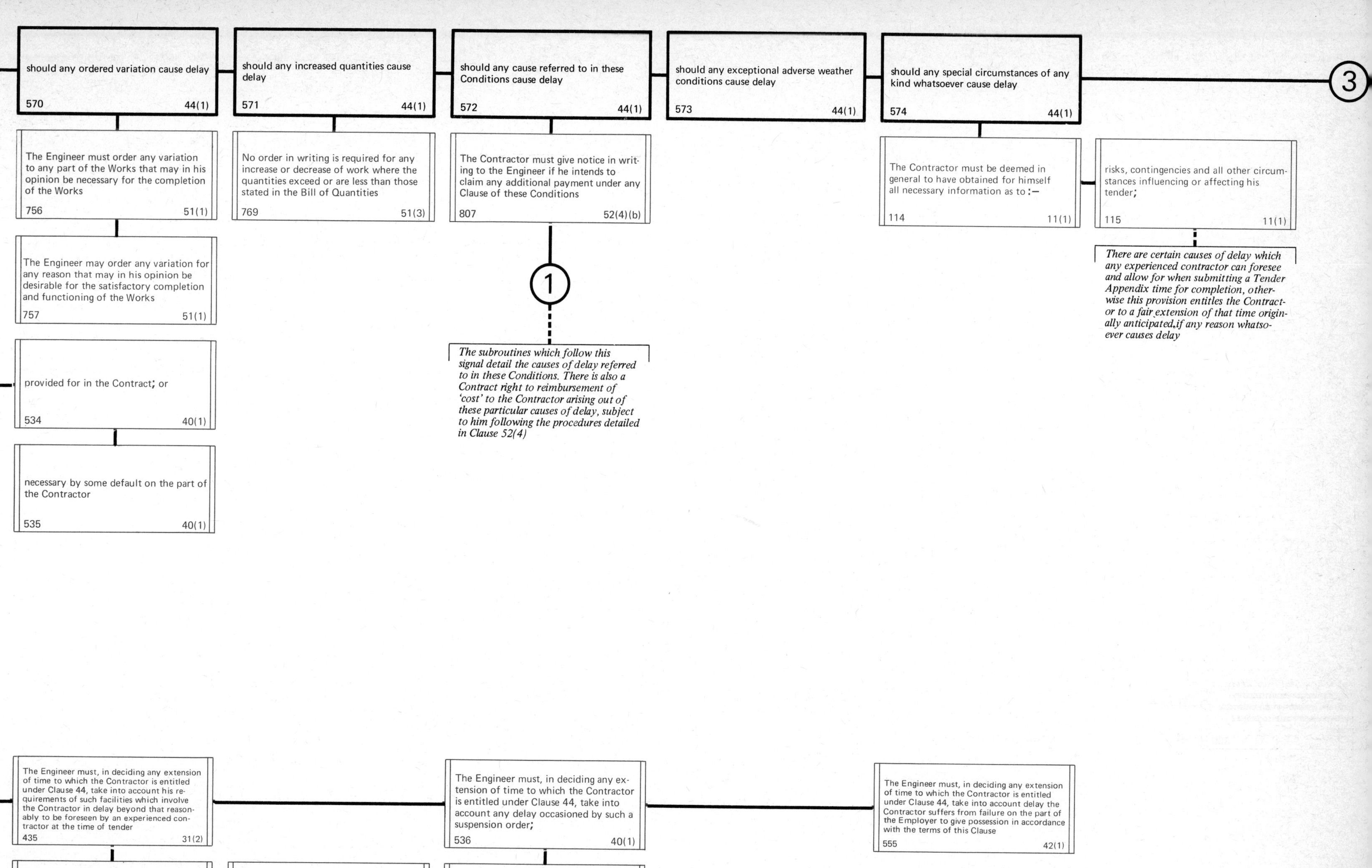
should any ordered variation cause delay
570 44(1)
should any increased quantities cause delay
571 44(1)
should any cause referred to in these Conditions cause delay
572 44(1)
should any exceptional adverse weather conditions cause delay
573 44(1)
should any special circumstances of any kind whatsoever cause delay
574 44(1)
3
The Engineer must order any variation to any part of the Works that may in his opinion be necessary for the completion of the Works
756 51(1)
No order in writing is required for any increase or decrease of work where the quantities exceed or are less than those stated in the Bill of Quantities
769 51(3)
The Contractor must give notice in writing to the Engineer if he intends to claim any additional payment under any Clause of these Conditions
807 52(4)(b)
The Contractor must be deemed in general to have obtained for himself all necessary information as to :—
114 11(1)
risks, contingencies and all other circumstances influencing or affecting his tender;
115 11(1)
There are certain causes of delay which any experienced contractor can foresee and allow for when submitting a Tender Appendix time for completion, otherwise this provision entitles the Contractor to a fair extension of that time originally anticipated, if any reason whatsoever causes delay
The Engineer may order any variation for any reason that may in his opinion be desirable for the satisfactory completion and functioning of the Works
757 51(1)
1
The subroutines which follow this signal detail the causes of delay referred to in these Conditions. There is also a Contract right to reimbursement of 'cost' to the Contractor arising out of these particular causes of delay, subject to him following the procedures detailed in Clause 52(4)
provided for in the Contract; or
534 40(1)
necessary by some default on the part of the Contractor
535 40(1)
The Engineer must, in deciding any extension of time to which the Contractor is entitled under Clause 44, take into account his requirements of such facilities which involve the Contractor in delay beyond that reasonably to be foreseen by an experienced contractor at the time of tender
435 31(2)
The Engineer must, in deciding any extension of time to which the Contractor is entitled under Clause 44, take into account any delay occasioned by such a suspension order;
536 40(1)
The Engineer must, in deciding any extension of time to which the Contractor is entitled under Clause 44, take into account delay the Contractor suffers from failure on the part of the Employer to give possession in accordance with the terms of this Clause
555 42(1)
The Contractor must at once notify the Engineer in writing if any event arises which in the opinion of the Contractor would entitle the Contractor to exercise his right under the Forfeiture Clause; or
1003 59B(2)
The Engineer must take into account any delay to the completion of the Works in deciding any extension of time to which the Contractor is entitled
1061 59B(4)(b)
including that arising from any act or default of the Engineer or the Employer
537 40(1)

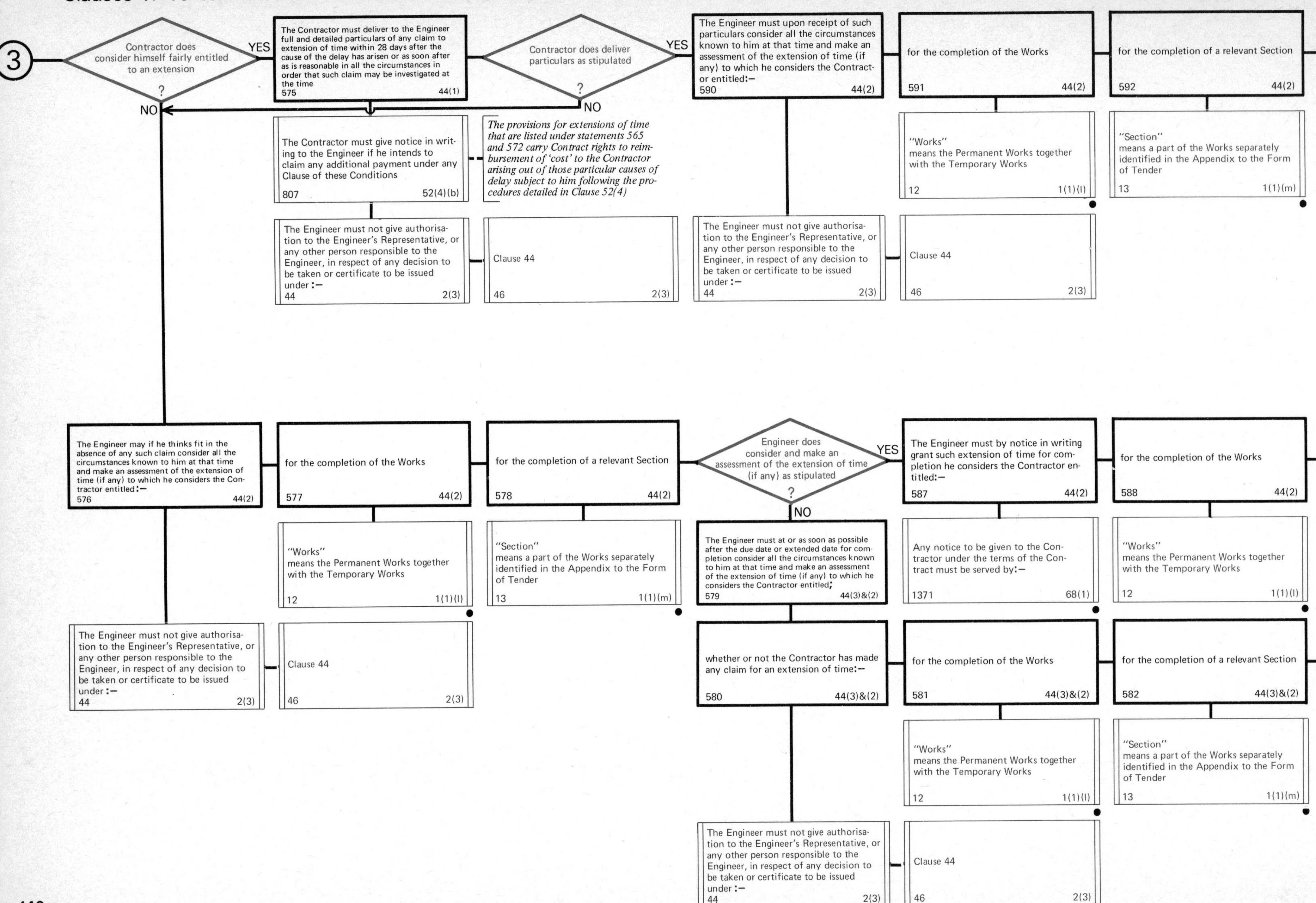
3
Contractor does consider himself fairly entitled to an extension ?
YES
NO
The Contractor must deliver to the Engineer full and detailed particulars of any claim to extension of time within 28 days after the cause of the delay has arisen or as soon after as is reasonable in all the circumstances in order that such claim may be investigated at the time
575 44(1)
Contractor does deliver particulars as stipulated ?
YES
NO
The Engineer must upon receipt of such particulars consider all the circumstances known to him at that time and make an assessment of the extension of time (if any) to which he considers the Contractor entitled:—
590 44(2)
for the completion of the Works
591 44(2)
for the completion of a relevant Section
592 44(2)
The Contractor must give notice in writing to the Engineer if he intends to claim any additional payment under any Clause of these Conditions
807 52(4)(b)
The provisions for extensions of time that are listed under statements 565 and 572 carry Contract rights to reimbursement of 'cost' to the Contractor arising out of those particular causes of delay subject to him following the procedures detailed in Clause 52(4)
"Works" means the Permanent Works together with the Temporary Works
12 1(1)(l)
"Section" means a part of the Works separately identified in the Appendix to the Form of Tender
13 1(1)(m)
The Engineer must not give authorisation to the Engineer's Representative, or any other person responsible to the Engineer, in respect of any decision to be taken or certificate to be issued under:—
44 2(3)
Clause 44
46 2(3)
The Engineer must not give authorisation to the Engineer's Representative, or any other person responsible to the Engineer, in respect of any decision to be taken or certificate to be issued under:—
44 2(3)
Clause 44
46 2(3)
The Engineer may if he thinks fit in the absence of any such claim consider all the circumstances known to him at that time and make an assessment of the extension of time (if any) to which he considers the Contractor entitled:—
576 44(2)
for the completion of the Works
577 44(2)
for the completion of a relevant Section
578 44(2)
Engineer does consider and make an assessment of the extension of time (if any) as stipulated ?
YES
NO
The Engineer must by notice in writing grant such extension of time for completion he considers the Contractor entitled:—
587 44(2)
for the completion of the Works
588 44(2)
"Works" means the Permanent Works together with the Temporary Works
12 1(1)(l)
"Section" means a part of the Works separately identified in the Appendix to the Form of Tender
13 1(1)(m)
The Engineer must at or as soon as possible after the due date or extended date for completion consider all the circumstances known to him at that time and make an assessment of the extension of time (if any) to which he considers the Contractor entitled;
579 44(3)&(2)
Any notice to be given to the Contractor under the terms of the Contract must be served by:—
1371 68(1)
"Works" means the Permanent Works together with the Temporary Works
12 1(1)(l)
The Engineer must not give authorisation to the Engineer's Representative, or any other person responsible to the Engineer, in respect of any decision to be taken or certificate to be issued under:—
44 2(3)
Clause 44
46 2(3)
whether or not the Contractor has made any claim for an extension of time:—
580 44(3)&(2)
for the completion of the Works
581 44(3)&(2)
for the completion of a relevant Section
582 44(3)&(2)
"Works" means the Permanent Works together with the Temporary Works
12 1(1)(l)
"Section" means a part of the Works separately identified in the Appendix to the Form of Tender
13 1(1)(m)
The Engineer must not give authorisation to the Engineer's Representative, or any other person responsible to the Engineer, in respect of any decision to be taken or certificate to be issued under:—
44 2(3)
Clause 44
46 2(3)

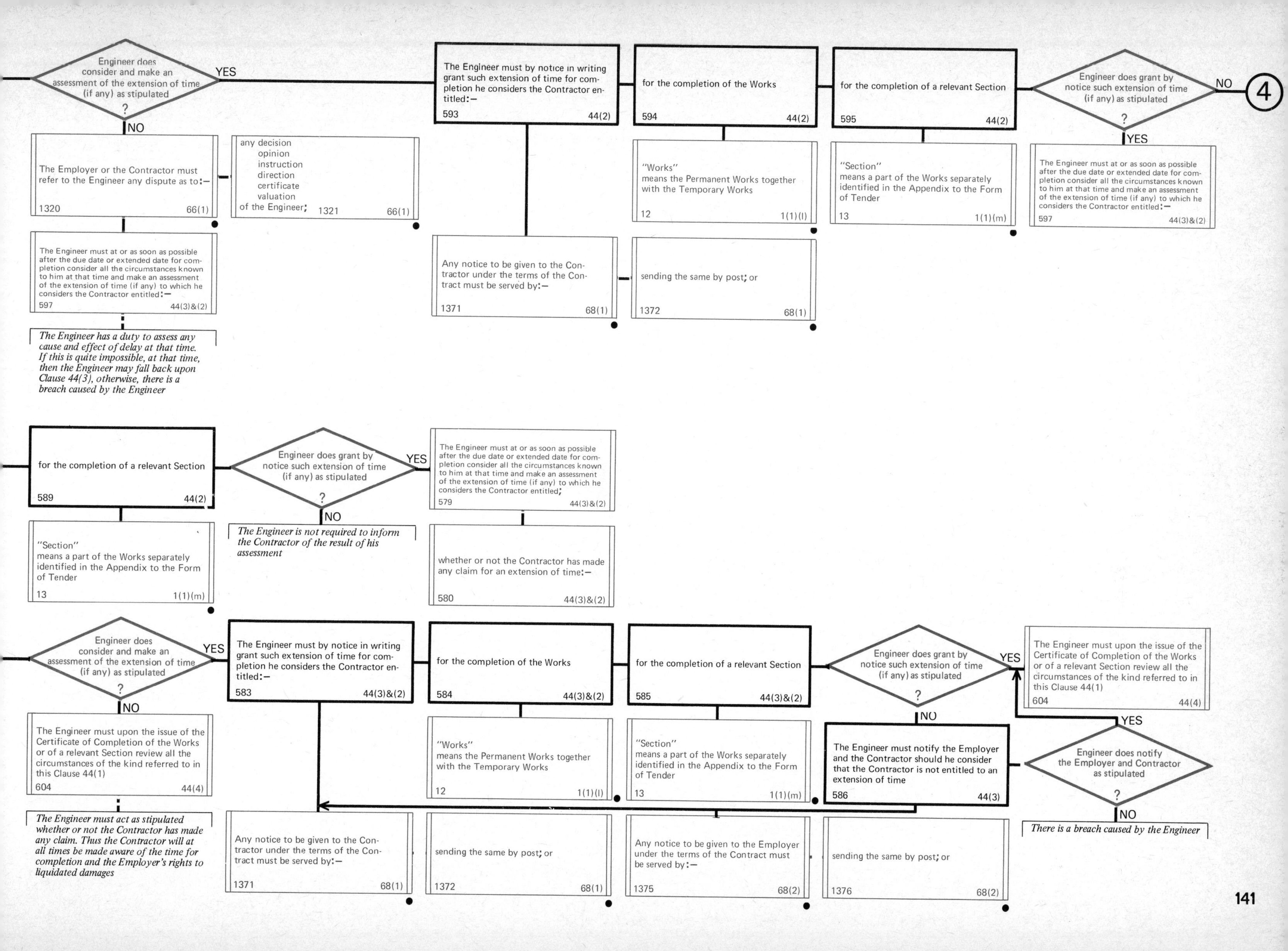
Engineer does consider and make an assessment of the extension of time (if any) as stipulated ?
YES
NO
The Engineer must by notice in writing grant such extension of time for completion he considers the Contractor entitled:—
593 44(2)
for the completion of the Works
594 44(2)
for the completion of a relevant Section
595 44(2)
Engineer does grant by notice such extension of time (if any) as stipulated ?
NO
4
YES
The Employer or the Contractor must refer to the Engineer any dispute as to:—
1320 66(1)
any decision opinion instruction direction certificate valuation of the Engineer;
1321 66(1)
"Works" means the Permanent Works together with the Temporary Works
12 1(1)(l)
"Section" means a part of the Works separately identified in the Appendix to the Form of Tender
13 1(1)(m)
The Engineer must at or as soon as possible after the due date or extended date for completion consider all the circumstances known to him at that time and make an assessment of the extension of time (if any) to which he considers the Contractor entitled:—
597 44(3)&(2)
The Engineer must at or as soon as possible after the due date or extended date for completion consider all the circumstances known to him at that time and make an assessment of the extension of time (if any) to which he considers the Contractor entitled:—
597 44(3)&(2)
Any notice to be given to the Contractor under the terms of the Contract must be served by:—
1371 68(1)
sending the same by post; or
1372 68(1)
The Engineer has a duty to assess any cause and effect of delay at that time. If this is quite impossible, at that time, then the Engineer may fall back upon Clause 44(3), otherwise, there is a breach caused by the Engineer
for the completion of a relevant Section
589 44(2)
Engineer does grant by notice such extension of time (if any) as stipulated ?
YES
NO
The Engineer must at or as soon as possible after the due date or extended date for completion consider all the circumstances known to him at that time and make an assessment of the extension of time (if any) to which he considers the Contractor entitled;
579 44(3)&(2)
"Section" means a part of the Works separately identified in the Appendix to the Form of Tender
13 1(1)(m)
The Engineer is not required to inform the Contractor of the result of his assessment
whether or not the Contractor has made any claim for an extension of time:—
580 44(3)&(2)
Engineer does consider and make an assessment of the extension of time (if any) as stipulated ?
YES
NO
The Engineer must by notice in writing grant such extension of time for completion he considers the Contractor entitled:—
583 44(3)&(2)
for the completion of the Works
584 44(3)&(2)
for the completion of a relevant Section
585 44(3)&(2)
Engineer does grant by notice such extension of time (if any) as stipulated ?
YES
NO
The Engineer must upon the issue of the Certificate of Completion of the Works or of a relevant Section review all the circumstances of the kind referred to in this Clause 44(1)
604 44(4)
The Engineer must upon the issue of the Certificate of Completion of the Works or of a relevant Section review all the circumstances of the kind referred to in this Clause 44(1)
604 44(4)
"Works" means the Permanent Works together with the Temporary Works
12 1(1)(l)
"Section" means a part of the Works separately identified in the Appendix to the Form of Tender
13 1(1)(m)
The Engineer must notify the Employer and the Contractor should he consider that the Contractor is not entitled to an extension of time
586 44(3)
YES
Engineer does notify the Employer and Contractor as stipulated ?
NO
There is a breach caused by the Engineer
The Engineer must act as stipulated whether or not the Contractor has made any claim. Thus the Contractor will at all times be made aware of the time for completion and the Employer's rights to liquidated damages
Any notice to be given to the Contractor under the terms of the Contract must be served by:—
1371 68(1)
sending the same by post; or
1372 68(1)
Any notice to be given to the Employer under the terms of the Contract must be served by:—
1375 68(2)
sending the same by post; or
1376 68(2)

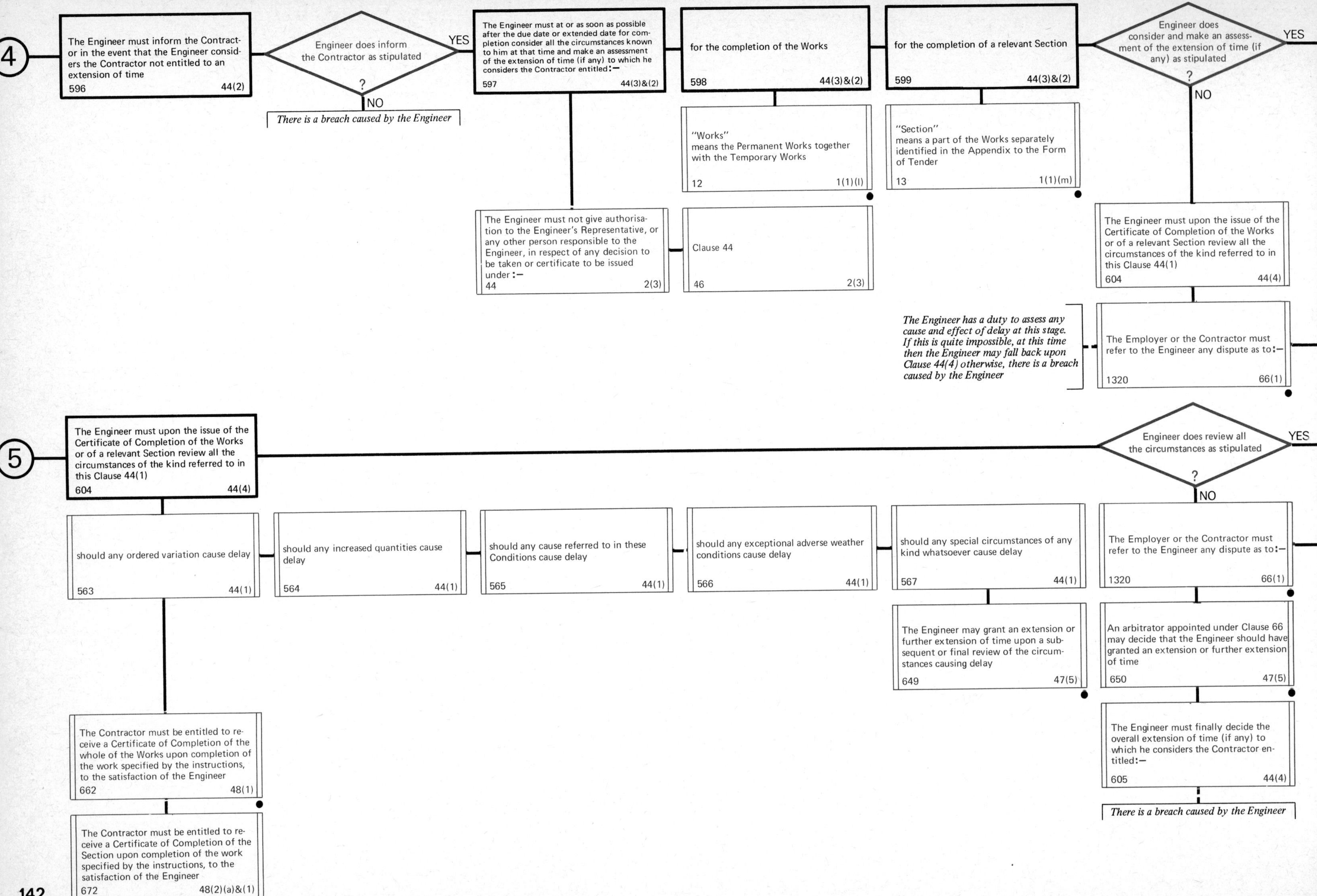
4
The Engineer must inform the Contractor in the event that the Engineer considers the Contractor not entitled to an extension of time
596 44(2)
Engineer does inform the Contractor as stipulated
?
YES
NO
There is a breach caused by the Engineer
The Engineer must at or as soon as possible after the due date or extended date for completion consider all the circumstances known to him at that time and make an assessment of the extension of time (if any) to which he considers the Contractor entitled:—
597 44(3)&(2)
for the completion of the Works
598 44(3)&(2)
for the completion of a relevant Section
599 44(3)&(2)
Engineer does consider and make an assessment of the extension of time (if any) as stipulated
?
YES
NO
"Works"
means the Permanent Works together with the Temporary Works
12 1(1)(l)
"Section"
means a part of the Works separately identified in the Appendix to the Form of Tender
13 1(1)(m)
The Engineer must not give authorisation to the Engineer's Representative, or any other person responsible to the Engineer, in respect of any decision to be taken or certificate to be issued under:—
44 2(3)
Clause 44
46 2(3)
The Engineer must upon the issue of the Certificate of Completion of the Works or of a relevant Section review all the circumstances of the kind referred to in this Clause 44(1)
604 44(4)
The Engineer has a duty to assess any cause and effect of delay at this stage. If this is quite impossible, at this time then the Engineer may fall back upon Clause 44(4) otherwise, there is a breach caused by the Engineer
The Employer or the Contractor must refer to the Engineer any dispute as to:—
1320 66(1)
5
The Engineer must upon the issue of the Certificate of Completion of the Works or of a relevant Section review all the circumstances of the kind referred to in this Clause 44(1)
604 44(4)
Engineer does review all the circumstances as stipulated
?
YES
NO
should any ordered variation cause delay
563 44(1)
should any increased quantities cause delay
564 44(1)
should any cause referred to in these Conditions cause delay
565 44(1)
should any exceptional adverse weather conditions cause delay
566 44(1)
should any special circumstances of any kind whatsoever cause delay
567 44(1)
The Employer or the Contractor must refer to the Engineer any dispute as to:—
1320 66(1)
The Engineer may grant an extension or further extension of time upon a subsequent or final review of the circumstances causing delay
649 47(5)
An arbitrator appointed under Clause 66 may decide that the Engineer should have granted an extension or further extension of time
650 47(5)
The Engineer must finally decide the overall extension of time (if any) to which he considers the Contractor entitled:—
605 44(4)
There is a breach caused by the Engineer
The Contractor must be entitled to receive a Certificate of Completion of the whole of the Works upon completion of the work specified by the instructions, to the satisfaction of the Engineer
662 48(1)
The Contractor must be entitled to receive a Certificate of Completion of the Section upon completion of the work specified by the instructions, to the satisfaction of the Engineer
672 48(2)(a)&(1)

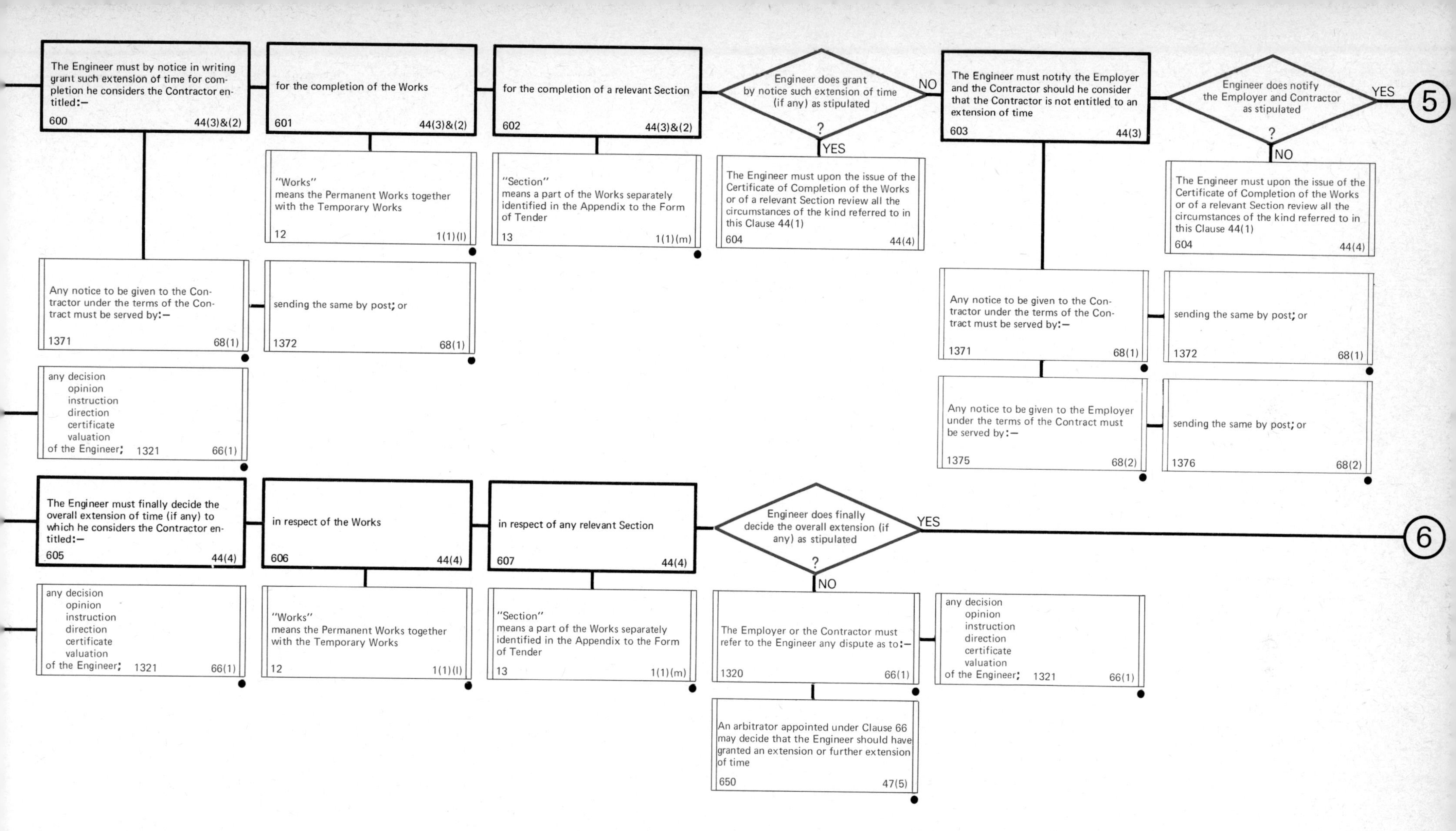
The Engineer must by notice in writing grant such extension of time for completion he considers the Contractor entitled:— 600 44(3)&(2)
for the completion of the Works 601 44(3)&(2)
for the completion of a relevant Section 602 44(3)&(2)
Engineer does grant by notice such extension of time (if any) as stipulated ?
NO
YES
The Engineer must notify the Employer and the Contractor should he consider that the Contractor is not entitled to an extension of time 603 44(3)
Engineer does notify the Employer and Contractor as stipulated ?
YES
NO
5
"Works" means the Permanent Works together with the Temporary Works 12 1(1)(l)
"Section" means a part of the Works separately identified in the Appendix to the Form of Tender 13 1(1)(m)
The Engineer must upon the issue of the Certificate of Completion of the Works or of a relevant Section review all the circumstances of the kind referred to in this Clause 44(1) 604 44(4)
The Engineer must upon the issue of the Certificate of Completion of the Works or of a relevant Section review all the circumstances of the kind referred to in this Clause 44(1) 604 44(4)
Any notice to be given to the Contractor under the terms of the Contract must be served by:— 1371 68(1)
sending the same by post; or 1372 68(1)
Any notice to be given to the Contractor under the terms of the Contract must be served by:— 1371 68(1)
sending the same by post; or 1372 68(1)
any decision opinion instruction direction certificate valuation of the Engineer; 1321 66(1)
Any notice to be given to the Employer under the terms of the Contract must be served by:— 1375 68(2)
sending the same by post; or 1376 68(2)
The Engineer must finally decide the overall extension of time (if any) to which he considers the Contractor entitled:— 605 44(4)
in respect of the Works 606 44(4)
in respect of any relevant Section 607 44(4)
Engineer does finally decide the overall extension (if any) as stipulated ?
YES
NO
6
any decision opinion instruction direction certificate valuation of the Engineer; 1321 66(1)
"Works" means the Permanent Works together with the Temporary Works 12 1(1)(l)
"Section" means a part of the Works separately identified in the Appendix to the Form of Tender 13 1(1)(m)
The Employer or the Contractor must refer to the Engineer any dispute as to:— 1320 66(1)
any decision opinion instruction direction certificate valuation of the Engineer; 1321 66(1)
An arbitrator appointed under Clause 66 may decide that the Engineer should have granted an extension or further extension of time 650 47(5)

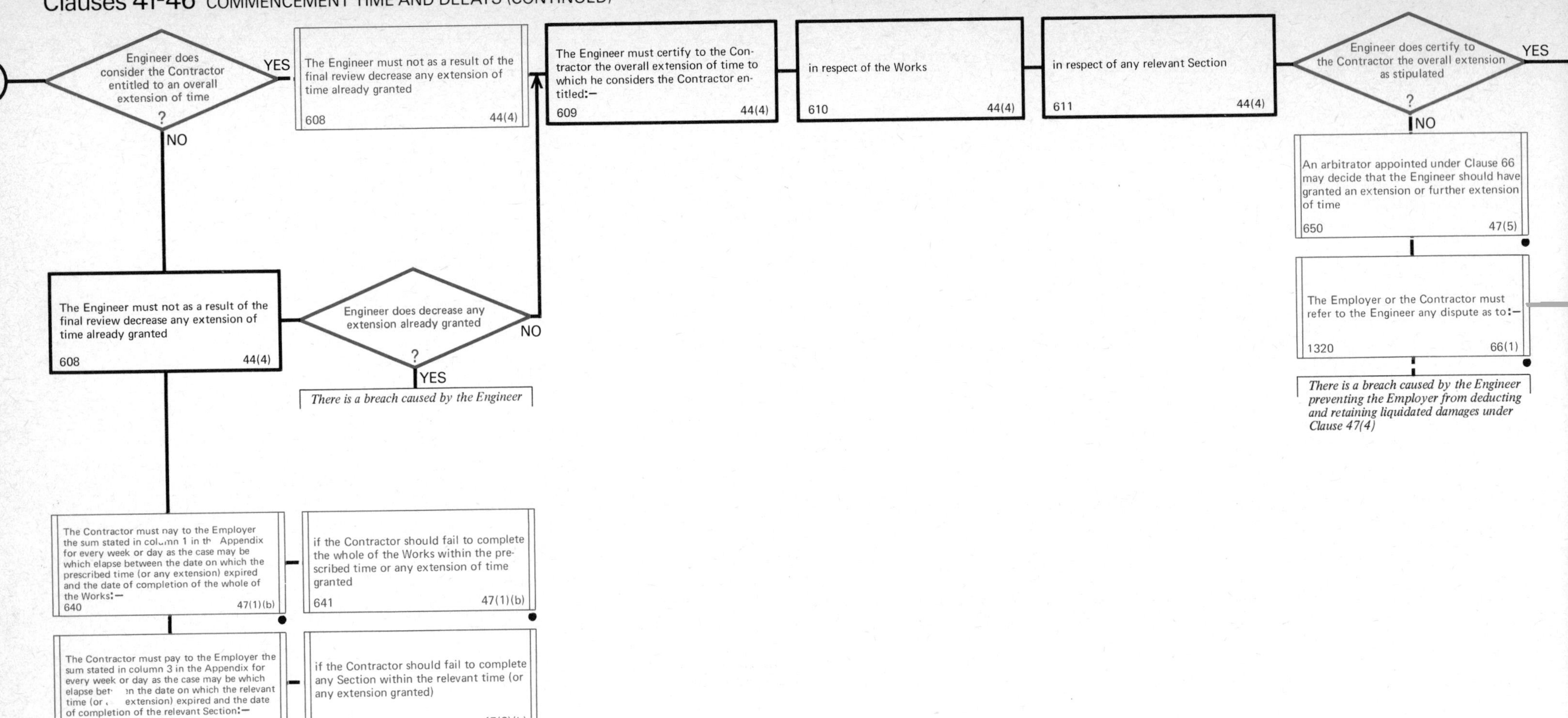
6
Engineer does consider the Contractor entitled to an overall extension of time ?
YES
NO
The Engineer must not as a result of the final review decrease any extension of time already granted 608 44(4)
The Engineer must certify to the Contractor the overall extension of time to which he considers the Contractor entitled:— 609 44(4)
in respect of the Works 610 44(4)
in respect of any relevant Section 611 44(4)
Engineer does certify to the Contractor the overall extension as stipulated ?
YES
NO
An arbitrator appointed under Clause 66 may decide that the Engineer should have granted an extension or further extension of time 650 47(5)
The Employer or the Contractor must refer to the Engineer any dispute as to:— 1320 66(1)
There is a breach caused by the Engineer preventing the Employer from deducting and retaining liquidated damages under Clause 47(4)
The Engineer must not as a result of the final review decrease any extension of time already granted 608 44(4)
Engineer does decrease any extension already granted ?
NO
YES
There is a breach caused by the Engineer
The Contractor must pay to the Employer the sum stated in column 1 in the Appendix for every week or day as the case may be which elapse between the date on which the prescribed time (or any extension) expired and the date of completion of the whole of the Works:— 640 47(1)(b)
if the Contractor should fail to complete the whole of the Works within the prescribed time or any extension of time granted 641 47(1)(b)
The Contractor must pay to the Employer the sum stated in column 3 in the Appendix for every week or day as the case may be which elapse between the date on which the relevant time (or any extension) expired and the date of completion of the relevant Section:— 632 47(2)(b)
if the Contractor should fail to complete any Section within the relevant time (or any extension granted) 633 47(2)(b)

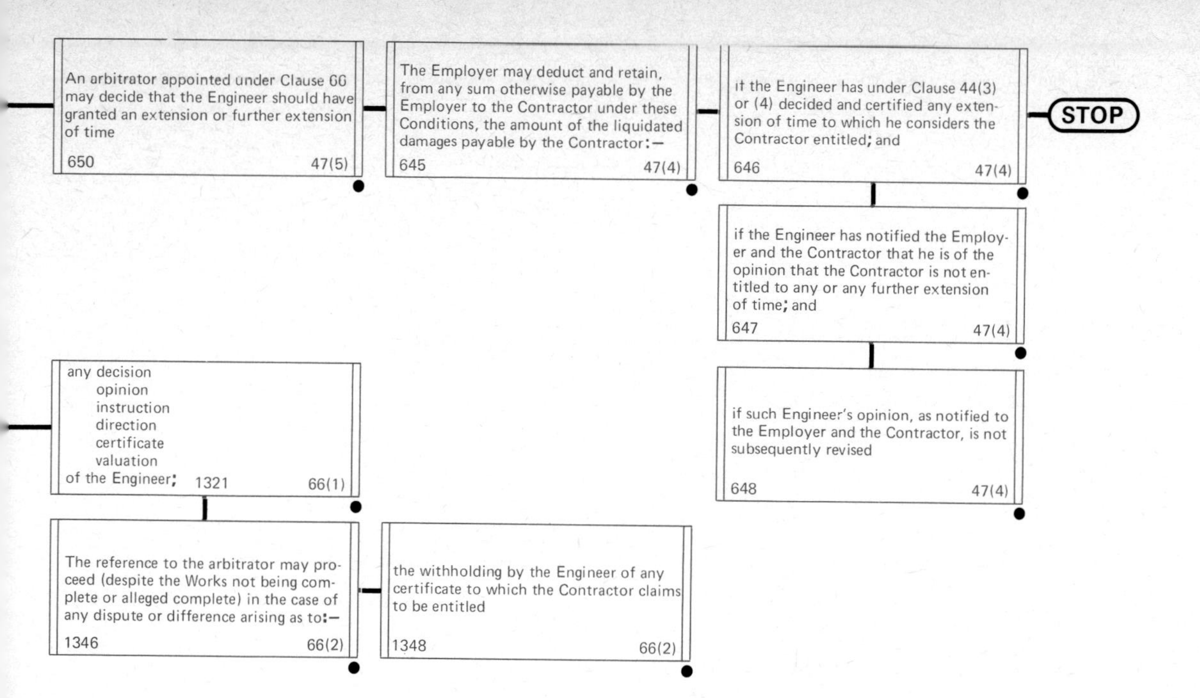
An arbitrator appointed under Clause 66 may decide that the Engineer should have granted an extension or further extension of time
650
47(5)
The Employer may deduct and retain, from any sum otherwise payable by the Employer to the Contractor under these Conditions, the amount of the liquidated damages payable by the Contractor:—
645
47(4)
if the Engineer has under Clause 44(3) or (4) decided and certified any extension of time to which he considers the Contractor entitled; and
646
47(4)
STOP
if the Engineer has notified the Employer and the Contractor that he is of the opinion that the Contractor is not entitled to any or any further extension of time; and
647
47(4)
if such Engineer's opinion, as notified to the Employer and the Contractor, is not subsequently revised
648
47(4)
any decision
opinion
instruction
direction
certificate
valuation
of the Engineer;
1321
66(1)
The reference to the arbitrator may proceed (despite the Works not being complete or alleged complete) in the case of any dispute or difference arising as to:—
1346
66(2)
the withholding by the Engineer of any certificate to which the Contractor claims to be entitled
1348
66(2)

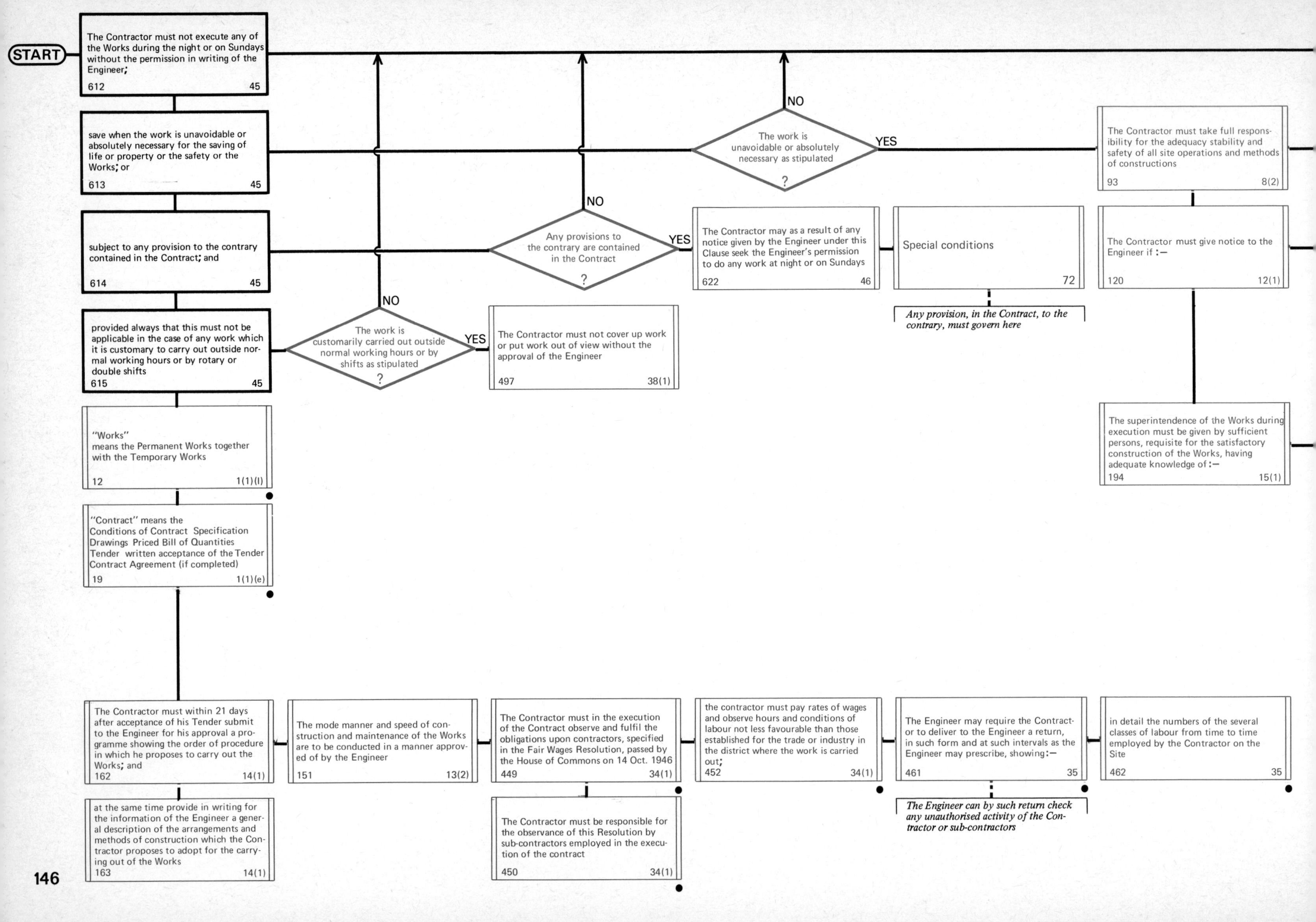
START
The Contractor must not execute any of the Works during the night or on Sundays without the permission in writing of the Engineer;
612 45
save when the work is unavoidable or absolutely necessary for the saving of life or property or the safety or the Works; or
613 45
subject to any provision to the contrary contained in the Contract; and
614 45
provided always that this must not be applicable in the case of any work which it is customary to carry out outside normal working hours or by rotary or double shifts
615 45
"Works" means the Permanent Works together with the Temporary Works
12 1(1)(l)
"Contract" means the Conditions of Contract Specification Drawings Priced Bill of Quantities Tender written acceptance of the Tender Contract Agreement (if completed)
19 1(1)(e)
The work is customarily carried out outside normal working hours or by shifts as stipulated ?
YES
NO
The Contractor must not cover up work or put work out of view without the approval of the Engineer
497 38(1)
Any provisions to the contrary are contained in the Contract ?
YES
NO
The Contractor may as a result of any notice given by the Engineer under this Clause seek the Engineer's permission to do any work at night or on Sundays
622 46
The work is unavoidable or absolutely necessary as stipulated ?
YES
NO
Special conditions
72
Any provision, in the Contract, to the contrary, must govern here
The Contractor must take full responsibility for the adequacy stability and safety of all site operations and methods of constructions
93 8(2)
The Contractor must give notice to the Engineer if :—
120 12(1)
The superintendence of the Works during execution must be given by sufficient persons, requisite for the satisfactory construction of the Works, having adequate knowledge of :—
194 15(1)
The Contractor must within 21 days after acceptance of his Tender submit to the Engineer for his approval a programme showing the order of procedure in which he proposes to carry out the Works; and
162 14(1)
at the same time provide in writing for the information of the Engineer a general description of the arrangements and methods of construction which the Contractor proposes to adopt for the carrying out of the Works
163 14(1)
The mode manner and speed of construction and maintenance of the Works are to be conducted in a manner approved of by the Engineer
151 13(2)
The Contractor must in the execution of the Contract observe and fulfil the obligations upon contractors, specified in the Fair Wages Resolution, passed by the House of Commons on 14 Oct. 1946
449 34(1)
The Contractor must be responsible for the observance of this Resolution by sub-contractors employed in the execution of the contract
450 34(1)
the contractor must pay rates of wages and observe hours and conditions of labour not less favourable than those established for the trade or industry in the district where the work is carried out;
452 34(1)
The Engineer may require the Contractor to deliver to the Engineer a return, in such form and at such intervals as the Engineer may prescribe, showing:—
461 35
The Engineer can by such return check any unauthorised activity of the Contractor or sub-contractors
in detail the numbers of the several classes of labour from time to time employed by the Contractor on the Site
462 35

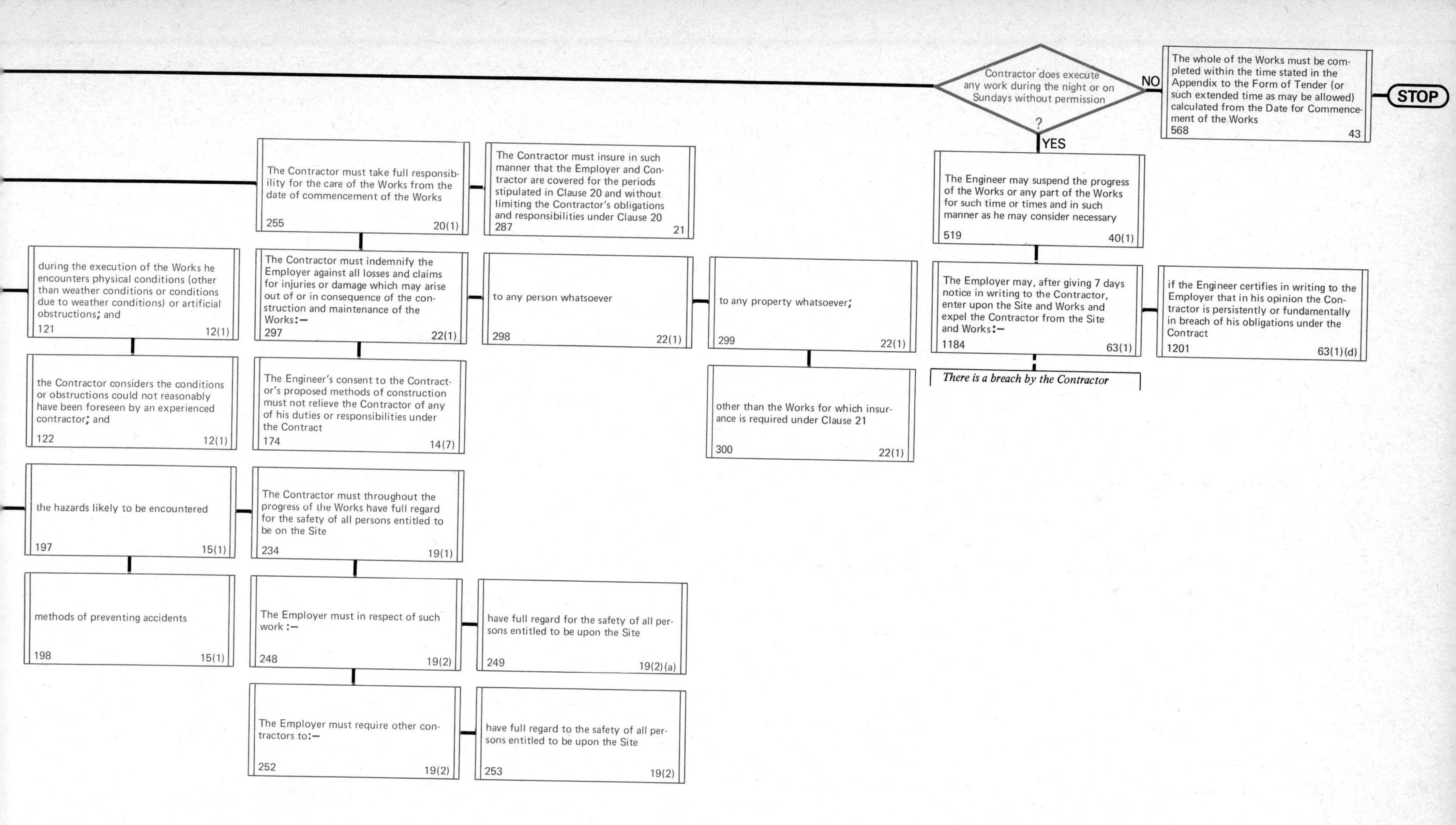

Contractor does execute any work during the night or on Sundays without permission ?
NO
YES
The whole of the Works must be completed within the time stated in the Appendix to the Form of Tender (or such extended time as may be allowed) calculated from the Date for Commencement of the Works
568 43
STOP
The Contractor must take full responsibility for the care of the Works from the date of commencement of the Works
255 20(1)
The Contractor must insure in such manner that the Employer and Contractor are covered for the periods stipulated in Clause 20 and without limiting the Contractor's obligations and responsibilities under Clause 20
287 21
The Engineer may suspend the progress of the Works or any part of the Works for such time or times and in such manner as he may consider necessary
519 40(1)
during the execution of the Works he encounters physical conditions (other than weather conditions or conditions due to weather conditions) or artificial obstructions; and
121 12(1)
The Contractor must indemnify the Employer against all losses and claims for injuries or damage which may arise out of or in consequence of the construction and maintenance of the Works:—
297 22(1)
to any person whatsoever
298 22(1)
to any property whatsoever;
299 22(1)
The Employer may, after giving 7 days notice in writing to the Contractor, enter upon the Site and Works and expel the Contractor from the Site and Works:—
1184 63(1)
if the Engineer certifies in writing to the Employer that in his opinion the Contractor is persistently or fundamentally in breach of his obligations under the Contract
1201 63(1)(d)
There is a breach by the Contractor
the Contractor considers the conditions or obstructions could not reasonably have been foreseen by an experienced contractor; and
122 12(1)
The Engineer's consent to the Contractor's proposed methods of construction must not relieve the Contractor of any of his duties or responsibilities under the Contract
174 14(7)
other than the Works for which insurance is required under Clause 21
300 22(1)
the hazards likely to be encountered
197 15(1)
The Contractor must throughout the progress of the Works have full regard for the safety of all persons entitled to be on the Site
234 19(1)
methods of preventing accidents
198 15(1)
The Employer must in respect of such work :—
248 19(2)
have full regard for the safety of all persons entitled to be upon the Site
249 19(2)(a)
The Employer must require other contractors to:—
252 19(2)
have full regard to the safety of all persons entitled to be upon the Site
253 19(2)

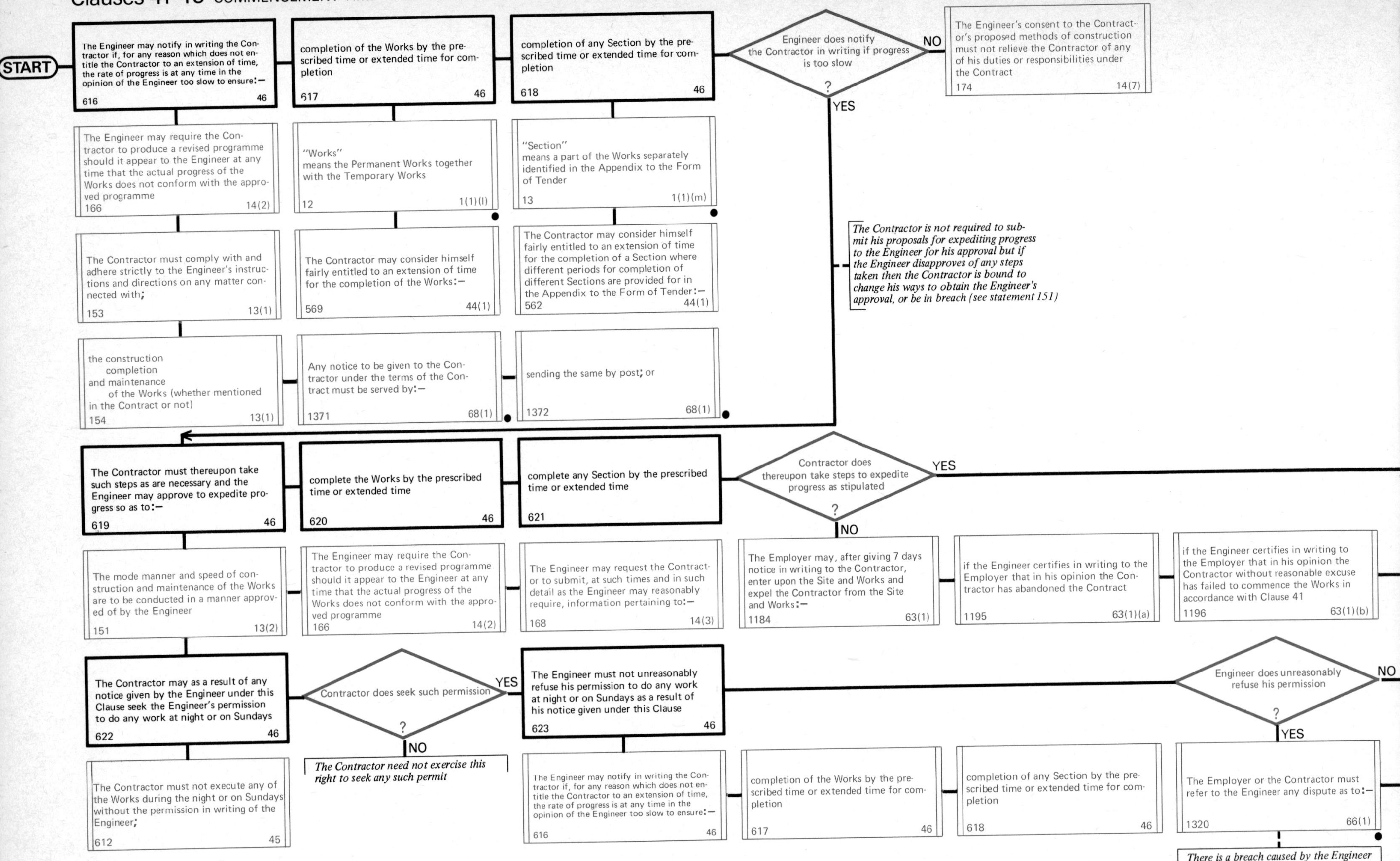
START
The Engineer may notify in writing the Contractor if, for any reason which does not entitle the Contractor to an extension of time, the rate of progress is at any time in the opinion of the Engineer too slow to ensure:— 616 46
completion of the Works by the prescribed time or extended time for completion 617 46
completion of any Section by the prescribed time or extended time for completion 618 46
Engineer does notify the Contractor in writing if progress is too slow ?
NO
YES
The Engineer's consent to the Contractor's proposed methods of construction must not relieve the Contractor of any of his duties or responsibilities under the Contract 174 14(7)
The Engineer may require the Contractor to produce a revised programme should it appear to the Engineer at any time that the actual progress of the Works does not conform with the approved programme 166 14(2)
"Works" means the Permanent Works together with the Temporary Works 12 1(1)(l)
"Section" means a part of the Works separately identified in the Appendix to the Form of Tender 13 1(1)(m)
The Contractor must comply with and adhere strictly to the Engineer's instructions and directions on any matter connected with; 153 13(1)
The Contractor may consider himself fairly entitled to an extension of time for the completion of the Works:— 569 44(1)
The Contractor may consider himself fairly entitled to an extension of time for the completion of a Section where different periods for completion of different Sections are provided for in the Appendix to the Form of Tender:— 562 44(1)
The Contractor is not required to submit his proposals for expediting progress to the Engineer for his approval but if the Engineer disapproves of any steps taken then the Contractor is bound to change his ways to obtain the Engineer's approval, or be in breach (see statement 151)
the construction completion and maintenance of the Works (whether mentioned in the Contract or not) 154 13(1)
Any notice to be given to the Contractor under the terms of the Contract must be served by:— 1371 68(1)
sending the same by post; or 1372 68(1)
The Contractor must thereupon take such steps as are necessary and the Engineer may approve to expedite progress so as to:— 619 46
complete the Works by the prescribed time or extended time 620 46
complete any Section by the prescribed time or extended time 621
Contractor does thereupon take steps to expedite progress as stipulated ?
YES
NO
The mode manner and speed of construction and maintenance of the Works are to be conducted in a manner approved of by the Engineer 151 13(2)
The Engineer may require the Contractor to produce a revised programme should it appear to the Engineer at any time that the actual progress of the Works does not conform with the approved programme 166 14(2)
The Engineer may request the Contractor to submit, at such times and in such detail as the Engineer may reasonably require, information pertaining to:— 168 14(3)
The Employer may, after giving 7 days notice in writing to the Contractor, enter upon the Site and Works and expel the Contractor from the Site and Works:— 1184 63(1)
if the Engineer certifies in writing to the Employer that in his opinion the Contractor has abandoned the Contract 1195 63(1)(a)
if the Engineer certifies in writing to the Employer that in his opinion the Contractor without reasonable excuse has failed to commence the Works in accordance with Clause 41 1196 63(1)(b)
The Contractor may as a result of any notice given by the Engineer under this Clause seek the Engineer's permission to do any work at night or on Sundays 622 46
Contractor does seek such permission ?
YES
NO
The Contractor need not exercise this right to seek any such permit
The Engineer must not unreasonably refuse his permission to do any work at night or on Sundays as a result of his notice given under this Clause 623 46
Engineer does unreasonably refuse his permission ?
NO
YES
The Contractor must not execute any of the Works during the night or on Sundays without the permission in writing of the Engineer; 612 45
The Engineer may notify in writing the Contractor if, for any reason which does not entitle the Contractor to an extension of time, the rate of progress is at any time in the opinion of the Engineer too slow to ensure:— 616 46
completion of the Works by the prescribed time or extended time for completion 617 46
completion of any Section by the prescribed time or extended time for completion 618 46
The Employer or the Contractor must refer to the Engineer any dispute as to:— 1320 66(1)
There is a breach caused by the Engineer

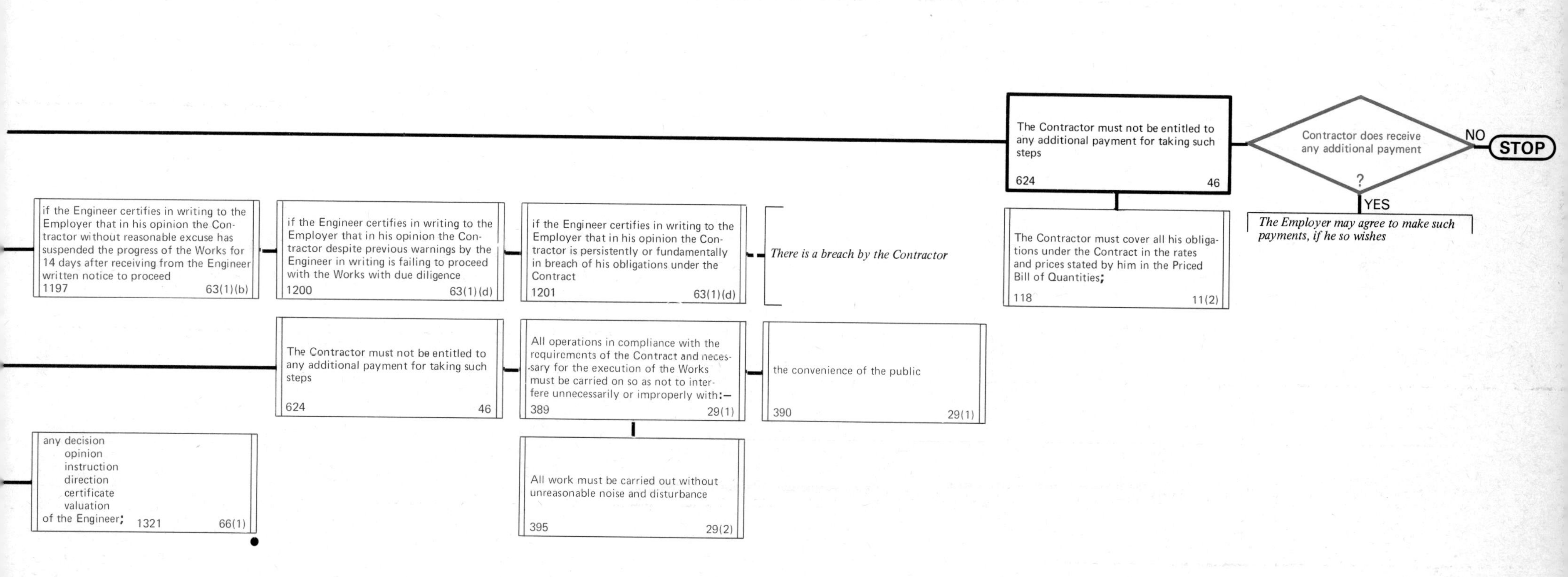
The Contractor must not be entitled to any additional payment for taking such steps
624
46
Contractor does receive any additional payment
?
NO
STOP
YES
The Employer may agree to make such payments, if he so wishes
if the Engineer certifies in writing to the Employer that in his opinion the Contractor without reasonable excuse has suspended the progress of the Works for 14 days after receiving from the Engineer written notice to proceed
1197
63(1)(b)
if the Engineer certifies in writing to the Employer that in his opinion the Contractor despite previous warnings by the Engineer in writing is failing to proceed with the Works with due diligence
1200
63(1)(d)
if the Engineer certifies in writing to the Employer that in his opinion the Contractor is persistently or fundamentally in breach of his obligations under the Contract
1201
63(1)(d)
There is a breach by the Contractor
The Contractor must cover all his obligations under the Contract in the rates and prices stated by him in the Priced Bill of Quantities;
118
11(2)
The Contractor must not be entitled to any additional payment for taking such steps
624
46
All operations in compliance with the requirements of the Contract and necessary for the execution of the Works must be carried on so as not to interfere unnecessarily or improperly with:—
389
29(1)
the convenience of the public
390
29(1)
any decision
opinion
instruction
direction
certificate
valuation
of the Engineer;
1321
66(1)
All work must be carried out without unreasonable noise and disturbance
395
29(2)

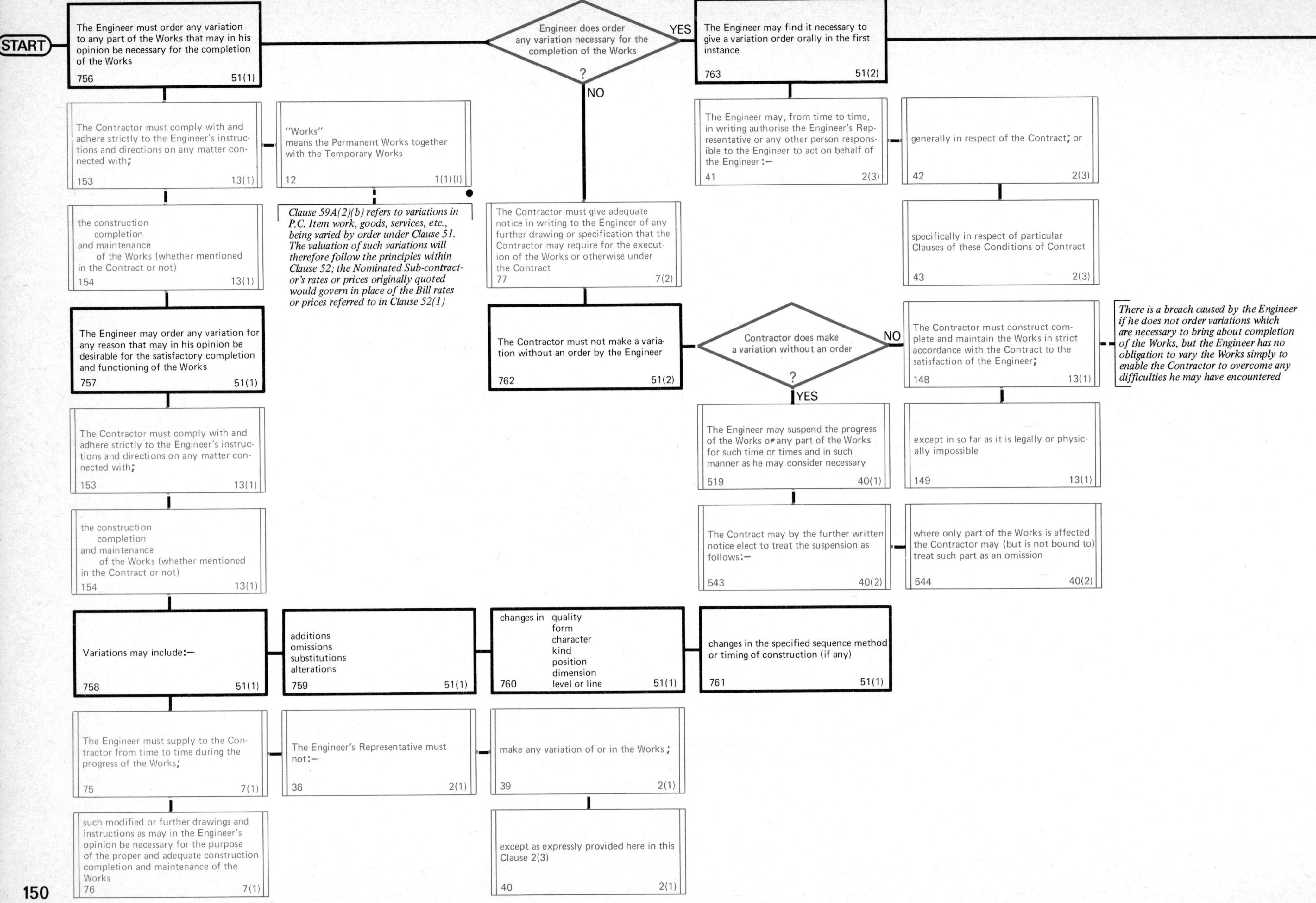
START
The Engineer must order any variation to any part of the Works that may in his opinion be necessary for the completion of the Works
756 51(1)
Engineer does order any variation necessary for the completion of the Works ?
YES
NO
The Engineer may find it necessary to give a variation order orally in the first instance
763 51(2)
The Contractor must comply with and adhere strictly to the Engineer's instructions and directions on any matter connected with;
153 13(1)
"Works" means the Permanent Works together with the Temporary Works
12 1(1)(l)
The Engineer may, from time to time, in writing authorise the Engineer's Representative or any other person responsible to the Engineer to act on behalf of the Engineer:—
41 2(3)
generally in respect of the Contract; or
42 2(3)
the construction completion and maintenance of the Works (whether mentioned in the Contract or not)
154 13(1)
Clause 59A(2)(b) refers to variations in P.C. Item work, goods, services, etc., being varied by order under Clause 51. The valuation of such variations will therefore follow the principles within Clause 52; the Nominated Sub-contractor's rates or prices originally quoted would govern in place of the Bill rates or prices referred to in Clause 52(1)
The Contractor must give adequate notice in writing to the Engineer of any further drawing or specification that the Contractor may require for the execution of the Works or otherwise under the Contract
77 7(2)
specifically in respect of particular Clauses of these Conditions of Contract
43 2(3)
The Engineer may order any variation for any reason that may in his opinion be desirable for the satisfactory completion and functioning of the Works
757 51(1)
The Contractor must not make a variation without an order by the Engineer
762 51(2)
Contractor does make a variation without an order ?
NO
YES
The Contractor must construct complete and maintain the Works in strict accordance with the Contract to the satisfaction of the Engineer;
148 13(1)
There is a breach caused by the Engineer if he does not order variations which are necessary to bring about completion of the Works, but the Engineer has no obligation to vary the Works simply to enable the Contractor to overcome any difficulties he may have encountered
The Contractor must comply with and adhere strictly to the Engineer's instructions and directions on any matter connected with;
153 13(1)
The Engineer may suspend the progress of the Works or any part of the Works for such time or times and in such manner as he may consider necessary
519 40(1)
except in so far as it is legally or physically impossible
149 13(1)
the construction completion and maintenance of the Works (whether mentioned in the Contract or not)
154 13(1)
The Contract may by the further written notice elect to treat the suspension as follows:—
543 40(2)
where only part of the Works is affected the Contractor may (but is not bound to) treat such part as an omission
544 40(2)
Variations may include:—
758 51(1)
additions omissions substitutions alterations
759 51(1)
changes in quality form character kind position dimension level or line
760 51(1)
changes in the specified sequence method or timing of construction (if any)
761 51(1)
The Engineer must supply to the Contractor from time to time during the progress of the Works;
75 7(1)
The Engineer's Representative must not:—
36 2(1)
make any variation of or in the Works;
39 2(1)
such modified or further drawings and instructions as may in the Engineer's opinion be necessary for the purpose of the proper and adequate construction completion and maintenance of the Works
76 7(1)
except as expressly provided here in this Clause 2(3)
40 2(1)

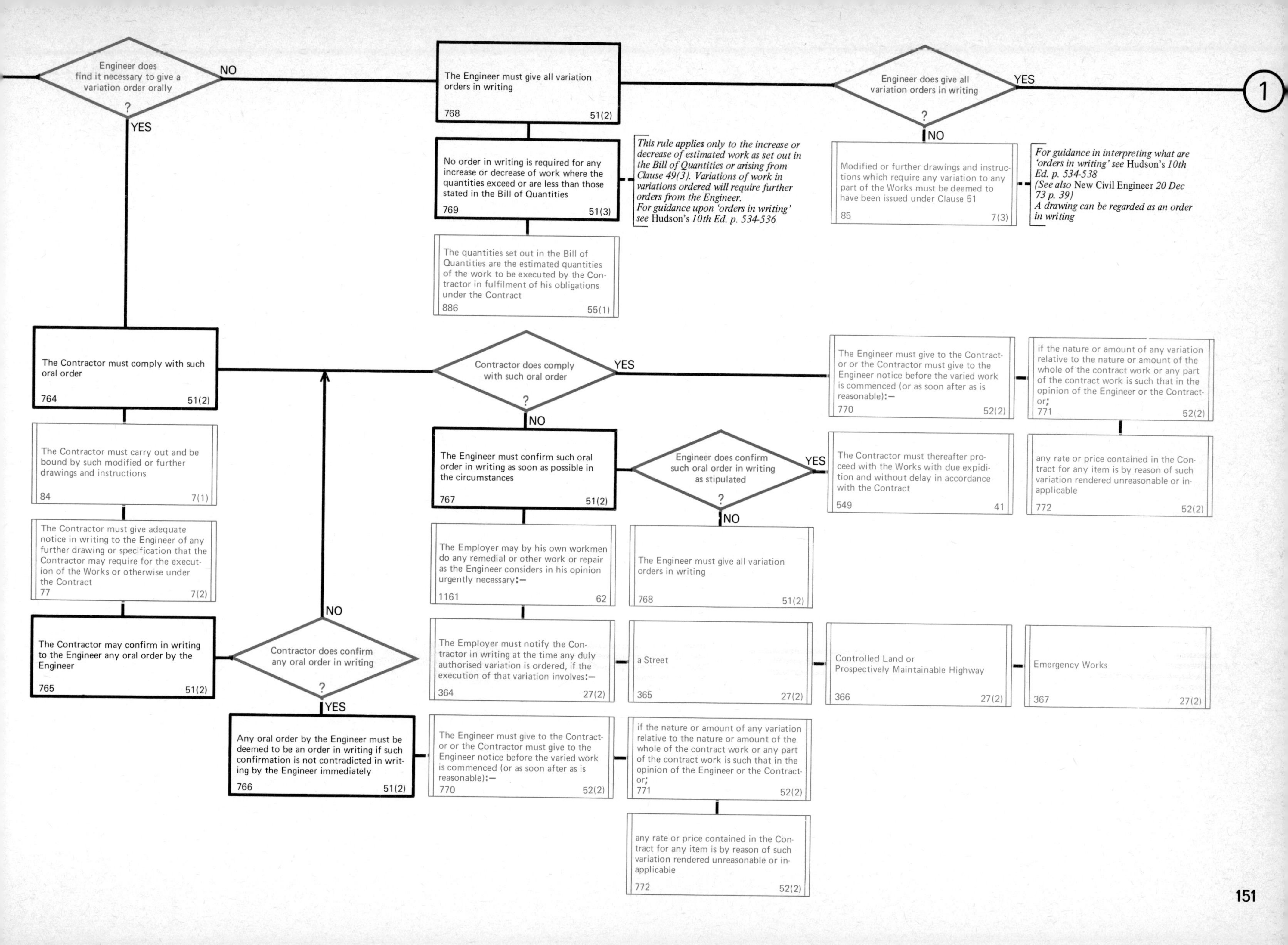

Engineer does find it necessary to give a variation order orally ?
NO
YES
The Engineer must give all variation orders in writing
768 51(2)
Engineer does give all variation orders in writing ?
YES
NO
1
No order in writing is required for any increase or decrease of work where the quantities exceed or are less than those stated in the Bill of Quantities
769 51(3)
This rule applies only to the increase or decrease of estimated work as set out in the Bill of Quantities or arising from Clause 49(3). Variations of work in variations ordered will require further orders from the Engineer.
For guidance upon 'orders in writing' see Hudson's 10th Ed. p. 534-536
Modified or further drawings and instructions which require any variation to any part of the Works must be deemed to have been issued under Clause 51
85 7(3)
For guidance in interpreting what are 'orders in writing' see Hudson's 10th Ed. p. 534-538
(See also New Civil Engineer 20 Dec 73 p. 39)
A drawing can be regarded as an order in writing
The quantities set out in the Bill of Quantities are the estimated quantities of the work to be executed by the Contractor in fulfilment of his obligations under the Contract
886 55(1)
The Contractor must comply with such oral order
764 51(2)
Contractor does comply with such oral order ?
YES
NO
The Engineer must give to the Contractor or the Contractor must give to the Engineer notice before the varied work is commenced (or as soon after as is reasonable):—
770 52(2)
if the nature or amount of any variation relative to the nature or amount of the whole of the contract work or any part of the contract work is such that in the opinion of the Engineer or the Contractor or;
771 52(2)
The Contractor must carry out and be bound by such modified or further drawings and instructions
84 7(1)
The Engineer must confirm such oral order in writing as soon as possible in the circumstances
767 51(2)
Engineer does confirm such oral order in writing as stipulated ?
YES
NO
The Contractor must thereafter proceed with the Works with due expidition and without delay in accordance with the Contract
549 41
any rate or price contained in the Contract for any item is by reason of such variation rendered unreasonable or inapplicable
772 52(2)
The Contractor must give adequate notice in writing to the Engineer of any further drawing or specification that the Contractor may require for the execution of the Works or otherwise under the Contract
77 7(2)
The Employer may by his own workmen do any remedial or other work or repair as the Engineer considers in his opinion urgently necessary:—
1161 62
The Engineer must give all variation orders in writing
768 51(2)
NO
The Contractor may confirm in writing to the Engineer any oral order by the Engineer
765 51(2)
Contractor does confirm any oral order in writing ?
YES
The Employer must notify the Contractor in writing at the time any duly authorised variation is ordered, if the execution of that variation involves:—
364 27(2)
a Street
365 27(2)
Controlled Land or Prospectively Maintainable Highway
366 27(2)
Emergency Works
367 27(2)
Any oral order by the Engineer must be deemed to be an order in writing if such confirmation is not contradicted in writing by the Engineer immediately
766 51(2)
The Engineer must give to the Contractor or the Contractor must give to the Engineer notice before the varied work is commenced (or as soon after as is reasonable):—
770 52(2)
if the nature or amount of any variation relative to the nature or amount of the whole of the contract work or any part of the contract work is such that in the opinion of the Engineer or the Contractor or;
771 52(2)
any rate or price contained in the Contract for any item is by reason of such variation rendered unreasonable or inapplicable
772 52(2)

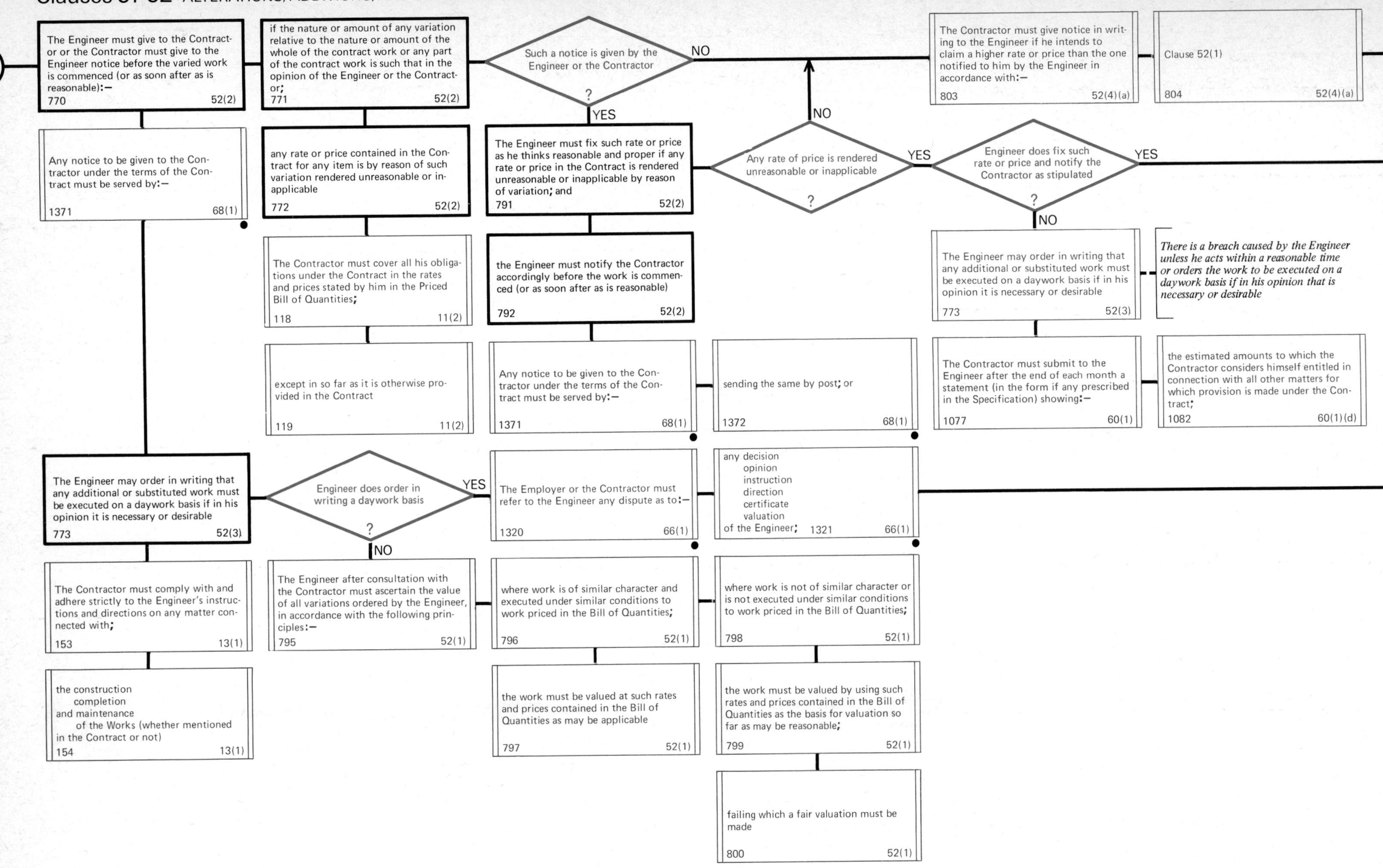
1
The Engineer must give to the Contractor or the Contractor must give to the Engineer notice before the varied work is commenced (or as soon after as is reasonable):— 770 52(2)
if the nature or amount of any variation relative to the nature or amount of the whole of the contract work or any part of the contract work is such that in the opinion of the Engineer or the Contractor; 771 52(2)
Such a notice is given by the Engineer or the Contractor ?
NO
YES
The Contractor must give notice in writing to the Engineer if he intends to claim a higher rate or price than the one notified to him by the Engineer in accordance with:— 803 52(4)(a)
Clause 52(1) 804 52(4)(a)
Any notice to be given to the Contractor under the terms of the Contract must be served by:— 1371 68(1)
any rate or price contained in the Contract for any item is by reason of such variation rendered unreasonable or inapplicable 772 52(2)
The Engineer must fix such rate or price as he thinks reasonable and proper if any rate or price in the Contract is rendered unreasonable or inapplicable by reason of variation; and 791 52(2)
Any rate of price is rendered unreasonable or inapplicable ?
NO
YES
Engineer does fix such rate or price and notify the Contractor as stipulated ?
YES
NO
The Contractor must cover all his obligations under the Contract in the rates and prices stated by him in the Priced Bill of Quantities; 118 11(2)
the Engineer must notify the Contractor accordingly before the work is commenced (or as soon after as is reasonable) 792 52(2)
The Engineer may order in writing that any additional or substituted work must be executed on a daywork basis if in his opinion it is necessary or desirable 773 52(3)
There is a breach caused by the Engineer unless he acts within a reasonable time or orders the work to be executed on a daywork basis if in his opinion that is necessary or desirable
except in so far as it is otherwise provided in the Contract 119 11(2)
Any notice to be given to the Contractor under the terms of the Contract must be served by:— 1371 68(1)
sending the same by post; or 1372 68(1)
The Contractor must submit to the Engineer after the end of each month a statement (in the form if any prescribed in the Specification) showing:— 1077 60(1)
the estimated amounts to which the Contractor considers himself entitled in connection with all other matters for which provision is made under the Contract; 1082 60(1)(d)
The Engineer may order in writing that any additional or substituted work must be executed on a daywork basis if in his opinion it is necessary or desirable 773 52(3)
Engineer does order in writing a daywork basis ?
YES
NO
The Employer or the Contractor must refer to the Engineer any dispute as to:— 1320 66(1)
any decision opinion instruction direction certificate valuation of the Engineer; 1321 66(1)
The Contractor must comply with and adhere strictly to the Engineer's instructions and directions on any matter connected with; 153 13(1)
The Engineer after consultation with the Contractor must ascertain the value of all variations ordered by the Engineer, in accordance with the following principles:— 795 52(1)
where work is of similar character and executed under similar conditions to work priced in the Bill of Quantities; 796 52(1)
where work is not of similar character or is not executed under similar conditions to work priced in the Bill of Quantities; 798 52(1)
the construction completion and maintenance of the Works (whether mentioned in the Contract or not) 154 13(1)
the work must be valued at such rates and prices contained in the Bill of Quantities as may be applicable 797 52(1)
the work must be valued by using such rates and prices contained in the Bill of Quantities as the basis for valuation so far as may be reasonable; 799 52(1)
failing which a fair valuation must be made 800 52(1)

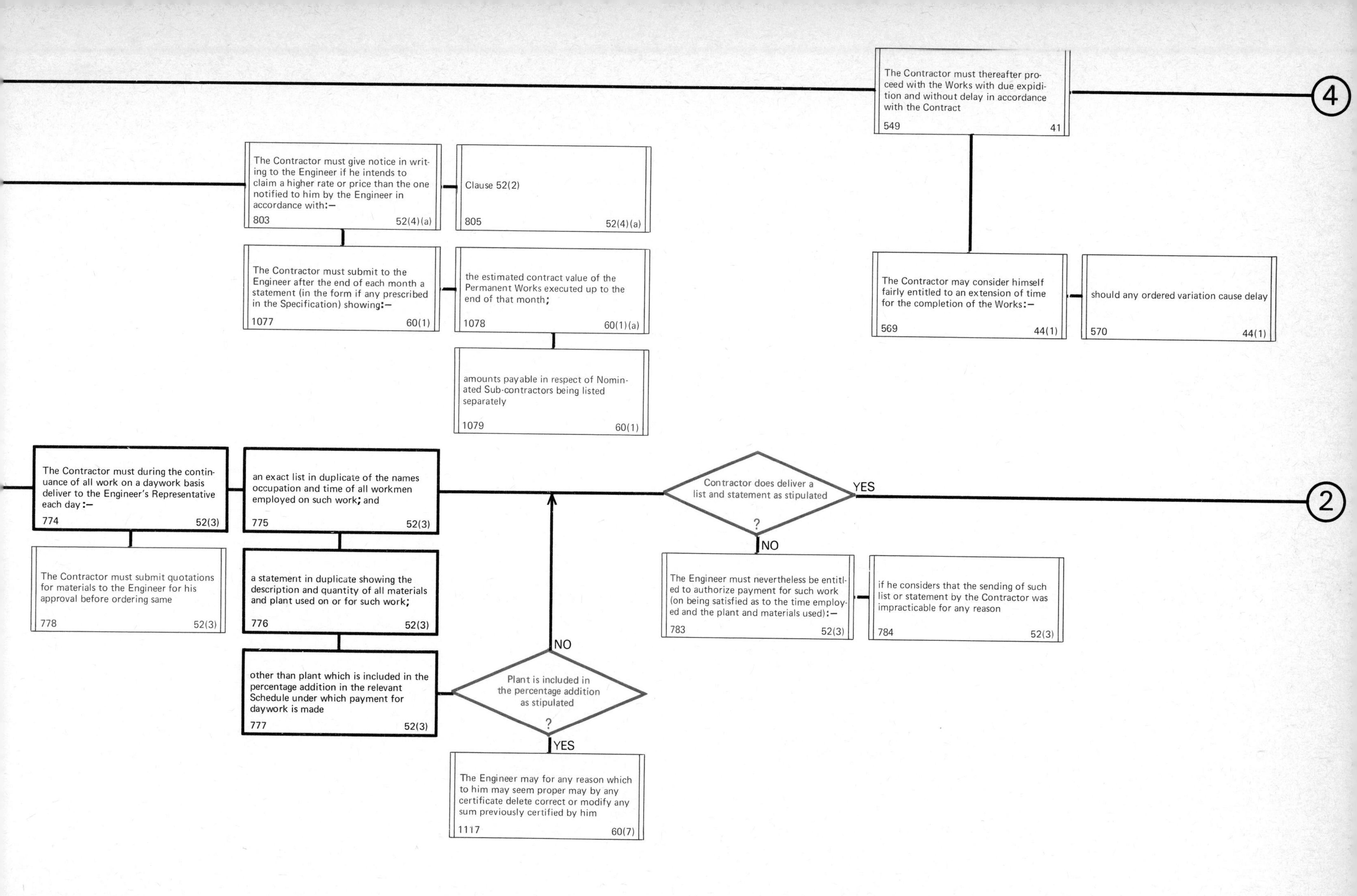

The Contractor must thereafter proceed with the Works with due expidition and without delay in accordance with the Contract
549 41
4
The Contractor must give notice in writing to the Engineer if he intends to claim a higher rate or price than the one notified to him by the Engineer in accordance with:—
803 52(4)(a)
Clause 52(2)
805 52(4)(a)
The Contractor must submit to the Engineer after the end of each month a statement (in the form if any prescribed in the Specification) showing:—
1077 60(1)
the estimated contract value of the Permanent Works executed up to the end of that month;
1078 60(1)(a)
amounts payable in respect of Nominated Sub-contractors being listed separately
1079 60(1)
The Contractor may consider himself fairly entitled to an extension of time for the completion of the Works:—
569 44(1)
should any ordered variation cause delay
570 44(1)
The Contractor must during the continuance of all work on a daywork basis deliver to the Engineer's Representative each day:—
774 52(3)
an exact list in duplicate of the names occupation and time of all workmen employed on such work; and
775 52(3)
Contractor does deliver a list and statement as stipulated
?
YES
NO
2
The Contractor must submit quotations for materials to the Engineer for his approval before ordering same
778 52(3)
a statement in duplicate showing the description and quantity of all materials and plant used on or for such work;
776 52(3)
The Engineer must nevertheless be entitled to authorize payment for such work (on being satisfied as to the time employed and the plant and materials used):—
783 52(3)
if he considers that the sending of such list or statement by the Contractor was impracticable for any reason
784 52(3)
NO
other than plant which is included in the percentage addition in the relevant Schedule under which payment for daywork is made
777 52(3)
Plant is included in the percentage addition as stipulated
?
YES
The Engineer may for any reason which to him may seem proper may by any certificate delete correct or modify any sum previously certified by him
1117 60(7)

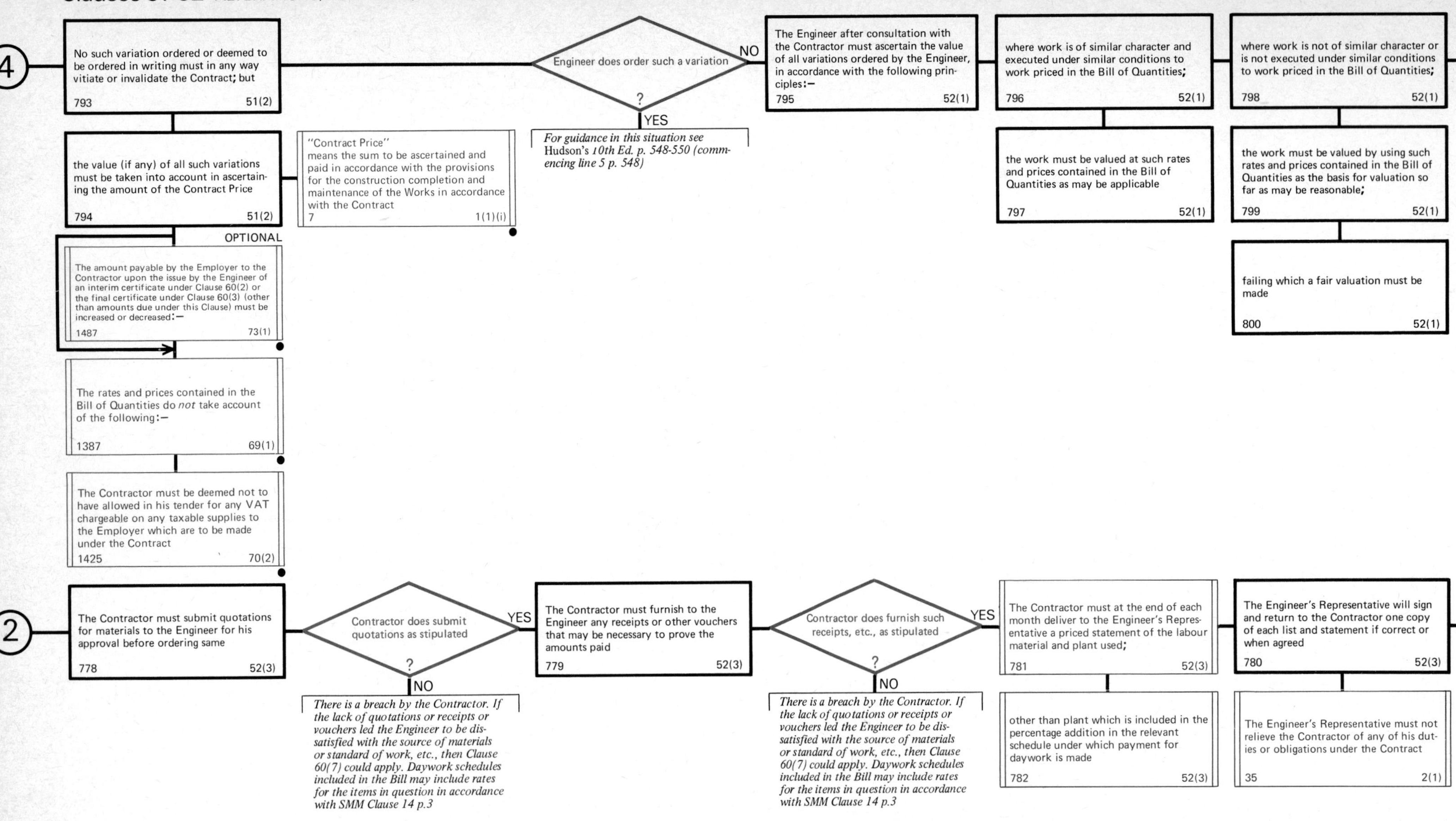
4
No such variation ordered or deemed to be ordered in writing must in any way vitiate or invalidate the Contract; but
793 51(2)
Engineer does order such a variation
?
NO
YES
For guidance in this situation see Hudson's 10th Ed. p. 548-550 (commencing line 5 p. 548)
The Engineer after consultation with the Contractor must ascertain the value of all variations ordered by the Engineer, in accordance with the following principles:—
795 52(1)
where work is of similar character and executed under similar conditions to work priced in the Bill of Quantities;
796 52(1)
where work is not of similar character or is not executed under similar conditions to work priced in the Bill of Quantities;
798 52(1)
the value (if any) of all such variations must be taken into account in ascertaining the amount of the Contract Price
794 51(2)
"Contract Price" means the sum to be ascertained and paid in accordance with the provisions for the construction completion and maintenance of the Works in accordance with the Contract
7 1(1)(i)
the work must be valued at such rates and prices contained in the Bill of Quantities as may be applicable
797 52(1)
the work must be valued by using such rates and prices contained in the Bill of Quantities as the basis for valuation so far as may be reasonable;
799 52(1)
failing which a fair valuation must be made
800 52(1)
OPTIONAL
The amount payable by the Employer to the Contractor upon the issue by the Engineer of an interim certificate under Clause 60(2) or the final certificate under Clause 60(3) (other than amounts due under this Clause) must be increased or decreased:—
1487 73(1)
The rates and prices contained in the Bill of Quantities do not take account of the following:—
1387 69(1)
The Contractor must be deemed not to have allowed in his tender for any VAT chargeable on any taxable supplies to the Employer which are to be made under the Contract
1425 70(2)
2
The Contractor must submit quotations for materials to the Engineer for his approval before ordering same
778 52(3)
Contractor does submit quotations as stipulated
?
YES
NO
There is a breach by the Contractor. If the lack of quotations or receipts or vouchers led the Engineer to be dissatisfied with the source of materials or standard of work, etc., then Clause 60(7) could apply. Daywork schedules included in the Bill may include rates for the items in question in accordance with SMM Clause 14 p.3
The Contractor must furnish to the Engineer any receipts or other vouchers that may be necessary to prove the amounts paid
779 52(3)
Contractor does furnish such receipts, etc., as stipulated
?
YES
NO
There is a breach by the Contractor. If the lack of quotations or receipts or vouchers led the Engineer to be dissatisfied with the source of materials or standard of work, etc., then Clause 60(7) could apply. Daywork schedules included in the Bill may include rates for the items in question in accordance with SMM Clause 14 p.3
The Contractor must at the end of each month deliver to the Engineer's Representative a priced statement of the labour material and plant used;
781 52(3)
other than plant which is included in the percentage addition in the relevant schedule under which payment for daywork is made
782 52(3)
The Engineer's Representative will sign and return to the Contractor one copy of each list and statement if correct or when agreed
780 52(3)
The Engineer's Representative must not relieve the Contractor of any of his duties or obligations under the Contract
35 2(1)

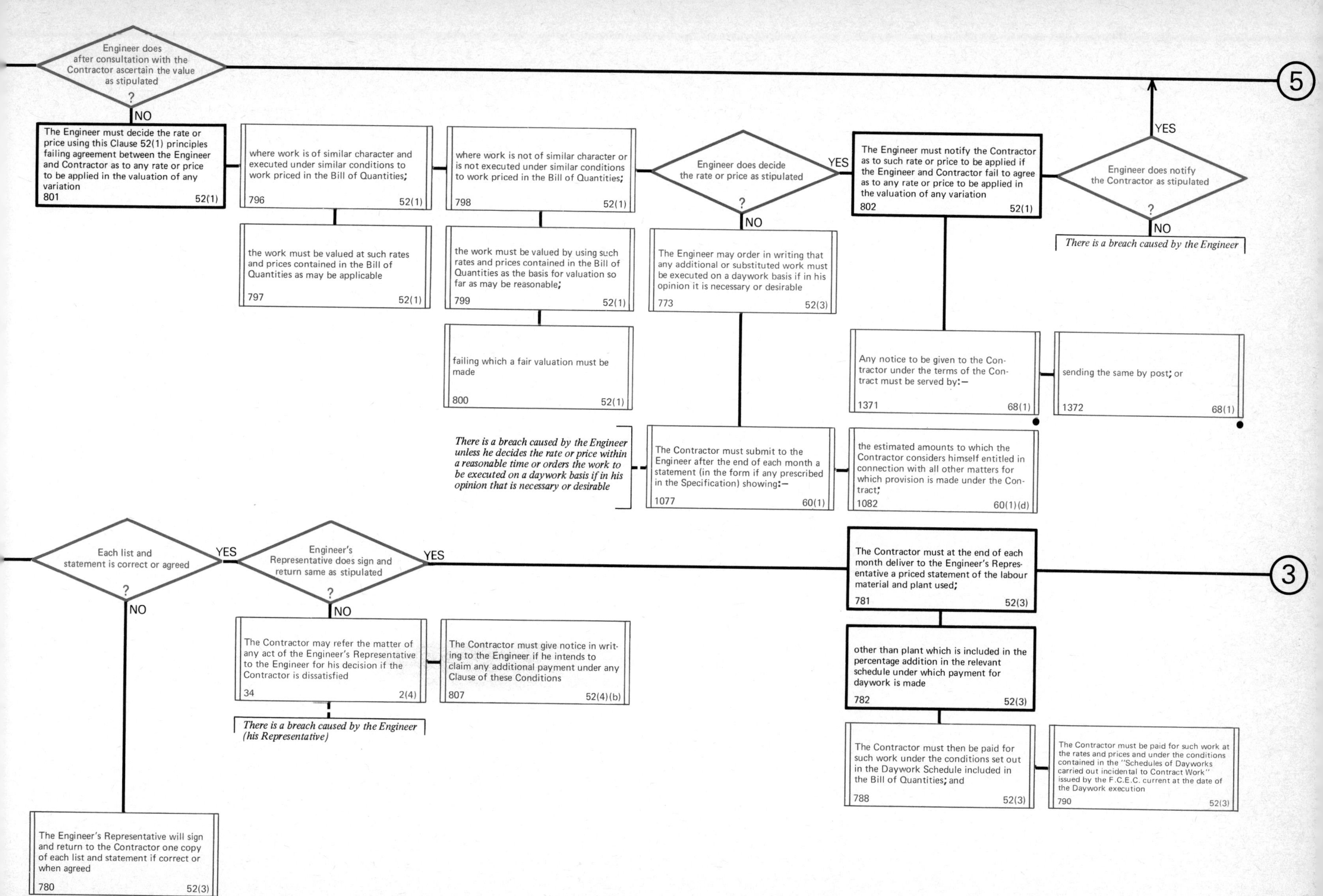
Engineer does after consultation with the Contractor ascertain the value as stipulated ?
NO
YES
5
The Engineer must decide the rate or price using this Clause 52(1) principles failing agreement between the Engineer and Contractor as to any rate or price to be applied in the valuation of any variation 801 52(1)
where work is of similar character and executed under similar conditions to work priced in the Bill of Quantities; 796 52(1)
where work is not of similar character or is not executed under similar conditions to work priced in the Bill of Quantities; 798 52(1)
Engineer does decide the rate or price as stipulated ?
The Engineer must notify the Contractor as to such rate or price to be applied if the Engineer and Contractor fail to agree as to any rate or price to be applied in the valuation of any variation 802 52(1)
Engineer does notify the Contractor as stipulated ?
There is a breach caused by the Engineer
the work must be valued at such rates and prices contained in the Bill of Quantities as may be applicable 797 52(1)
the work must be valued by using such rates and prices contained in the Bill of Quantities as the basis for valuation so far as may be reasonable; 799 52(1)
The Engineer may order in writing that any additional or substituted work must be executed on a daywork basis if in his opinion it is necessary or desirable 773 52(3)
failing which a fair valuation must be made 800 52(1)
Any notice to be given to the Contractor under the terms of the Contract must be served by:— 1371 68(1)
sending the same by post; or 1372 68(1)
There is a breach caused by the Engineer unless he decides the rate or price within a reasonable time or orders the work to be executed on a daywork basis if in his opinion that is necessary or desirable
The Contractor must submit to the Engineer after the end of each month a statement (in the form if any prescribed in the Specification) showing:— 1077 60(1)
the estimated amounts to which the Contractor considers himself entitled in connection with all other matters for which provision is made under the Contract; 1082 60(1)(d)
Each list and statement is correct or agreed ?
Engineer's Representative does sign and return same as stipulated ?
The Contractor must at the end of each month deliver to the Engineer's Representative a priced statement of the labour material and plant used; 781 52(3)
3
The Contractor may refer the matter of any act of the Engineer's Representative to the Engineer for his decision if the Contractor is dissatisfied 34 2(4)
The Contractor must give notice in writing to the Engineer if he intends to claim any additional payment under any Clause of these Conditions 807 52(4)(b)
There is a breach caused by the Engineer (his Representative)
other than plant which is included in the percentage addition in the relevant schedule under which payment for daywork is made 782 52(3)
The Contractor must then be paid for such work under the conditions set out in the Daywork Schedule included in the Bill of Quantities; and 788 52(3)
The Contractor must be paid for such work at the rates and prices and under the conditions contained in the "Schedules of Dayworks carried out incidental to Contract Work" issued by the F.C.E.C. current at the date of the Daywork execution 790 52(3)
The Engineer's Representative will sign and return to the Contractor one copy of each list and statement if correct or when agreed 780 52(3)

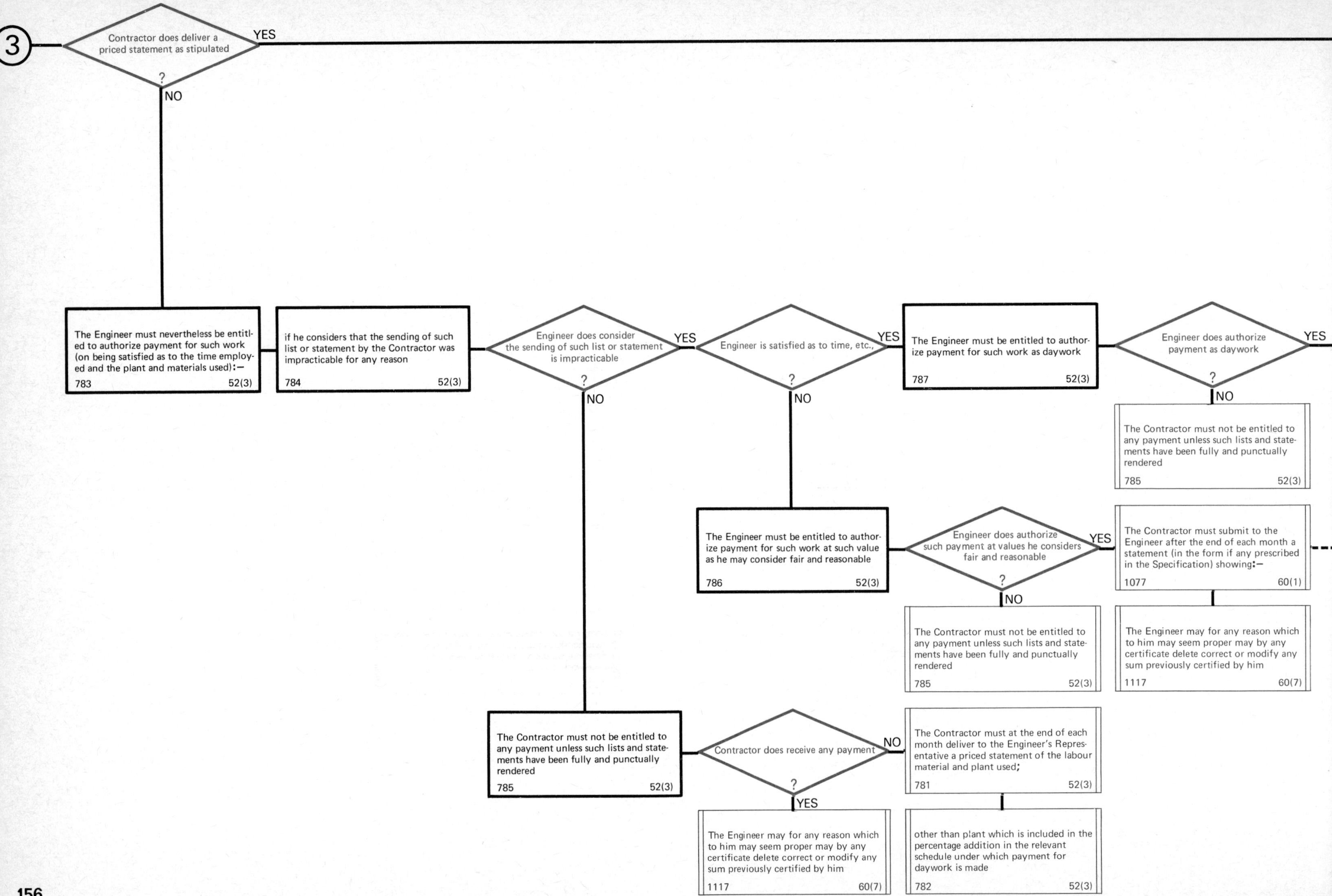
3
Contractor does deliver a priced statement as stipulated ?
YES
NO
The Engineer must nevertheless be entitled to authorize payment for such work (on being satisfied as to the time employed and the plant and materials used):—
783 52(3)
if he considers that the sending of such list or statement by the Contractor was impracticable for any reason
784 52(3)
Engineer does consider the sending of such list or statement is impracticable ?
YES
NO
Engineer is satisfied as to time, etc., ?
YES
NO
The Engineer must be entitled to authorize payment for such work as daywork
787 52(3)
Engineer does authorize payment as daywork ?
YES
NO
The Contractor must not be entitled to any payment unless such lists and statements have been fully and punctually rendered
785 52(3)
The Engineer must be entitled to authorize payment for such work at such value as he may consider fair and reasonable
786 52(3)
Engineer does authorize such payment at values he considers fair and reasonable ?
YES
NO
The Contractor must submit to the Engineer after the end of each month a statement (in the form if any prescribed in the Specification) showing:—
1077 60(1)
The Contractor must not be entitled to any payment unless such lists and statements have been fully and punctually rendered
785 52(3)
The Engineer may for any reason which to him may seem proper may by any certificate delete correct or modify any sum previously certified by him
1117 60(7)
The Contractor must not be entitled to any payment unless such lists and statements have been fully and punctually rendered
785 52(3)
Contractor does receive any payment ?
NO
YES
The Contractor must at the end of each month deliver to the Engineer's Representative a priced statement of the labour material and plant used;
781 52(3)
The Engineer may for any reason which to him may seem proper may by any certificate delete correct or modify any sum previously certified by him
1117 60(7)
other than plant which is included in the percentage addition in the relevant schedule under which payment for daywork is made
782 52(3)

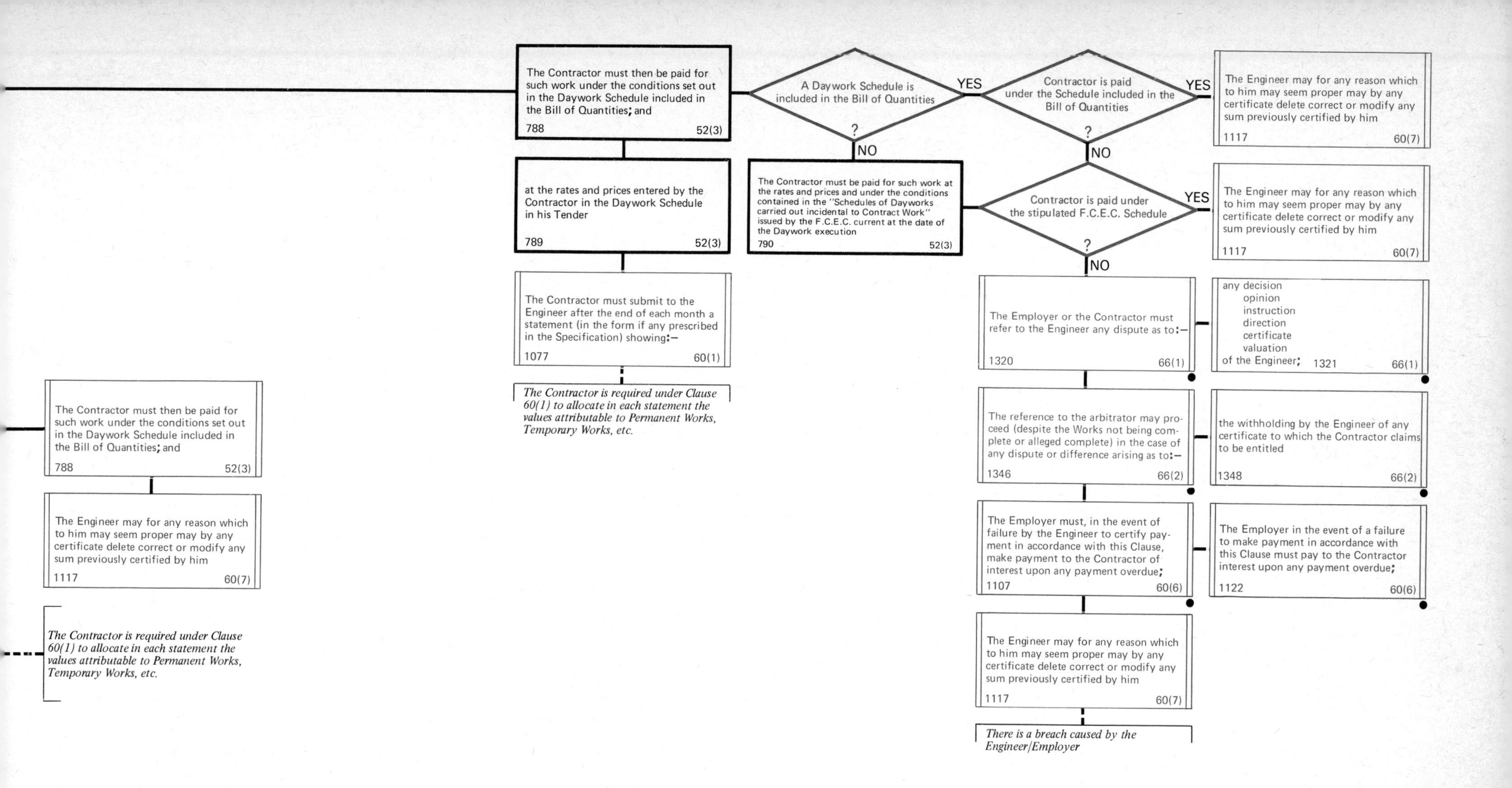

The Contractor must then be paid for such work under the conditions set out in the Daywork Schedule included in the Bill of Quantities; and
788 52(3)
at the rates and prices entered by the Contractor in the Daywork Schedule in his Tender
789 52(3)
The Contractor must submit to the Engineer after the end of each month a statement (in the form if any prescribed in the Specification) showing:–
1077 60(1)
The Contractor is required under Clause 60(1) to allocate in each statement the values attributable to Permanent Works, Temporary Works, etc.
A Daywork Schedule is included in the Bill of Quantities
?
YES
NO
The Contractor must be paid for such work at the rates and prices and under the conditions contained in the "Schedules of Dayworks carried out incidental to Contract Work" issued by the F.C.E.C. current at the date of the Daywork execution
790 52(3)
Contractor is paid under the Schedule included in the Bill of Quantities
?
YES
NO
The Engineer may for any reason which to him may seem proper may by any certificate delete correct or modify any sum previously certified by him
1117 60(7)
Contractor is paid under the stipulated F.C.E.C. Schedule
?
YES
NO
The Engineer may for any reason which to him may seem proper may by any certificate delete correct or modify any sum previously certified by him
1117 60(7)
The Employer or the Contractor must refer to the Engineer any dispute as to:–
1320 66(1)
any decision
opinion
instruction
direction
certificate
valuation
of the Engineer; 1321 66(1)
The reference to the arbitrator may proceed (despite the Works not being complete or alleged complete) in the case of any dispute or difference arising as to:–
1346 66(2)
the withholding by the Engineer of any certificate to which the Contractor claims to be entitled
1348 66(2)
The Employer must, in the event of failure by the Engineer to certify payment in accordance with this Clause, make payment to the Contractor of interest upon any payment overdue;
1107 60(6)
The Employer in the event of a failure to make payment in accordance with this Clause must pay to the Contractor interest upon any payment overdue;
1122 60(6)
The Engineer may for any reason which to him may seem proper may by any certificate delete correct or modify any sum previously certified by him
1117 60(7)
There is a breach caused by the Engineer/Employer
The Contractor must then be paid for such work under the conditions set out in the Daywork Schedule included in the Bill of Quantities; and
788 52(3)
The Engineer may for any reason which to him may seem proper may by any certificate delete correct or modify any sum previously certified by him
1117 60(7)
The Contractor is required under Clause 60(1) to allocate in each statement the values attributable to Permanent Works, Temporary Works, etc.

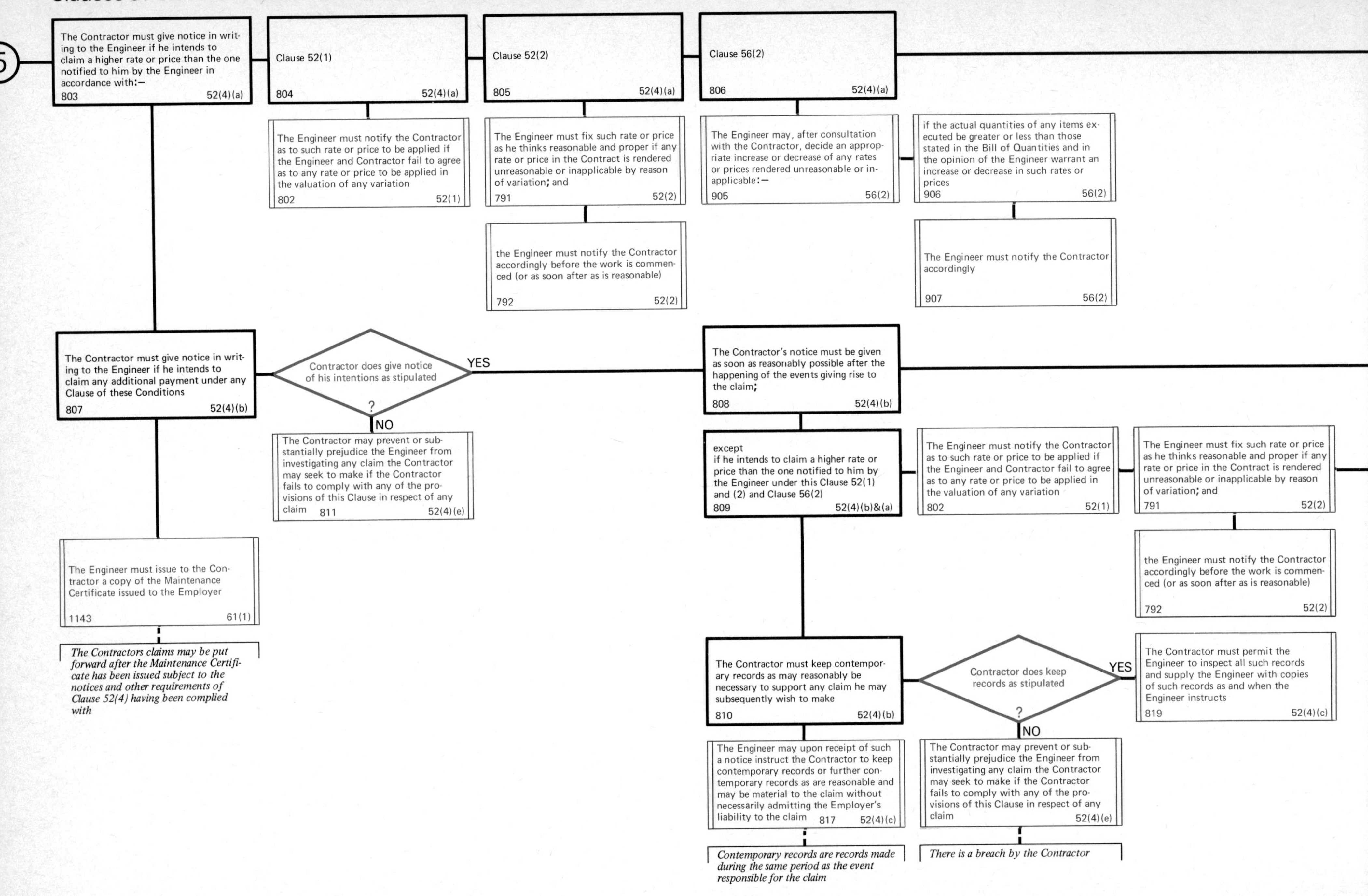
5
The Contractor must give notice in writing to the Engineer if he intends to claim a higher rate or price than the one notified to him by the Engineer in accordance with:—
803 52(4)(a)
Clause 52(1)
804 52(4)(a)
Clause 52(2)
805 52(4)(a)
Clause 56(2)
806 52(4)(a)
The Engineer must notify the Contractor as to such rate or price to be applied if the Engineer and Contractor fail to agree as to any rate or price to be applied in the valuation of any variation
802 52(1)
The Engineer must fix such rate or price as he thinks reasonable and proper if any rate or price in the Contract is rendered unreasonable or inapplicable by reason of variation; and
791 52(2)
the Engineer must notify the Contractor accordingly before the work is commenced (or as soon after as is reasonable)
792 52(2)
The Engineer may, after consultation with the Contractor, decide an appropriate increase or decrease of any rates or prices rendered unreasonable or inapplicable:—
905 56(2)
if the actual quantities of any items executed be greater or less than those stated in the Bill of Quantities and in the opinion of the Engineer warrant an increase or decrease in such rates or prices
906 56(2)
The Engineer must notify the Contractor accordingly
907 56(2)
The Contractor must give notice in writing to the Engineer if he intends to claim any additional payment under any Clause of these Conditions
807 52(4)(b)
Contractor does give notice of his intentions as stipulated
?
YES
NO
The Contractor may prevent or substantially prejudice the Engineer from investigating any claim the Contractor may seek to make if the Contractor fails to comply with any of the provisions of this Clause in respect of any claim
811 52(4)(e)
The Contractor's notice must be given as soon as reasonably possible after the happening of the events giving rise to the claim;
808 52(4)(b)
except if he intends to claim a higher rate or price than the one notified to him by the Engineer under this Clause 52(1) and (2) and Clause 56(2)
809 52(4)(b)&(a)
The Engineer must notify the Contractor as to such rate or price to be applied if the Engineer and Contractor fail to agree as to any rate or price to be applied in the valuation of any variation
802 52(1)
The Engineer must fix such rate or price as he thinks reasonable and proper if any rate or price in the Contract is rendered unreasonable or inapplicable by reason of variation; and
791 52(2)
the Engineer must notify the Contractor accordingly before the work is commenced (or as soon after as is reasonable)
792 52(2)
The Engineer must issue to the Contractor a copy of the Maintenance Certificate issued to the Employer
1143 61(1)
The Contractors claims may be put forward after the Maintenance Certificate has been issued subject to the notices and other requirements of Clause 52(4) having been complied with
The Contractor must keep contemporary records as may reasonably be necessary to support any claim he may subsequently wish to make
810 52(4)(b)
Contractor does keep records as stipulated
?
YES
NO
The Contractor must permit the Engineer to inspect all such records and supply the Engineer with copies of such records as and when the Engineer instructs
819 52(4)(c)
The Engineer may upon receipt of such a notice instruct the Contractor to keep contemporary records or further contemporary records as are reasonable and may be material to the claim without necessarily admitting the Employer's liability to the claim
817 52(4)(c)
The Contractor may prevent or substantially prejudice the Engineer from investigating any claim the Contractor may seek to make if the Contractor fails to comply with any of the provisions of this Clause in respect of any claim
52(4)(e)
Contemporary records are records made during the same period as the event responsible for the claim
There is a breach by the Contractor

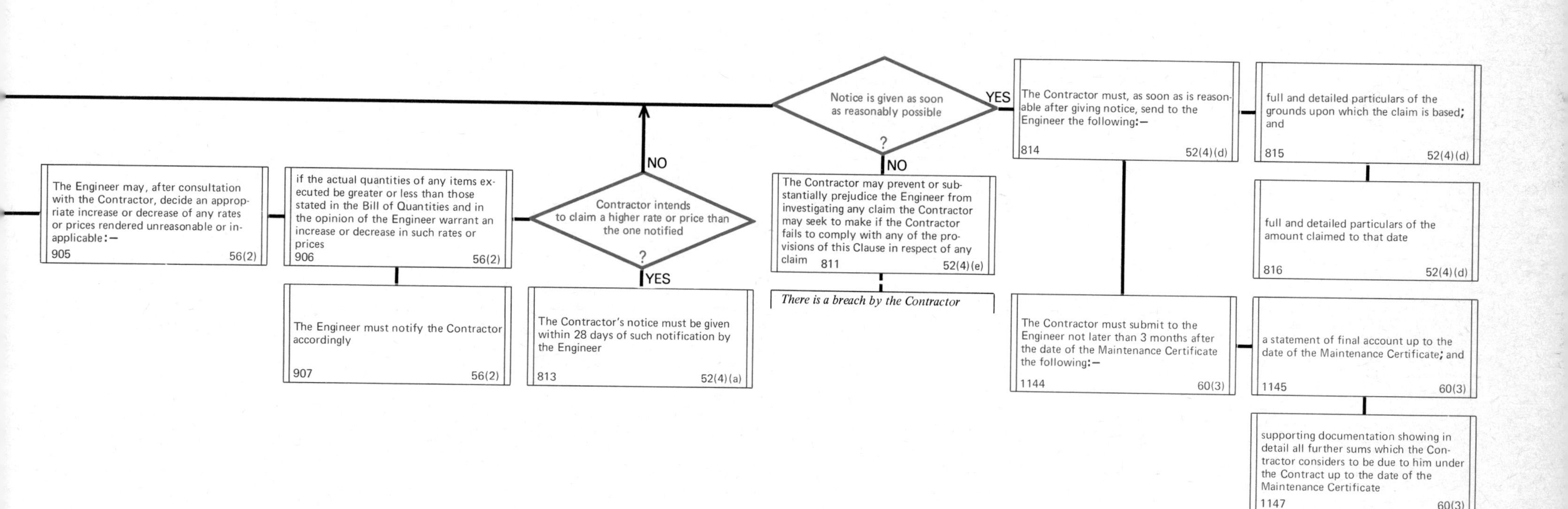
6
The Engineer may, after consultation with the Contractor, decide an appropriate increase or decrease of any rates or prices rendered unreasonable or inapplicable:—
905 56(2)
if the actual quantities of any items executed be greater or less than those stated in the Bill of Quantities and in the opinion of the Engineer warrant an increase or decrease in such rates or prices
906 56(2)
The Engineer must notify the Contractor accordingly
907 56(2)
Contractor intends to claim a higher rate or price than the one notified ?
NO
YES
The Contractor's notice must be given within 28 days of such notification by the Engineer
813 52(4)(a)
Notice is given as soon as reasonably possible ?
YES
NO
The Contractor may prevent or substantially prejudice the Engineer from investigating any claim the Contractor may seek to make if the Contractor fails to comply with any of the provisions of this Clause in respect of any claim 811 52(4)(e)
There is a breach by the Contractor
The Contractor must, as soon as is reasonable after giving notice, send to the Engineer the following:—
814 52(4)(d)
full and detailed particulars of the grounds upon which the claim is based; and
815 52(4)(d)
full and detailed particulars of the amount claimed to that date
816 52(4)(d)
The Contractor must submit to the Engineer not later than 3 months after the date of the Maintenance Certificate the following:—
1144 60(3)
a statement of final account up to the date of the Maintenance Certificate; and
1145 60(3)
supporting documentation showing in detail all further sums which the Contractor considers to be due to him under the Contract up to the date of the Maintenance Certificate
1147 60(3)

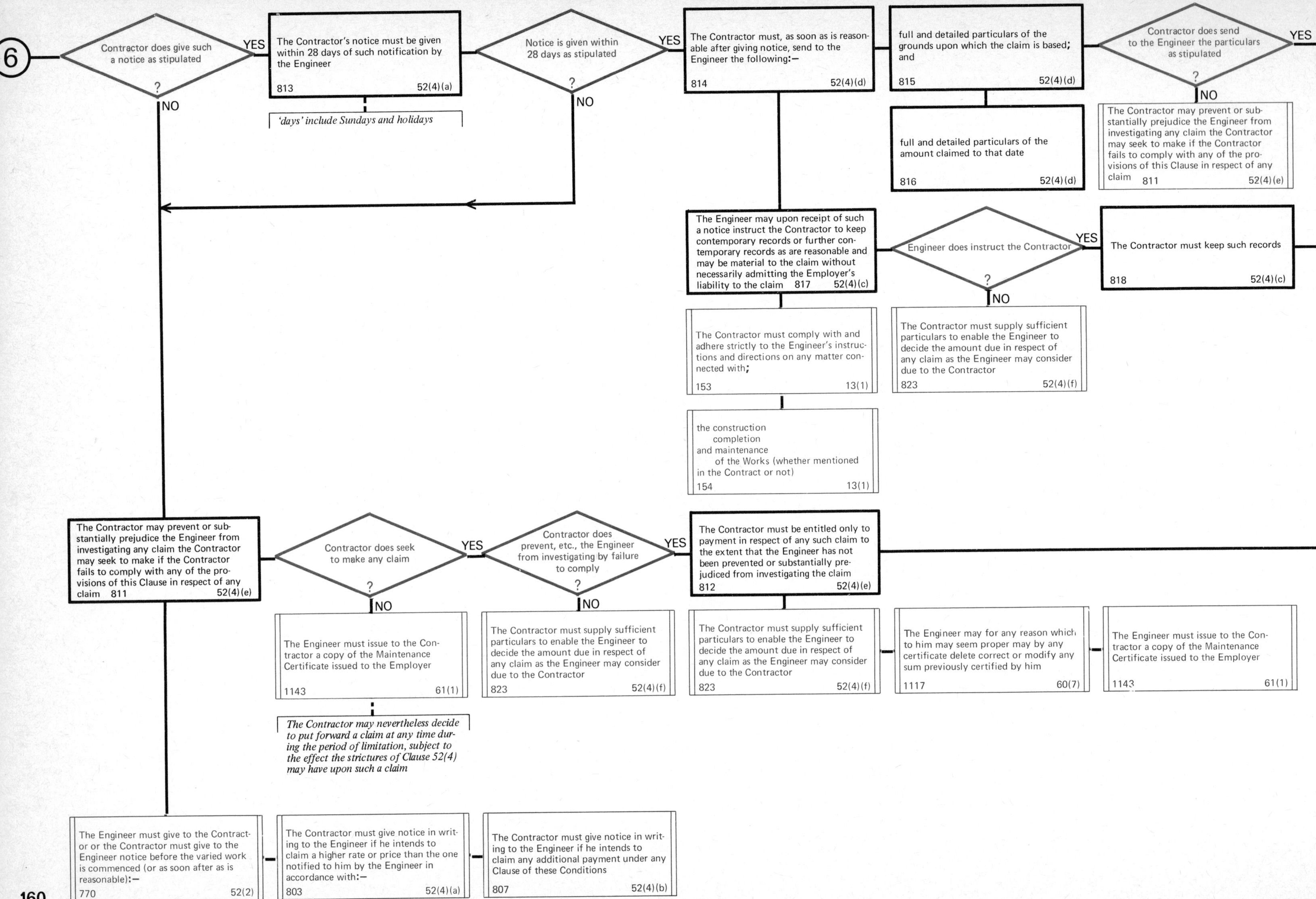
6
Contractor does give such a notice as stipulated ?
YES
NO
The Contractor's notice must be given within 28 days of such notification by the Engineer 813 52(4)(a)
'days' include Sundays and holidays
Notice is given within 28 days as stipulated ?
YES
NO
The Contractor must, as soon as is reasonable after giving notice, send to the Engineer the following:— 814 52(4)(d)
full and detailed particulars of the grounds upon which the claim is based; and 815 52(4)(d)
full and detailed particulars of the amount claimed to that date 816 52(4)(d)
Contractor does send to the Engineer the particulars as stipulated ?
YES
NO
The Contractor may prevent or substantially prejudice the Engineer from investigating any claim the Contractor may seek to make if the Contractor fails to comply with any of the provisions of this Clause in respect of any claim 811 52(4)(e)
The Engineer may upon receipt of such a notice instruct the Contractor to keep contemporary records or further contemporary records as are reasonable and may be material to the claim without necessarily admitting the Employer's liability to the claim 817 52(4)(c)
Engineer does instruct the Contractor ?
YES
NO
The Contractor must keep such records 818 52(4)(c)
The Contractor must comply with and adhere strictly to the Engineer's instructions and directions on any matter connected with; 153 13(1)
the construction completion and maintenance of the Works (whether mentioned in the Contract or not) 154 13(1)
The Contractor must supply sufficient particulars to enable the Engineer to decide the amount due in respect of any claim as the Engineer may consider due to the Contractor 823 52(4)(f)
The Contractor may prevent or substantially prejudice the Engineer from investigating any claim the Contractor may seek to make if the Contractor fails to comply with any of the provisions of this Clause in respect of any claim 811 52(4)(e)
Contractor does seek to make any claim ?
YES
NO
Contractor does prevent, etc., the Engineer from investigating by failure to comply ?
YES
NO
The Contractor must be entitled only to payment in respect of any such claim to the extent that the Engineer has not been prevented or substantially prejudiced from investigating the claim 812 52(4)(e)
The Engineer must issue to the Contractor a copy of the Maintenance Certificate issued to the Employer 1143 61(1)
The Contractor may nevertheless decide to put forward a claim at any time during the period of limitation, subject to the effect the strictures of Clause 52(4) may have upon such a claim
The Contractor must supply sufficient particulars to enable the Engineer to decide the amount due in respect of any claim as the Engineer may consider due to the Contractor 823 52(4)(f)
The Contractor must supply sufficient particulars to enable the Engineer to decide the amount due in respect of any claim as the Engineer may consider due to the Contractor 823 52(4)(f)
The Engineer may for any reason which to him may seem proper may by any certificate delete correct or modify any sum previously certified by him 1117 60(7)
The Engineer must issue to the Contractor a copy of the Maintenance Certificate issued to the Employer 1143 61(1)
The Engineer must give to the Contractor or the Contractor must give to the Engineer notice before the varied work is commenced (or as soon after as is reasonable):— 770 52(2)
The Contractor must give notice in writing to the Engineer if he intends to claim a higher rate or price than the one notified to him by the Engineer in accordance with:— 803 52(4)(a)
The Contractor must give notice in writing to the Engineer if he intends to claim any additional payment under any Clause of these Conditions 807 52(4)(b)

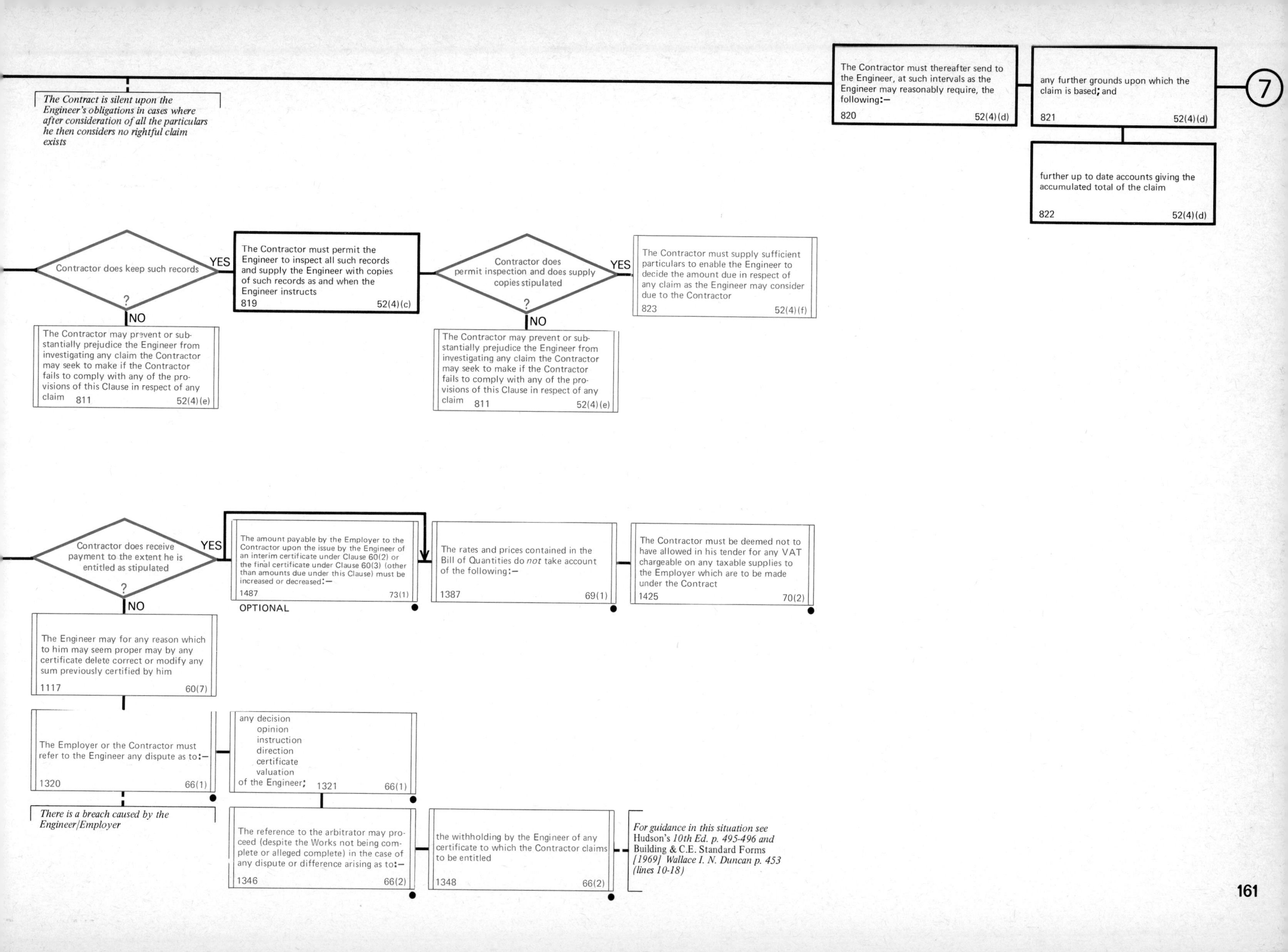
The Contract is silent upon the Engineer's obligations in cases where after consideration of all the particulars he then considers no rightful claim exists
The Contractor must thereafter send to the Engineer, at such intervals as the Engineer may reasonably require, the following:– 820 52(4)(d)
any further grounds upon which the claim is based; and 821 52(4)(d)
7
further up to date accounts giving the accumulated total of the claim 822 52(4)(d)
Contractor does keep such records ?
YES
NO
The Contractor must permit the Engineer to inspect all such records and supply the Engineer with copies of such records as and when the Engineer instructs 819 52(4)(c)
Contractor does permit inspection and does supply copies stipulated ?
YES
NO
The Contractor must supply sufficient particulars to enable the Engineer to decide the amount due in respect of any claim as the Engineer may consider due to the Contractor 823 52(4)(f)
The Contractor may prevent or substantially prejudice the Engineer from investigating any claim the Contractor may seek to make if the Contractor fails to comply with any of the provisions of this Clause in respect of any claim 811 52(4)(e)
The Contractor may prevent or substantially prejudice the Engineer from investigating any claim the Contractor may seek to make if the Contractor fails to comply with any of the provisions of this Clause in respect of any claim 811 52(4)(e)
Contractor does receive payment to the extent he is entitled as stipulated ?
YES
NO
The amount payable by the Employer to the Contractor upon the issue by the Engineer of an interim certificate under Clause 60(2) or the final certificate under Clause 60(3) (other than amounts due under this Clause) must be increased or decreased:– 1487 73(1)
OPTIONAL
The rates and prices contained in the Bill of Quantities do not take account of the following:– 1387 69(1)
The Contractor must be deemed not to have allowed in his tender for any VAT chargeable on any taxable supplies to the Employer which are to be made under the Contract 1425 70(2)
The Engineer may for any reason which to him may seem proper may by any certificate delete correct or modify any sum previously certified by him 1117 60(7)
The Employer or the Contractor must refer to the Engineer any dispute as to:– 1320 66(1)
any decision opinion instruction direction certificate valuation of the Engineer; 1321 66(1)
There is a breach caused by the Engineer/Employer
The reference to the arbitrator may proceed (despite the Works not being complete or alleged complete) in the case of any dispute or difference arising as to:– 1346 66(2)
the withholding by the Engineer of any certificate to which the Contractor claims to be entitled 1348 66(2)
For guidance in this situation see Hudson's 10th Ed. p. 495-496 and Building & C.E. Standard Forms [1969] Wallace I. N. Duncan p. 453 (lines 10-18)

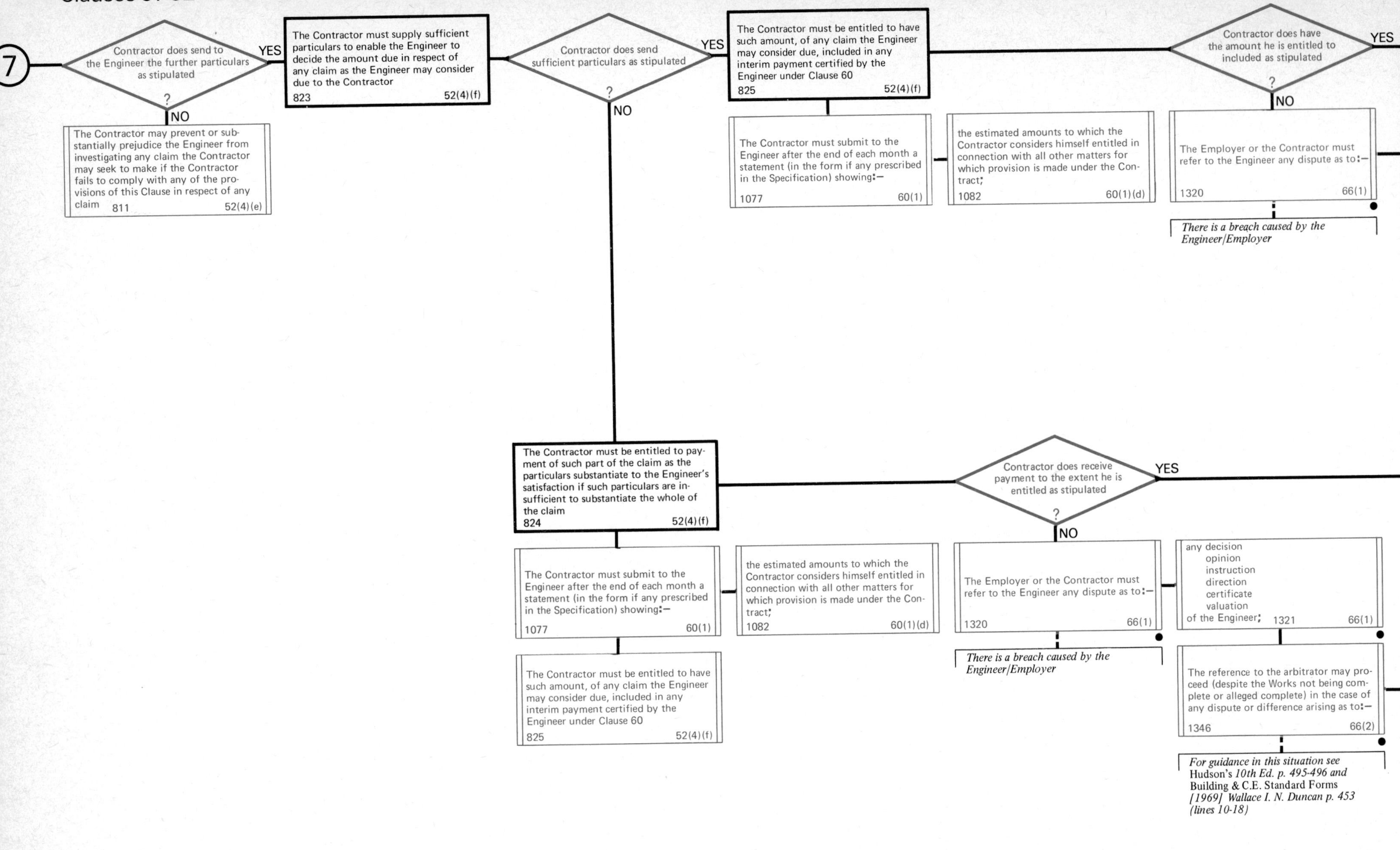
7
Contractor does send to the Engineer the further particulars as stipulated ?
YES
NO
The Contractor may prevent or substantially prejudice the Engineer from investigating any claim the Contractor may seek to make if the Contractor fails to comply with any of the provisions of this Clause in respect of any claim 811 52(4)(e)
The Contractor must supply sufficient particulars to enable the Engineer to decide the amount due in respect of any claim as the Engineer may consider due to the Contractor 823 52(4)(f)
Contractor does send sufficient particulars as stipulated ?
YES
NO
The Contractor must be entitled to have such amount, of any claim the Engineer may consider due, included in any interim payment certified by the Engineer under Clause 60 825 52(4)(f)
The Contractor must submit to the Engineer after the end of each month a statement (in the form if any prescribed in the Specification) showing:— 1077 60(1)
the estimated amounts to which the Contractor considers himself entitled in connection with all other matters for which provision is made under the Contract; 1082 60(1)(d)
Contractor does have the amount he is entitled to included as stipulated ?
YES
NO
The Employer or the Contractor must refer to the Engineer any dispute as to:— 1320 66(1)
There is a breach caused by the Engineer/Employer
The Contractor must be entitled to payment of such part of the claim as the particulars substantiate to the Engineer's satisfaction if such particulars are insufficient to substantiate the whole of the claim 824 52(4)(f)
The Contractor must submit to the Engineer after the end of each month a statement (in the form if any prescribed in the Specification) showing:— 1077 60(1)
the estimated amounts to which the Contractor considers himself entitled in connection with all other matters for which provision is made under the Contract; 1082 60(1)(d)
The Contractor must be entitled to have such amount, of any claim the Engineer may consider due, included in any interim payment certified by the Engineer under Clause 60 825 52(4)(f)
Contractor does receive payment to the extent he is entitled as stipulated ?
YES
NO
The Employer or the Contractor must refer to the Engineer any dispute as to:— 1320 66(1)
There is a breach caused by the Engineer/Employer
any decision opinion instruction direction certificate valuation of the Engineer; 1321 66(1)
The reference to the arbitrator may proceed (despite the Works not being complete or alleged complete) in the case of any dispute or difference arising as to:— 1346 66(2)
For guidance in this situation see Hudson's 10th Ed. p. 495-496 and Building & C.E. Standard Forms [1969] Wallace I. N. Duncan p. 453 (lines 10-18)

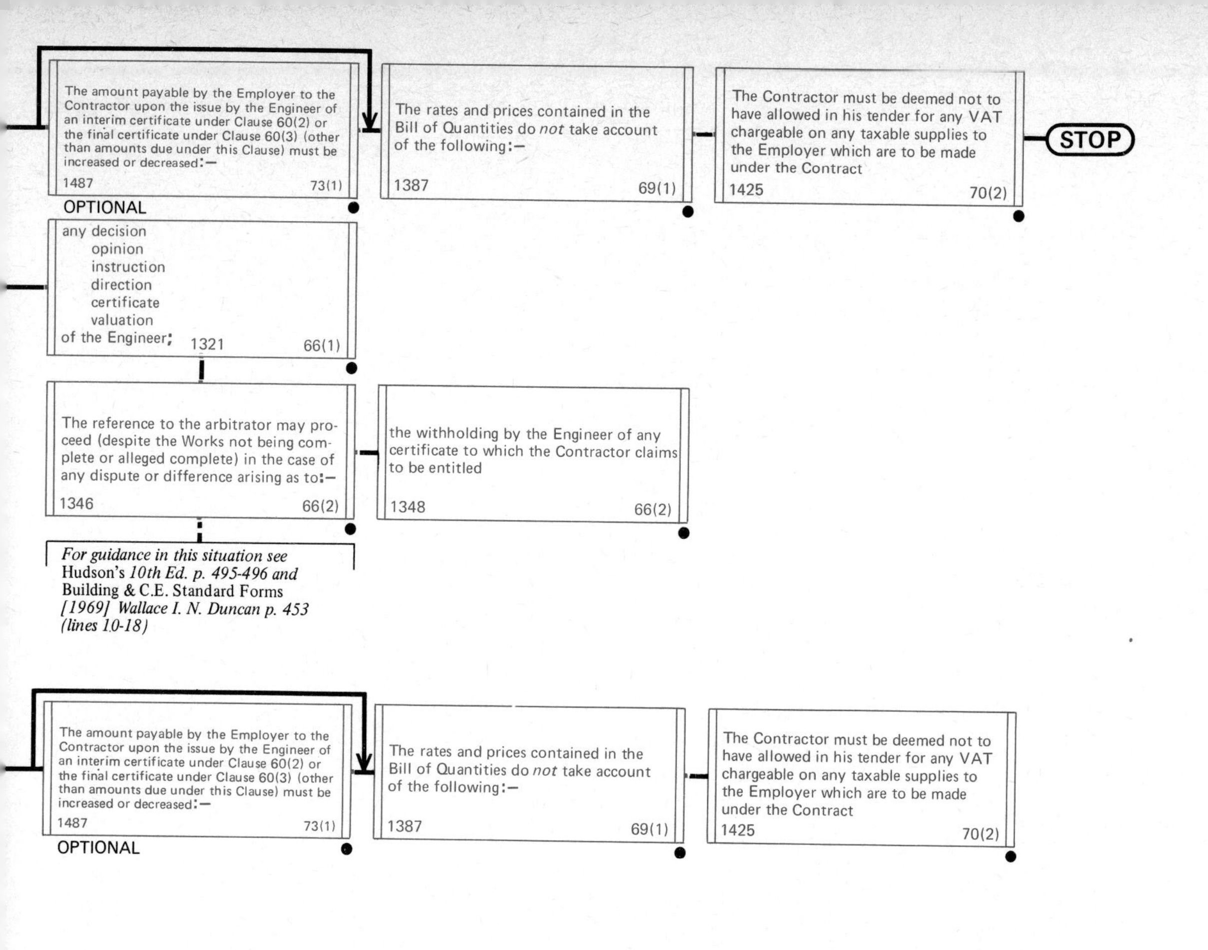

the withholding by the Engineer of any certificate to which the Contractor claims to be entitled

1348 66(2)

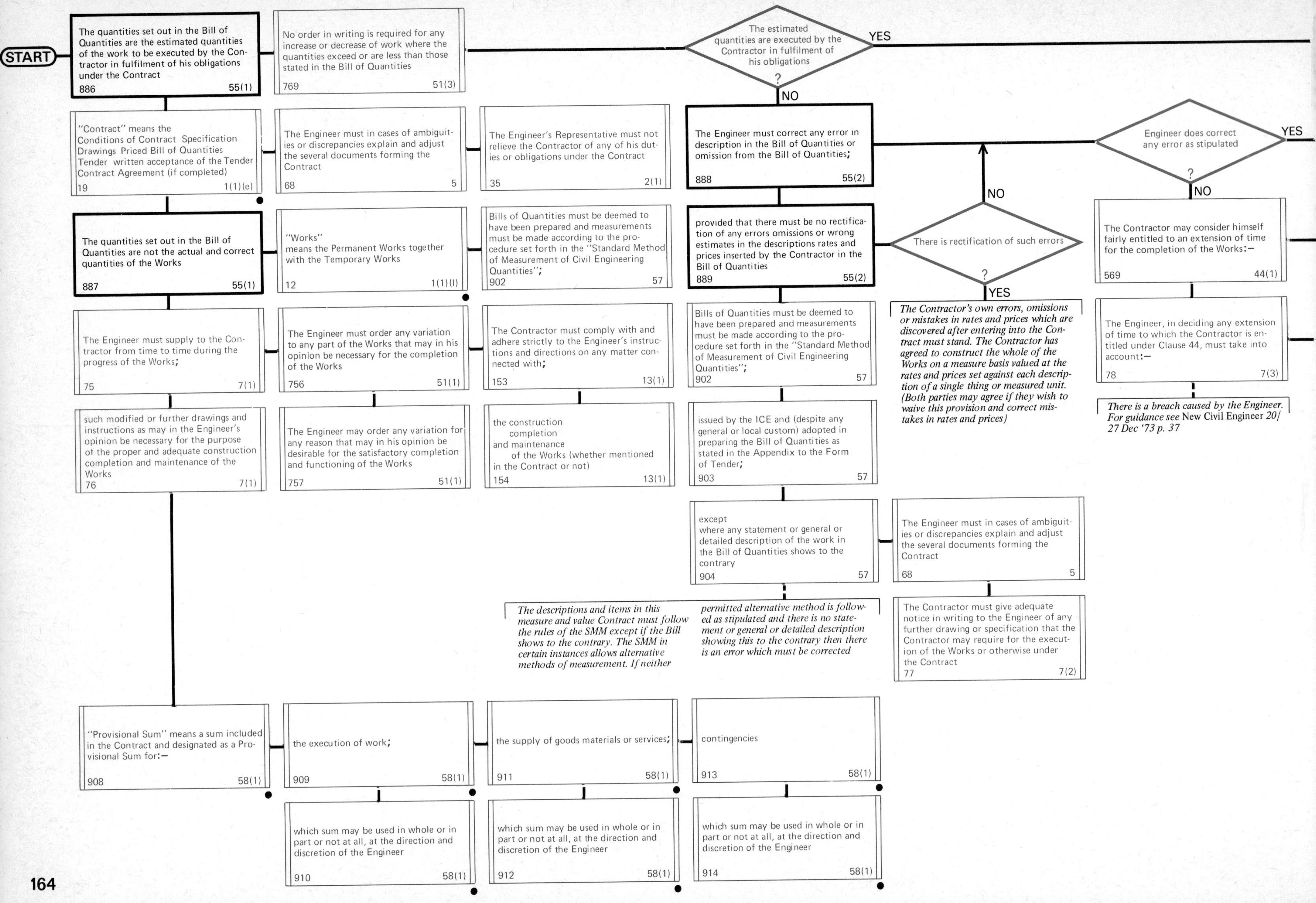
START
The quantities set out in the Bill of Quantities are the estimated quantities of the work to be executed by the Contractor in fulfilment of his obligations under the Contract
886 55(1)
No order in writing is required for any increase or decrease of work where the quantities exceed or are less than those stated in the Bill of Quantities
769 51(3)
The estimated quantities are executed by the Contractor in fulfilment of his obligations ?
YES
NO
"Contract" means the Conditions of Contract Specification Drawings Priced Bill of Quantities Tender written acceptance of the Tender Contract Agreement (if completed)
19 1(1)(e)
The Engineer must in cases of ambiguities or discrepancies explain and adjust the several documents forming the Contract
68 5
The Engineer's Representative must not relieve the Contractor of any of his duties or obligations under the Contract
35 2(1)
The Engineer must correct any error in description in the Bill of Quantities or omission from the Bill of Quantities;
888 55(2)
Engineer does correct any error as stipulated ?
YES
NO
The quantities set out in the Bill of Quantities are not the actual and correct quantities of the Works
887 55(1)
"Works" means the Permanent Works together with the Temporary Works
12 1(1)(l)
Bills of Quantities must be deemed to have been prepared and measurements must be made according to the procedure set forth in the "Standard Method of Measurement of Civil Engineering Quantities";
902 57
provided that there must be no rectification of any errors omissions or wrong estimates in the descriptions rates and prices inserted by the Contractor in the Bill of Quantities
889 55(2)
There is rectification of such errors ?
NO
YES
The Contractor may consider himself fairly entitled to an extension of time for the completion of the Works:—
569 44(1)
The Engineer must supply to the Contractor from time to time during the progress of the Works;
75 7(1)
The Engineer must order any variation to any part of the Works that may in his opinion be necessary for the completion of the Works
756 51(1)
The Contractor must comply with and adhere strictly to the Engineer's instructions and directions on any matter connected with;
153 13(1)
Bills of Quantities must be deemed to have been prepared and measurements must be made according to the procedure set forth in the "Standard Method of Measurement of Civil Engineering Quantities";
902 57
The Contractor's own errors, omissions or mistakes in rates and prices which are discovered after entering into the Contract must stand. The Contractor has agreed to construct the whole of the Works on a measure basis valued at the rates and prices set against each description of a single thing or measured unit. (Both parties may agree if they wish to waive this provision and correct mistakes in rates and prices)
The Engineer, in deciding any extension of time to which the Contractor is entitled under Clause 44, must take into account:—
78 7(3)
There is a breach caused by the Engineer. For guidance see New Civil Engineer 20/27 Dec '73 p. 37
such modified or further drawings and instructions as may in the Engineer's opinion be necessary for the purpose of the proper and adequate construction completion and maintenance of the Works
76 7(1)
The Engineer may order any variation for any reason that may in his opinion be desirable for the satisfactory completion and functioning of the Works
757 51(1)
the construction completion and maintenance of the Works (whether mentioned in the Contract or not)
154 13(1)
issued by the ICE and (despite any general or local custom) adopted in preparing the Bill of Quantities as stated in the Appendix to the Form of Tender;
903 57
except where any statement or general or detailed description of the work in the Bill of Quantities shows to the contrary
904 57
The Engineer must in cases of ambiguities or discrepancies explain and adjust the several documents forming the Contract
68 5
The descriptions and items in this measure and value Contract must follow the rules of the SMM except if the Bill shows to the contrary. The SMM in certain instances allows alternative methods of measurement. If neither permitted alternative method is followed as stipulated and there is no statement or general or detailed description showing this to the contrary then there is an error which must be corrected
The Contractor must give adequate notice in writing to the Engineer of any further drawing or specification that the Contractor may require for the execution of the Works or otherwise under the Contract
77 7(2)
"Provisional Sum" means a sum included in the Contract and designated as a Provisional Sum for:—
908 58(1)
the execution of work;
909 58(1)
the supply of goods materials or services;
911 58(1)
contingencies
913 58(1)
which sum may be used in whole or in part or not at all, at the direction and discretion of the Engineer
910 58(1)
which sum may be used in whole or in part or not at all, at the direction and discretion of the Engineer
912 58(1)
which sum may be used in whole or in part or not at all, at the direction and discretion of the Engineer
914 58(1)

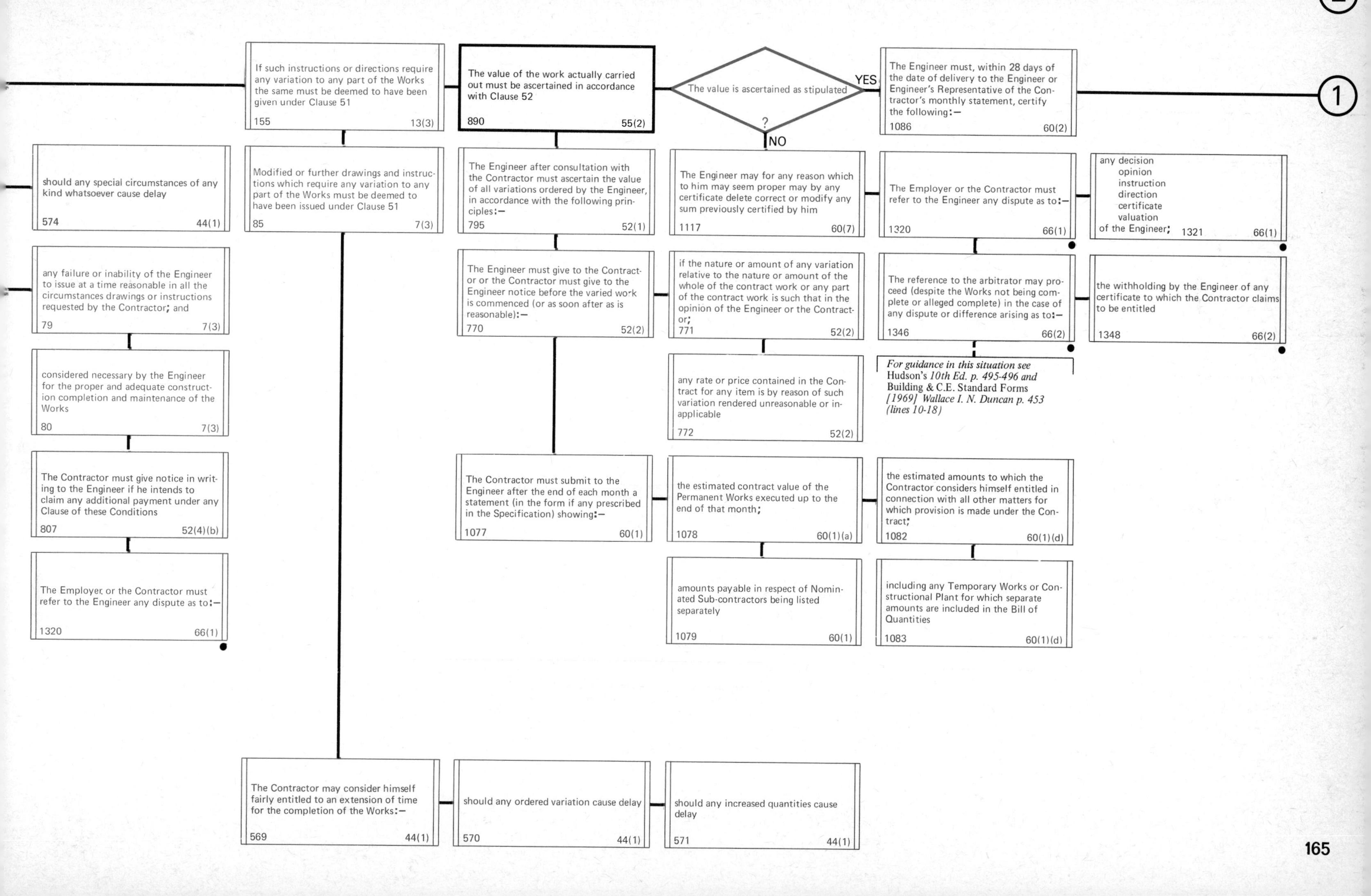

2
1
If such instructions or directions require any variation to any part of the Works the same must be deemed to have been given under Clause 51
155 13(3)
The value of the work actually carried out must be ascertained in accordance with Clause 52
890 55(2)
The value is ascertained as stipulated ?
YES
NO
The Engineer must, within 28 days of the date of delivery to the Engineer or Engineer's Representative of the Contractor's monthly statement, certify the following:—
1086 60(2)
should any special circumstances of any kind whatsoever cause delay
574 44(1)
Modified or further drawings and instructions which require any variation to any part of the Works must be deemed to have been issued under Clause 51
85 7(3)
The Engineer after consultation with the Contractor must ascertain the value of all variations ordered by the Engineer, in accordance with the following principles:—
795 52(1)
The Engineer may for any reason which to him may seem proper by any certificate delete correct or modify any sum previously certified by him
1117 60(7)
The Employer or the Contractor must refer to the Engineer any dispute as to:—
1320 66(1)
any decision opinion instruction direction certificate valuation of the Engineer;
1321 66(1)
any failure or inability of the Engineer to issue at a time reasonable in all the circumstances drawings or instructions requested by the Contractor; and
79 7(3)
The Engineer must give to the Contractor or the Contractor must give to the Engineer notice before the varied work is commenced (or as soon after as is reasonable):—
770 52(2)
if the nature or amount of any variation relative to the nature or amount of the whole of the contract work or any part of the contract work is such that in the opinion of the Engineer or the Contractor;
771 52(2)
The reference to the arbitrator may proceed (despite the Works not being complete or alleged complete) in the case of any dispute or difference arising as to:—
1346 66(2)
the withholding by the Engineer of any certificate to which the Contractor claims to be entitled
1348 66(2)
For guidance in this situation see Hudson's 10th Ed. p. 495-496 and Building & C.E. Standard Forms [1969] Wallace I. N. Duncan p. 453 (lines 10-18)
considered necessary by the Engineer for the proper and adequate construction completion and maintenance of the Works
80 7(3)
any rate or price contained in the Contract for any item is by reason of such variation rendered unreasonable or inapplicable
772 52(2)
The Contractor must give notice in writing to the Engineer if he intends to claim any additional payment under any Clause of these Conditions
807 52(4)(b)
The Contractor must submit to the Engineer after the end of each month a statement (in the form if any prescribed in the Specification) showing:—
1077 60(1)
the estimated contract value of the Permanent Works executed up to the end of that month;
1078 60(1)(a)
the estimated amounts to which the Contractor considers himself entitled in connection with all other matters for which provision is made under the Contract;
1082 60(1)(d)
The Employer or the Contractor must refer to the Engineer any dispute as to:—
1320 66(1)
amounts payable in respect of Nominated Sub-contractors being listed separately
1079 60(1)
including any Temporary Works or Constructional Plant for which separate amounts are included in the Bill of Quantities
1083 60(1)(d)
The Contractor may consider himself fairly entitled to an extension of time for the completion of the Works:—
569 44(1)
should any ordered variation cause delay
570 44(1)
should any increased quantities cause delay
571 44(1)

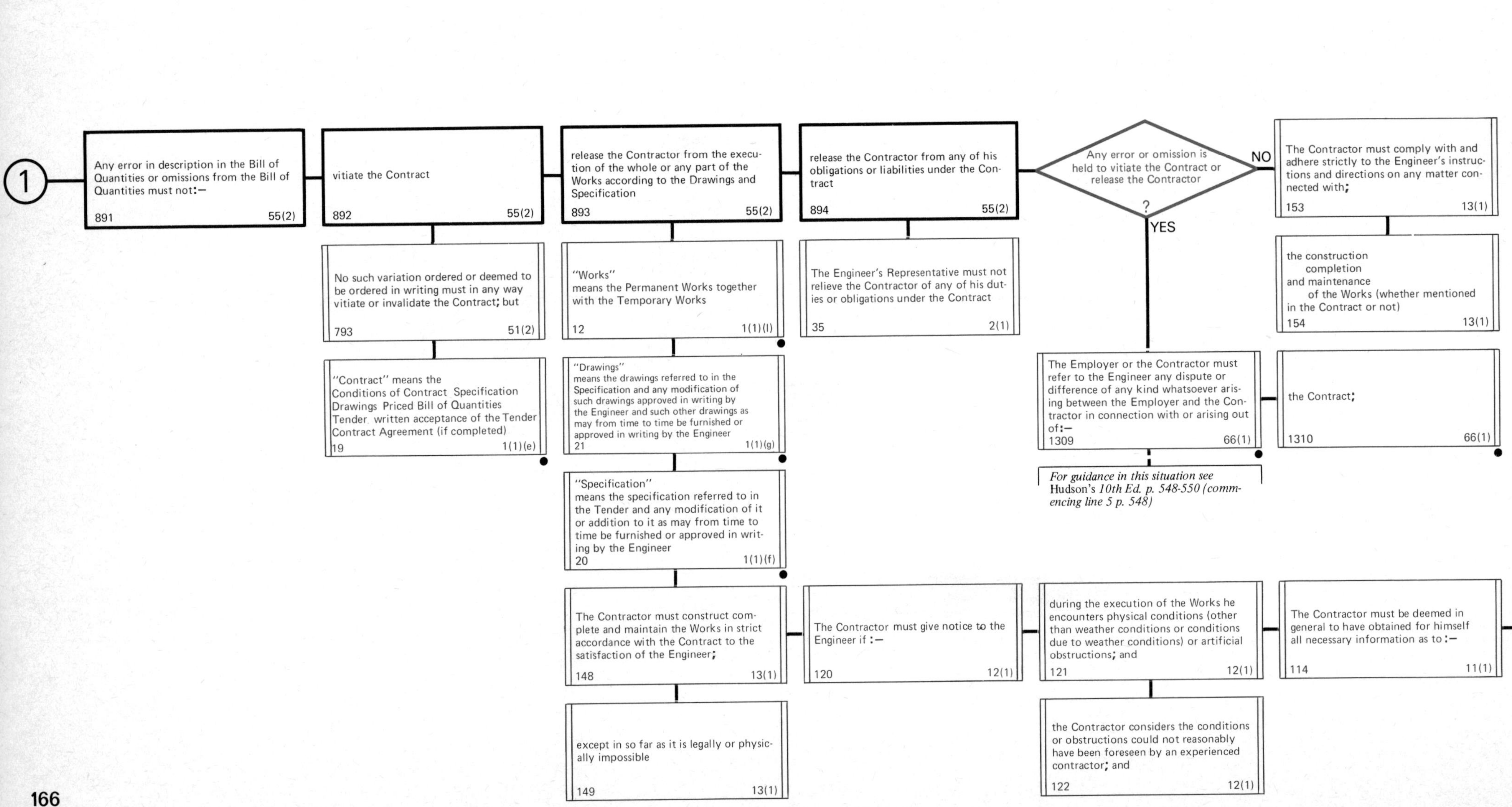
2
1
Any error in description in the Bill of Quantities or omissions from the Bill of Quantities must not:—
891 55(2)
vitiate the Contract
892 55(2)
release the Contractor from the execution of the whole or any part of the Works according to the Drawings and Specification
893 55(2)
release the Contractor from any of his obligations or liabilities under the Contract
894 55(2)
Any error or omission is held to vitiate the Contract or release the Contractor
?
NO
YES
The Contractor must comply with and adhere strictly to the Engineer's instructions and directions on any matter connected with;
153 13(1)
the construction completion and maintenance of the Works (whether mentioned in the Contract or not)
154 13(1)
No such variation ordered or deemed to be ordered in writing must in any way vitiate or invalidate the Contract; but
793 51(2)
"Works" means the Permanent Works together with the Temporary Works
12 1(1)(l)
The Engineer's Representative must not relieve the Contractor of any of his duties or obligations under the Contract
35 2(1)
"Contract" means the Conditions of Contract Specification Drawings Priced Bill of Quantities Tender written acceptance of the Tender Contract Agreement (if completed)
19 1(1)(e)
"Drawings" means the drawings referred to in the Specification and any modification of such drawings approved in writing by the Engineer and such other drawings as may from time to time be furnished or approved in writing by the Engineer
21 1(1)(g)
The Employer or the Contractor must refer to the Engineer any dispute or difference of any kind whatsoever arising between the Employer and the Contractor in connection with or arising out of:—
1309 66(1)
the Contract;
1310 66(1)
For guidance in this situation see Hudson's 10th Ed. p. 548-550 (commencing line 5 p. 548)
"Specification" means the specification referred to in the Tender and any modification of it or addition to it as may from time to time be furnished or approved in writing by the Engineer
20 1(1)(f)
The Contractor must construct complete and maintain the Works in strict accordance with the Contract to the satisfaction of the Engineer;
148 13(1)
The Contractor must give notice to the Engineer if :—
120 12(1)
during the execution of the Works he encounters physical conditions (other than weather conditions or conditions due to weather conditions) or artificial obstructions; and
121 12(1)
The Contractor must be deemed in general to have obtained for himself all necessary information as to:—
114 11(1)
except in so far as it is legally or physically impossible
149 13(1)
the Contractor considers the conditions or obstructions could not reasonably have been foreseen by an experienced contractor; and
122 12(1)

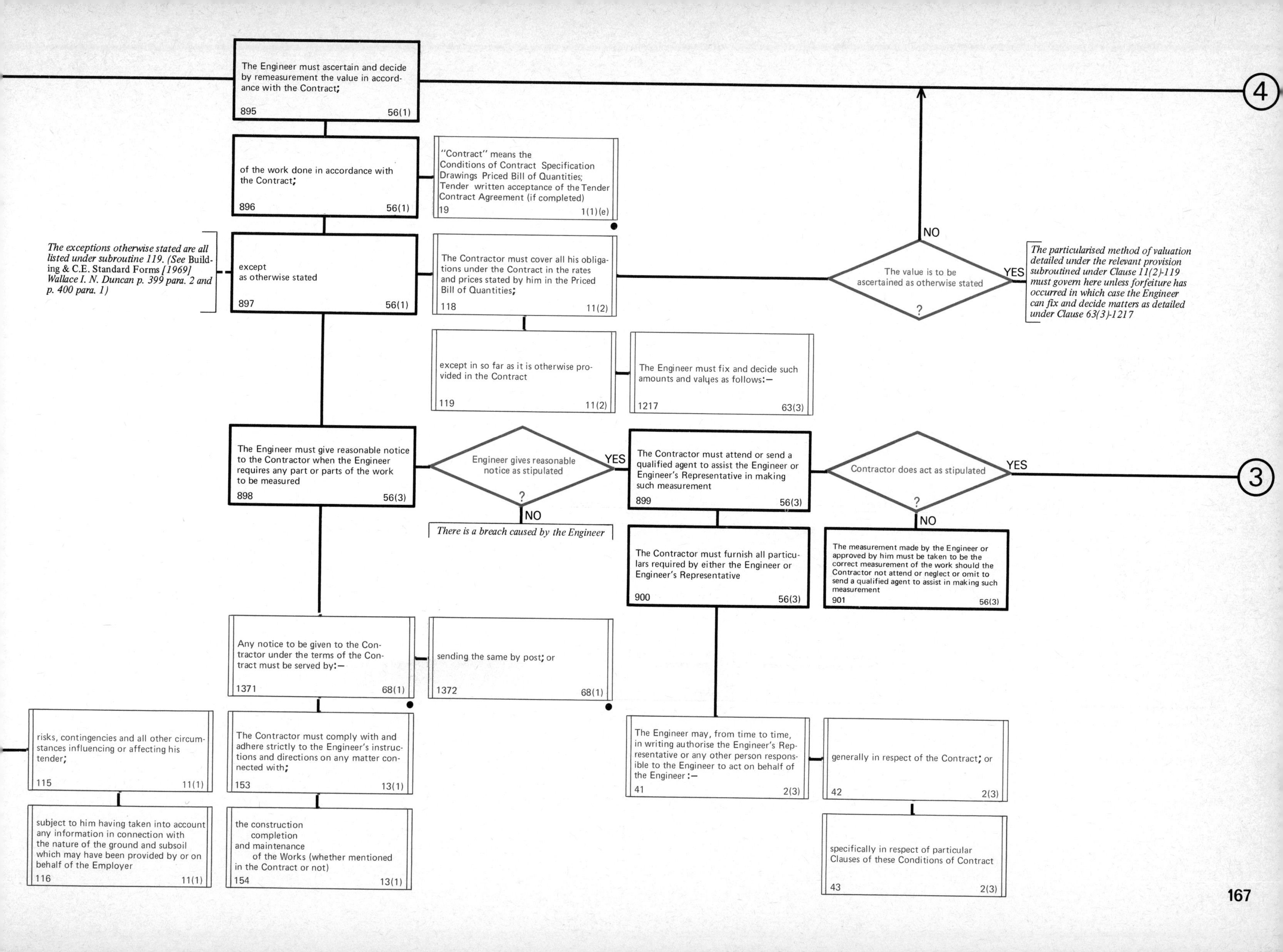
The Engineer must ascertain and decide by remeasurement the value in accordance with the Contract;
895 56(1)
of the work done in accordance with the Contract;
896 56(1)
"Contract" means the Conditions of Contract Specification Drawings Priced Bill of Quantities; Tender written acceptance of the Tender Contract Agreement (if completed)
19 1(1)(e)
The exceptions otherwise stated are all listed under subroutine 119. (See Building & C.E. Standard Forms [1969] Wallace I. N. Duncan p. 399 para. 2 and p. 400 para. 1)
except as otherwise stated
897 56(1)
The Contractor must cover all his obligations under the Contract in the rates and prices stated by him in the Priced Bill of Quantities;
118 11(2)
The value is to be ascertained as otherwise stated ?
NO
YES
The particularised method of valuation detailed under the relevant provision subroutined under Clause 11(2)-119 must govern here unless forfeiture has occurred in which case the Engineer can fix and decide matters as detailed under Clause 63(3)-1217
except in so far as it is otherwise provided in the Contract
119 11(2)
The Engineer must fix and decide such amounts and values as follows:—
1217 63(3)
The Engineer must give reasonable notice to the Contractor when the Engineer requires any part or parts of the work to be measured
898 56(3)
Engineer gives reasonable notice as stipulated ?
YES
NO
There is a breach caused by the Engineer
The Contractor must attend or send a qualified agent to assist the Engineer or Engineer's Representative in making such measurement
899 56(3)
Contractor does act as stipulated ?
YES
NO
The Contractor must furnish all particulars required by either the Engineer or Engineer's Representative
900 56(3)
The measurement made by the Engineer or approved by him must be taken to be the correct measurement of the work should the Contractor not attend or neglect or omit to send a qualified agent to assist in making such measurement
901 56(3)
Any notice to be given to the Contractor under the terms of the Contract must be served by:—
1371 68(1)
sending the same by post; or
1372 68(1)
risks, contingencies and all other circumstances influencing or affecting his tender;
115 11(1)
subject to him having taken into account any information in connection with the nature of the ground and subsoil which may have been provided by or on behalf of the Employer
116 11(1)
The Contractor must comply with and adhere strictly to the Engineer's instructions and directions on any matter connected with;
153 13(1)
the construction completion and maintenance of the Works (whether mentioned in the Contract or not)
154 13(1)
The Engineer may, from time to time, in writing authorise the Engineer's Representative or any other person responsible to the Engineer to act on behalf of the Engineer:—
41 2(3)
generally in respect of the Contract; or
42 2(3)
specifically in respect of particular Clauses of these Conditions of Contract
43 2(3)
4
3

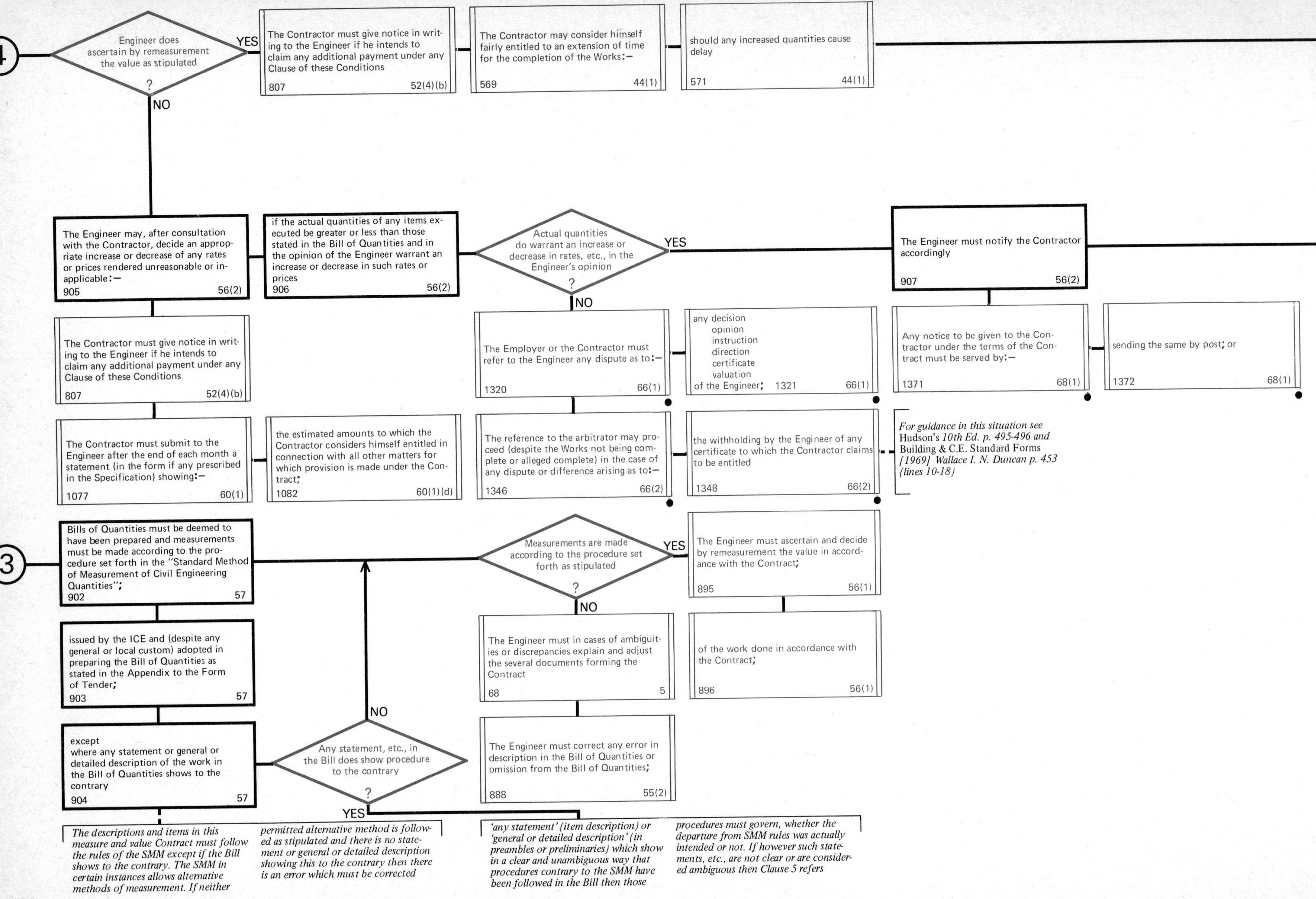
4
Engineer does ascertain by remeasurement the value as stipulated ?
YES
NO
The Contractor must give notice in writing to the Engineer if he intends to claim any additional payment under any Clause of these Conditions 807 52(4)(b)
The Contractor may consider himself fairly entitled to an extension of time for the completion of the Works:— 569 44(1)
should any increased quantities cause delay 571 44(1)
The Engineer may, after consultation with the Contractor, decide an appropriate increase or decrease of any rates or prices rendered unreasonable or inapplicable:— 905 56(2)
if the actual quantities of any items executed be greater or less than those stated in the Bill of Quantities and in the opinion of the Engineer warrant an increase or decrease in such rates or prices 906 56(2)
Actual quantities do warrant an increase or decrease in rates, etc., in the Engineer's opinion ?
YES
NO
The Engineer must notify the Contractor accordingly 907 56(2)
The Contractor must give notice in writing to the Engineer if he intends to claim any additional payment under any Clause of these Conditions 807 52(4)(b)
The Employer or the Contractor must refer to the Engineer any dispute as to:— 1320 66(1)
any decision opinion instruction direction certificate valuation of the Engineer; 1321 66(1)
Any notice to be given to the Contractor under the terms of the Contract must be served by:— 1371 68(1)
sending the same by post; or 1372 68(1)
The Contractor must submit to the Engineer after the end of each month a statement (in the form if any prescribed in the Specification) showing:— 1077 60(1)
the estimated amounts to which the Contractor considers himself entitled in connection with all other matters for which provision is made under the Contract; 1082 60(1)(d)
The reference to the arbitrator may proceed (despite the Works not being complete or alleged complete) in the case of any dispute or difference arising as to:— 1346 66(2)
the withholding by the Engineer of any certificate to which the Contractor claims to be entitled 1348 66(2)
For guidance in this situation see Hudson's 10th Ed. p. 495-496 and Building & C.E. Standard Forms [1969] Wallace I. N. Duncan p. 453 (lines 10-18)
3
Bills of Quantities must be deemed to have been prepared and measurements must be made according to the procedure set forth in the "Standard Method of Measurement of Civil Engineering Quantities"; 902 57
Measurements are made according to the procedure set forth as stipulated ?
YES
NO
The Engineer must ascertain and decide by remeasurement the value in accordance with the Contract; 895 56(1)
issued by the ICE and (despite any general or local custom) adopted in preparing the Bill of Quantities as stated in the Appendix to the Form of Tender; 903 57
The Engineer must in cases of ambiguities or discrepancies explain and adjust the several documents forming the Contract 68 5
of the work done in accordance with the Contract; 896 56(1)
NO
except where any statement or general or detailed description of the work in the Bill of Quantities shows to the contrary 904 57
Any statement, etc., in the Bill does show procedure to the contrary ?
YES
The Engineer must correct any error in description in the Bill of Quantities or omission from the Bill of Quantities; 888 55(2)
The descriptions and items in this measure and value Contract must follow the rules of the SMM except if the Bill shows to the contrary. The SMM in certain instances allows alternative methods of measurement. If neither permitted alternative method is followed as stipulated and there is no statement or general or detailed description showing this to the contrary then there is an error which must be corrected
'any statement' (item description) or 'general or detailed description' (in preambles or preliminaries) which show in a clear and unambiguous way that procedures contrary to the SMM have been followed in the Bill then those procedures must govern, whether the departure from SMM rules was actually intended or not. If however such statements, etc., are not clear or are considered ambiguous then Clause 5 refers

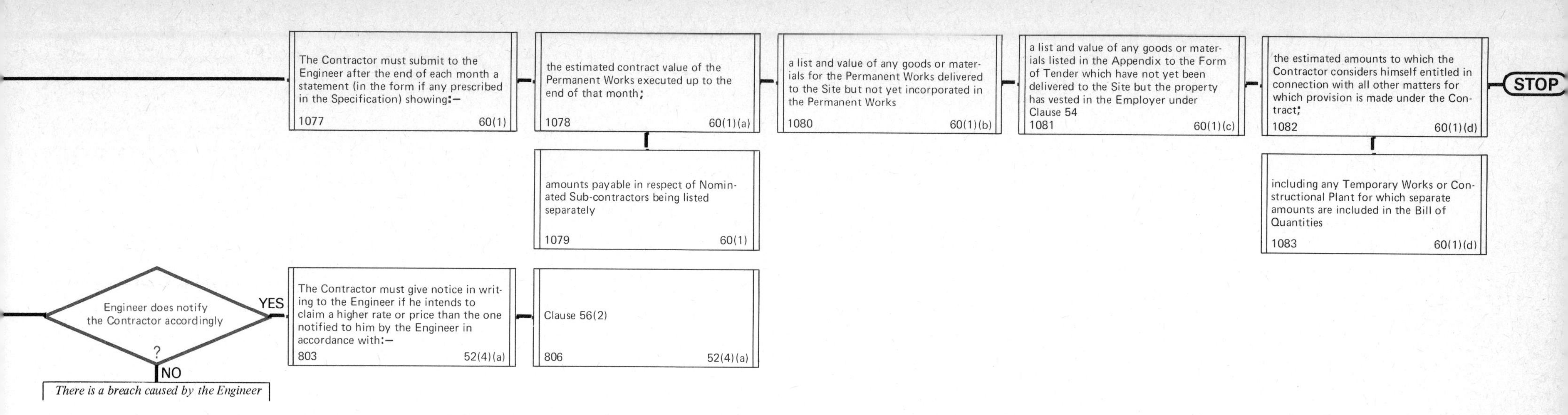
The Contractor must submit to the Engineer after the end of each month a statement (in the form if any prescribed in the Specification) showing:–
1077 60(1)
the estimated contract value of the Permanent Works executed up to the end of that month;
1078 60(1)(a)
amounts payable in respect of Nominated Sub-contractors being listed separately
1079 60(1)
a list and value of any goods or materials for the Permanent Works delivered to the Site but not yet incorporated in the Permanent Works
1080 60(1)(b)
a list and value of any goods or materials listed in the Appendix to the Form of Tender which have not yet been delivered to the Site but the property has vested in the Employer under Clause 54
1081 60(1)(c)
the estimated amounts to which the Contractor considers himself entitled in connection with all other matters for which provision is made under the Contract;
1082 60(1)(d)
including any Temporary Works or Constructional Plant for which separate amounts are included in the Bill of Quantities
1083 60(1)(d)
STOP
Engineer does notify the Contractor accordingly
?
YES
NO
There is a breach caused by the Engineer
The Contractor must give notice in writing to the Engineer if he intends to claim a higher rate or price than the one notified to him by the Engineer in accordance with:–
803 52(4)(a)
Clause 56(2)
806 52(4)(a)

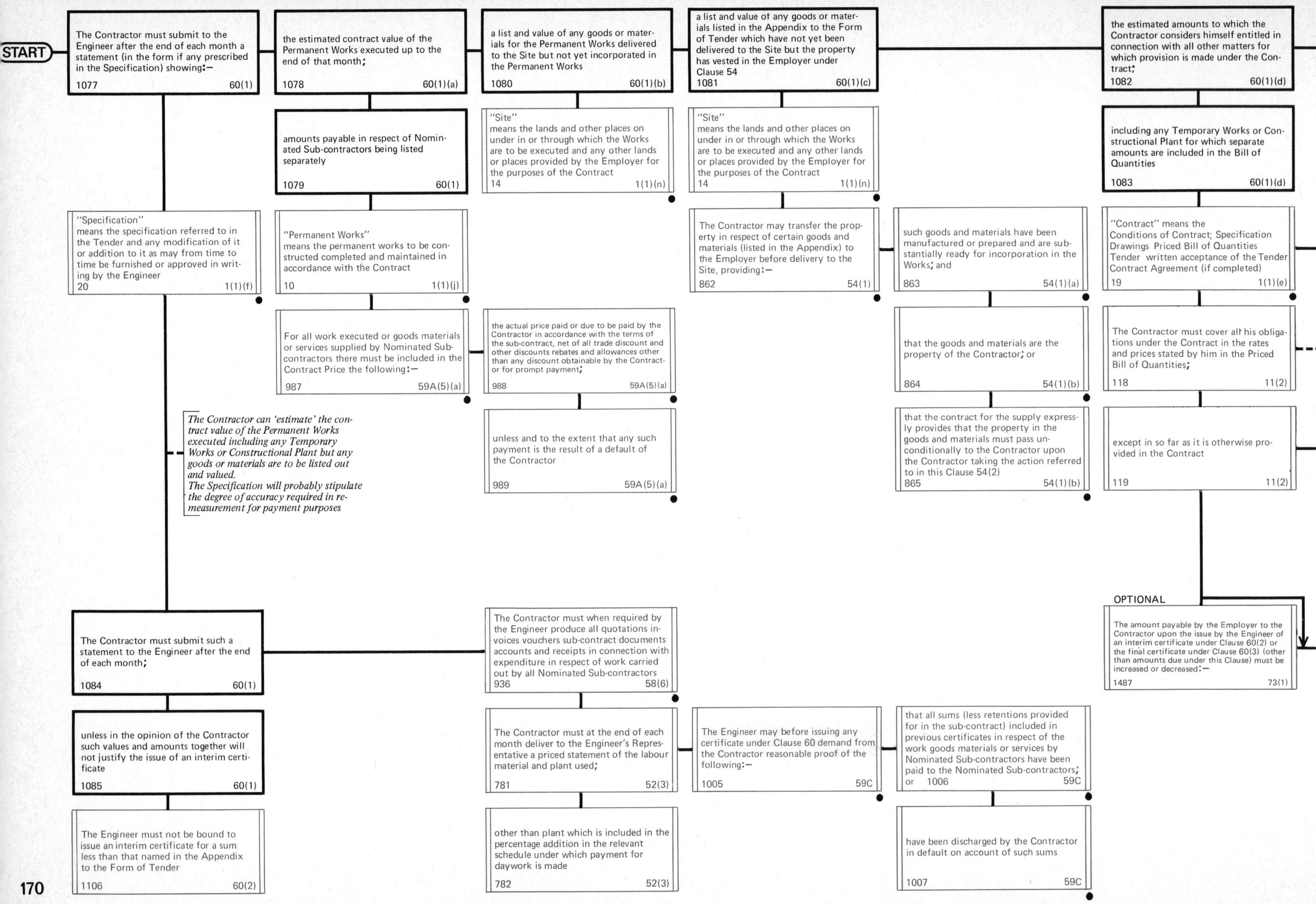
START
The Contractor must submit to the Engineer after the end of each month a statement (in the form if any prescribed in the Specification) showing:— 1077 60(1)
the estimated contract value of the Permanent Works executed up to the end of that month; 1078 60(1)(a)
a list and value of any goods or materials for the Permanent Works delivered to the Site but not yet incorporated in the Permanent Works 1080 60(1)(b)
a list and value of any goods or materials listed in the Appendix to the Form of Tender which have not yet been delivered to the Site but the property has vested in the Employer under Clause 54 1081 60(1)(c)
the estimated amounts to which the Contractor considers himself entitled in connection with all other matters for which provision is made under the Contract; 1082 60(1)(d)
amounts payable in respect of Nominated Sub-contractors being listed separately 1079 60(1)
"Site" means the lands and other places on under in or through which the Works are to be executed and any other lands or places provided by the Employer for the purposes of the Contract 14 1(1)(n)
"Site" means the lands and other places on under in or through which the Works are to be executed and any other lands or places provided by the Employer for the purposes of the Contract 14 1(1)(n)
including any Temporary Works or Constructional Plant for which separate amounts are included in the Bill of Quantities 1083 60(1)(d)
"Specification" means the specification referred to in the Tender and any modification of it or addition to it as may from time to time be furnished or approved in writing by the Engineer 20 1(1)(f)
"Permanent Works" means the permanent works to be constructed completed and maintained in accordance with the Contract 10 1(1)(j)
The Contractor may transfer the property in respect of certain goods and materials (listed in the Appendix) to the Employer before delivery to the Site, providing:— 862 54(1)
such goods and materials have been manufactured or prepared and are substantially ready for incorporation in the Works; and 863 54(1)(a)
"Contract" means the Conditions of Contract; Specification Drawings Priced Bill of Quantities Tender written acceptance of the Tender Contract Agreement (if completed) 19 1(1)(e)
For all work executed or goods materials or services supplied by Nominated Sub-contractors there must be included in the Contract Price the following:— 987 59A(5)(a)
the actual price paid or due to be paid by the Contractor in accordance with the terms of the sub-contract, net of all trade discount and other discounts rebates and allowances other than any discount obtainable by the Contractor for prompt payment; 988 59A(5)(a)
that the goods and materials are the property of the Contractor; or 864 54(1)(b)
The Contractor must cover all his obligations under the Contract in the rates and prices stated by him in the Priced Bill of Quantities; 118 11(2)
The Contractor can 'estimate' the contract value of the Permanent Works executed including any Temporary Works or Constructional Plant but any goods or materials are to be listed out and valued.
The Specification will probably stipulate the degree of accuracy required in remeasurement for payment purposes
unless and to the extent that any such payment is the result of a default of the Contractor 989 59A(5)(a)
that the contract for the supply expressly provides that the property in the goods and materials must pass unconditionally to the Contractor upon the Contractor taking the action referred to in this Clause 54(2) 865 54(1)(b)
except in so far as it is otherwise provided in the Contract 119 11(2)
OPTIONAL
The amount payable by the Employer to the Contractor upon the issue by the Engineer of an interim certificate under Clause 60(2) or the final certificate under Clause 60(3) (other than amounts due under this Clause) must be increased or decreased:— 1487 73(1)
The Contractor must submit such a statement to the Engineer after the end of each month; 1084 60(1)
The Contractor must when required by the Engineer produce all quotations invoices vouchers sub-contract documents accounts and receipts in connection with expenditure in respect of work carried out by all Nominated Sub-contractors 936 58(6)
unless in the opinion of the Contractor such values and amounts together will not justify the issue of an interim certificate 1085 60(1)
The Contractor must at the end of each month deliver to the Engineer's Representative a priced statement of the labour material and plant used; 781 52(3)
The Engineer may before issuing any certificate under Clause 60 demand from the Contractor reasonable proof of the following:— 1005 59C
that all sums (less retentions provided for in the sub-contract) included in previous certificates in respect of the work goods materials or services by Nominated Sub-contractors have been paid to the Nominated Sub-contractors; or 1006 59C
The Engineer must not be bound to issue an interim certificate for a sum less than that named in the Appendix to the Form of Tender 1106 60(2)
other than plant which is included in the percentage addition in the relevant schedule under which payment for daywork is made 782 52(3)
have been discharged by the Contractor in default on account of such sums 1007 59C

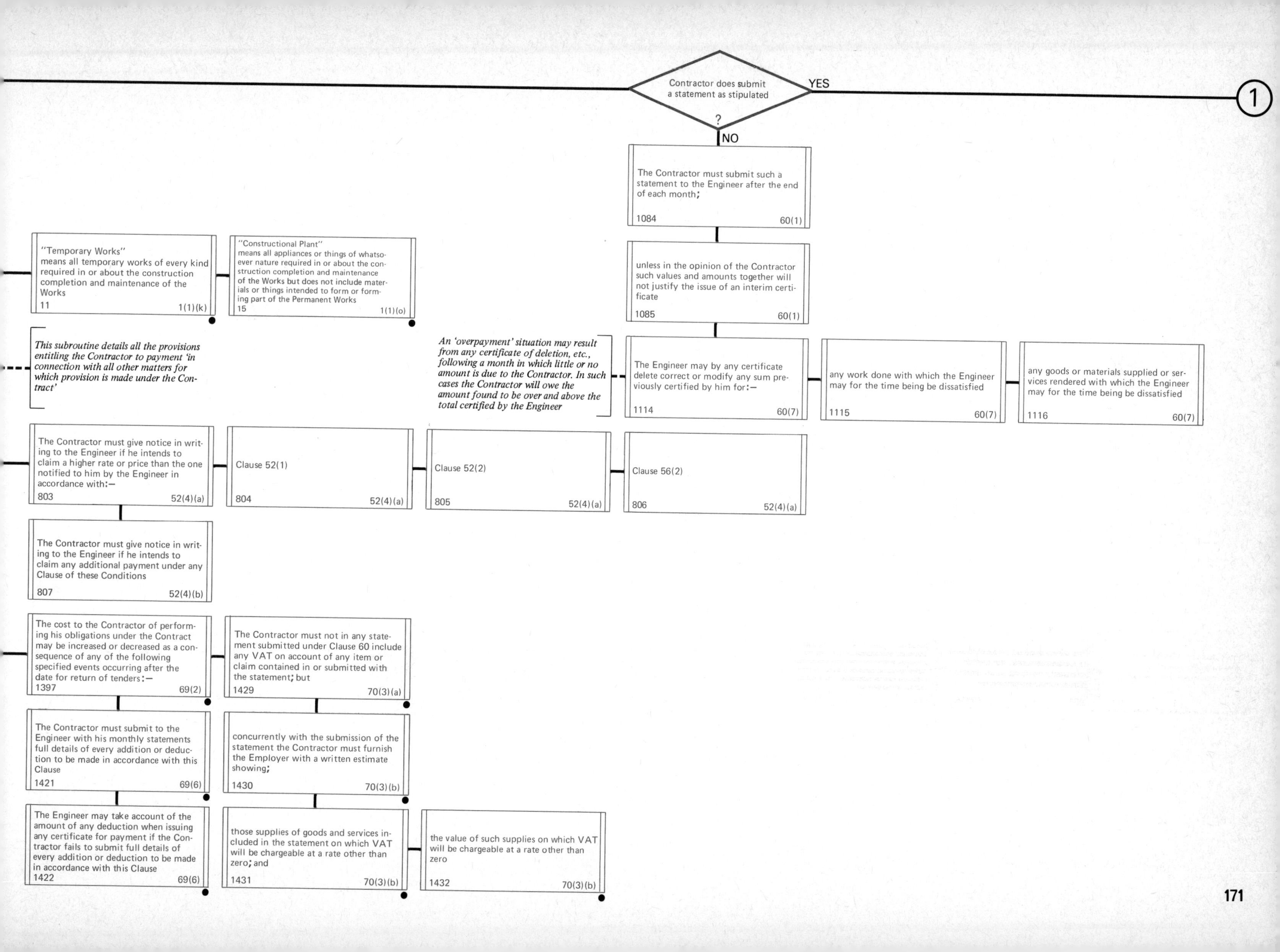

Contractor does submit a statement as stipulated ?
YES
NO
1
The Contractor must submit such a statement to the Engineer after the end of each month;
1084 60(1)
unless in the opinion of the Contractor such values and amounts together will not justify the issue of an interim certificate
1085 60(1)
"Temporary Works" means all temporary works of every kind required in or about the construction completion and maintenance of the Works
11 1(1)(k)
"Constructional Plant" means all appliances or things of whatsoever nature required in or about the construction completion and maintenance of the Works but does not include materials or things intended to form or forming part of the Permanent Works
15 1(1)(o)
This subroutine details all the provisions entitling the Contractor to payment 'in connection with all other matters for which provision is made under the Contract'
An 'overpayment' situation may result from any certificate of deletion, etc., following a month in which little or no amount is due to the Contractor. In such cases the Contractor will owe the amount found to be over and above the total certified by the Engineer
The Engineer may by any certificate delete correct or modify any sum previously certified by him for:—
1114 60(7)
any work done with which the Engineer may for the time being be dissatisfied
1115 60(7)
any goods or materials supplied or services rendered with which the Engineer may for the time being be dissatisfied
1116 60(7)
The Contractor must give notice in writing to the Engineer if he intends to claim a higher rate or price than the one notified to him by the Engineer in accordance with:—
803 52(4)(a)
Clause 52(1)
804 52(4)(a)
Clause 52(2)
805 52(4)(a)
Clause 56(2)
806 52(4)(a)
The Contractor must give notice in writing to the Engineer if he intends to claim any additional payment under any Clause of these Conditions
807 52(4)(b)
The cost to the Contractor of performing his obligations under the Contract may be increased or decreased as a consequence of any of the following specified events occurring after the date for return of tenders:—
1397 69(2)
The Contractor must not in any statement submitted under Clause 60 include any VAT on account of any item or claim contained in or submitted with the statement; but
1429 70(3)(a)
The Contractor must submit to the Engineer with his monthly statements full details of every addition or deduction to be made in accordance with this Clause
1421 69(6)
concurrently with the submission of the statement the Contractor must furnish the Employer with a written estimate showing;
1430 70(3)(b)
The Engineer may take account of the amount of any deduction when issuing any certificate for payment if the Contractor fails to submit full details of every addition or deduction to be made in accordance with this Clause
1422 69(6)
those supplies of goods and services included in the statement on which VAT will be chargeable at a rate other than zero; and
1431 70(3)(b)
the value of such supplies on which VAT will be chargeable at a rate other than zero
1432 70(3)(b)

INTERIM CERTIFICATES

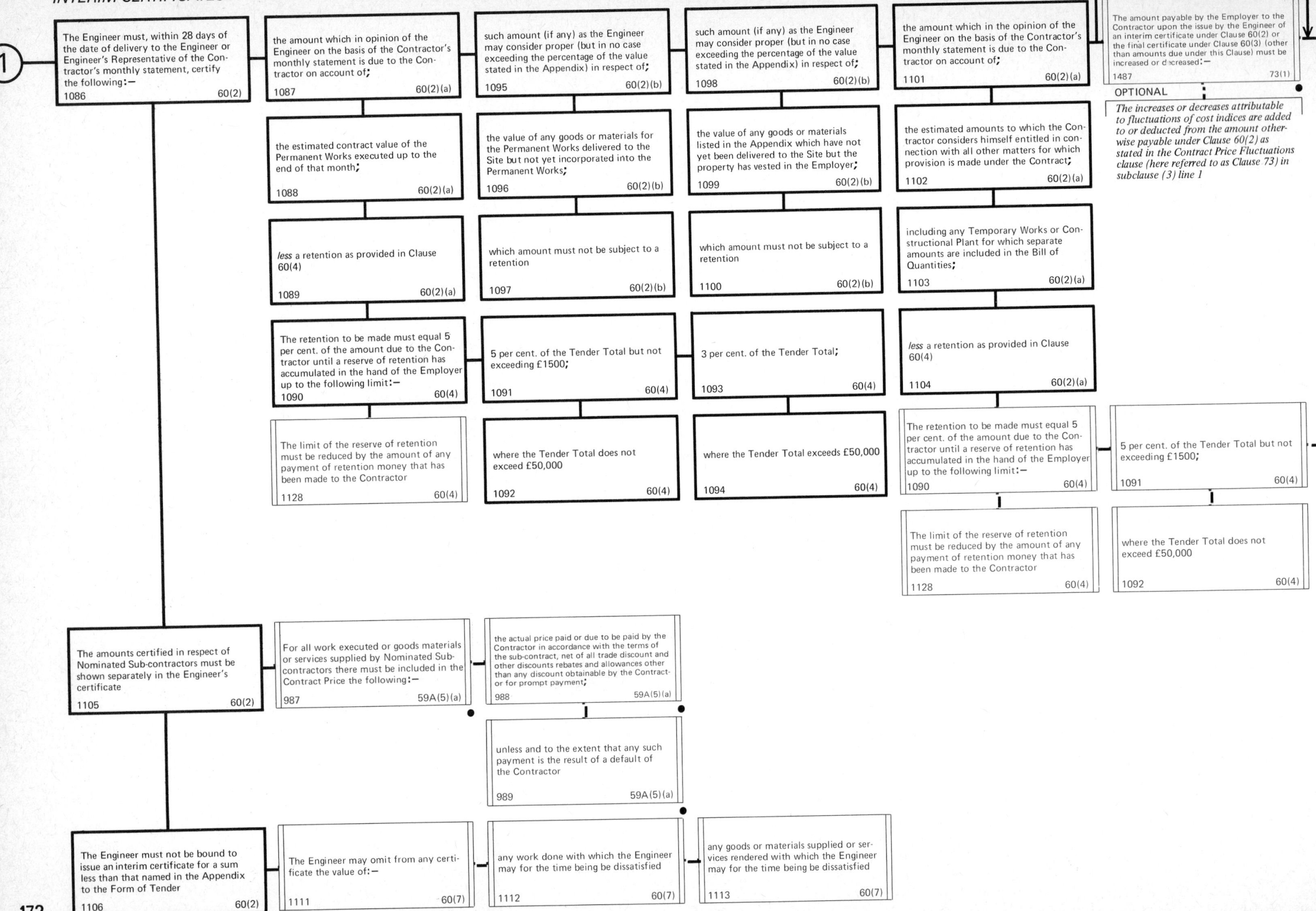

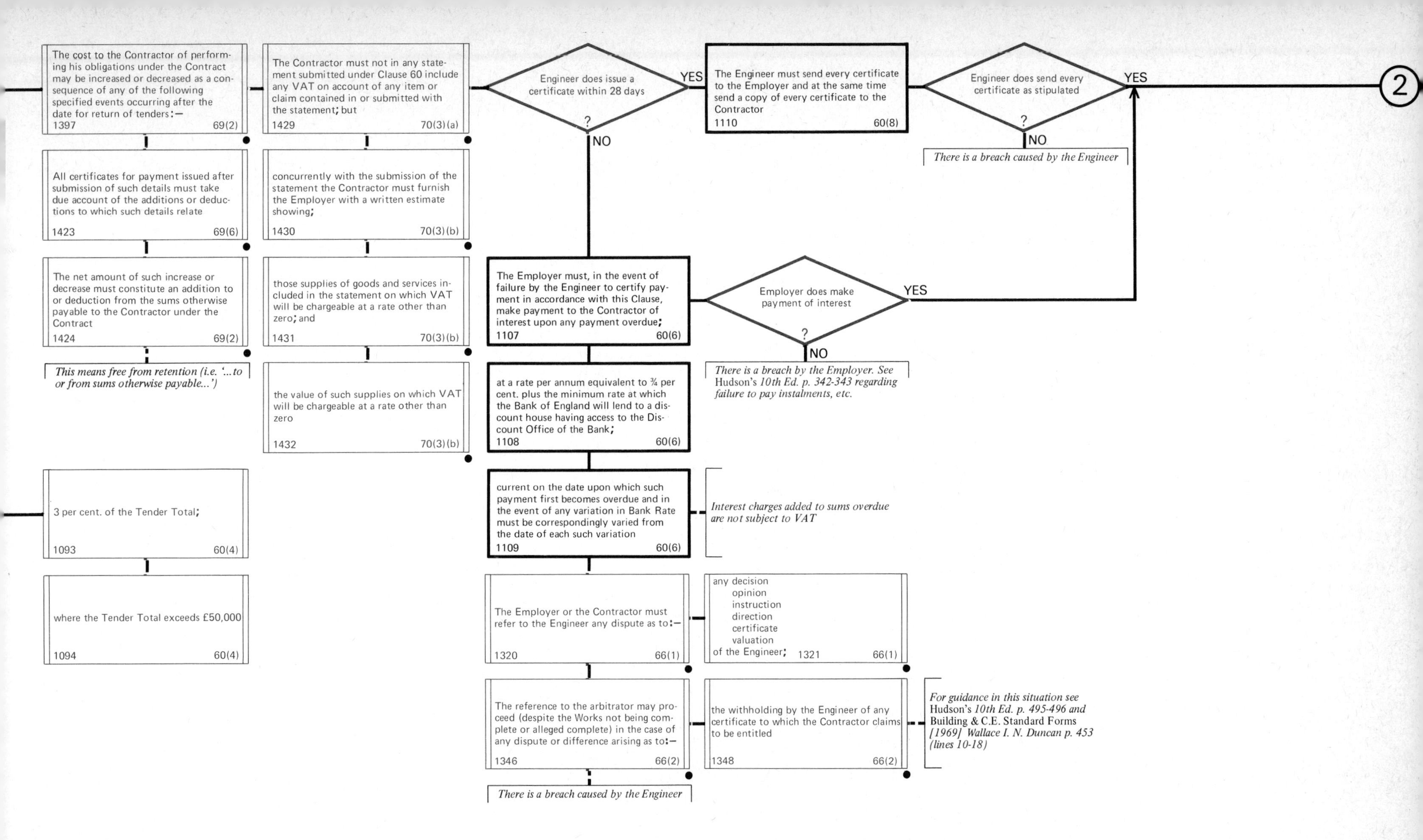
The cost to the Contractor of performing his obligations under the Contract may be increased or decreased as a consequence of any of the following specified events occurring after the date for return of tenders:—
1397 69(2)
All certificates for payment issued after submission of such details must take due account of the additions or deductions to which such details relate
1423 69(6)
The net amount of such increase or decrease must constitute an addition to or deduction from the sums otherwise payable to the Contractor under the Contract
1424 69(2)
This means free from retention (i.e. '...to or from sums otherwise payable...')
3 per cent. of the Tender Total;
1093 60(4)
where the Tender Total exceeds £50,000
1094 60(4)
The Contractor must not in any statement submitted under Clause 60 include any VAT on account of any item or claim contained in or submitted with the statement; but
1429 70(3)(a)
concurrently with the submission of the statement the Contractor must furnish the Employer with a written estimate showing;
1430 70(3)(b)
those supplies of goods and services included in the statement on which VAT will be chargeable at a rate other than zero; and
1431 70(3)(b)
the value of such supplies on which VAT will be chargeable at a rate other than zero
1432 70(3)(b)
Engineer does issue a certificate within 28 days
?
YES
NO
The Engineer must send every certificate to the Employer and at the same time send a copy of every certificate to the Contractor
1110 60(8)
Engineer does send every certificate as stipulated
?
YES
NO
There is a breach caused by the Engineer
2
The Employer must, in the event of failure by the Engineer to certify payment in accordance with this Clause, make payment to the Contractor of interest upon any payment overdue;
1107 60(6)
Employer does make payment of interest
?
YES
NO
There is a breach by the Employer. See Hudson's 10th Ed. p. 342-343 regarding failure to pay instalments, etc.
at a rate per annum equivalent to ¾ per cent. plus the minimum rate at which the Bank of England will lend to a discount house having access to the Discount Office of the Bank;
1108 60(6)
current on the date upon which such payment first becomes overdue and in the event of any variation in Bank Rate must be correspondingly varied from the date of each such variation
1109 60(6)
Interest charges added to sums overdue are not subject to VAT
The Employer or the Contractor must refer to the Engineer any dispute as to:—
1320 66(1)
any decision
opinion
instruction
direction
certificate
valuation
of the Engineer; 1321 66(1)
The reference to the arbitrator may proceed (despite the Works not being complete or alleged complete) in the case of any dispute or difference arising as to:—
1346 66(2)
the withholding by the Engineer of any certificate to which the Contractor claims to be entitled
1348 66(2)
For guidance in this situation see Hudson's 10th Ed. p. 495-496 and Building & C.E. Standard Forms [1969] Wallace I. N. Duncan p. 453 (lines 10-18)
There is a breach caused by the Engineer

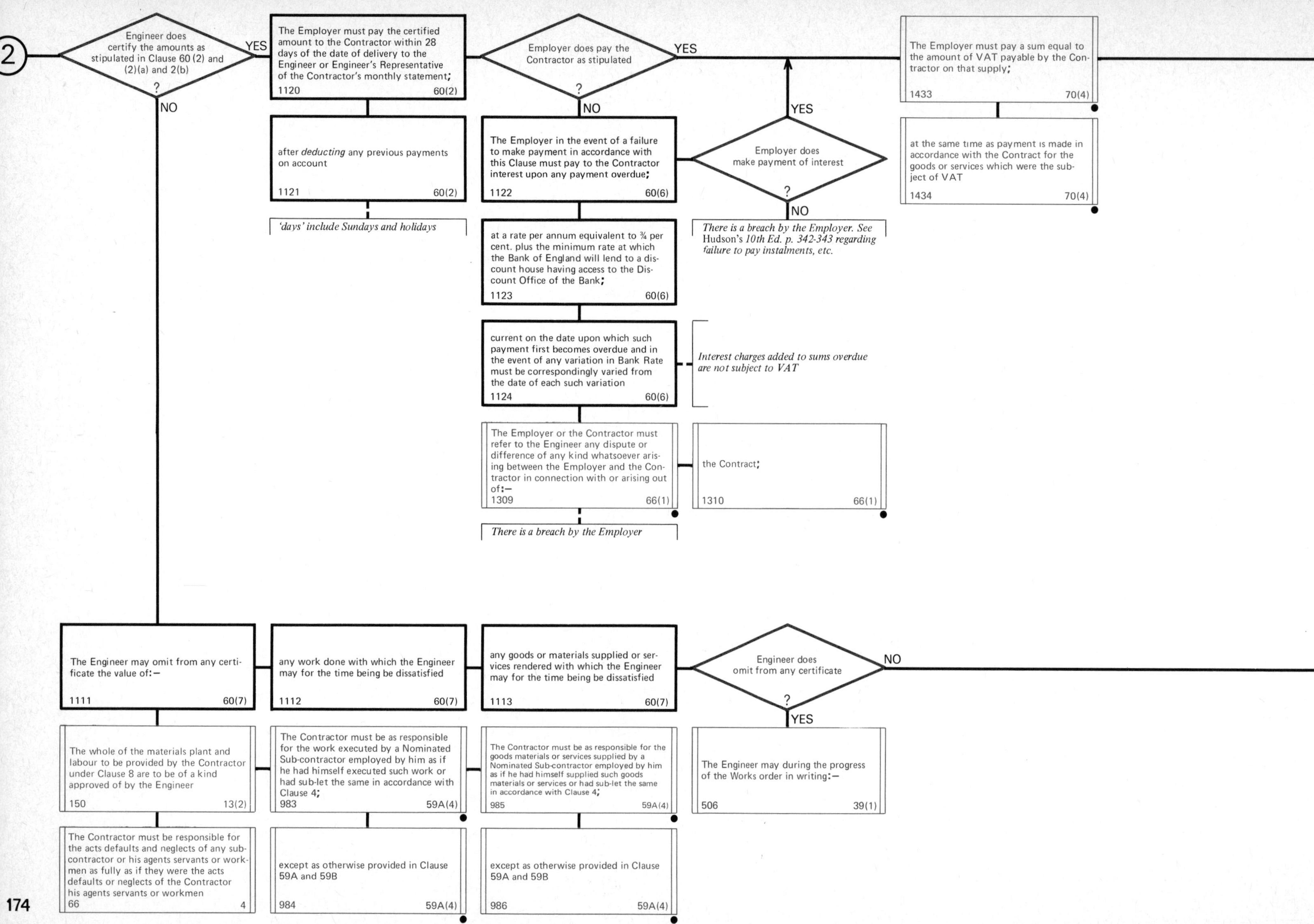
2
Engineer does certify the amounts as stipulated in Clause 60 (2) and (2)(a) and 2(b) ?
YES
NO
The Employer must pay the certified amount to the Contractor within 28 days of the date of delivery to the Engineer or Engineer's Representative of the Contractor's monthly statement;
1120 60(2)
after *deducting* any previous payments on account
1121 60(2)
'days' include Sundays and holidays
Employer does pay the Contractor as stipulated ?
YES
NO
The Employer in the event of a failure to make payment in accordance with this Clause must pay to the Contractor interest upon any payment overdue;
1122 60(6)
at a rate per annum equivalent to ¾ per cent. plus the minimum rate at which the Bank of England will lend to a discount house having access to the Discount Office of the Bank;
1123 60(6)
current on the date upon which such payment first becomes overdue and in the event of any variation in Bank Rate must be correspondingly varied from the date of each such variation
1124 60(6)
Interest charges added to sums overdue are not subject to VAT
The Employer or the Contractor must refer to the Engineer any dispute or difference of any kind whatsoever arising between the Employer and the Contractor in connection with or arising out of:—
1309 66(1)
the Contract;
1310 66(1)
There is a breach by the Employer
Employer does make payment of interest ?
YES
NO
There is a breach by the Employer. See Hudson's 10th Ed. p. 342-343 regarding failure to pay instalments, etc.
The Employer must pay a sum equal to the amount of VAT payable by the Contractor on that supply;
1433 70(4)
at the same time as payment is made in accordance with the Contract for the goods or services which were the subject of VAT
1434 70(4)
The Engineer may omit from any certificate the value of:—
1111 60(7)
any work done with which the Engineer may for the time being be dissatisfied
1112 60(7)
any goods or materials supplied or services rendered with which the Engineer may for the time being be dissatisfied
1113 60(7)
Engineer does omit from any certificate ?
NO
YES
The Engineer may during the progress of the Works order in writing:—
506 39(1)
The whole of the materials plant and labour to be provided by the Contractor under Clause 8 are to be of a kind approved of by the Engineer
150 13(2)
The Contractor must be responsible for the acts defaults and neglects of any sub-contractor or his agents servants or workmen as fully as if they were the acts defaults or neglects of the Contractor his agents servants or workmen
66 4
The Contractor must be as responsible for the work executed by a Nominated Sub-contractor employed by him as if he had himself executed such work or had sub-let the same in accordance with Clause 4;
983 59A(4)
except as otherwise provided in Clause 59A and 59B
984 59A(4)
The Contractor must be as responsible for the goods materials or services supplied by a Nominated Sub-contractor employed by him as if he had himself supplied such goods materials or services or had sub-let the same in accordance with Clause 4;
985 59A(4)
except as otherwise provided in Clause 59A and 59B
986 59A(4)

PAYMENT OF RETENTION MONEY

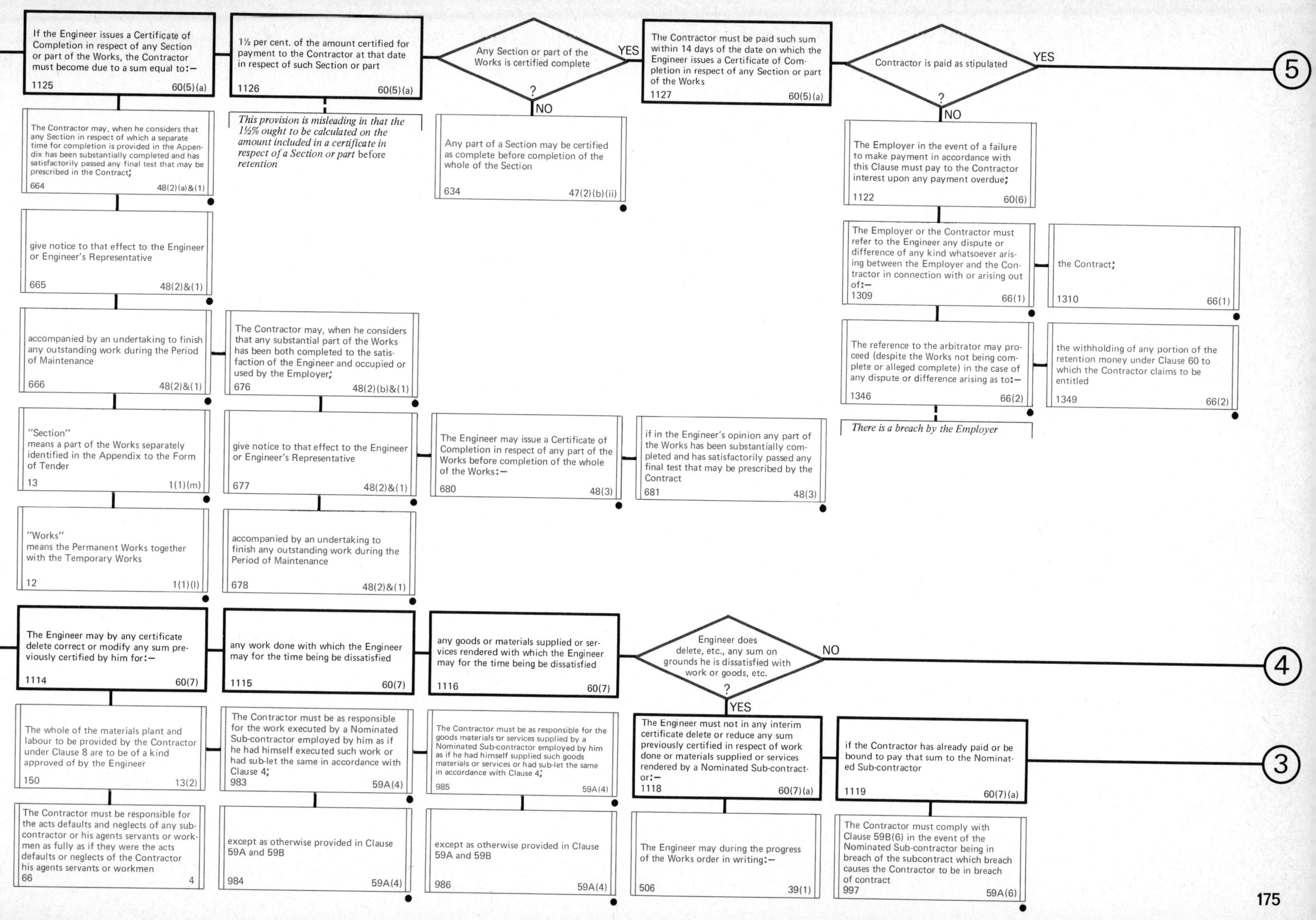

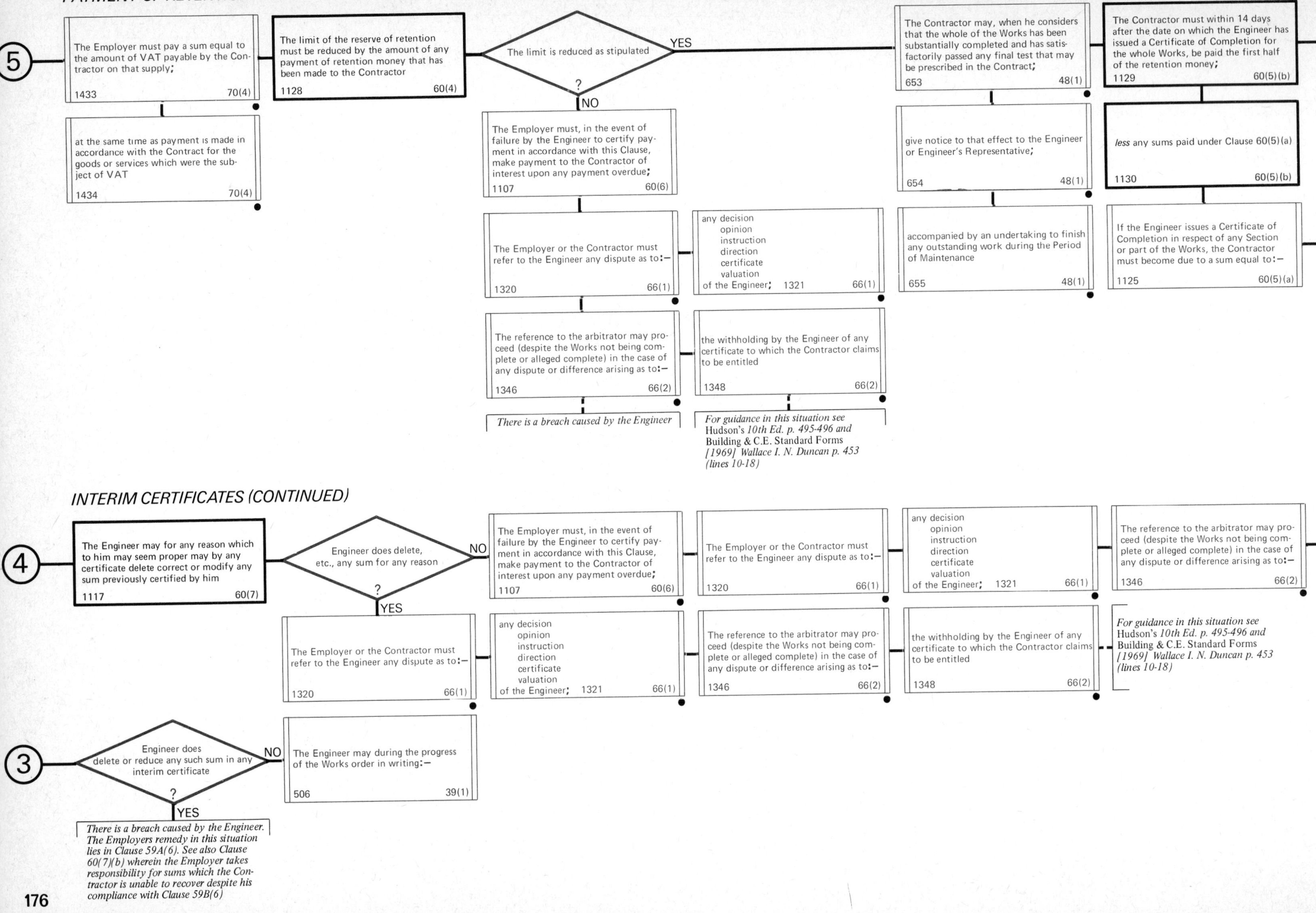
PAYMENT OF RETENTION MONEY (CONTINUED)
5
The Employer must pay a sum equal to the amount of VAT payable by the Contractor on that supply; 1433 70(4)
at the same time as payment is made in accordance with the Contract for the goods or services which were the subject of VAT 1434 70(4)
The limit of the reserve of retention must be reduced by the amount of any payment of retention money that has been made to the Contractor 1128 60(4)
The limit is reduced as stipulated ?
YES
NO
The Employer must, in the event of failure by the Engineer to certify payment in accordance with this Clause, make payment to the Contractor of interest upon any payment overdue; 1107 60(6)
The Employer or the Contractor must refer to the Engineer any dispute as to:– 1320 66(1)
any decision opinion instruction direction certificate valuation of the Engineer; 1321 66(1)
The reference to the arbitrator may proceed (despite the Works not being complete or alleged complete) in the case of any dispute or difference arising as to:– 1346 66(2)
the withholding by the Engineer of any certificate to which the Contractor claims to be entitled 1348 66(2)
There is a breach caused by the Engineer
For guidance in this situation see Hudson's 10th Ed. p. 495-496 and Building & C.E. Standard Forms [1969] Wallace I. N. Duncan p. 453 (lines 10-18)
The Contractor may, when he considers that the whole of the Works has been substantially completed and has satisfactorily passed any final test that may be prescribed in the Contract; 653 48(1)
give notice to that effect to the Engineer or Engineer's Representative; 654 48(1)
accompanied by an undertaking to finish any outstanding work during the Period of Maintenance 655 48(1)
The Contractor must within 14 days after the date on which the Engineer has issued a Certificate of Completion for the whole Works, be paid the first half of the retention money; 1129 60(5)(b)
less any sums paid under Clause 60(5)(a) 1130 60(5)(b)
If the Engineer issues a Certificate of Completion in respect of any Section or part of the Works, the Contractor must become due to a sum equal to:– 1125 60(5)(a)
INTERIM CERTIFICATES (CONTINUED)
4
The Engineer may for any reason which to him may seem proper may by any certificate delete correct or modify any sum previously certified by him 1117 60(7)
Engineer does delete, etc., any sum for any reason ?
NO
YES
The Employer must, in the event of failure by the Engineer to certify payment in accordance with this Clause, make payment to the Contractor of interest upon any payment overdue; 1107 60(6)
The Employer or the Contractor must refer to the Engineer any dispute as to:– 1320 66(1)
any decision opinion instruction direction certificate valuation of the Engineer; 1321 66(1)
The reference to the arbitrator may proceed (despite the Works not being complete or alleged complete) in the case of any dispute or difference arising as to:– 1346 66(2)
The Employer or the Contractor must refer to the Engineer any dispute as to:– 1320 66(1)
any decision opinion instruction direction certificate valuation of the Engineer; 1321 66(1)
The reference to the arbitrator may proceed (despite the Works not being complete or alleged complete) in the case of any dispute or difference arising as to:– 1346 66(2)
the withholding by the Engineer of any certificate to which the Contractor claims to be entitled 1348 66(2)
For guidance in this situation see Hudson's 10th Ed. p. 495-496 and Building & C.E. Standard Forms [1969] Wallace I. N. Duncan p. 453 (lines 10-18)
3
Engineer does delete or reduce any such sum in any interim certificate ?
NO
YES
The Engineer may during the progress of the Works order in writing:– 506 39(1)
There is a breach caused by the Engineer. The Employers remedy in this situation lies in Clause 59A(6). See also Clause 60(7)(b) wherein the Employer takes responsibility for sums which the Contractor is unable to recover despite his compliance with Clause 59B(6)

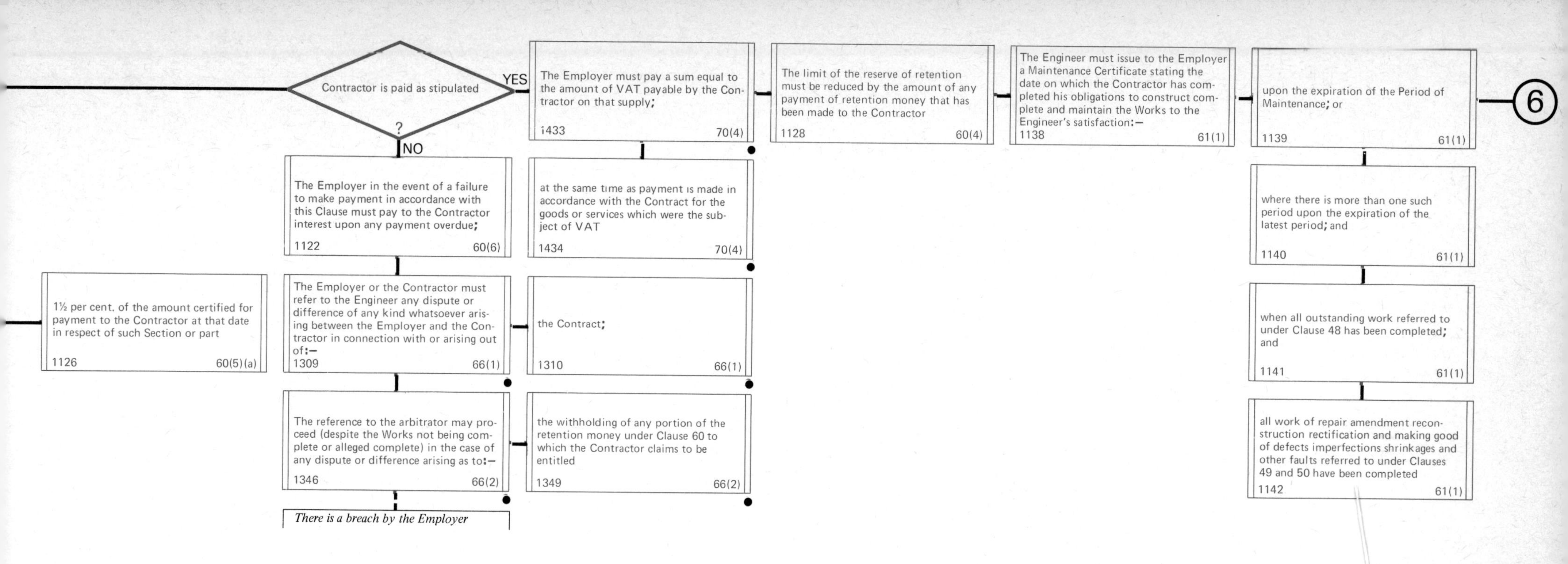

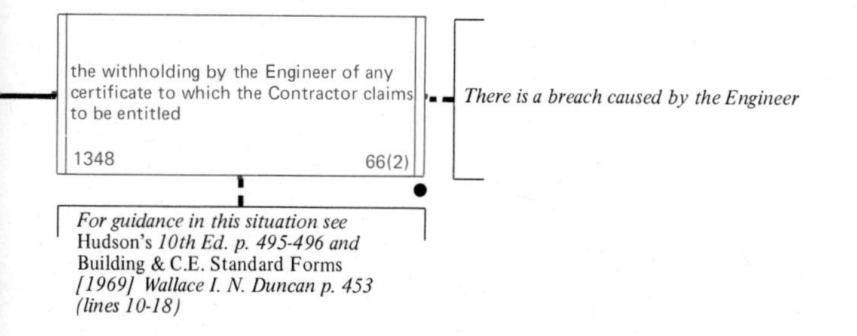

For guidance in this situation see Hudson's *10th Ed. p. 495-496 and* Building & C.E. Standard Forms *[1969] Wallace I. N. Duncan p. 453 (lines 10-18)*

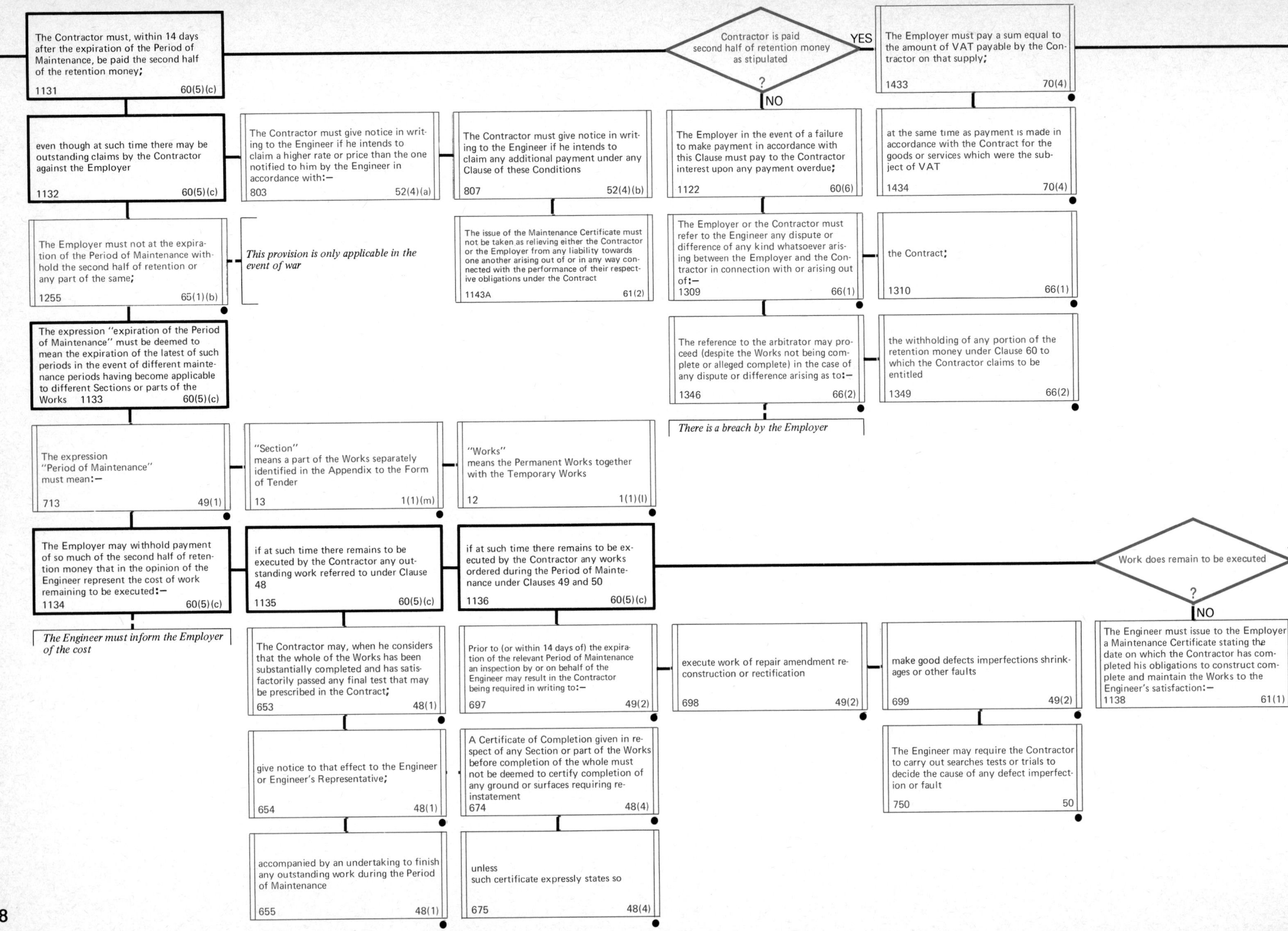
6
The Contractor must, within 14 days after the expiration of the Period of Maintenance, be paid the second half of the retention money;
1131 60(5)(c)
Contractor is paid second half of retention money as stipulated ?
YES
NO
The Employer must pay a sum equal to the amount of VAT payable by the Contractor on that supply;
1433 70(4)
even though at such time there may be outstanding claims by the Contractor against the Employer
1132 60(5)(c)
The Contractor must give notice in writing to the Engineer if he intends to claim a higher rate or price than the one notified to him by the Engineer in accordance with:—
803 52(4)(a)
The Contractor must give notice in writing to the Engineer if he intends to claim any additional payment under any Clause of these Conditions
807 52(4)(b)
The Employer in the event of a failure to make payment in accordance with this Clause must pay to the Contractor interest upon any payment overdue;
1122 60(6)
at the same time as payment is made in accordance with the Contract for the goods or services which were the subject of VAT
1434 70(4)
The Employer must not at the expiration of the Period of Maintenance withhold the second half of retention or any part of the same;
1255 65(1)(b)
This provision is only applicable in the event of war
The issue of the Maintenance Certificate must not be taken as relieving either the Contractor or the Employer from any liability towards one another arising out of or in any way connected with the performance of their respective obligations under the Contract
1143A 61(2)
The Employer or the Contractor must refer to the Engineer any dispute or difference of any kind whatsoever arising between the Employer and the Contractor in connection with or arising out of:—
1309 66(1)
the Contract;
1310 66(1)
The expression "expiration of the Period of Maintenance" must be deemed to mean the expiration of the latest of such periods in the event of different maintenance periods having become applicable to different Sections or parts of the Works 1133 60(5)(c)
The reference to the arbitrator may proceed (despite the Works not being complete or alleged complete) in the case of any dispute or difference arising as to:—
1346 66(2)
the withholding of any portion of the retention money under Clause 60 to which the Contractor claims to be entitled
1349 66(2)
There is a breach by the Employer
The expression "Period of Maintenance" must mean:—
713 49(1)
"Section" means a part of the Works separately identified in the Appendix to the Form of Tender
13 1(1)(m)
"Works" means the Permanent Works together with the Temporary Works
12 1(1)(l)
The Employer may withhold payment of so much of the second half of retention money that in the opinion of the Engineer represent the cost of work remaining to be executed:—
1134 60(5)(c)
if at such time there remains to be executed by the Contractor any outstanding work referred to under Clause 48
1135 60(5)(c)
if at such time there remains to be executed by the Contractor any works ordered during the Period of Maintenance under Clauses 49 and 50
1136 60(5)(c)
Work does remain to be executed ?
YES
NO
The Engineer must inform the Employer of the cost
The Contractor may, when he considers that the whole of the Works has been substantially completed and has satisfactorily passed any final test that may be prescribed in the Contract;
653 48(1)
Prior to (or within 14 days of) the expiration of the relevant Period of Maintenance an inspection by or on behalf of the Engineer may result in the Contractor being required in writing to:—
697 49(2)
execute work of repair amendment reconstruction or rectification
698 49(2)
make good defects imperfections shrinkages or other faults
699 49(2)
The Engineer must issue to the Employer a Maintenance Certificate stating the date on which the Contractor has completed his obligations to construct complete and maintain the Works to the Engineer's satisfaction:—
1138 61(1)
give notice to that effect to the Engineer or Engineer's Representative;
654 48(1)
A Certificate of Completion given in respect of any Section or part of the Works before completion of the whole must not be deemed to certify completion of any ground or surfaces requiring reinstatement
674 48(4)
The Engineer may require the Contractor to carry out searches tests or trials to decide the cause of any defect imperfection or fault
750 50
accompanied by an undertaking to finish any outstanding work during the Period of Maintenance
655 48(1)
unless such certificate expressly states so
675 48(4)

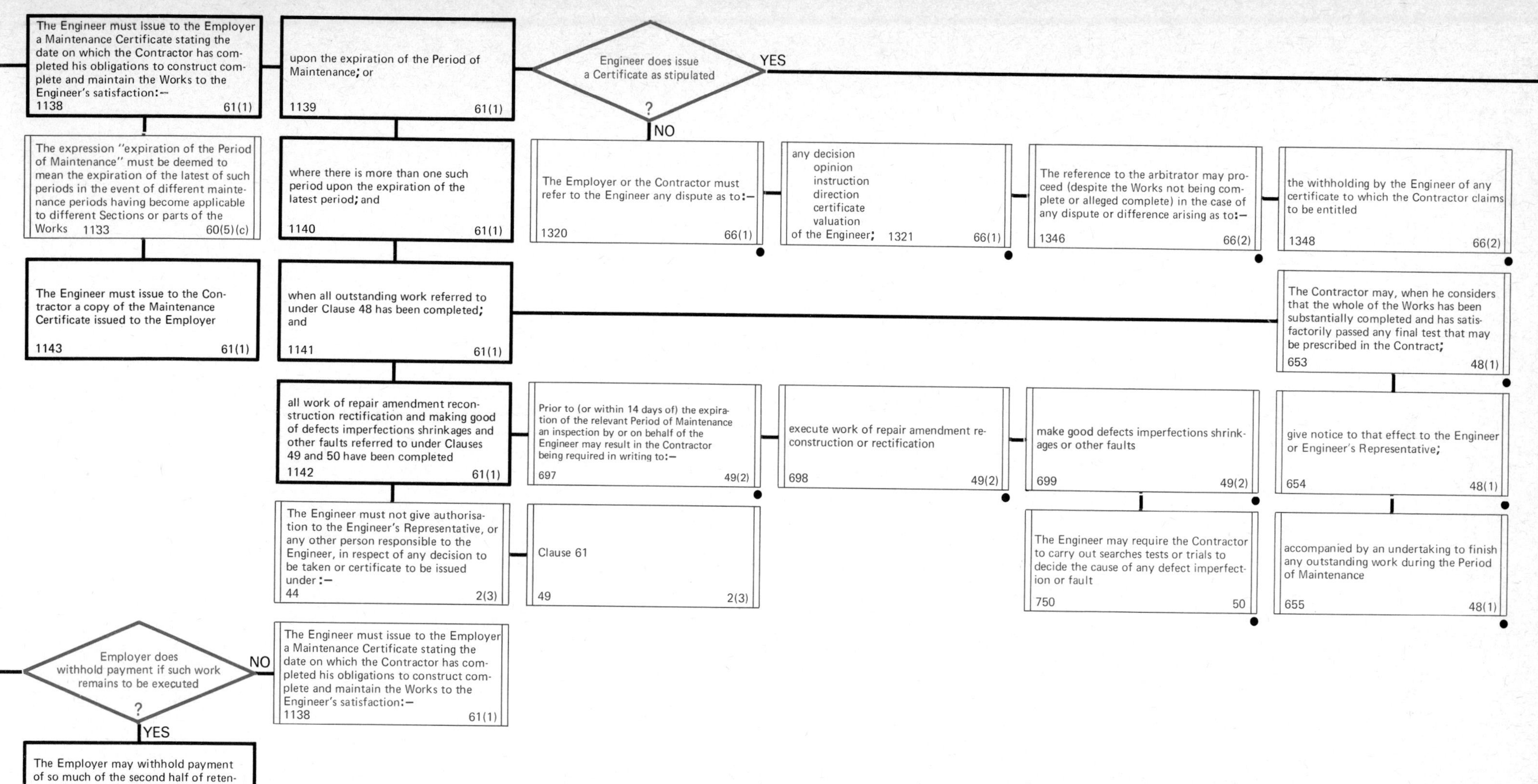
The Engineer must issue to the Employer a Maintenance Certificate stating the date on which the Contractor has completed his obligations to construct complete and maintain the Works to the Engineer's satisfaction:— 1138 61(1)
upon the expiration of the Period of Maintenance; or 1139 61(1)
Engineer does issue a Certificate as stipulated ?
YES
7
NO
The expression "expiration of the Period of Maintenance" must be deemed to mean the expiration of the latest of such periods in the event of different maintenance periods having become applicable to different Sections or parts of the Works 1133 60(5)(c)
where there is more than one such period upon the expiration of the latest period; and 1140 61(1)
The Employer or the Contractor must refer to the Engineer any dispute as to:— 1320 66(1)
any decision opinion instruction direction certificate valuation of the Engineer; 1321 66(1)
The reference to the arbitrator may proceed (despite the Works not being complete or alleged complete) in the case of any dispute or difference arising as to:— 1346 66(2)
the withholding by the Engineer of any certificate to which the Contractor claims to be entitled 1348 66(2)
The Engineer must issue to the Contractor a copy of the Maintenance Certificate issued to the Employer 1143 61(1)
when all outstanding work referred to under Clause 48 has been completed; and 1141 61(1)
The Contractor may, when he considers that the whole of the Works has been substantially completed and has satisfactorily passed any final test that may be prescribed in the Contract; 653 48(1)
all work of repair amendment reconstruction rectification and making good of defects imperfections shrinkages and other faults referred to under Clauses 49 and 50 have been completed 1142 61(1)
Prior to (or within 14 days of) the expiration of the relevant Period of Maintenance an inspection by or on behalf of the Engineer may result in the Contractor being required in writing to:— 697 49(2)
execute work of repair amendment reconstruction or rectification 698 49(2)
make good defects imperfections shrinkages or other faults 699 49(2)
give notice to that effect to the Engineer or Engineer's Representative; 654 48(1)
The Engineer must not give authorisation to the Engineer's Representative, or any other person responsible to the Engineer, in respect of any decision to be taken or certificate to be issued under:— 44 2(3)
Clause 61 49 2(3)
The Engineer may require the Contractor to carry out searches tests or trials to decide the cause of any defect imperfection or fault 750 50
accompanied by an undertaking to finish any outstanding work during the Period of Maintenance 655 48(1)
Employer does withhold payment if such work remains to be executed ?
NO
YES
The Engineer must issue to the Employer a Maintenance Certificate stating the date on which the Contractor has completed his obligations to construct complete and maintain the Works to the Engineer's satisfaction:— 1138 61(1)
The Employer may withhold payment of so much of the second half of retention money that in the opinion of the Engineer represent the cost of the works so remaining to be executed 1137 60(5)(c)

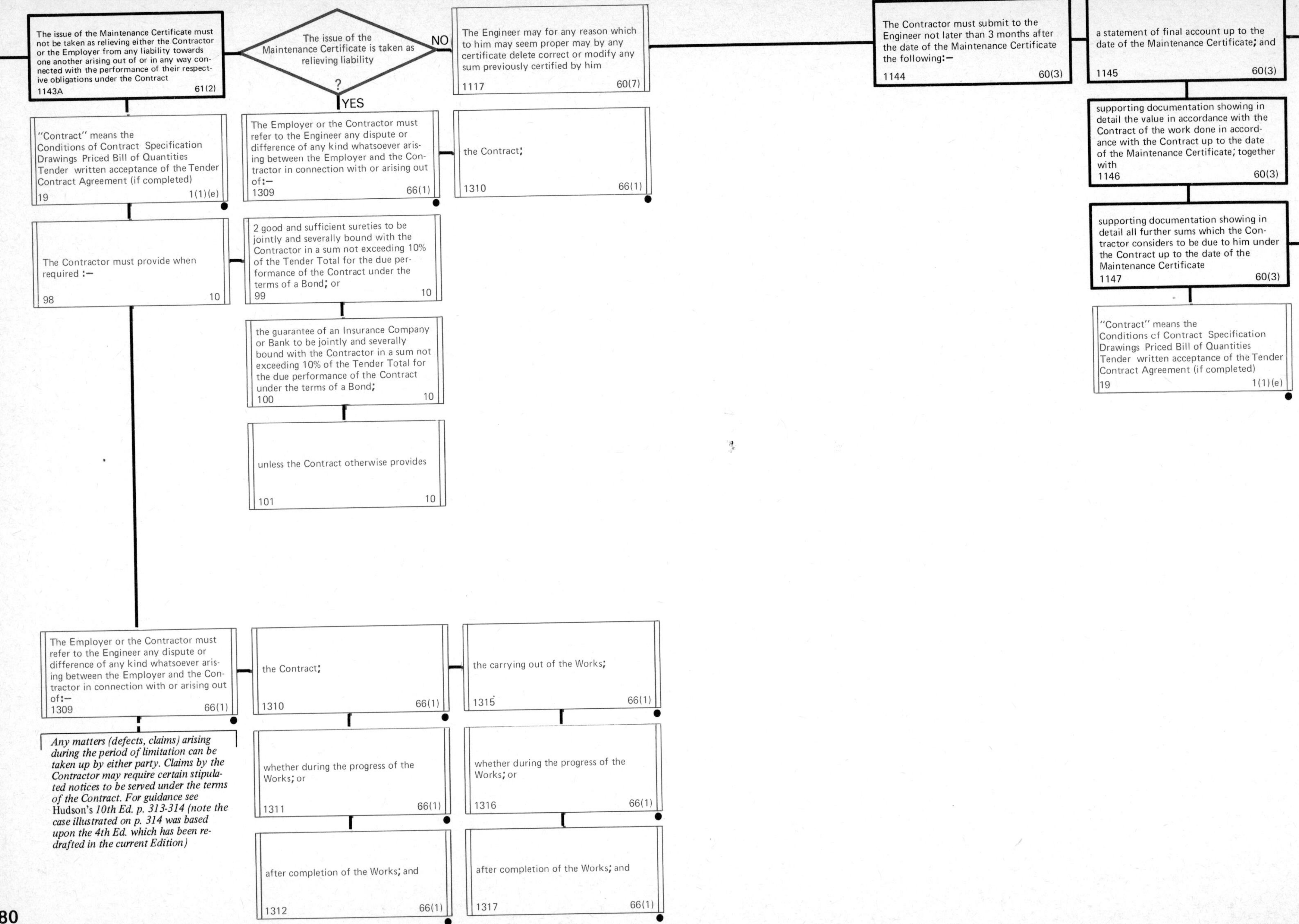
7
The issue of the Maintenance Certificate must not be taken as relieving either the Contractor or the Employer from any liability towards one another arising out of or in any way connected with the performance of their respective obligations under the Contract
1143A 61(2)
The issue of the Maintenance Certificate is taken as relieving liability ?
NO
YES
The Engineer may for any reason which to him may seem proper may by any certificate delete correct or modify any sum previously certified by him
1117 60(7)
"Contract" means the Conditions of Contract Specification Drawings Priced Bill of Quantities Tender written acceptance of the Tender Contract Agreement (if completed)
19 1(1)(e)
The Employer or the Contractor must refer to the Engineer any dispute or difference of any kind whatsoever arising between the Employer and the Contractor in connection with or arising out of:—
1309 66(1)
the Contract;
1310 66(1)
The Contractor must provide when required :—
98 10
2 good and sufficient sureties to be jointly and severally bound with the Contractor in a sum not exceeding 10% of the Tender Total for the due performance of the Contract under the terms of a Bond; or
99 10
the guarantee of an Insurance Company or Bank to be jointly and severally bound with the Contractor in a sum not exceeding 10% of the Tender Total for the due performance of the Contract under the terms of a Bond;
100 10
unless the Contract otherwise provides
101 10
The Employer or the Contractor must refer to the Engineer any dispute or difference of any kind whatsoever arising between the Employer and the Contractor in connection with or arising out of:—
1309 66(1)
the Contract;
1310 66(1)
the carrying out of the Works;
1315 66(1)
whether during the progress of the Works; or
1311 66(1)
whether during the progress of the Works; or
1316 66(1)
after completion of the Works; and
1312 66(1)
after completion of the Works; and
1317 66(1)
Any matters (defects, claims) arising during the period of limitation can be taken up by either party. Claims by the Contractor may require certain stipulated notices to be served under the terms of the Contract. For guidance see Hudson's 10th Ed. p. 313-314 (note the case illustrated on p. 314 was based upon the 4th Ed. which has been re-drafted in the current Edition)
FINAL ACCOUNT
The Contractor must submit to the Engineer not later than 3 months after the date of the Maintenance Certificate the following:—
1144 60(3)
a statement of final account up to the date of the Maintenance Certificate; and
1145 60(3)
supporting documentation showing in detail the value in accordance with the Contract of the work done in accordance with the Contract up to the date of the Maintenance Certificate; together with
1146 60(3)
supporting documentation showing in detail all further sums which the Contractor considers to be due to him under the Contract up to the date of the Maintenance Certificate
1147 60(3)
"Contract" means the Conditions of Contract Specification Drawings Priced Bill of Quantities Tender written acceptance of the Tender Contract Agreement (if completed)
19 1(1)(e)

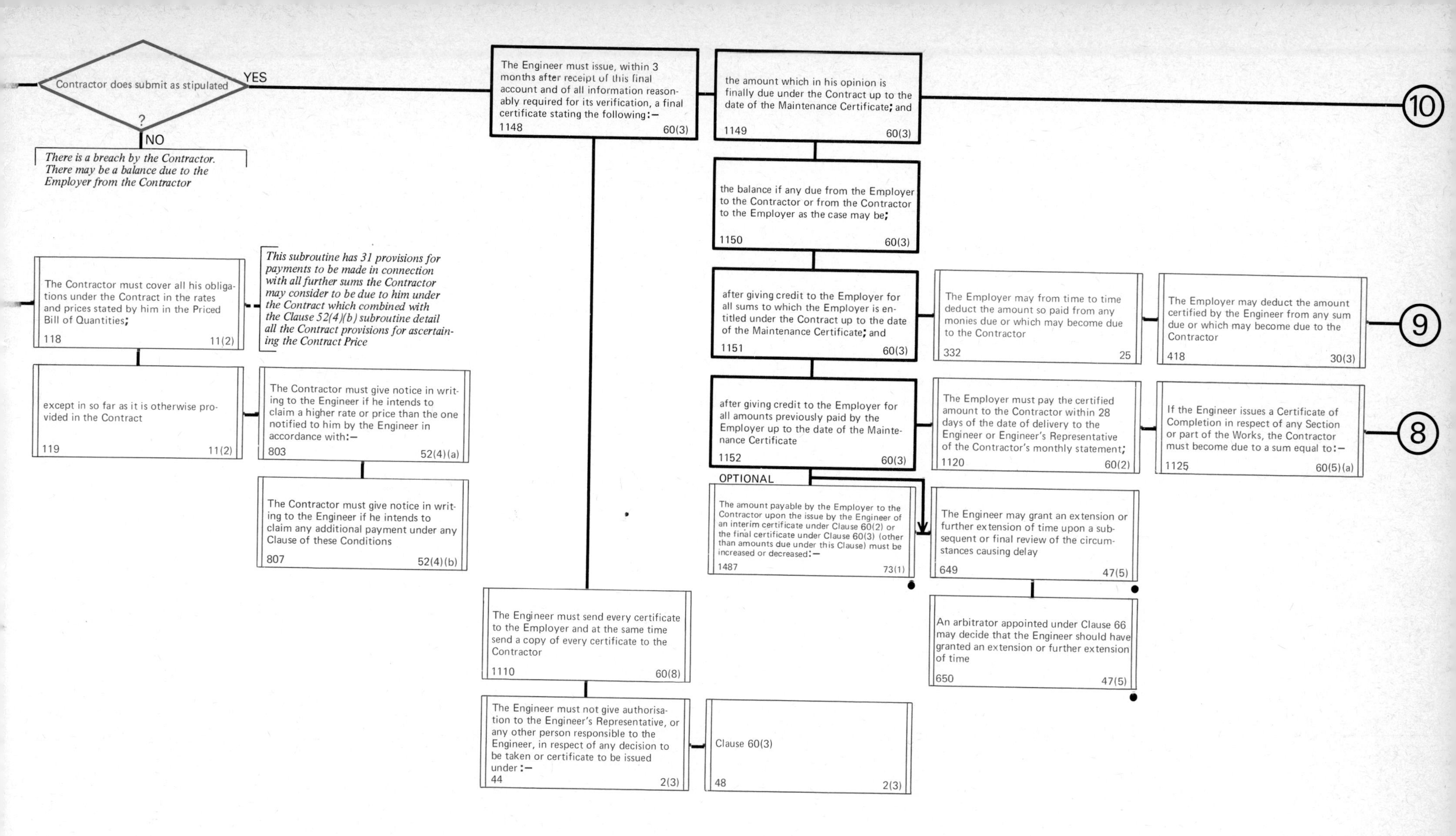
Contractor does submit as stipulated
?
YES
NO
There is a breach by the Contractor. There may be a balance due to the Employer from the Contractor
The Engineer must issue, within 3 months after receipt of this final account and of all information reasonably required for its verification, a final certificate stating the following:—
1148 60(3)
the amount which in his opinion is finally due under the Contract up to the date of the Maintenance Certificate; and
1149 60(3)
10
the balance if any due from the Employer to the Contractor or from the Contractor to the Employer as the case may be;
1150 60(3)
The Contractor must cover all his obligations under the Contract in the rates and prices stated by him in the Priced Bill of Quantities;
118 11(2)
This subroutine has 31 provisions for payments to be made in connection with all further sums the Contractor may consider to be due to him under the Contract which combined with the Clause 52(4)(b) subroutine detail all the Contract provisions for ascertaining the Contract Price
after giving credit to the Employer for all sums to which the Employer is entitled under the Contract up to the date of the Maintenance Certificate; and
1151 60(3)
The Employer may from time to time deduct the amount so paid from any monies due or which may become due to the Contractor
332 25
The Employer may deduct the amount certified by the Engineer from any sum due or which may become due to the Contractor
418 30(3)
9
except in so far as it is otherwise provided in the Contract
119 11(2)
The Contractor must give notice in writing to the Engineer if he intends to claim a higher rate or price than the one notified to him by the Engineer in accordance with:—
803 52(4)(a)
after giving credit to the Employer for all amounts previously paid by the Employer up to the date of the Maintenance Certificate
1152 60(3)
The Employer must pay the certified amount to the Contractor within 28 days of the date of delivery to the Engineer or Engineer's Representative of the Contractor's monthly statement;
1120 60(2)
If the Engineer issues a Certificate of Completion in respect of any Section or part of the Works, the Contractor must become due to a sum equal to:—
1125 60(5)(a)
8
OPTIONAL
The Contractor must give notice in writing to the Engineer if he intends to claim any additional payment under any Clause of these Conditions
807 52(4)(b)
The amount payable by the Employer to the Contractor upon the issue by the Engineer of an interim certificate under Clause 60(2) or the final certificate under Clause 60(3) (other than amounts due under this Clause) must be increased or decreased:—
1487 73(1)
The Engineer may grant an extension or further extension of time upon a subsequent or final review of the circumstances causing delay
649 47(5)
The Engineer must send every certificate to the Employer and at the same time send a copy of every certificate to the Contractor
1110 60(8)
An arbitrator appointed under Clause 66 may decide that the Engineer should have granted an extension or further extension of time
650 47(5)
The Engineer must not give authorisation to the Engineer's Representative, or any other person responsible to the Engineer, in respect of any decision to be taken or certificate to be issued under:—
44 2(3)
Clause 60(3)
48 2(3)

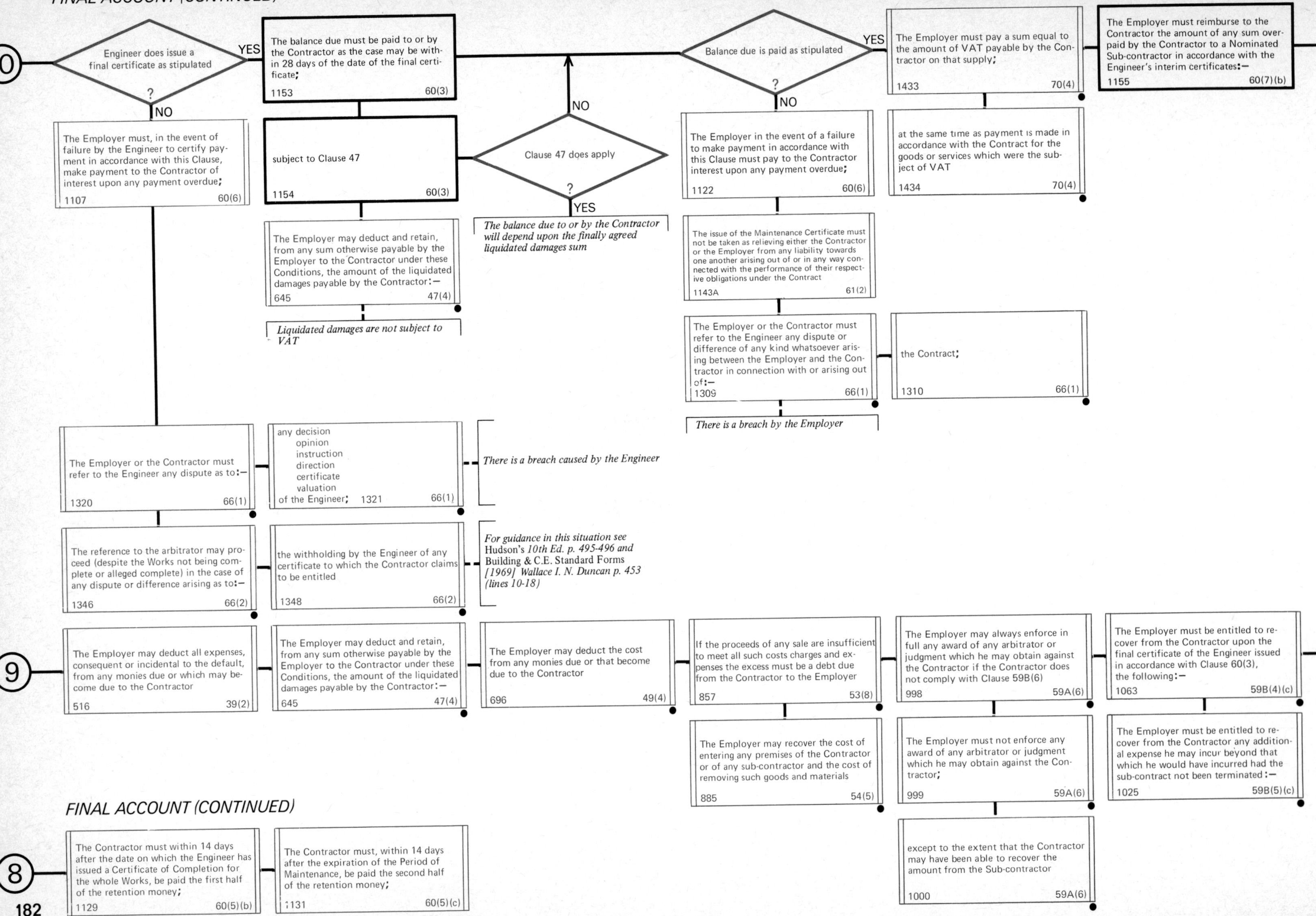
10
Engineer does issue a final certificate as stipulated
YES
NO
?
The balance due must be paid to or by the Contractor as the case may be within 28 days of the date of the final certificate;
1153 60(3)
subject to Clause 47
1154 60(3)
Clause 47 does apply
NO
?
YES
The balance due to or by the Contractor will depend upon the finally agreed liquidated damages sum
Balance due is paid as stipulated
?
YES
NO
The Employer must pay a sum equal to the amount of VAT payable by the Contractor on that supply;
1433 70(4)
The Employer must reimburse to the Contractor the amount of any sum overpaid by the Contractor to a Nominated Sub-contractor in accordance with the Engineer's interim certificates:—
1155 60(7)(b)
The Employer must, in the event of failure by the Engineer to certify payment in accordance with this Clause, make payment to the Contractor of interest upon any payment overdue;
1107 60(6)
The Employer may deduct and retain, from any sum otherwise payable by the Employer to the Contractor under these Conditions, the amount of the liquidated damages payable by the Contractor:—
645 47(4)
Liquidated damages are not subject to VAT
The Employer in the event of a failure to make payment in accordance with this Clause must pay to the Contractor interest upon any payment overdue;
1122 60(6)
at the same time as payment is made in accordance with the Contract for the goods or services which were the subject of VAT
1434 70(4)
The issue of the Maintenance Certificate must not be taken as relieving either the Contractor or the Employer from any liability towards one another arising out of or in any way connected with the performance of their respective obligations under the Contract
1143A 61(2)
The Employer or the Contractor must refer to the Engineer any dispute or difference of any kind whatsoever arising between the Employer and the Contractor in connection with or arising out of:—
1309 66(1)
the Contract;
1310 66(1)
There is a breach by the Employer
The Employer or the Contractor must refer to the Engineer any dispute as to:—
1320 66(1)
any decision
opinion
instruction
direction
certificate
valuation
of the Engineer; 1321 66(1)
There is a breach caused by the Engineer
The reference to the arbitrator may proceed (despite the Works not being complete or alleged complete) in the case of any dispute or difference arising as to:—
1346 66(2)
the withholding by the Engineer of any certificate to which the Contractor claims to be entitled
1348 66(2)
For guidance in this situation see Hudson's 10th Ed. p. 495-496 and Building & C.E. Standard Forms [1969] Wallace I. N. Duncan p. 453 (lines 10-18)
9
The Employer may deduct all expenses, consequent or incidental to the default, from any monies due or which may become due to the Contractor
516 39(2)
The Employer may deduct and retain, from any sum otherwise payable by the Employer to the Contractor under these Conditions, the amount of the liquidated damages payable by the Contractor:—
645 47(4)
The Employer may deduct the cost from any monies due or that become due to the Contractor
696 49(4)
If the proceeds of any sale are insufficient to meet all such costs charges and expenses the excess must be a debt due from the Contractor to the Employer
857 53(8)
The Employer may always enforce in full any award of any arbitrator or judgment which he may obtain against the Contractor if the Contractor does not comply with Clause 59B(6)
998 59A(6)
The Employer must be entitled to recover from the Contractor upon the final certificate of the Engineer issued in accordance with Clause 60(3), the following:—
1063 59B(4)(c)
The Employer may recover the cost of entering any premises of the Contractor or of any sub-contractor and the cost of removing such goods and materials
885 54(5)
The Employer must not enforce any award of any arbitrator or judgment which he may obtain against the Contractor;
999 59A(6)
The Employer must be entitled to recover from the Contractor any additional expense he may incur beyond that which he would have incurred had the sub-contract not been terminated:—
1025 59B(5)(c)
except to the extent that the Contractor may have been able to recover the amount from the Sub-contractor
1000 59A(6)
FINAL ACCOUNT (CONTINUED)
8
The Contractor must within 14 days after the date on which the Engineer has issued a Certificate of Completion for the whole Works, be paid the first half of the retention money;
1129 60(5)(b)
The Contractor must, within 14 days after the expiration of the Period of Maintenance, be paid the second half of the retention money;
1131 60(5)(c)

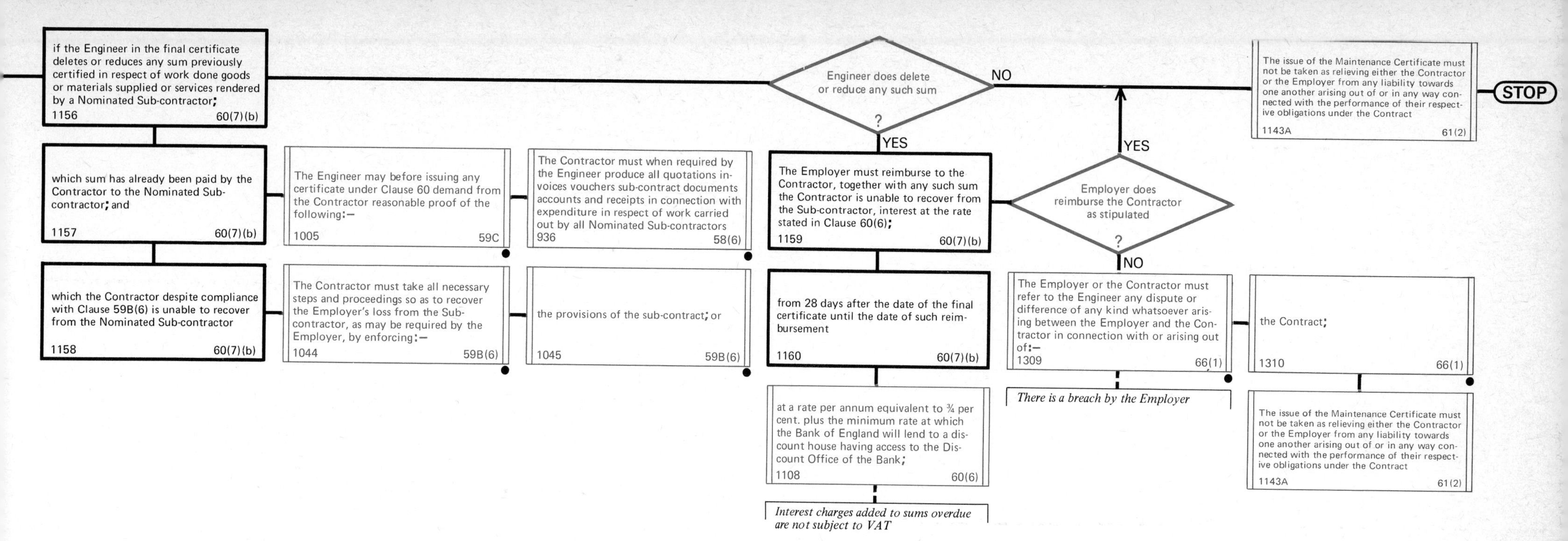

if the Engineer in the final certificate deletes or reduces any sum previously certified in respect of work done goods or materials supplied or services rendered by a Nominated Sub-contractor;
1156 60(7)(b)
Engineer does delete or reduce any such sum ?
NO
YES
The issue of the Maintenance Certificate must not be taken as relieving either the Contractor or the Employer from any liability towards one another arising out of or in any way connected with the performance of their respective obligations under the Contract
1143A 61(2)
STOP
which sum has already been paid by the Contractor to the Nominated Sub-contractor; and
1157 60(7)(b)
The Engineer may before issuing any certificate under Clause 60 demand from the Contractor reasonable proof of the following:—
1005 59C
The Contractor must when required by the Engineer produce all quotations invoices vouchers sub-contract documents accounts and receipts in connection with expenditure in respect of work carried out by all Nominated Sub-contractors
936 58(6)
The Employer must reimburse to the Contractor, together with any such sum the Contractor is unable to recover from the Sub-contractor, interest at the rate stated in Clause 60(6);
1159 60(7)(b)
Employer does reimburse the Contractor as stipulated ?
YES
NO
which the Contractor despite compliance with Clause 59B(6) is unable to recover from the Nominated Sub-contractor
1158 60(7)(b)
The Contractor must take all necessary steps and proceedings so as to recover the Employer's loss from the Sub-contractor, as may be required by the Employer, by enforcing:—
1044 59B(6)
the provisions of the sub-contract; or
1045 59B(6)
from 28 days after the date of the final certificate until the date of such reimbursement
1160 60(7)(b)
The Employer or the Contractor must refer to the Engineer any dispute or difference of any kind whatsoever arising between the Employer and the Contractor in connection with or arising out of:—
1309 66(1)
the Contract;
1310 66(1)
There is a breach by the Employer
at a rate per annum equivalent to ¾ per cent. plus the minimum rate at which the Bank of England will lend to a discount house having access to the Discount Office of the Bank;
1108 60(6)
The issue of the Maintenance Certificate must not be taken as relieving either the Contractor or the Employer from any liability towards one another arising out of or in any way connected with the performance of their respective obligations under the Contract
1143A 61(2)
Interest charges added to sums overdue are not subject to VAT

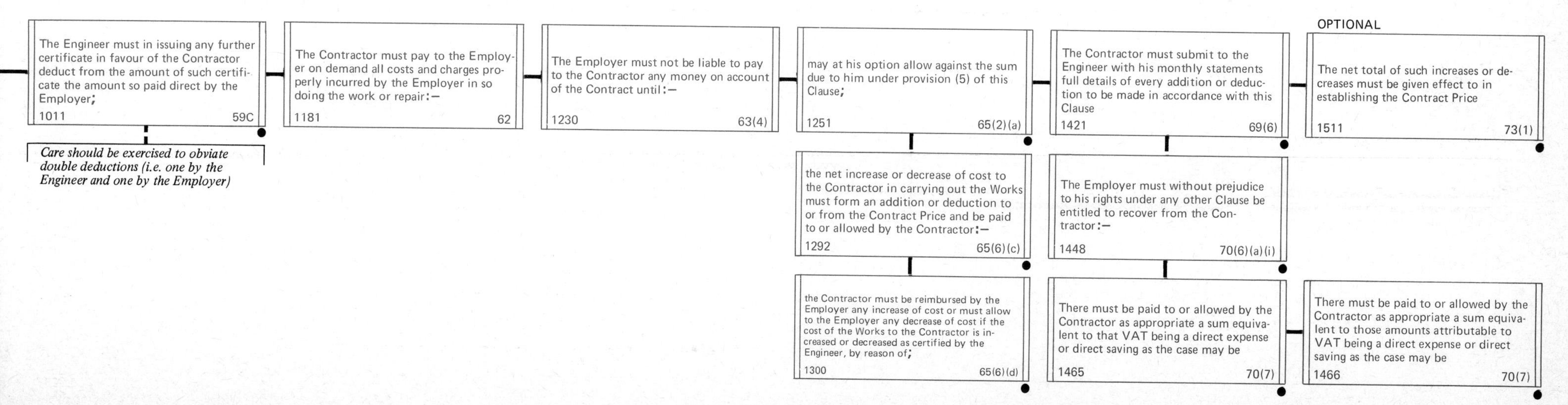

The Engineer must in issuing any further certificate in favour of the Contractor deduct from the amount of such certificate the amount so paid direct by the Employer;
1011 59C
Care should be exercised to obviate double deductions (i.e. one by the Engineer and one by the Employer)
The Contractor must pay to the Employer on demand all costs and charges properly incurred by the Employer in so doing the work or repair:—
1181 62
The Employer must not be liable to pay to the Contractor any money on account of the Contract until:—
1230 63(4)
may at his option allow against the sum due to him under provision (5) of this Clause;
1251 65(2)(a)
The Contractor must submit to the Engineer with his monthly statements full details of every addition or deduction to be made in accordance with this Clause
1421 69(6)
OPTIONAL
The net total of such increases or decreases must be given effect to in establishing the Contract Price
1511 73(1)
the net increase or decrease of cost to the Contractor in carrying out the Works must form an addition or deduction to or from the Contract Price and be paid to or allowed by the Contractor:—
1292 65(6)(c)
The Employer must without prejudice to his rights under any other Clause be entitled to recover from the Contractor:—
1448 70(6)(a)(i)
the Contractor must be reimbursed by the Employer any increase of cost or must allow to the Employer any decrease of cost if the cost of the Works to the Contractor is increased or decreased as certified by the Engineer, by reason of;
1300 65(6)(d)
There must be paid to or allowed by the Contractor as appropriate a sum equivalent to that VAT being a direct expense or direct saving as the case may be
1465 70(7)
There must be paid to or allowed by the Contractor as appropriate a sum equivalent to those amounts attributable to VAT being a direct expense or direct saving as the case may be
1466 70(7)

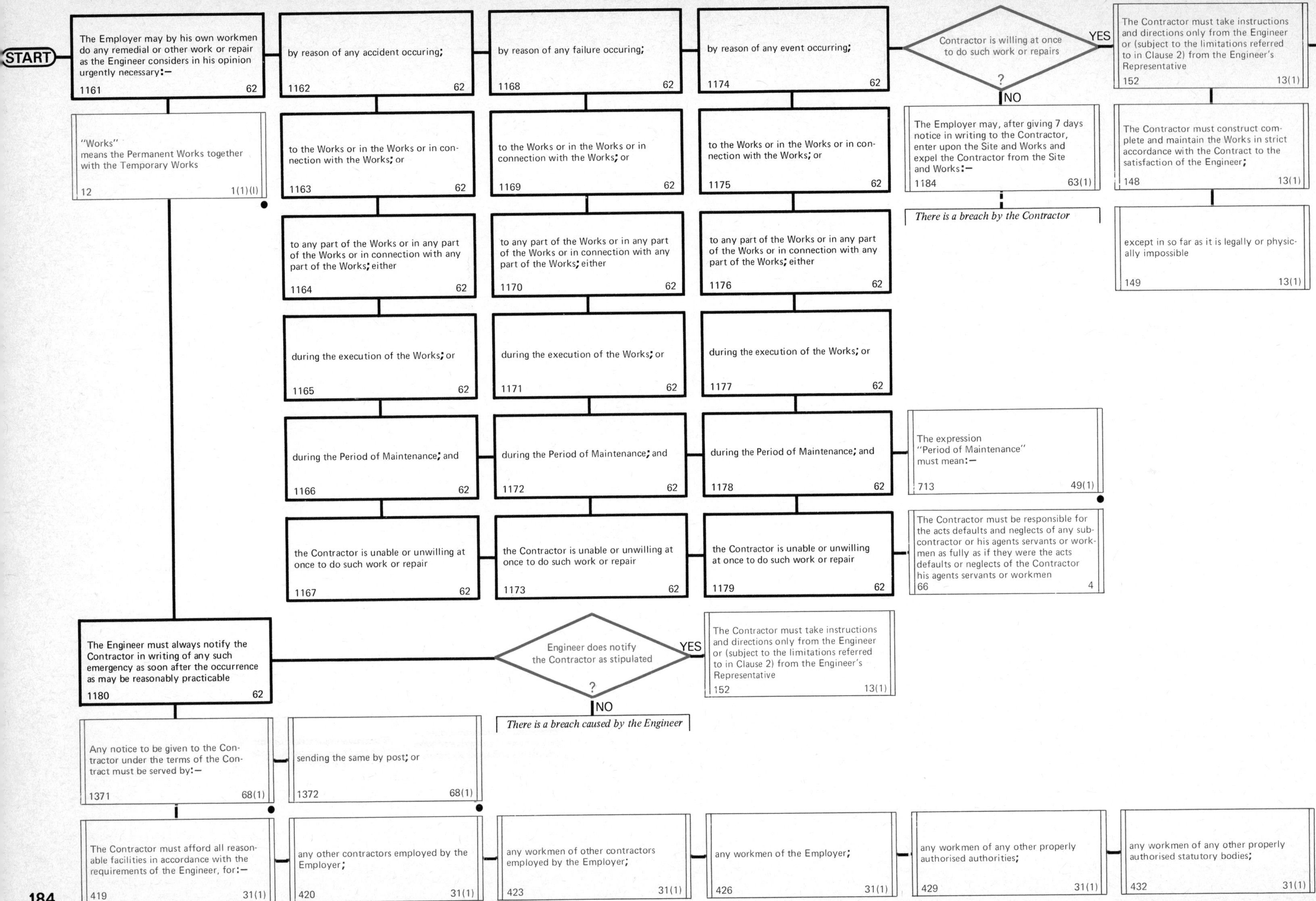
START
The Employer may by his own workmen do any remedial or other work or repair as the Engineer considers in his opinion urgently necessary:— 1161 62
by reason of any accident occuring; 1162 62
by reason of any failure occuring; 1168 62
by reason of any event occurring; 1174 62
Contractor is willing at once to do such work or repairs ?
YES
NO
The Contractor must take instructions and directions only from the Engineer or (subject to the limitations referred to in Clause 2) from the Engineer's Representative 152 13(1)
"Works" means the Permanent Works together with the Temporary Works 12 1(1)(l)
to the Works or in the Works or in connection with the Works; or 1163 62
to the Works or in the Works or in connection with the Works; or 1169 62
to the Works or in the Works or in connection with the Works; or 1175 62
The Employer may, after giving 7 days notice in writing to the Contractor, enter upon the Site and Works and expel the Contractor from the Site and Works:— 1184 63(1)
There is a breach by the Contractor
The Contractor must construct complete and maintain the Works in strict accordance with the Contract to the satisfaction of the Engineer; 148 13(1)
to any part of the Works or in any part of the Works or in connection with any part of the Works; either 1164 62
to any part of the Works or in any part of the Works or in connection with any part of the Works; either 1170 62
to any part of the Works or in any part of the Works or in connection with any part of the Works; either 1176 62
except in so far as it is legally or physically impossible 149 13(1)
during the execution of the Works; or 1165 62
during the execution of the Works; or 1171 62
during the execution of the Works; or 1177 62
during the Period of Maintenance; and 1166 62
during the Period of Maintenance; and 1172 62
during the Period of Maintenance; and 1178 62
The expression "Period of Maintenance" must mean:— 713 49(1)
the Contractor is unable or unwilling at once to do such work or repair 1167 62
the Contractor is unable or unwilling at once to do such work or repair 1173 62
the Contractor is unable or unwilling at once to do such work or repair 1179 62
The Contractor must be responsible for the acts defaults and neglects of any sub-contractor or his agents servants or workmen as fully as if they were the acts defaults or neglects of the Contractor his agents servants or workmen 66 4
The Engineer must always notify the Contractor in writing of any such emergency as soon after the occurrence as may be reasonably practicable 1180 62
Engineer does notify the Contractor as stipulated ?
YES
NO
The Contractor must take instructions and directions only from the Engineer or (subject to the limitations referred to in Clause 2) from the Engineer's Representative 152 13(1)
There is a breach caused by the Engineer
Any notice to be given to the Contractor under the terms of the Contract must be served by:— 1371 68(1)
sending the same by post; or 1372 68(1)
The Contractor must afford all reasonable facilities in accordance with the requirements of the Engineer, for:— 419 31(1)
any other contractors employed by the Employer; 420 31(1)
any workmen of other contractors employed by the Employer; 423 31(1)
any workmen of the Employer; 426 31(1)
any workmen of any other properly authorised authorities; 429 31(1)
any workmen of any other properly authorised statutory bodies; 432 31(1)

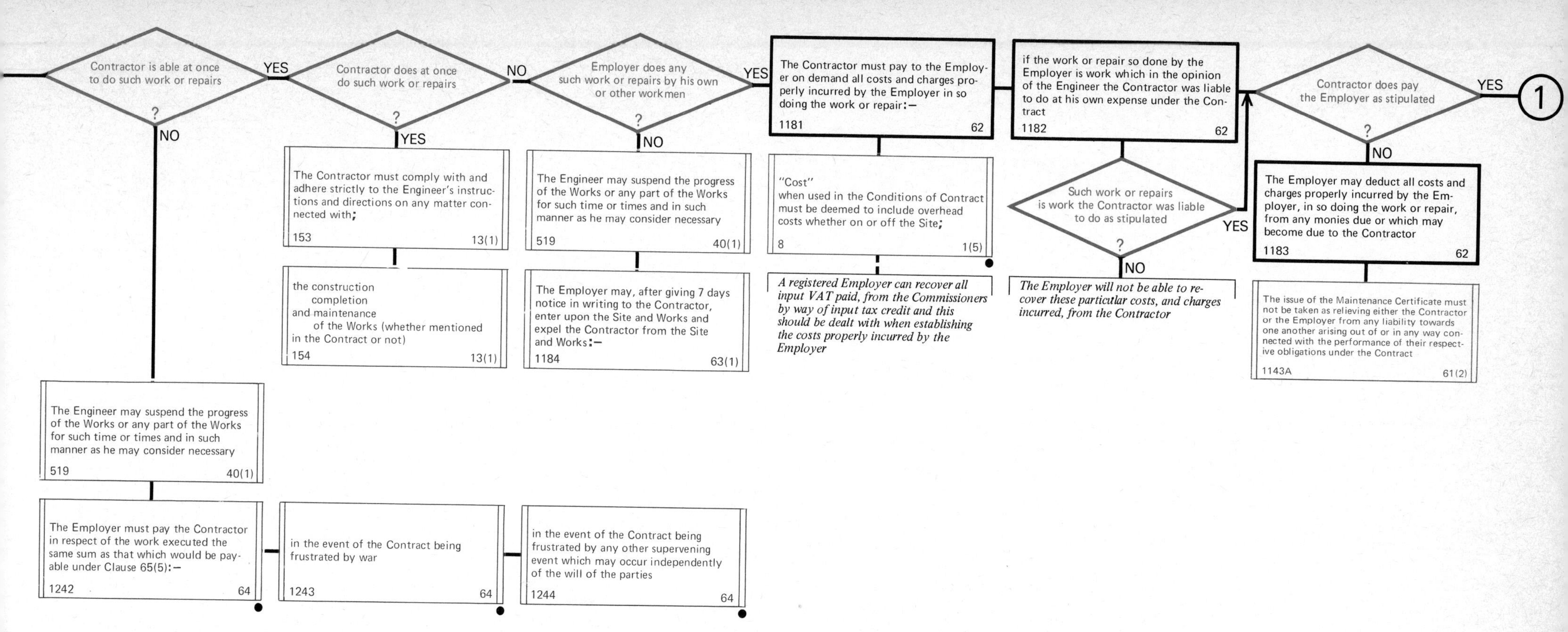

Contractor is able at once to do such work or repairs ?
YES
NO
Contractor does at once do such work or repairs ?
NO
YES
Employer does any such work or repairs by his own or other workmen ?
YES
NO
The Contractor must pay to the Employer on demand all costs and charges properly incurred by the Employer in so doing the work or repair:–
1181 62
if the work or repair so done by the Employer is work which in the opinion of the Engineer the Contractor was liable to do at his own expense under the Contract
1182 62
Contractor does pay the Employer as stipulated ?
YES
NO
1
The Contractor must comply with and adhere strictly to the Engineer's instructions and directions on any matter connected with;
153 13(1)
the construction completion and maintenance of the Works (whether mentioned in the Contract or not)
154 13(1)
The Engineer may suspend the progress of the Works or any part of the Works for such time or times and in such manner as he may consider necessary
519 40(1)
The Employer may, after giving 7 days notice in writing to the Contractor, enter upon the Site and Works and expel the Contractor from the Site and Works:–
1184 63(1)
"Cost" when used in the Conditions of Contract must be deemed to include overhead costs whether on or off the Site;
8 1(5)
A registered Employer can recover all input VAT paid, from the Commissioners by way of input tax credit and this should be dealt with when establishing the costs properly incurred by the Employer
Such work or repairs is work the Contractor was liable to do as stipulated ?
YES
NO
The Employer will not be able to recover these particular costs, and charges incurred, from the Contractor
The Employer may deduct all costs and charges properly incurred by the Employer, in so doing the work or repair, from any monies due or which may become due to the Contractor
1183 62
The issue of the Maintenance Certificate must not be taken as relieving either the Contractor or the Employer from any liability towards one another arising out of or in any way connected with the performance of their respective obligations under the Contract
1143A 61(2)
The Engineer may suspend the progress of the Works or any part of the Works for such time or times and in such manner as he may consider necessary
519 40(1)
The Employer must pay the Contractor in respect of the work executed the same sum as that which would be payable under Clause 65(5):–
1242 64
in the event of the Contract being frustrated by war
1243 64
in the event of the Contract being frustrated by any other supervening event which may occur independently of the will of the parties
1244 64

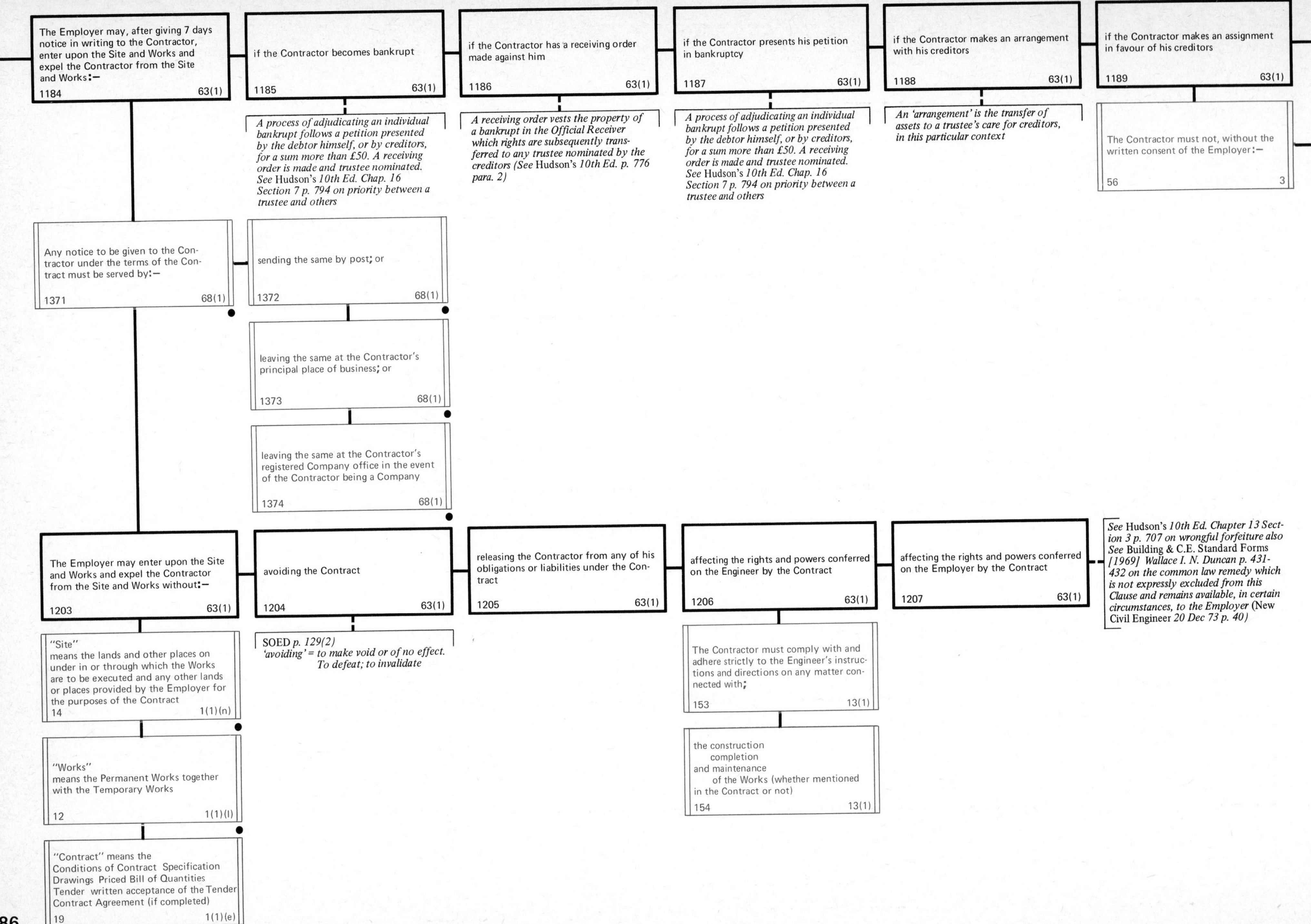
1
The Employer may, after giving 7 days notice in writing to the Contractor, enter upon the Site and Works and expel the Contractor from the Site and Works:— 1184 63(1)
if the Contractor becomes bankrupt 1185 63(1)
if the Contractor has a receiving order made against him 1186 63(1)
if the Contractor presents his petition in bankruptcy 1187 63(1)
if the Contractor makes an arrangement with his creditors 1188 63(1)
if the Contractor makes an assignment in favour of his creditors 1189 63(1)
A process of adjudicating an individual bankrupt follows a petition presented by the debtor himself, or by creditors, for a sum more than £50. A receiving order is made and trustee nominated. See Hudson's 10th Ed. Chap. 16 Section 7 p. 794 on priority between a trustee and others
A receiving order vests the property of a bankrupt in the Official Receiver which rights are subsequently transferred to any trustee nominated by the creditors (See Hudson's 10th Ed. p. 776 para. 2)
A process of adjudicating an individual bankrupt follows a petition presented by the debtor himself, or by creditors, for a sum more than £50. A receiving order is made and trustee nominated. See Hudson's 10th Ed. Chap. 16 Section 7 p. 794 on priority between a trustee and others
An 'arrangement' is the transfer of assets to a trustee's care for creditors, in this particular context
The Contractor must not, without the written consent of the Employer:— 56 3
Any notice to be given to the Contractor under the terms of the Contract must be served by:— 1371 68(1)
sending the same by post; or 1372 68(1)
leaving the same at the Contractor's principal place of business; or 1373 68(1)
leaving the same at the Contractor's registered Company office in the event of the Contractor being a Company 1374 68(1)
The Employer may enter upon the Site and Works and expel the Contractor from the Site and Works without:— 1203 63(1)
avoiding the Contract 1204 63(1)
releasing the Contractor from any of his obligations or liabilities under the Contract 1205 63(1)
affecting the rights and powers conferred on the Engineer by the Contract 1206 63(1)
affecting the rights and powers conferred on the Employer by the Contract 1207 63(1)
See Hudson's 10th Ed. Chapter 13 Section 3 p. 707 on wrongful forfeiture also See Building & C.E. Standard Forms [1969] Wallace I. N. Duncan p. 431-432 on the common law remedy which is not expressly excluded from this Clause and remains available, in certain circumstances, to the Employer (New Civil Engineer 20 Dec 73 p. 40)
SOED p. 129(2) 'avoiding' = to make void or of no effect. To defeat; to invalidate
"Site" means the lands and other places on under in or through which the Works are to be executed and any other lands or places provided by the Employer for the purposes of the Contract 14 1(1)(n)
"Works" means the Permanent Works together with the Temporary Works 12 1(1)(l)
"Contract" means the Conditions of Contract Specification Drawings Priced Bill of Quantities Tender written acceptance of the Tender Contract Agreement (if completed) 19 1(1)(e)
The Contractor must comply with and adhere strictly to the Engineer's instructions and directions on any matter connected with; 153 13(1)
the construction completion and maintenance of the Works (whether mentioned in the Contract or not) 154 13(1)

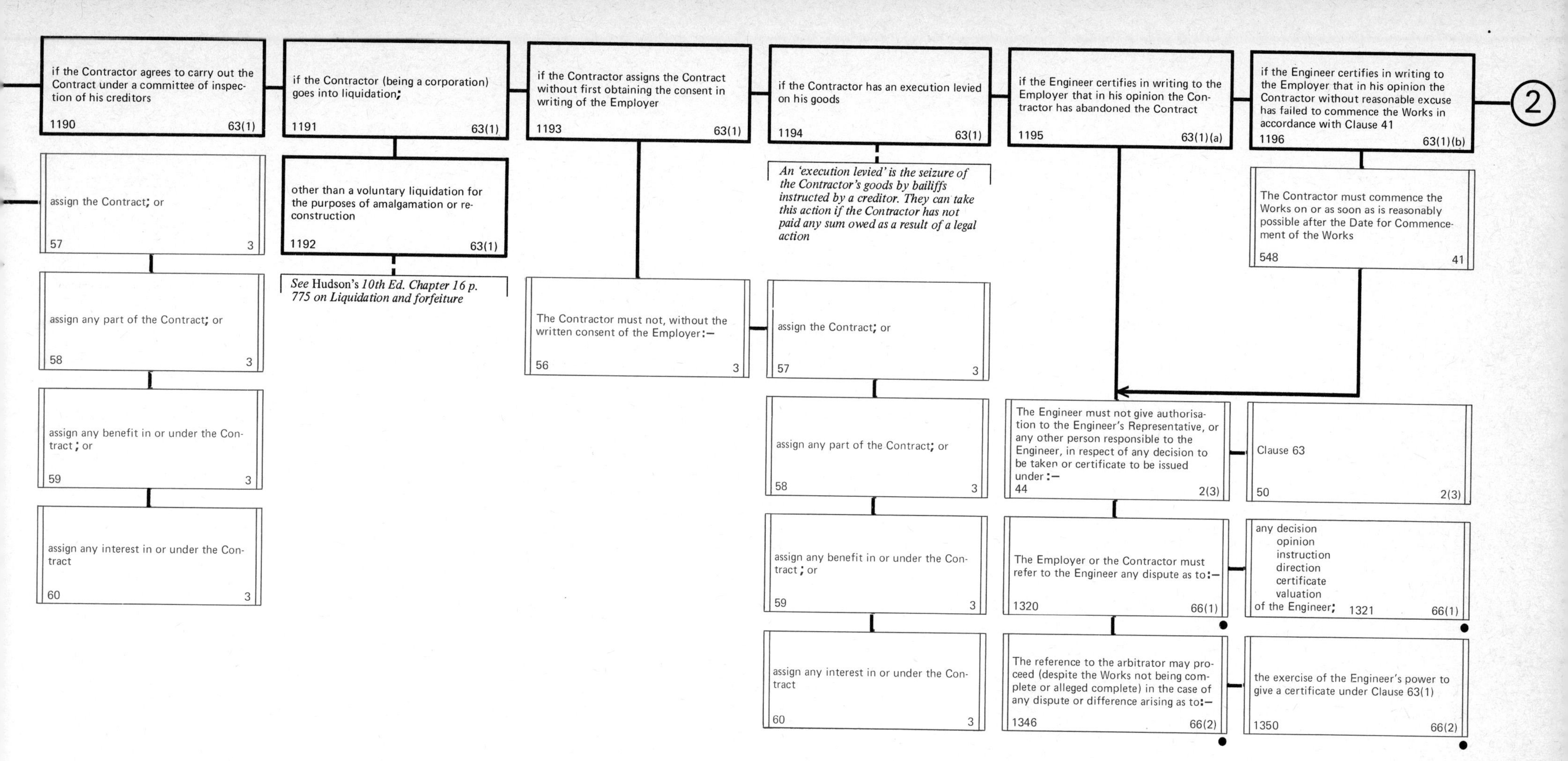
if the Contractor agrees to carry out the Contract under a committee of inspection of his creditors
1190 63(1)
if the Contractor (being a corporation) goes into liquidation;
1191 63(1)
if the Contractor assigns the Contract without first obtaining the consent in writing of the Employer
1193 63(1)
if the Contractor has an execution levied on his goods
1194 63(1)
if the Engineer certifies in writing to the Employer that in his opinion the Contractor has abandoned the Contract
1195 63(1)(a)
if the Engineer certifies in writing to the Employer that in his opinion the Contractor without reasonable excuse has failed to commence the Works in accordance with Clause 41
1196 63(1)(b)
2
assign the Contract; or
57 3
other than a voluntary liquidation for the purposes of amalgamation or reconstruction
1192 63(1)
See Hudson's 10th Ed. Chapter 16 p. 775 on Liquidation and forfeiture
An 'execution levied' is the seizure of the Contractor's goods by bailiffs instructed by a creditor. They can take this action if the Contractor has not paid any sum owed as a result of a legal action
The Contractor must commence the Works on or as soon as is reasonably possible after the Date for Commencement of the Works
548 41
assign any part of the Contract; or
58 3
The Contractor must not, without the written consent of the Employer:—
56 3
assign the Contract; or
57 3
assign any benefit in or under the Contract; or
59 3
assign any part of the Contract; or
58 3
The Engineer must not give authorisation to the Engineer's Representative, or any other person responsible to the Engineer, in respect of any decision to be taken or certificate to be issued under:—
44 2(3)
Clause 63
50 2(3)
assign any interest in or under the Contract
60 3
assign any benefit in or under the Contract; or
59 3
The Employer or the Contractor must refer to the Engineer any dispute as to:—
1320 66(1)
any decision
opinion
instruction
direction
certificate
valuation
of the Engineer; 1321 66(1)
assign any interest in or under the Contract
60 3
The reference to the arbitrator may proceed (despite the Works not being complete or alleged complete) in the case of any dispute or difference arising as to:—
1346 66(2)
the exercise of the Engineer's power to give a certificate under Clause 63(1)
1350 66(2)

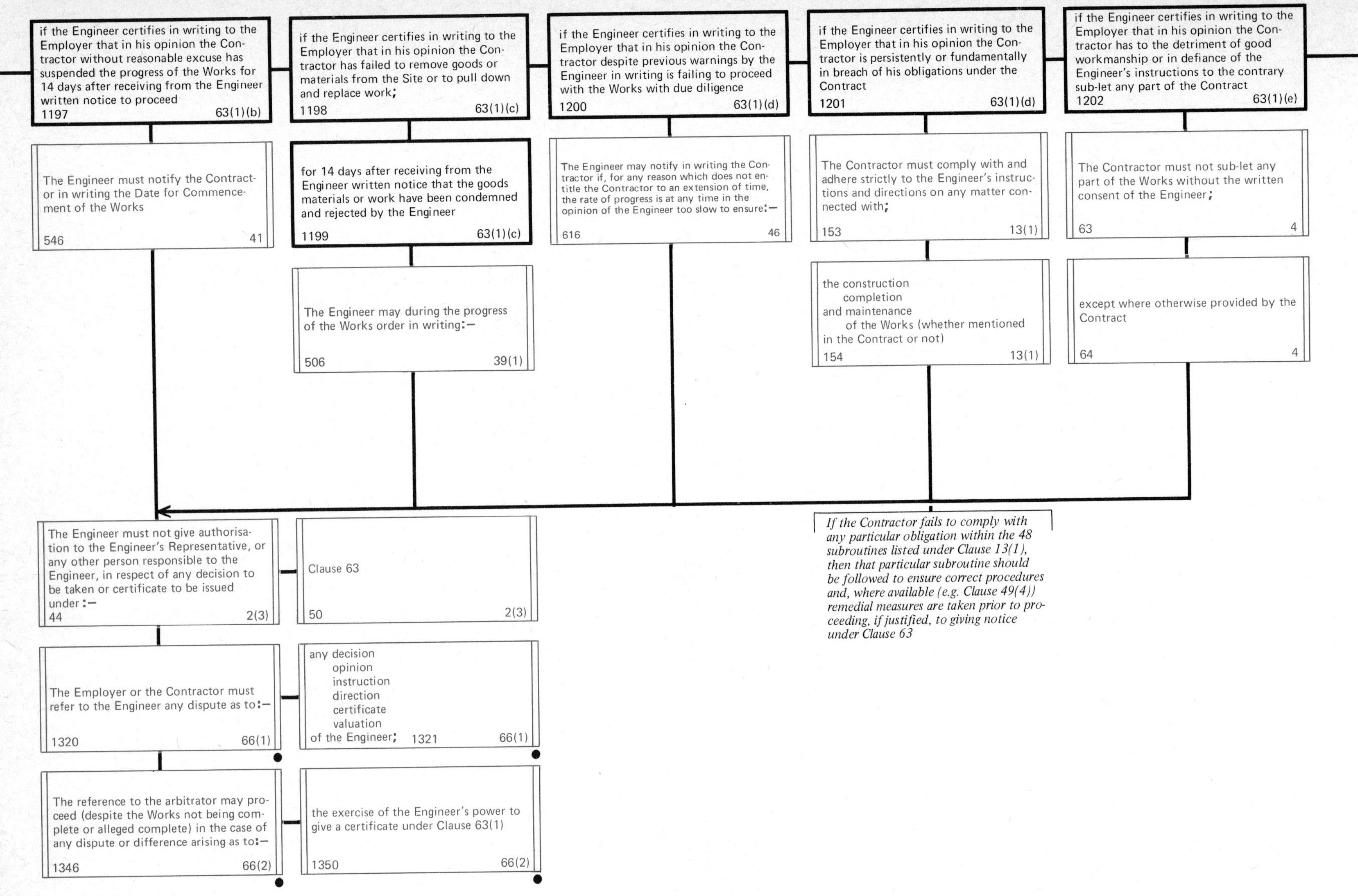
2
if the Engineer certifies in writing to the Employer that in his opinion the Contractor without reasonable excuse has suspended the progress of the Works for 14 days after receiving from the Engineer written notice to proceed 1197 63(1)(b)
if the Engineer certifies in writing to the Employer that in his opinion the Contractor has failed to remove goods or materials from the Site or to pull down and replace work; 1198 63(1)(c)
if the Engineer certifies in writing to the Employer that in his opinion the Contractor despite previous warnings by the Engineer in writing is failing to proceed with the Works with due diligence 1200 63(1)(d)
if the Engineer certifies in writing to the Employer that in his opinion the Contractor is persistently or fundamentally in breach of his obligations under the Contract 1201 63(1)(d)
if the Engineer certifies in writing to the Employer that in his opinion the Contractor has to the detriment of good workmanship or in defiance of the Engineer's instructions to the contrary sub-let any part of the Contract 1202 63(1)(e)
The Engineer must notify the Contractor in writing the Date for Commencement of the Works 546 41
for 14 days after receiving from the Engineer written notice that the goods materials or work have been condemned and rejected by the Engineer 1199 63(1)(c)
The Engineer may notify in writing the Contractor if, for any reason which does not entitle the Contractor to an extension of time, the rate of progress is at any time in the opinion of the Engineer too slow to ensure:— 616 46
The Contractor must comply with and adhere strictly to the Engineer's instructions and directions on any matter connected with; 153 13(1)
The Contractor must not sub-let any part of the Works without the written consent of the Engineer; 63 4
The Engineer may during the progress of the Works order in writing:— 506 39(1)
the construction completion and maintenance of the Works (whether mentioned in the Contract or not) 154 13(1)
except where otherwise provided by the Contract 64 4
The Engineer must not give authorisation to the Engineer's Representative, or any other person responsible to the Engineer, in respect of any decision to be taken or certificate to be issued under:— 44 2(3)
Clause 63 50 2(3)
The Employer or the Contractor must refer to the Engineer any dispute as to:— 1320 66(1)
any decision opinion instruction direction certificate valuation of the Engineer; 1321 66(1)
The reference to the arbitrator may proceed (despite the Works not being complete or alleged complete) in the case of any dispute or difference arising as to:— 1346 66(2)
the exercise of the Engineer's power to give a certificate under Clause 63(1) 1350 66(2)
If the Contractor fails to comply with any particular obligation within the 48 subroutines listed under Clause 13(1), then that particular subroutine should be followed to ensure correct procedures and, where available (e.g. Clause 49(4)) remedial measures are taken prior to proceeding, if justified, to giving notice under Clause 63

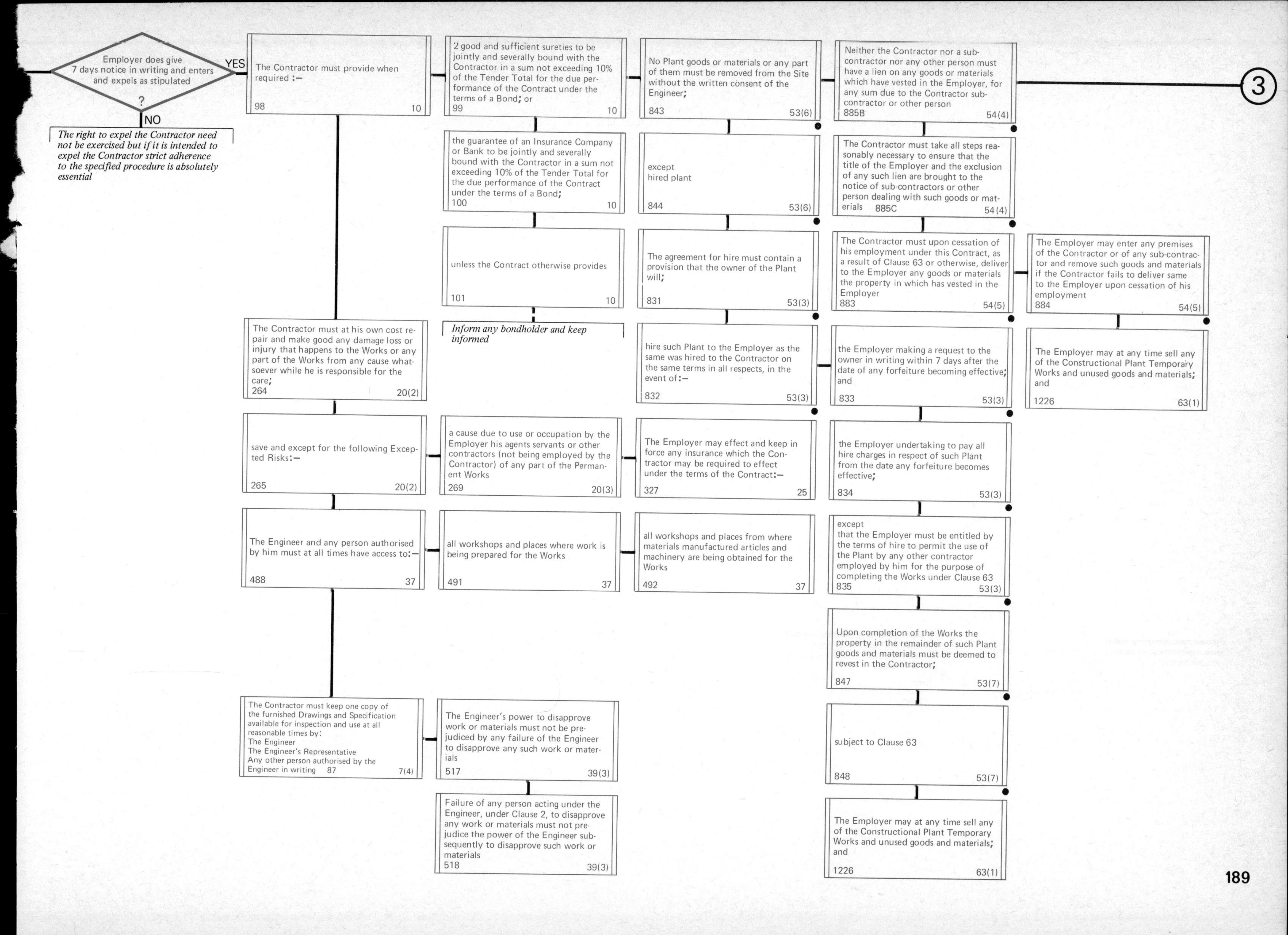
Employer does give 7 days notice in writing and enters and expels as stipulated ?
YES
NO
The right to expel the Contractor need not be exercised but if it is intended to expel the Contractor strict adherence to the specified procedure is absolutely essential
The Contractor must provide when required :— 98 10
2 good and sufficient sureties to be jointly and severally bound with the Contractor in a sum not exceeding 10% of the Tender Total for the due performance of the Contract under the terms of a Bond; or 99 10
the guarantee of an Insurance Company or Bank to be jointly and severally bound with the Contractor in a sum not exceeding 10% of the Tender Total for the due performance of the Contract under the terms of a Bond; 100 10
unless the Contract otherwise provides 101 10
Inform any bondholder and keep informed
No Plant goods or materials or any part of them must be removed from the Site without the written consent of the Engineer; 843 53(6)
except hired plant 844 53(6)
The agreement for hire must contain a provision that the owner of the Plant will; 831 53(3)
hire such Plant to the Employer as the same was hired to the Contractor on the same terms in all respects, in the event of:— 832 53(3)
Neither the Contractor nor a sub-contractor nor any other person must have a lien on any goods or materials which have vested in the Employer, for any sum due to the Contractor sub-contractor or other person 885B 54(4)
The Contractor must take all steps reasonably necessary to ensure that the title of the Employer and the exclusion of any such lien are brought to the notice of sub-contractors or other person dealing with such goods or materials 885C 54(4)
The Contractor must upon cessation of his employment under this Contract, as a result of Clause 63 or otherwise, deliver to the Employer any goods or materials the property in which has vested in the Employer 883 54(5)
The Employer may enter any premises of the Contractor or of any sub-contractor and remove such goods and materials if the Contractor fails to deliver same to the Employer upon cessation of his employment 884 54(5)
The Employer may at any time sell any of the Constructional Plant Temporary Works and unused goods and materials; and 1226 63(1)
3
the Employer making a request to the owner in writing within 7 days after the date of any forfeiture becoming effective; and 833 53(3)
the Employer undertaking to pay all hire charges in respect of such Plant from the date any forfeiture becomes effective; 834 53(3)
except that the Employer must be entitled by the terms of hire to permit the use of the Plant by any other contractor employed by him for the purpose of completing the Works under Clause 63 835 53(3)
Upon completion of the Works the property in the remainder of such Plant goods and materials must be deemed to revest in the Contractor; 847 53(7)
subject to Clause 63 848 53(7)
The Employer may at any time sell any of the Constructional Plant Temporary Works and unused goods and materials; and 1226 63(1)
The Contractor must at his own cost repair and make good any damage loss or injury that happens to the Works or any part of the Works from any cause whatsoever while he is responsible for the care; 264 20(2)
save and except for the following Excepted Risks:— 265 20(2)
a cause due to use or occupation by the Employer his agents servants or other contractors (not being employed by the Contractor) of any part of the Permanent Works 269 20(3)
The Employer may effect and keep in force any insurance which the Contractor may be required to effect under the terms of the Contract:— 327 25
The Engineer and any person authorised by him must at all times have access to:— 488 37
all workshops and places where work is being prepared for the Works 491 37
all workshops and places from where materials manufactured articles and machinery are being obtained for the Works 492 37
The Contractor must keep one copy of the furnished Drawings and Specification available for inspection and use at all reasonable times by:
The Engineer
The Engineer's Representative
Any other person authorised by the Engineer in writing 87 7(4)
The Engineer's power to disapprove work or materials must not be prejudiced by any failure of the Engineer to disapprove any such work or materials 517 39(3)
Failure of any person acting under the Engineer, under Clause 2, to disapprove any work or materials must not prejudice the power of the Engineer subsequently to disapprove such work or materials 518 39(3)

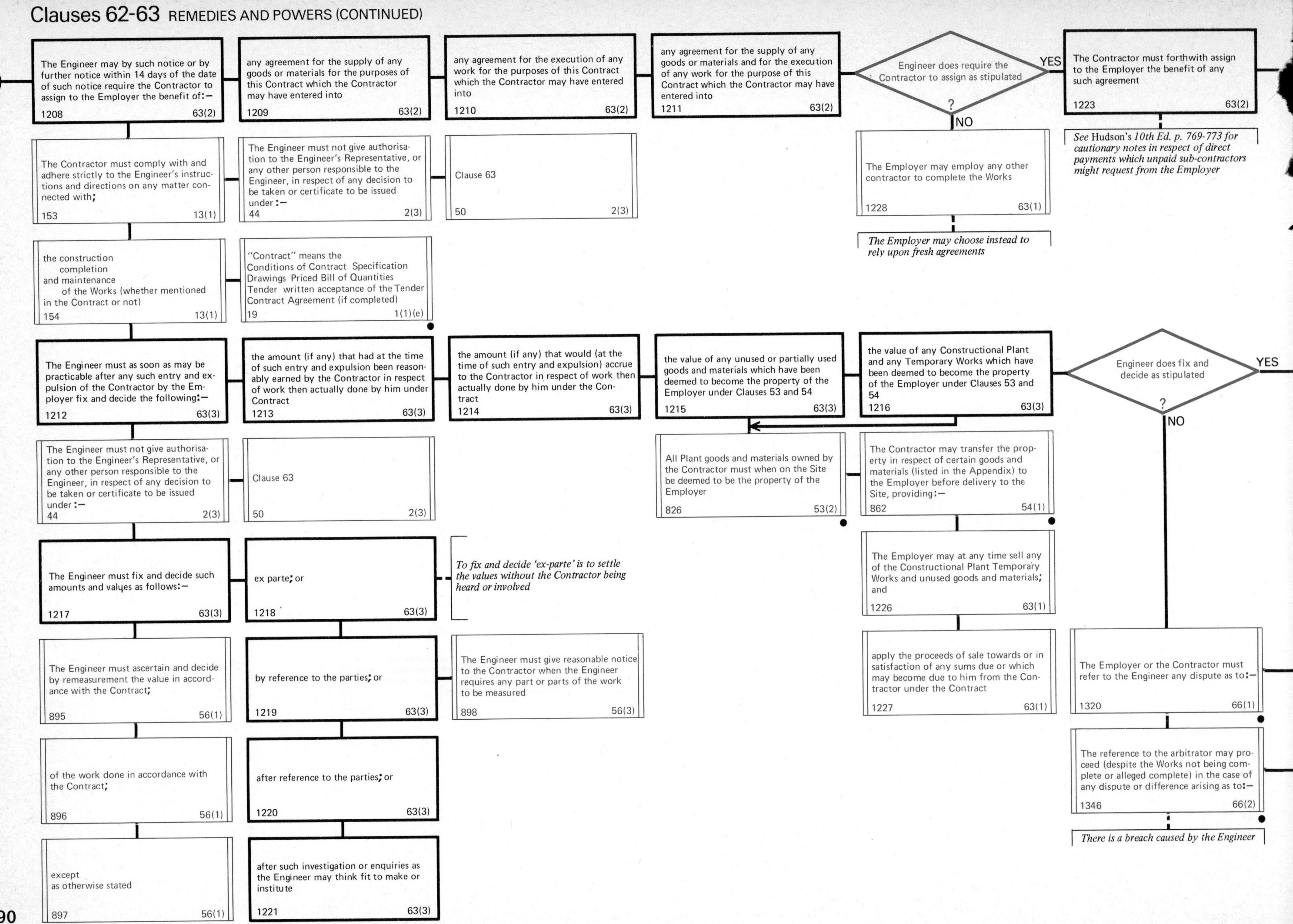
3
The Engineer may by such notice or by further notice within 14 days of the date of such notice require the Contractor to assign to the Employer the benefit of:— 1208 63(2)
any agreement for the supply of any goods or materials for the purposes of this Contract which the Contractor may have entered into 1209 63(2)
any agreement for the execution of any work for the purposes of this Contract which the Contractor may have entered into 1210 63(2)
any agreement for the supply of any goods or materials and for the execution of any work for the purpose of this Contract which the Contractor may have entered into 1211 63(2)
Engineer does require the Contractor to assign as stipulated ?
YES
NO
The Contractor must forthwith assign to the Employer the benefit of any such agreement 1223 63(2)
See Hudson's 10th Ed. p. 769-773 for cautionary notes in respect of direct payments which unpaid sub-contractors might request from the Employer
The Contractor must comply with and adhere strictly to the Engineer's instructions and directions on any matter connected with; 153 13(1)
The Engineer must not give authorisation to the Engineer's Representative, or any other person responsible to the Engineer, in respect of any decision to be taken or certificate to be issued under :— 44 2(3)
Clause 63 50 2(3)
The Employer may employ any other contractor to complete the Works 1228 63(1)
The Employer may choose instead to rely upon fresh agreements
the construction completion and maintenance of the Works (whether mentioned in the Contract or not) 154 13(1)
"Contract" means the Conditions of Contract Specification Drawings Priced Bill of Quantities Tender written acceptance of the Tender Contract Agreement (if completed) 19 1(1)(e)
The Engineer must as soon as may be practicable after any such entry and expulsion of the Contractor by the Employer fix and decide the following:— 1212 63(3)
the amount (if any) that had at the time of such entry and expulsion been reasonably earned by the Contractor in respect of work then actually done by him under Contract 1213 63(3)
the amount (if any) that would (at the time of such entry and expulsion) accrue to the Contractor in respect of work then actually done by him under the Contract 1214 63(3)
the value of any unused or partially used goods and materials which have been deemed to become the property of the Employer under Clauses 53 and 54 1215 63(3)
the value of any Constructional Plant and any Temporary Works which have been deemed to become the property of the Employer under Clauses 53 and 54 1216 63(3)
Engineer does fix and decide as stipulated ?
YES
NO
The Engineer must not give authorisation to the Engineer's Representative, or any other person responsible to the Engineer, in respect of any decision to be taken or certificate to be issued under :— 44 2(3)
Clause 63 50 2(3)
All Plant goods and materials owned by the Contractor must when on the Site be deemed to be the property of the Employer 826 53(2)
The Contractor may transfer the property in respect of certain goods and materials (listed in the Appendix) to the Employer before delivery to the Site, providing:— 862 54(1)
The Engineer must fix and decide such amounts and values as follows:— 1217 63(3)
ex parte; or 1218 63(3)
To fix and decide 'ex-parte' is to settle the values without the Contractor being heard or involved
The Employer may at any time sell any of the Constructional Plant Temporary Works and unused goods and materials; and 1226 63(1)
The Engineer must ascertain and decide by remeasurement the value in accordance with the Contract; 895 56(1)
by reference to the parties; or 1219 63(3)
The Engineer must give reasonable notice to the Contractor when the Engineer requires any part or parts of the work to be measured 898 56(3)
apply the proceeds of sale towards or in satisfaction of any sums due or which may become due to him from the Contractor under the Contract 1227 63(1)
The Employer or the Contractor must refer to the Engineer any dispute as to:— 1320 66(1)
of the work done in accordance with the Contract; 896 56(1)
after reference to the parties; or 1220 63(3)
The reference to the arbitrator may proceed (despite the Works not being complete or alleged complete) in the case of any dispute or difference arising as to:— 1346 66(2)
There is a breach caused by the Engineer
except as otherwise stated 897 56(1)
after such investigation or enquiries as the Engineer may think fit to make or institute 1221 63(3)

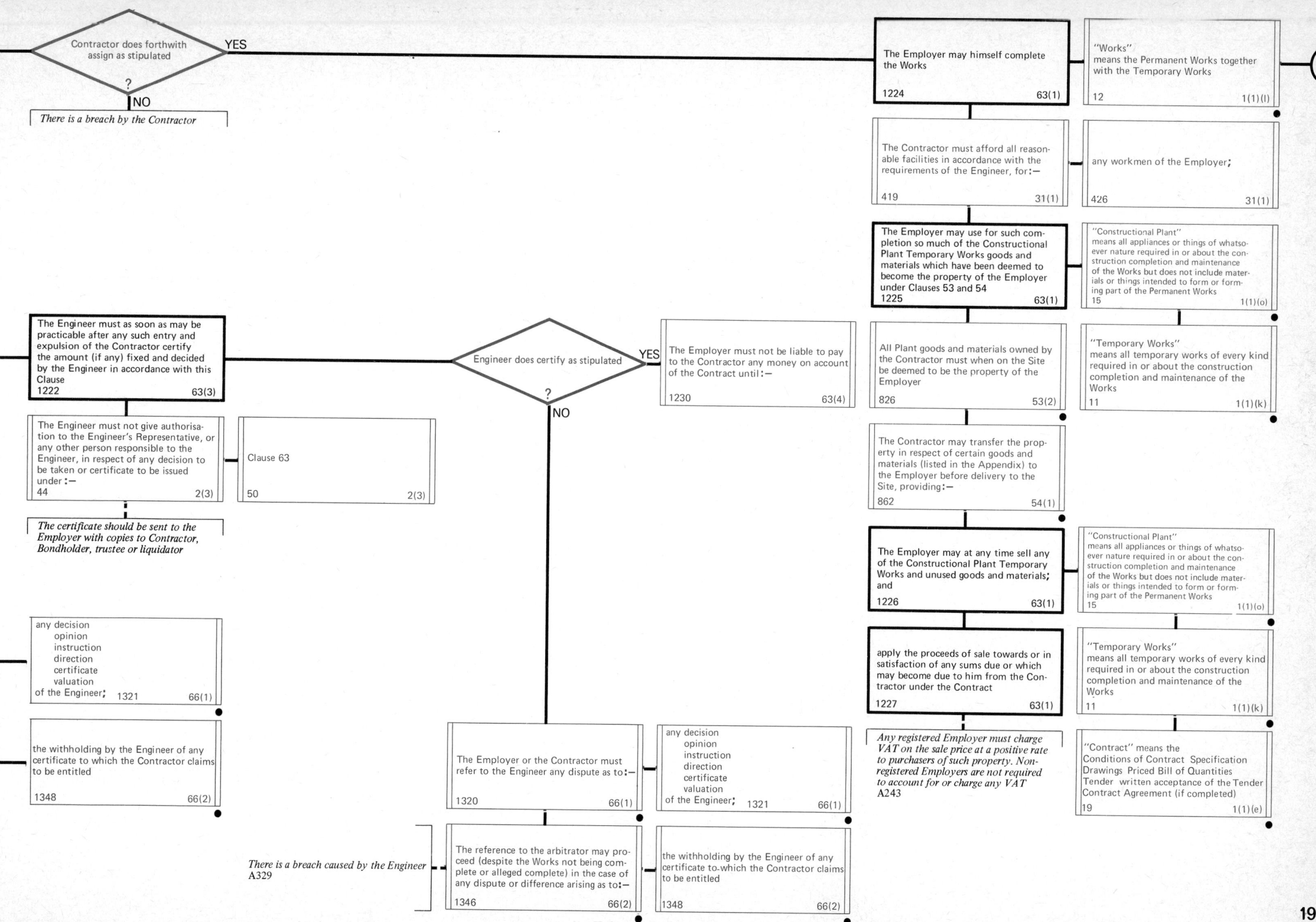

Contractor does forthwith assign as stipulated ?
YES
NO
There is a breach by the Contractor
The Employer may himself complete the Works 1224 63(1)
"Works" means the Permanent Works together with the Temporary Works 12 1(1)(l)
4
The Contractor must afford all reasonable facilities in accordance with the requirements of the Engineer, for:— 419 31(1)
any workmen of the Employer; 426 31(1)
The Employer may use for such completion so much of the Constructional Plant Temporary Works goods and materials which have been deemed to become the property of the Employer under Clauses 53 and 54 1225 63(1)
"Constructional Plant" means all appliances or things of whatsoever nature required in or about the construction completion and maintenance of the Works but does not include materials or things intended to form or forming part of the Permanent Works 15 1(1)(o)
The Engineer must as soon as may be practicable after any such entry and expulsion of the Contractor certify the amount (if any) fixed and decided by the Engineer in accordance with this Clause 1222 63(3)
Engineer does certify as stipulated ?
YES
NO
The Employer must not be liable to pay to the Contractor any money on account of the Contract until:— 1230 63(4)
All Plant goods and materials owned by the Contractor must when on the Site be deemed to be the property of the Employer 826 53(2)
"Temporary Works" means all temporary works of every kind required in or about the construction completion and maintenance of the Works 11 1(1)(k)
The Engineer must not give authorisation to the Engineer's Representative, or any other person responsible to the Engineer, in respect of any decision to be taken or certificate to be issued under:— 44 2(3)
Clause 63 50 2(3)
The certificate should be sent to the Employer with copies to Contractor, Bondholder, trustee or liquidator
The Contractor may transfer the property in respect of certain goods and materials (listed in the Appendix) to the Employer before delivery to the Site, providing:— 862 54(1)
The Employer may at any time sell any of the Constructional Plant Temporary Works and unused goods and materials; and 1226 63(1)
"Constructional Plant" means all appliances or things of whatsoever nature required in or about the construction completion and maintenance of the Works but does not include materials or things intended to form or forming part of the Permanent Works 15 1(1)(o)
any decision opinion instruction direction certificate valuation of the Engineer; 1321 66(1)
apply the proceeds of sale towards or in satisfaction of any sums due or which may become due to him from the Contractor under the Contract 1227 63(1)
"Temporary Works" means all temporary works of every kind required in or about the construction completion and maintenance of the Works 11 1(1)(k)
the withholding by the Engineer of any certificate to which the Contractor claims to be entitled 1348 66(2)
The Employer or the Contractor must refer to the Engineer any dispute as to:— 1320 66(1)
any decision opinion instruction direction certificate valuation of the Engineer; 1321 66(1)
Any registered Employer must charge VAT on the sale price at a positive rate to purchasers of such property. Non-registered Employers are not required to account for or charge any VAT A243
"Contract" means the Conditions of Contract Specification Drawings Priced Bill of Quantities Tender written acceptance of the Tender Contract Agreement (if completed) 19 1(1)(e)
There is a breach caused by the Engineer A329
The reference to the arbitrator may proceed (despite the Works not being complete or alleged complete) in the case of any dispute or difference arising as to:— 1346 66(2)
the withholding by the Engineer of any certificate to which the Contractor claims to be entitled 1348 66(2)

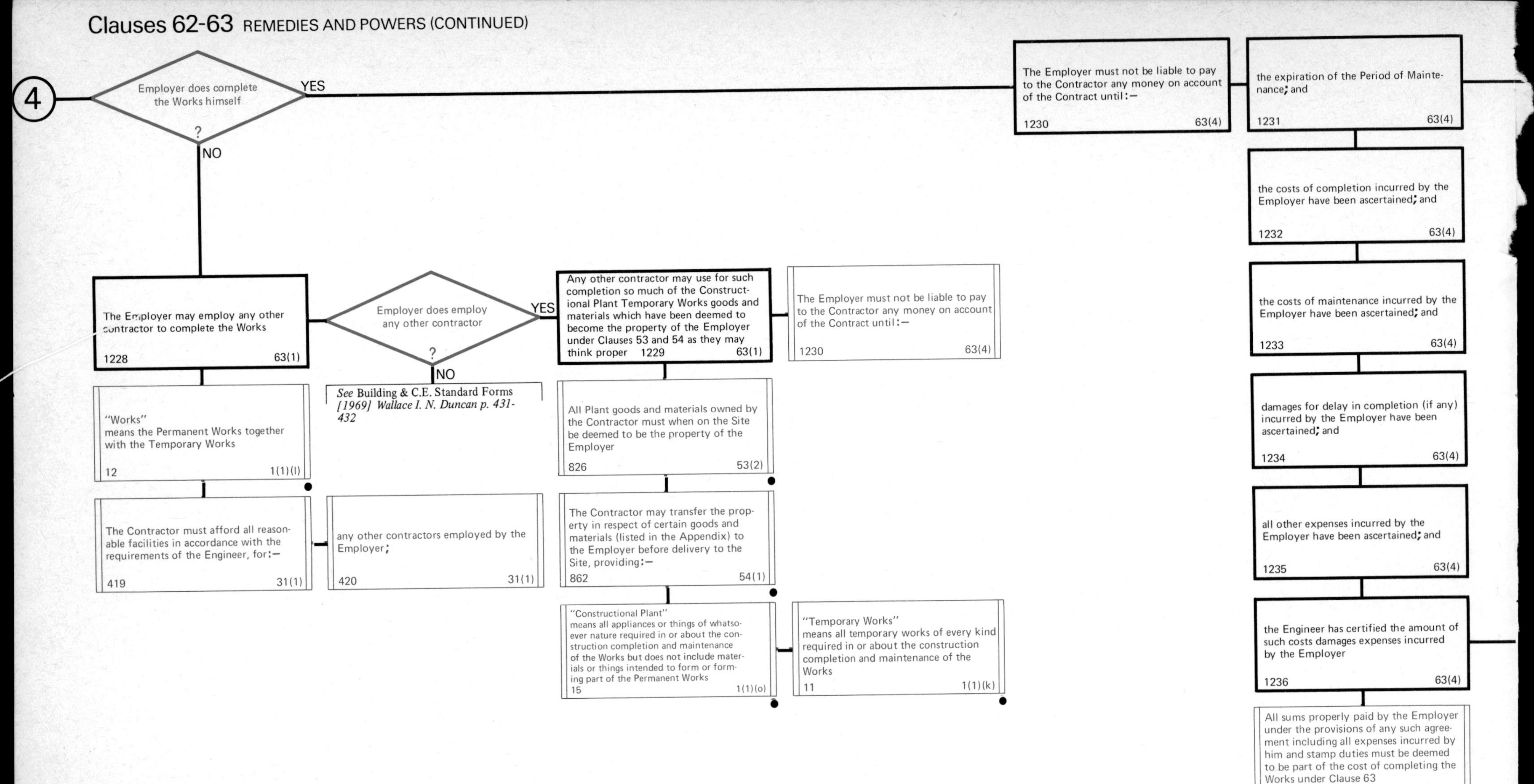
4
Employer does complete the Works himself ?
YES
NO
The Employer must not be liable to pay to the Contractor any money on account of the Contract until:— 1230 63(4)
the expiration of the Period of Maintenance; and 1231 63(4)
the costs of completion incurred by the Employer have been ascertained; and 1232 63(4)
the costs of maintenance incurred by the Employer have been ascertained; and 1233 63(4)
damages for delay in completion (if any) incurred by the Employer have been ascertained; and 1234 63(4)
all other expenses incurred by the Employer have been ascertained; and 1235 63(4)
the Engineer has certified the amount of such costs damages expenses incurred by the Employer 1236 63(4)
All sums properly paid by the Employer under the provisions of any such agreement including all expenses incurred by him and stamp duties must be deemed to be part of the cost of completing the Works under Clause 63 837 53(4)
The Employer may employ any other contractor to complete the Works 1228 63(1)
Employer does employ any other contractor ?
YES
NO
See Building & C.E. Standard Forms [1969] Wallace I. N. Duncan p. 431-432
Any other contractor may use for such completion so much of the Constructional Plant Temporary Works goods and materials which have been deemed to become the property of the Employer under Clauses 53 and 54 as they may think proper 1229 63(1)
The Employer must not be liable to pay to the Contractor any money on account of the Contract until:— 1230 63(4)
"Works" means the Permanent Works together with the Temporary Works 12 1(1)(l)
The Contractor must afford all reasonable facilities in accordance with the requirements of the Engineer, for:— 419 31(1)
any other contractors employed by the Employer; 420 31(1)
All Plant goods and materials owned by the Contractor must when on the Site be deemed to be the property of the Employer 826 53(2)
The Contractor may transfer the property in respect of certain goods and materials (listed in the Appendix) to the Employer before delivery to the Site, providing:— 862 54(1)
"Constructional Plant" means all appliances or things of whatsoever nature required in or about the construction completion and maintenance of the Works but does not include materials or things intended to form or forming part of the Permanent Works 15 1(1)(o)
"Temporary Works" means all temporary works of every kind required in or about the construction completion and maintenance of the Works 11 1(1)(k)

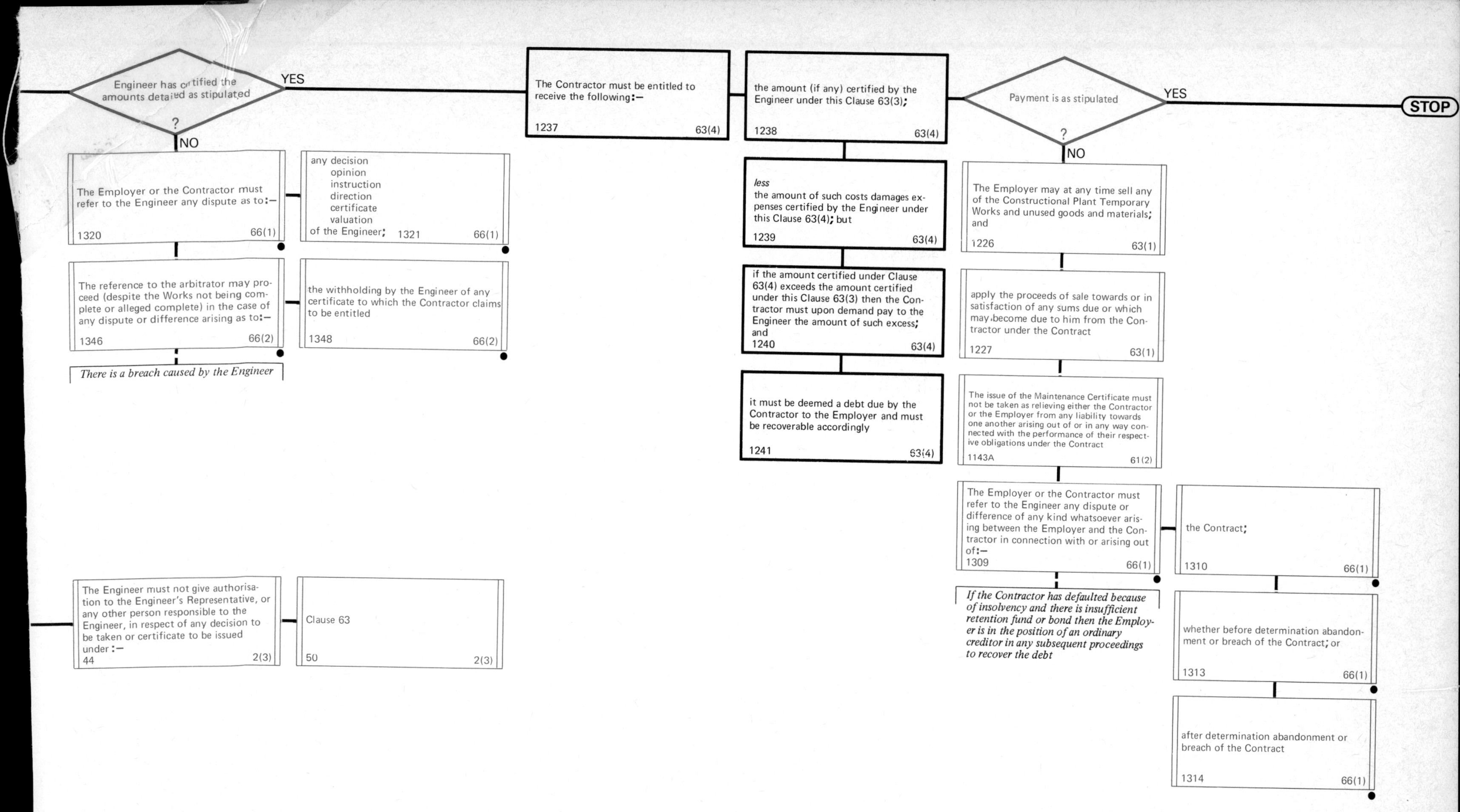
Engineer has certified the amounts detailed as stipulated
?
YES
NO
The Employer or the Contractor must refer to the Engineer any dispute as to:—
1320 66(1)
any decision
opinion
instruction
direction
certificate
valuation
of the Engineer; 1321 66(1)
The reference to the arbitrator may proceed (despite the Works not being complete or alleged complete) in the case of any dispute or difference arising as to:—
1346 66(2)
the withholding by the Engineer of any certificate to which the Contractor claims to be entitled
1348 66(2)
There is a breach caused by the Engineer
The Engineer must not give authorisation to the Engineer's Representative, or any other person responsible to the Engineer, in respect of any decision to be taken or certificate to be issued under:—
44 2(3)
Clause 63
50 2(3)
The Contractor must be entitled to receive the following:—
1237 63(4)
the amount (if any) certified by the Engineer under this Clause 63(3);
1238 63(4)
less
the amount of such costs damages expenses certified by the Engineer under this Clause 63(4); but
1239 63(4)
if the amount certified under Clause 63(4) exceeds the amount certified under this Clause 63(3) then the Contractor must upon demand pay to the Engineer the amount of such excess; and
1240 63(4)
it must be deemed a debt due by the Contractor to the Employer and must be recoverable accordingly
1241 63(4)
Payment is as stipulated
?
YES
NO
STOP
The Employer may at any time sell any of the Constructional Plant Temporary Works and unused goods and materials; and
1226 63(1)
apply the proceeds of sale towards or in satisfaction of any sums due or which may become due to him from the Contractor under the Contract
1227 63(1)
The issue of the Maintenance Certificate must not be taken as relieving either the Contractor or the Employer from any liability towards one another arising out of or in any way connected with the performance of their respective obligations under the Contract
1143A 61(2)
The Employer or the Contractor must refer to the Engineer any dispute or difference of any kind whatsoever arising between the Employer and the Contractor in connection with or arising out of:—
1309 66(1)
the Contract;
1310 66(1)
whether before determination abandonment or breach of the Contract; or
1313 66(1)
after determination abandonment or breach of the Contract
1314 66(1)
If the Contractor has defaulted because of insolvency and there is insufficient retention fund or bond then the Employer is in the position of an ordinary creditor in any subsequent proceedings to recover the debt

Bibliography

Chase, S., *The Tyranny of Words* (New York: Harcourt Brace & World Inc., 1937)

Cherry, C., *On Human Communication* (Massachusetts: MIT Press, 1957).

Duncan Wallace, I. N., *Building and Civil Engineering Standard Forms* (London: Sweet & Maxwell, 1969).

— *Hudson's Building and Civil Engineering Contracts* (tenth edition) (London: Sweet and Maxwell, 1970).

Empson, W., *Seven Types of Ambiguity* (London: Penguin Press, 1973).

Gowers, Sir Ernest (rev. Sir Bruce Fraser), *The Complete Plain Words* (London: Penguin Press).

Jones, Glyn P., *A New Approach to the Standard Form of Building Contract* (Lancaster: MTP, 1972).

Lewis, B. N., *Decision Logic Tables for Algorithms and Logical Trees* (London: HMSO, 1970).

— Horabin, I. S., and Gane, C. P., *Flow Charts, Logical Trees and Algorithms for Rules and Regulations* (London: HMSO, 1967).

Miller, G. A., *The Psychology of Communication* (London: Penguin Press, 1968).

Onions, C. J. (ed. and rev.) *The Shorter Oxford English Dictionary* (third edition) (Oxford: Clarendon Press, 1972).

Piesse, E. L., and Gilchrist-Smith, J., *The Elements of Drafting* (London: Stevens and Sons, 1965).

Young, J., *Information Theory* (London: Butterworths, 1971).